Studies in the History of Civil Engineering

Volume 5

Water-Supply and Public Health Engineering

STUDIES IN THE HISTORY OF CIVIL ENGINEERING

General Editor: Joyce Brown

with the assistance of M.M. Chrimes, A.W. Skempton, N.A.F. Smith and R.J.M. Sutherland

---

Twelve volumes are to appear from the Autumn of 1997:

1. **The Engineering of Medieval Cathedrals**
   Edited by LYNN T. COURTENAY, University of Wisconsin

2. **Masonry Bridges, Viaducts and Aquaducts**
   Edited by TED RUDDOCK, University of Edinburgh

3. **Land Drainage and Irrigation**
   Edited by SALVATORE CIRIACONO, University of Padova

4. **Dams**
   Edited by DONALD C. JACKSON, Lafayette College

5. **Water-Supply and Public Health Engineering**
   Edited by DENIS SMITH, formerly University of East London

6. **Port and Harbour Engineering**
   Edited by ADRIAN JARVIS, Merseyside Maritime Museum and Liverpool University

7. **The Civil Engineering of Canals and Railways before 1850**
   Edited by MIKE CHRIMES, Institution of Civil Engineers

8. **The Development of Timber as a Structural Material**
   Edited by DAVID T. YEOMANS, Liverpool University

9. **Structural Iron, 1750–1850**
   Edited by R.J.M. SUTHERLAND, Consultant, Harris and Sutherland, London

10. **Structural Iron and Steel, 1850–1900**
    Edited by ROBERT THORNE, Editor, *Construction History Journal*

11. **Early Reinforced Concrete**
    Edited by FRANK NEWBY, formerly F.J. Samuely and Partners, London

12. **Structural and Civil Engineering Design**
    Edited by WILLIAM ADDIS, University of Reading

Studies in the History of Civil Engineering
General Editor: Joyce Brown

Volume 5

Water-Supply and Public Health Engineering

edited by
Denis Smith

Ashgate
VARIORUM

Aldershot • Brookfield USA • Singapore • Sydney

Published in the series **Studies in the History of Civil Engineering** by

Ashgate Publishing Limited
Gower House, Croft Road
Aldershot, Hampshire GU11 3HR
Great Britain

Ashgate Publishing Company
Old Post Road
Brookfield, Vermont 05036–9704
USA

ISBN 0–86078–754–0

British Library CIP data
Water-Supply and Public Health Engineering
(Studies in the History of Civil Engineering; v. 5).
1. Sanitary engineering–History. 2. Water-supply engineering–History. 3. Water-supply–History.
I. Smith, Denis
628.1'09

US Library of Congress CIP data
Library of Congress Catalog Card Number preassigned as: 98–074001

This book is printed on acid free paper.

Printed in Great Britain by Galliards, Great Yarmouth

Contents

Acknowledgements vii–ix

General Editor's Preface xi–xii

Introduction xiii–xxxiii

URBAN WATER-SUPPLY

1 Attitudes to Roman engineering and the question of the inverted siphon
Norman A.F. Smith 1

2 Our debt to Roman engineering: the water supply of Lincoln to the present day
M.J.T. Lewis 29

3 Sir Hugh Myddelton and the New River
G.C. Berry 46

4 George Sorocold of Derby: a pioneer of water supply
F. Williamson 79

5 The old water-supply of Seville
George Higgin 115

6 Portsmouth's water supply, 1800–1860
Mary Hallett 131

7 The impounding reservoirs of the Newcastle and Gateshead Water Company, 1845–1905
R.W. Rennison 155

8 The old Croton Aqueduct
George H. Rappole 183

9 Bringing water to the Crescent City: Benjamin Latrobe and the New Orleans waterworks system
Gary A. Donaldson 195

10 Urban sanitation in preindustrial Japan
Susan B. Hanley 211

SANITARY REFORM

11 The development of Victorian infrastructures: the example of Portsmouth
Robert. A. Otter 237

12 Edwin Chadwick and the engineers, 1842–1854: systems and antisystems in the pipe-and-brick sewers war
Christopher Hamlin 259

13 The separate vs. combined sewer problem: a case study in urban technology and design choice
Joel A. Tarr 289

14 Sir Joseph William Bazalgette (1819–1891): engineer to the Metropolitan Board of Works
Denis Smith 321

15 Eugène Belgrand (1810–1878): civil engineer, geologist and pioneer hydrologist
George Atkinson 345

Index 377

Acknowledgements

The chapters in this volume are taken from the sources listed below, for which the editors and publishers wish to thank their authors, original publishers or other copyright holders for permission to use their material as follows:

Chapter 1: Norman A.F. Smith, 'Attitudes to Roman engineering and the question of the inverted syphon', *History of Technology*, 1 (1976), pp. 45–71. Copyright © 1976 by the contributors. Reproduced by permission of Mansell, a Casell imprint, Wellington House, 125 Strand, London, WC2R 0BB, England.

Chapter 2: M.J.T. Lewis, 'Our debt to Roman engineering: the water supply of Lincoln to the present day', *Industrial Archaeology Review*, 7, no. 1 (1984), pp. 57–73. Copyright © 1984 by the Association for Industrial Archaeology. Reprinted by permission of the editors, Marilyn Palmer and Peter A. Neaverson.

Chapter 3: G.C. Berry, 'Sir Hugh Myddelton and the New River', *Transactions of the Honourable Society of Cymmrodorion* (1957), pp. 17–45. Copyright © 1957 by the Honourable Society of Cymmrodorion.

Chapter 4: F. Williamson, 'George Sorocold of Derby: a pioneer of water supply', *Journal of Derbyshire Archaeological and Natural History Society*, NS 10, no. 57 (1936), pp. 43–47, 65–93. Reproduced by kind permission of the Derbyshire Archaeological Society.

Chapter 5: George Higgin, 'The old water-supply of Seville', *Minutes of Proceedings of the Institution of Civil Engineers*, 78, pt. 4 (1983–4), pp. 334–345. Copyright © 1983–4 by the Institution of Civil Engineers.

Chapter 6: Mary Hallett, 'Portsmouth's water supply, 1800–1860', *Portsmouth Papers*, no. 12 (1971), pp. 3–25. Copyright © 1971 by Mary Hallett.

Chapter 7: R.W. Rennison, 'The impounding reservoirs of the Newcastle and Gateshead Water Company, 1845–1905', *Transactions of the Newcomen Society*, 54 (1982–83), pp. 27–54. Copyright © 1982–83 by the Newcomen Society, London.

Chapter 8: George H. Rappole, 'The old Croton Aqueduct', *Journal of the Society for Industrial Archeology*, 4, no. 1 (1978), pp. 15–25. Copyright © 1978 by the Society for Industrial Archeology, Houghton, MI.

Chapter 9: Gary A. Donaldson, 'Bringing water to the Crescent City: Benjamin Latrobe and the New Orleans waterworks system', *Louisiana History*, 28, no. 4 (1987), pp. 381–396. Copyright © 1987 by the Louisiana Historical Association, Lafayette, LA.

Chapter 10: Susan B. Hanley, 'Urban sanitation in pre-industrial Japan', *Journal of Interdisciplinary History*, 18, no. 1 (1987), pp. 1–26. Copyright © 1987 by the Massachusetts Institute of Technology and the editors of the *Journal of Interdisciplinary History*. Reprinted by permission of the editors and the MIT Press, Cambridge, MA.

Chapter 11: Robert A. Otter, 'The development of Victorian infrastructures: the example of Portsmouth', *Proceedings of the ICE, North Western Association Centenary Conference: Infrastructure Renovation and Waste Control*, ed. D.C. Proctor and I.L. Whyte, (Manchester, 8–10 April 1986), pp. 109–119. Copyright © 1987 by the Institution of Civil Engineers.

Chapter 12: Christopher Hamlin, 'Edwin Chadwick and the engineers, 1842–1854: the systems and antisystems in the pipe-and-brick sewers war', *Technology and Culture*, 33, no. 4 (1992), pp. 680–709. Copyright © 1992 by the Society for the History of Technology. Reprinted by permission of the University of Chicago Press.

Chapter 13: The separate vs. combined sewer problem: a case study in urban technology and design choice', *Journal of Urban History*, 5, no. 3 (1979), pp. 308–339. Copyright © 1979 by Sage Publications, Inc. Reprinted by permission of Sage Publications, Inc.

Chapter 14: Denis Smith, 'Sir Joseph William Bazalgette (1819–1891): engineer to the Metropolitan Board of Works', *Transactions of the Newcomen Society*, 58 (1986–87), pp. 89–111. Copyright © 1986–87 by the Newcomen Society, London.

Chapter 15: George Atkinson, 'Eugène Belgrand (1810–1878): civil engineer, geologist and pioneer hydrologist', *Transactions of the Newcomen Society*, 69, no. 1 (1997–98), pp. 97–127. Copyright © 1997–98 by the Newcomen Society, London.

***

Chapters 7, 14 and 15 are reproduced from the *Transactions of the Newcomen Society* by permission of the Society's Council. The Newcomen Society promotes, encourages and co-ordinates the study of the history of all types of engineering and technology from the earliest times to the present day. Details from the Executive Secretary, The Newcomen Society, The Science Museum, London SW7 2DD, Tel/Fax 0171–589–1793.

Every effort has been made to trace all the copyright holders, but if any have been inadvertently overlooked the publishers will be pleased to make the necessary arrangement at the first opportunity.

General Editor's Preface

Joyce Brown

Civil engineering has a long history. Although practitioners did not begin to describe themselves as 'civil engineers' until the eighteenth century, the origins of their work lie in the construction skills of the ancient world and in the works of military engineers. Shortly after British civil engineers had formed themselves into a professional body, the Institution of Civil Engineers in London, their role was described in 1828 by Thomas Tredgold in these words:

> Civil engineering is the art of directing the great sources of power in Nature for the use and convenience of man.[1]

He thus neatly encapsulated what the civil engineer's job is about – the provision of the means to give society what it wants for its survival.

As time has gone by, the civil engineer's skills have had to diffuse over an ever-widening field and at the same time become more specialised. In all periods he has had to be the ingenious 'fixer', but more than a mere artisan – a point observed in the Institution's motto, *scientia et ingenio*.

The history of civil engineering is a fascinating one. Certainly 'the great sources of power in Nature' are not always directed without a certain amount of opposition by them, and the provision of the basic requirements – habitation, water supply, main drainage, harbours bridges, places of worship, transport systems by land and water – has called for inventiveness and sometimes for daring and courage.

Yet interest in the history of civil engineering is not of long standing, and it seems fair to state that most of the important work has been done in the last forty years. With a few notable exceptions, most of the major books on the subject have appeared in that era; many specialist societies with their own transactions have been founded; and many of the journals in which articles might be published have come into existence. While archaeologists and architectural historians have long made their contribution to this subject, their ranks have been swelled by a large number of serious scholars – including engineers themselves – now interested in recording and evaluating the engineering achievements of earlier times. In particular, there is a lively interest in the study of our industrial past, at least in part because some of the evidence of it are already disappearing from the landscape.

Our intention in creating the series of which this volume forms part is to provide, through the reproduction of important contributions to this subject, an

---

[1] *Minutes of Council, Institution of Civil Engineers*, 4 January 1828.

invaluable reference collection for its study. The series encompasses many different branches of engineering from early times to the beginning of this century; its perspective is global; and the chosen articles have an international authorship, to the extent that this can be achieved for an essentially English-language series. The introduction to each volume has allowed the volume editor to set the pages selected in the context of the whole history of the subject and its historiography, while the provision of references and a bibliography will enable the reader to go further into the study of the topic.

This series will have succeeded if it gives civil engineers some unexpected insights into their own craft and other readers a new way of looking at engineering structures.

***

The volume editors were chosen for their particular knowledge and expertise in the field. They have worked with enthusiasm and good will to put together collections of articles intended to be both informative and stimulating, and I am grateful to them for their efforts. Where possible, they have incorporated the work of the best scholars, while at the same time giving preference to articles often obscurely published and not readily available. The learned journals are represented, as also are conference proceedings and essay chapters contributed to books.

In all of this, I have been much assisted by my Advisory Panel. Emeritus Professor A.W. Skempton (Imperial College) has made himself freely available to me for consultation on many aspects of this venture. Mike Chrimes (Librarian of the Institution of Civil Engineers), Dr. Norman Smith (Emeritus Reader, Imperial College) and James Sutherland (formerly a Senior Partner in Harris and Sutherland, London, consulting engineers) have all contributed valuable advice based on their extensive knowledge of particular fields. The idea for the series was developed in discussion with Dr John Smedley of Ashgate, and Publisher of the Variorum Collected Studies Series, who has given me much personal support and the benefit of his experience.

I am grateful also for the help given by the staff of several libraries, notably Kay Crooks and Susannah Parry in the library of the Department of Civil Engineering, Imperial College, and Mike Chrimes and staff of the library of the Institution of Civil Engineers, London. Many friends have given me encouragement and useful advice; in particular, I want to thank Paula Kahn and Mark Baldwin.

Formerly Department of Civil Engineering,
Imperial College, London

Introduction

Denis Smith

The term 'public health', as used today, has its origins in the nineteenth century and arose in response to the appalling conditions of urban life in large cities and towns. But many aspects of the design, construction, and operation of water-supply and drainage systems have a long pedigree, and the papers here have been selected to reflect this important fact and to reveal both the similarities and differences of approach in Europe, America and the Far East.

One of the principal characteristics that defines this area of civil engineering, is that the engineer was often working for national or local government, or other public bodies. The tackling of public health problems involves a complex group of related activities, embracing, certainly by the nineteenth century, the skills of scientists, physicians, politicians, lawyers, architects and engineers, and the involvement of many professional groups. Two areas of study were critical to developments in public health. Firstly, there was the research of Louis Pasteur (1822–95) on yeast fermentation for a Paris brewery in 1862, which laid the foundation for bacteriology which was to become crucially important in public health matters. Secondly, there was the work of Dr John Snow (1815–58), who was pre-eminent amongst the physicians studying the diffusion of cholera, and who demonstrated that an outbreak of cholera in Soho, London, was due to pollution by human excreta at a particular well and pump in Broad Street. This water-cycle link was the key to the transmission of cholera and made the subsequent engineering work inescapable. In his famous pamphlet of 1849, *On the Mode of Communication of Cholera*, he says: 'The sanitary measure most required in the metropolis is a supply of water ... from some source quite removed from the drains'. The fact that the promoter of the necessary engineering work was frequently a public body meant that political will had to be enlisted to find the funds to finance engineering projects. This is a subject where the social benefit of civil engineering is obvious – not always the case in other fields. The two principal strands of the topic which involve the civil engineer are water-supply and main drainage, though other municipal matters such as street cleansing, refuse collection, smoke prevention, building control, and even issues such as burial versus cremation, were to emerge as part of the effective administration of towns and cities. Although public health engineering covers such a variety of topics, the chosen papers examine the subject from its two main aspects, namely, urban water-supply and public health engineering in relation to sanitation.

Historiography

This volume deals with an important aspect of the history of civil engineering which has some special characteristics – one of which is that it has not received a great deal of attention from historians of engineering. Social and medical historians have, however, made interesting complementary contributions in their own fields.

In this subject, and in particular, in the period of Greek and Roman engineering, a great deal has been achieved through textual criticism by classical scholars and in fieldwork by archaeologists. The problem here is that those undertaking this research have often lacked an adequate knowledge of engineering to make sound judgements about technical systems or devices, an idea discussed by Smith (see chapter 1). Nevertheless, there is a large body of literature including reports of surveys where the site evidence no longer survives.[1]

Another aspect of the historiography of this subject area is the work done by historians primarily interested in studying the harnessing of energy to a variety of industrial purposes. For example, those interested in muscle power have studied devices operated by men and animals in raising water, where man has applied his ingenuity in a variety of ways.[2] Again, a great deal has been published on the water-wheel because of its intrinsic technical interest, whether it was applied to milling, water pumping or other industrial uses.[3] The same is true of historians interested in heat engines. Much has been written on Savery's pumping device of the late-seventeenth century,[4] of Newcomen's engine which dominated most of the eighteenth century,[5] and of James Watt's improved steam engine.[6] In all these cases a primary use of the engines has been in water-supply, and, in the second half of the nineteenth century, the steam engine was also extensively used in main drainage schemes. Historical studies of water and steam turbines, internal

---

[1] See White, *Greek and Roman Technology*. Chapter 8 deals with civil engineering and surveying, and chapter 12 with hydraulic engineering.

[2] See, for example, G.F. Zimmer, 'A chain of pots of the 6th and 7th century', *Transactions of the Newcomen Society*, 4 (1923–24), 30–31; H.A. Sandford, 'A 16th-century treadwheel for raising water', *Transactions of the Newcomen Society*, 4 (1923–24) 36–39; and J.K. Major, 'The horse engine in the 19th century', *Transactions of the Newcomen Society*, 60 (1988–89), 31–48.

[3] See Skempton, *John Smeaton*. Chapters 3 and 8 deal with the application of water-wheels and the Newcomen engine in eighteenth-century water-supply.

[4] See, for example R.L. Hills, 'A steam chimera: a review of the history of the Savery engine', *Transactions of the Newcomen Society*, 58 (1986–87), 27–44.

[5] See L.T.C. Rolt and J.S. Allen, *The Steam Engines of Thomas Newcomen* (Buxton, 1977).

[6] The literature on the development and application of the steam engine is extensive and accessible, but one interesting example where water-supply engineers successfully modified and applied a mine pumping engine was in the case of the Cornish beam engine. See: D.B. Barton, *The Cornish Beam Engine* (Truro, 1965); Binnie, *Early Victorian Water Engineers*, 73.

combustion engines and electric motors reveal their widespread use in the utilities industries.[7] So the literature of prime movers will contain much of interest to those studying the history of water-supply and public health engineering. There is also a considerable body of papers, often published locally, on the water-supply and drainage of particular towns.[8] Amongst the medical historians, Roy Porter of the Wellcome Institute has made a significant contribution to the study of public health.[9] A British engineering historian, N.A.F. Smith, has brought analysis and insight to illuminate many of the technical questions of ancient civil engineering (see Bibliography). However, a great deal remains to be done in this field by historians of engineering and technology.

Urban Water-Supply

The essential need for water to support the life of people, animals and crops[10] is, of course, as old as life on earth, but the role of engineers, and the level of technology involved has changed enormously. Being absolutely indispensible to life meant that water was not, until relatively recently, considered as being part of a qualitative concept of health. Without it there would be no life.

In engineering terms, there are three spheres of activity, namely, the 'capture', 'treatment', and 'distribution' of water. Naturally, capture is fundamental and involves the location of sources of water and may range from water-divining to scientific hydrology. Sources include springs, wells, and rivers. Having located the source, there is often the need to provide a means of storage in reservoirs. These reservoirs often also provide the first stage of water treatment. Water abstracted from a river often has material in suspension and the impounding reservoir acts as a stilling pond allowing the heavier material to settle by sedimentation. This was virtually the only form of water treatment until the early-nineteenth century when the slow sand filter-bed was devised by James Simpson.

[7] For the variety of pumping plant used, at various times, in one waterworks, see R.W. Rennison, 'The pumping stations of the Newcastle and Gateshead Water Company', *Transactions of the Newcomen Society*, 68 (1996–97), 137–64.

[8] See Alan Hardiman, 'Servicing the houses of Bath, 1714–1830: water supply', *Bristol Industrial Archaeology Society Journal*, 27 (1996), 11–18, and, Alan Hardiman, 'Servicing the houses of Bath, 1714–1830: sewage and rainwater disposal', *Bristol Industrial Archaeology Journal*, 29 (1996), 15–21. Other examples are Jeffrey Radley, 'York waterworks and other waterworks in the north before 1800', *Transactions of the Newcomen Society*, 39 (1966–67), 143–56, and John R. Hume, 'The water supply of New Lanark', *Industrial Archaeology*, 5, no.4 (1968), 384–7. In addition, see the Bibliography for Jardine on Edinburgh, Bateman on Manchester, Cecen on Istanbul, Dickinson on London, Gibson on Cambridge, and Jones on Bristol waterworks.

[9] Porter, *The Greatest Benefit...* especially his section 'Public Health', 405–15.

[10] See volume 3 in this series, *Land Drainage and Irrigation*, ed. Salvatore Ciriacono.

The construction of reservoir banks poses similar problems to those of earth dam construction,[11] and it was only during the last century that a greater scientific understanding of the stability of earth banks has been developed. The distribution of water has engaged the ingenuity of civil engineers over centuries, both in terms of pipe network systems and in pumping technology. The design of piped systems raises several technical issues, including choice of material, pipe-friction characteristics, and reduction of leakage. Over the centuries water-supply pipes have been made of lead (originally fabricated from flat sheet), timber (bored tree trunks), stone (bored), and, from the early-nineteenth century, cast iron.

The distribution of water has demanded continuously increased pump horsepower and over the centuries water engineers have adopted most forms of prime movers. From antiquity muscle-power was used, followed by the water-wheel with cranks and reciprocating pumps. Water-supply engineers were quick to adopt the Newcomen atmospheric engine and subsequently Watt's steam engine. The internal combustion engine, water and steam turbines have all been used. From the eighteenth century Montgolfier's hydraulic ram was adopted on remote country estates and some are still in use. Earlier in this century the water-supply industry was the first to adopt H.A. Humphrey's unique pistonless gas pump.[12] In pumping terms, if we look at the two main aspects of the water-cycle, then it is obvious that the energy demands on the water-supply side far exceed that of main drainage. It was the water-supply engineers who had to lift from wells, springs, and rivers to the customers' buildings, from whence it would flow under gravity to the drains. For example, in London in 1884 the water companies supplied a population of 4 million and employed an aggregate pump capacity of 16,490 horse power when, at the same time, the main drainage system employed only 3,520 horse power (a ratio of 4.68:1). In civil engineering design all decisions are site-specific, but the relative demands for horse power in the water-supply and drainage parts of the water-cycle would be somewhat similar in other situations.

The Classical Period

The location of sources of fresh water significantly shaped nomadic life and the settlement patterns of subsequent civilisations. The Egyptians, Greeks and Romans were all involved with elements of water-supply. Evidently, and necessarily, water-supply received attention over a very long period, and was one the earliest spurs to technical ingenuity in the development of water-lifting and pipe-network

[11] See volume 4 in this series, *Dams,* ed. Donald C. Jackson.

[12] See Denis Smith, 'The Humphrey pump and its inventor', *Transactions of the Newcomen Society,* 43 (1970–71), 67–92.

technologies. Marcus Vitruvius, an engineer working in the period of Augustus, is a rare (though not always reliable) source for some of the technical details of Roman water-supply distribution systems, saying, for example:

> There are three methods of conducting water, in channels through masonry conduits, or in lead pipes, or in pipes of baked clay. If in conduits, let the masonry be as solid as possible, and let the bed of the channel have a gradient of not less than a quarter of an inch for every hundred feet, and let the masonry structure be arched over, so that the sun does not strike the water at all. When it has reached the city, build a reservoir with a distribution tank in three compartments ... From this central tank, pipes will be laid to all the basins and fountains ... [13]

It is clear from another classic written source, that of Julius Sextus Frontinus, writing in first-century Rome, that the three elements which were to characterise the nineteenth-century public health movement, namely 'legislation', 'organisation', and 'engineering' were in place during his lifetime. The Roman Senate passed laws relating to public and private water rights and penalties for polluting water sources. The Senate also appointed a Water Commissioner to oversee the design, construction, management and maintenance of water-supply systems. When Frontinus was appointed Commissioner he revealed the status accorded to the post by saying that it was 'an office which concerns not merely the convenience but also the health and even the safety of the City, and which has always been administered by the most eminent men of our State'.[14] The Romans have left us a rich archaeological heritage of structures relating to their water-supply systems: massive aqueducts (involving bridges, tunnels, and inverted siphons), water-towers, baths, fountains, cisterns and pipe-networks from many parts of their Empire.[15] After describing the extensive system of aqueducts bringing water into Rome, Frontinus says, perhaps with justifiable pride: 'With such an array of indispensible structures carrying so many waters, compare, if you will, the idle pyramids, or the useless, though famous, works of the Greeks!'[16]

The concept that what they were doing in this respect was 'useful' is characteristic of the Roman attitude. The design of water-supply schemes was often subject to military considerations. Tunnelled lengths of aqueducts were preferred as being less likely to be damaged by an enemy, and wells would be sited inside defensible perimeters. A good example is the *Basilica Cistern* in Istanbul which was built in response to the major problem of ensuring a water-supply to the town during a siege. The magnificent underground cistern was built

[13] Vitruvius, *The Ten Books on Architecture,* 244.

[14] Frontinus, *Stratagems and Aqueducts...*, 331.

[15] For an excellent outline of Roman water-supply, see Smith, *Man and Water,* 80–93; and Hodge, *Roman Aqueducts...*

[16] Frontinus, *Stratagems and Aqueducts...* , 357.

during AD 306–337 in the reign of Constantine I. It is a rectangular chamber measuring 141 × 73 metres and has a grid of stone columns, 8 metres high, in twelve rows each with 28 columns (336 in total). The roof comprises brick groined vaults. It was restored and extended by Justinian during AD 527–565, and is now a tourist attraction.[17]

Certainly nothing as efficient as the achievements of the Roman engineers, or on so large a scale, was to be achieved again until the nineteenth century. Smith's fascinating paper (see chapter 1) provides an interesting critical assessment of some technical issues in Roman engineering, drawing on both archaeological and written sources. His analytical approach removes many previously held misconceptions about such matters as water pressure problems in pipes. Lewis (see chapter 2) gives an account of the water-supply, from Roman times, to Lincoln, a high city on a fenland plain, posing particularly interesting problems for the early water engineers. He discusses a range of technologies including pumping, aqueducts, and siphons and relates these to archaeological excavations and other site investigations.

The Middle Ages

After the classical period there were relatively few developments in water-supply or drainage, and the technological achievements of the Greeks and Romans seem largely to have been laid aside. There was instead a reliance on springs or wells for drinking water, accompanied by only rudimentary drainage. While it is generally thought that the Renaissance was the next important phase of technological development, Jean Gimpel cogently argues that there was an industrial revolution in western Europe during the middle ages.[18] Although power was largely derived from the harnessing of water, coal began to replace wood as a fuel for domestic and industrial purposes, thereby greatly increasing both air and river pollution.

The desire for clean drinking water manifested itself here and there in large cities through legislation to prevent the dumping of refuse and sewage in rivers.[19] But there are few examples of technological solutions being employed. Moslem Spain is unusual in this respect. Higgin's paper (see chapter 5) describes a twelfth-century system in Seville. For its date, it is a remarkably complex system for the collection and house-distribution of a water-supply. Higgin was, in the 1880s, site engineer to the then new Seville waterworks, whose construction would

[17] Described in Cecen, *Sinan's Water Supply...*

[18] Jean Gimpel, *The Medieval Machine* (London, 1976). See in particular chapter 4: 'Environment and Pollution'.

[19] Smith, *Man and Water*, 95–6.

destroy the evidence of the earlier works. Fortunately, he had the wit to make a record of what had been there.

Developments from the Seventeenth Century

In Britain, gradual growth of population prompted attempts to supplement water-supply from the seventeenth century onwards, with a gradual move away from reliance on wells and springs to the installation of waterworks and piped water. Hallet describes this process in Portsmouth (see chapter 6).

In financial terms, it was relatively easy to finance the construction and maintenance of water-supply works by private risk capital. As a result, a large proportion of water-supply undertakings, certainly from the seventeenth century, could be provided by private companies whose revenue was provided by a water-rate levied on consumers and which yielded a return to shareholders.

The growth of population in London placed an increasing strain on the Thames and its tributaries and prompted in 1606 the work by Hugh Myddelton to convey water from springs in Hertfordshire in a open channel about 10 feet wide and 40 miles long in a New River from Ware to Islington.[20] From there, water was distributed in a system of wooden pipes in an area around Newgate and Ludgate in the west of the City of London. Berry gives a very detailed account of this work (see chapter 3) and his paper is remarkable also for its range of sources, considering that the Company's records were largely destroyed in a fire in 1750.

By the sixteenth century, piston pumps were being applied in Germany, Spain, and France, as well as in England, to lift water to high levels,[21] whence it would be distributed by gravity flow. The Thames employed such a system from 1582 when Peter Morice built water-wheels in two of the arches of London Bridge. George Sorocold,[22] an obscure but important waterworks engineer, undertook the task of increasing the supply and over a number of years designed and installed in London Bridge four large water-wheels driving between them 52 piston pumps. He also designed waterworks for many towns in England during the last decade of the seventeenth century, as Williamson's paper (see chapter 4) reveals.

The eighteenth- and nineteenth-century water companies and local authorities usually employed fee-earning consulting engineers to advise on the initial design work of their proposed schemes. Once the system was commissioned, salaried staff engineers would be employed to maintain and design small improvement

---

[20] For a biography, see Gough, *Sir Hugh Myddelton*.

[21] Smith, *Man and Water*, 97–101.

[22] See also R. Jenkins, *The Collected Papers of Rhys Jenkins* (Cambridge, 1931), chapter 20, for an article on Sorocold.

schemes. In Victorian Britain these staff engineers were often permitted to act as external consulting engineers and to take pupils at a premium.[23]

Hallet, in the second part of her paper (see chapter 6), reviews the economic, legal, and technical issues surrounding the mid-eighteenth century and later schemes to supply water to Portsmouth. Rennison's thorough study (see chapter 7) of the impounding reservoirs of the Newcastle and Gateshead Water Company involves the work of three of the greatest water engineers of the nineteenth century: James Simpson, John F. Bateman, and Thomas Hawksley. The three engineers were employed, during the period 1845 to 1905, to design impounding reservoirs at Whittle Dean, Hallington, and Catcleugh, respectively.

The remaining chapters in this section explore two contrasting water-supply systems in the USA. Rappole (see chapter 8) deals with the problems posed by the solid rock base of Manhattan Island and the salinity of the adjacent river water, which led to massive engineering works for New York City's first municipal water-supply. Donaldson (see chapter 9) writes of Benjamin Latrobe and the scheme, interrupted by the war with England, for the water-supply to New Orleans at the mouth of the Mississippi. Latrobe was born in Yorkshire and trained as an architect under S.P. Cockerell. After the death of his wife he emigrated to the United States where, in 1803, he became surveyor of public buildings to the Federal Government, and was involved with both architectural and engineering projects.[24]

Public Health Engineering

Industrialisation, wherever it occurred in Europe, brought a rapid increase in population, over-crowding and appalling and insanitary housing conditions, with a high cost in mortality.

The selected papers (chapters 10–15) focus primarily on matters relating to sanitation and waste disposal. The paper by Hanley (see chapter 10) gives a fascinating insight into eighteenth-century Japan. Edo, renamed Tokyo in 1868, had one of the largest populations in the world at that date, perhaps, Hanley suggests, because economic, religious and cultural attitudes towards waste disposal made for a healthier environment than was usual in the west in the same period.

[23] James Simpson (1799–1869) is a good example. Whilst Engineer to the Chelsea Waterworks Company he built his first large slow sand filter-bed there in 1828, but was also consulted about water schemes for Bristol, Newcastle, Liverpool, Manchester, York, Cardiff, Reading, Cambridge, Carlisle, Exeter, Folkestone and Gravesend. Similarly, in main drainage works, Sir Joseph Bazalgette was widely consulted (see chapter 14).

[24] Other American cities have been studied, for example, Cain, 'Unfouling the public's nest...; Hoffecker, 'Water and sewage works....; O'Connell, 'Chicago's quest...'

Sanitary Reform in Britain

Reform was carried forward with particular success in Victorian Britain, where existing structures of local government could be built upon and where there was already a tradition of concern about the insanitary conditions of the poor. This evidenced itself in a number of publications, of which Engels' *Conditions of the Working Class in England* (1844) is probably the most famous.[25] It is worth looking in some detail at what was happening in Britain.

As we have seen, finance for water-supply undertakings was often readily available from private sources, but this was not the case for main drainage schemes. Customers were psychologically attuned to paying to have potable water delivered to their property; they were much less willing to pay for effluent to be removed. In the case of main drainage 'public health' was seen to be a matter of 'public wealth'. This led to the need to engage both local and national government, and therefore, political will, to find the funds for engineering work in this field.

The sanitary reform movement in Britain developed largely as a reaction to one dreadful disease, namely, cholera. Although its occurrence in epidemic form had been noted by physicians in the sixteenth century, it was its rapid spread in the early-nineteenth century that made the study of the disease inescapable and led to effective action. It is worth noting that there was also a strong religious and moral component in the motivation of the Victorian sanitary reformers. In 1848, for example, Lord Shaftesbury, in accepting an invitation to join the new General Board of Health, recorded in his diary:

> the immense and unparalleled value I have always attached, in public and private movement, to the sanitary question, as second only to the religious, and in some respects, inseparable from it.[26]

Lytton Strachey, in his well-known essay on Florence Nightingale, went further when remarking on her unorthodox conception of God, saying:

> She felt towards Him as she might have felt towards a glorified sanitary engineer; and in some of her speculations she seems hardly to distinguish between the Deity and the Drains.[27]

Again, Lord Shaftesbury, aware that public opinion was vitally important, said:

> Sanitary arrangements will never be carried into full effect over the length and breadth of this land, till the head of every family … is fully convinced that these

---

[25] Porter, *The Greatest Benefit...*, 408–10.

[26] J. Wesley Bready, *Lord Shaftesbury and Social-Industrial Progress* (London, 1926), 109.

[27] Lytton Strachey, *Eminent Victorians* (London, 1948), 181; first published, Chatto & Windus, 1918.

> are principles that are essentially necessary to his domestic comfort, to his domestic purity, to his physical condition, and to his moral strength.[28]

It was in 1817 that European physicians specifically directed their attention to the subject of cholera, when a violent outbreak of the disease occurred in Jessore in Bengal. It rapidly spread over a large part of British India, and for this reason became known as Asiatic cholera. The disease spread northwards through Europe and arrived in Britain in the 1830s.[29] The method of communicating the disease was not at first understood. It usually struck in late summer or the autumn, and for about half of those who contracted the disease it proved fatal. Faced with the awful scourge of cholera, the first step of those in the sanitary reform movement was to press for appropriate legislation.[30]

One of the first agitators for social reform was the British *utilitarian* philosopher Jeremy Bentham (1748–1832), who argued that society should be organised to secure 'The greatest happiness for the greatest number'. Edwin Chadwick (1800–1890) was introduced to Bentham in 1830 and indeed nursed him through his last illness in 1832. Chadwick, who was to become the figurehead of the sanitary reform movement, was a lawyer by training, having been admitted to the Middle Temple in 1823. His active involvement with public health matters began in 1834 when he was appointed Secretary to the Poor Law Commission. He did not trust engineers and was one of their severest critics. For example, writing to a local councillor in 1844, he said:

> For all purposes it would be of the greatest importance that you should get the advice of trustworthy engineers, of whom I am sorry to say there are marvellous few – a more ignorant or more jobbing set of men ... I have rarely met with.[31]

In turn, a pamphlet published in 1855, the year of the Metropolis Management Act, which established the Metropolitan Board of Works, summarised Chadwick thus:

---

[28] See note 26.

[29] For an excellent essay on the nature and spread of cholera, see *Encyclopaedia Britannica*, vol. V, ninth edition, 1875–89, 682–685.

[30] A brief chronology of the principal events and Acts of Parliament is as follows: 1834: Poor Law Act; 1842: Edwin Chadwick published his *Report on the Sanitary Condition of the Labouring Population of Great Britain*; 1848: First Public Health Act, established the General Board of Health, did not apply to London, and lasted only six years; 1849: Dr John Snow published his pamphlet *On the Mode of Communication of Cholera*; 1854: Board of Health dismantled; 1855: Metropolis Management Act, established (January 1856) the Metropolitan Board of Works with powers over 117 square miles of the Capital; 1875: First comprehensive Public Health Act, with jurisdiction over Drains, Sewers, Water-Supply, Inspection of Food, and Prevention of Epidemics; 1876: River Pollution Prevention Act.

[31] Letter dated 9 October, 1844: Chadwick to Councillor John Shuttleworth in Manchester. University College London Library, document 2181/5.

> He was a Lawyer more fond, as most Lawyers are, of victory than of truth, obstinate in his prejudices, intolerant of any opposition to his opinions, and as greedy of power as he is zealous for his mode of drainage and his mode of water supply. In a word, no one was to be clean except by Mr Chadwick's patent soap. There were persons who might have put up with a first-class engineer, but the Lawyer advocate amateur Commissioner would not do.[32]

Despite these difficulties, the reform movement did lead to legislation which began to establish the appropriate administrative apparatus. The 1848 Public Health Act was the first step.[33] This established the General Board of Health which included Lord Morpeth, Edwin Chadwick, Lord Ashley (Shaftesbury), and Dr Southwood Smith. The Board were authorised to make an Inquiry into the sanitary condition of a town either when the ratepayers petitioned for it or where the death rate was above 23 per thousand. Four Engineering Inspectors were appointed, including Robert Rawlinson and Edward Cresy, both of whom were lifelong friends of Henry Doulton, and who were as convinced as Doulton of the utility of earthenware pipes! Where the 1848 Act was applied, a Local Board of Health was established with powers to appoint Surveyors, Inspectors of Nuisances, and Medical Officers of Health. The first MOH to be appointed was Dr John Simon (1816–1904), who was appointed by the City of London.[34] The General Board of Health was dismantled in 1854, and after an interregnum, the first national comprehensive Public Health Act of 1875 was passed. This Bill covered Drains and Sewers (separately defined), Water-Supply, Inspection of Food, and Prevention of Epidemics.

The last group of papers in the present volume cover a range of political, economic, and technical issues which were debated publicly and widely in Parliament, Local Government, sanitary and health-lobbying groups, and the press. In chapter 11, Otter discusses the development of public health infrastructures in Victorian Portsmouth. His paper is a detailed illustration of the way in which local government structures developed during the period of sanitary reform.

Two of the technical debates of the day are presented in chapters 12 and 13. Hamlin explores the opposing views of Edwin Chadwick and the engineers on the relative merits of brick or pipe-sewers, a matter which was widely discussed at the time. Tarr examines another controversial technical issue in the American context, although it was also extensively discussed in this country: the question of whether sanitary sewers should also carry rainwater.

---

32 *Sanitary Reform and Sanitary Reformers; for the information of Metropolitan Vestries and Local Boards with an Appendix of the Metropolitan Local Management Act* (London, 1855), 80. ICE Tract., 8vo., 109 (1) held in the Library of the Institution of Civil Engineers, London.

33 This Act did not apply to London. The General Board of Health was established for five years, later extended to six.

34 Dr John Simon's Annual Reports to the City of London are a fascinating compilation covering a wide range of social conditions.

The last two chapters provide an opportunity for the comparison of the approach to public health problems in two European capitals: London and Paris. Chapter 14 deals with the career of Sir Joseph Bazalgette[35] and the interceptor sewer system in London, whilst Atkinson (see chapter 14) writes on Bazalgette's near contemporary, Eugène Belgrand, and his engineering work in Paris. The different education, training, and style of employment of the two engineers says a great deal about the different approaches to civil engineering in the two countries.

Associated Events

A significant piece of legislation which made an important contribution to the development of water-supply and public health engineering in Britain was the passing, in 1836, of the *Registration Act* covering births, deaths, and marriages. And most importantly for public health issues, Edwin Chadwick succeeded in obtaining an amendment requiring the 'cause' of death also to be registered from 1837. Sir Douglas Galton said that these statistics were 'the basis upon which all scientific sanitary investigation necessarily rests'.[36] It is also significant that The Statistical Society had been formed in 1834, gaining its Royal Charter in 1887.

The Society of Arts, founded in 1754, played an active role in promoting sanitation and water-supply, particularly in the period immediately after the first comprehensive Public Health Act of 1875 . This is perhaps not surprising as the Society had Sir Douglas Galton, Edwin Chadwick, and John Simon among its members. The Society ran a series of important national conferences between 1874 and 1884.[37]

Another body which contributed to the subject was The Sanitary Institute established in 1876 and Incorporated in 1888. The Institute held lectures, demonstrations and examinations three times a year in 'Sanitary Science as applied to Buildings and Public Works' for Inspectors of Nuisances and Smoke

[35] Bazalgette's Presidential Address to the Institution of Civil Engineers (8 January 1884) was concerned with 'engineering works which promote the health and comfort of the inhabitants of large cities', giving statistics of population and death-rate/1000 of 45 large cities. See Bibliography for full reference. See also J.W. Bazalgette, 'On the main drainage of London, and the interception of the sewage from the river Thames', *Minutes of Proceedings of the Institution of Civil Engineers,* 24 (14 March 1865), 280–358; J.W. Worth and W.S. Crimp, 'The main drainage of London', *Minutes of Proceedings of the Institution of Civil Engineers,* 129 (1897), 49–79.

[36] George Barclay Bruce, 'Presidential Address', *Minutes of Proceedings of the Institution of Civil Engineers,* 91 (8 November 1887), 23.

[37] See D. Hudson, and K.W. Luckhurst , *The Royal Society of Arts, 1754–1954* (London, 1954).

Inspectors amongst others. By 1919, these examinations had been extended to candidates in many parts of the British Empire.[38]

The Crimean War provides an interesting example of the interaction between civil and military engineers. In February 1855, I.K. Brunel was asked by the War Department to design and construct a portable hospital to accommodate 3,000 military patients. He duly designed a prefabricated building complete with all services, and an experimental ward was erected at Paddington Station. The building was shipped out to the Crimea by 23 vessels carrying a total of 11,500 tons. The first vessel arrived on 7 May 1855 and the last on 5 December. Construction on site at Renkoi began in May 1855 and the hospital was opened in stages and by March 1856 could house 2,200 patients. The buildings had a short life and having served their purpose were put up for sale in September 1856.[39] Other examples of collaboration of the civil and military are Sir Robert Rawlinson (1810–1898), a civil engineer specialising in sanitary engineering, visiting the Crimea in 1855 as part of a Government Commission; while a military engineer who was much involved with sanitary engineering was Captain Douglas Galton (1822–1899), whose lifelong friendship with Florence Nightingale began just after the Crimean War. He became one of her principal collaborators in the reform of army hospitals. In 1862, he was appointed Assistant Under-Secretary at the War Office in charge of military health matters and sanitary administration. His lectures on Sanitary Engineering at the School of Military Engineering at Chatham were published in 1877. Between 1869 and 1875 he was Director of Public Works and Buildings at the Government's Office of Works.

In medical education, St Thomas' Hospital in London was the first to establish a Lectureship in Public Health. The first appointment was Dr T.M. Greenhow[40] who gave his first Lecture in the hospital lecture theatre on 6 May 1857. A report of the lecture said:

> Dr. Greenhow had a quiet fling at lay sanitary reformers, some of whom he had invited to hear him, admitting that in certain departments they had effected

---

[38] *The Year Book of the Learned Societies of Great Britain and Ireland* (London, 1919), 312.

[39] For a good account (with drawings of the hospital) see, Isambard Brunel, *The Life of Isambard Kingdom Brunel* (London, 1870), 461–73. Brunel's Resident Engineer in the Crimea was John Brunton (1812–1899); for his own account see, *John Brunton's Book* (Cambridge, 1939), 47–75.

[40] Dr T.M. Greenhow, a Newcastle colleague of Dr John Snow, had published: *Cholera: its non-contagious nature, and the best means of arresting its progress shortly examined. In a letter addressed to the Right Worshipful Mayor of Newcastle* (Newcastle, 1831). Dr Greenhow was appointed Lecturer in Public Health in March 1856; see F.G. Parsons, *The History of St Thomas's Hospital*, 3 vols. (London, 1936), vol. 3, 136.

> good, but claiming for the medical profession the whole credit for any improvement which had been effected in public health ... .[41]

Nevertheless, William Cubitt, during his Presidential Address to the Institution of Civil Engineers had, characteristically, earlier claimed that:

> The engineers have always been the real sanitary reformers, as they are the originators of all onward movements; all their labours tend to the amelioration of their fellow-men.[42]

Professional Organisation

An interesting and significant consequence of the growth of public health engineering following the 1848 Public Health Act was the growing awareness of surveyors[43] and engineers of the need for a professional body representing their special sphere of work. As a result, on 2 May 1873, the Association of Municipal & Sanitary Engineers & Surveyors was formed and at this inaugural meeting[44] they decided that the Objects of the Association were to be:

a. The promotion and interchange among its Members of that species of knowledge and practice which falls within the department of an Engineer or Surveyor engaged in the discharge of the duties imposed by the Public Health, Local Government, and other Sanitary Acts.
b. The promotion of the professional interests of the Members.
c. The general promotion of the objects of Sanitary Science.

At this meeting the first President, Lewis Angell,[45] said that the office of Borough Engineer was no sinecure and, in an ironic mood, felt that their qualifications and duties might best be described in the language of the lengthy advertisements in the contemporary newspapers, signed by the Town Clerk, and headed:

> To Civil Engineers and – *others*
>
> The Corporation of Mudborough requires the services of an experienced person to fill the office of Borough Engineer and Surveyor. He must be a man of good

---

[41] 'Public Health at St Thomas's Hospital', *The Builder* (14 May 1857), 327.

[42] W. Cubitt, 'Presidential Address', *Minutes of Proceedings of the Institution of Civil Engineers*, 9 (8 January 1850), 143.

[43] There is a problem of terminology here. Some towns would have a Borough Surveyor, some a Borough Engineer, and others a Borough Engineer and Surveyor. For the surveyors see: F.M.L. Thompson, *Chartered Surveyors: the Growth of a Profession* (London, 1968).

[44] The Inaugural Meeting was held at The Institution of Civil Engineers.

[45] Borough Engineer of West Ham, London. He was the leading promoter of the Association.

> education and able to keep accounts ... competent to take charge of workmen and the direction of works ... acquainted with the value of building materials, to take out quantities, make estimates, specifications, and measure up works. ... competent to make surveys, take levels, and be a good draughtsman ... able to design and superintend public buildings ... and to enforce the Building Bye-Laws ... and advise the Corporation upon the best means of disposing of or utilizing sewage. He may also have to act as Engineer and Manager of the Water-works. ... He must be conversant with ... the Public Health, Local Government, Nuisances Removal, and various other Acts. ... (and) must produce testimonials of the highest character, find sureties, and be not more than 40 years of age.[46]

An amusing pastiche perhaps, but revealing something of the mood of the engineers. Salaried engineers working for local authorities, and whose proposed schemes had to be submitted for approval to annually re-elected lay committees, had special problems. A subsequent President of the Association said, 'we frequently lose those who best understand what requires to be done' which made it 'very difficult to carry on our work, do what we will, without fault-finding from some'.[47]

The new Institution gradually increased its membership and influence in the Town Halls throughout the country. In July 1889, the President's Annual Address contained a litany of unresolved issues, saying:

> We all only too well know and feel how far off perfection we still are .. in a solution of many of the great sanitary problems of the day.
> The question of the profitable disposal of water-carried town sewage still remains unsettled.
> The most effective method of sewer ventilation is still an active bone of contention.
> Whether we shall have public and private lighting by electricity, or remain contented with gas, has by no means yet been settled either by experts or by the *vox populi......*
> The practical working of many of the clauses of the Public Health and other sanitary Act is not all that could be desired ....
> The avoidance of that irritating and injurious noise caused by the great traffic of our streets; whether to be cured by improved pavements or improved wheels and horses' shoes, or both, or neither time will show.
> The better housing of the working classes has still a future ...
> That unostentatious but necessary work, the collection and disposal of house refuse, is by no means at present perfect.[48]

---

46 *Proceedings of the Association of Municipal & Sanitary Engineers & Surveyors*, 1 (1873–74), 18–19.

47 Presidential Address of A.W. Morant, MICE, Borough Engineer, Leeds. *Proceedings of the Association of Municipal & Sanitary Engineers & Surveyors,* 6 (May 1880), 135.

48 *Proceedings of the Association of Municipal & Sanitary Engineers & Surveyors*, 15 (1888–89), 97–8.

By the end of the nineteenth century the name was changed to The Incorporated Association of Municipal and County Engineers, and in July 1910 they became The Institution of Municipal and County Engineers. In 1948, they received their Royal Charter and became the The Institution of Municipal Engineers. The *Proceedings* of this body, under its various titles, provide an unrivalled source of information on the development of the specialised civil engineering works essential to the smooth running of large cities and towns. Other bodies include The Institute of Sanitary Engineers, founded in 1895, which, on Incorporation in 1916, became The Institution of Sanitary Engineers, and in 1955 was re-titled as The Institution of Public Health Engineers. Similarly, and indicative of the increasing fragmentation of professional engineering bodies, The British Association of Waterworks Engineers was formed in 1896, and became, in 1911, The Institution of Water Engineers.[49]

Concluding Comments

The field covered by this volume is that of service to society, of 'utilities', and the variety of professions that have contributed to its development. Scientific research led to bacteriology and a greatly improved understanding of the nature and transmission of infection. Medical research developed epidemiology based on statistics and a preventive approach to illness. Chemistry provided antiseptic surgery under anaesthetic, but also contributed to water purification and sewage treatment. Contributions to these fields were international.

The civil engineering profession in Britain had emerged from the great canal building programme of the late-eighteenth century and the unprecedented demands made on it by the construction of railways in the early-nineteenth century. In a sense the engineers were ready and waiting for their next challenge, and public health engineering was to provide it. But public health works did bring some new technical problems. Issues such as the best cross-section for sewers, the best materials for drain and sewer construction, the necessary fall per mile to make a sewer self-cleansing, the friction characteristics of large brick sewers, and questions regarding whether surface rainwater should be combined with the sewage. The treatment of sewage during the nineteenth century was largely one of sedimentation and the subsequent removal of sludge, treatment by aerobic bacteria only beginning at the very end of the nineteenth or early-twentieth century. An interesting piece of research work was undertaken by a civil engineer

---

[49] For a thorough study of the evolution of professional engineering bodies see R.A. Buchanan, *The Engineers: A History of the Engineering Profession in Britain, 1750–1914* (London, 1989). Chapter 5, 'The Fragmentation of the Profession' is of particular interest. See also, H. Percy Boulnois, *Reminiscences of a Municipal Engineer* (London, 1920).

during the construction of London's Main Drainage in the 1860s. Bazalgette's Assistant Engineer, John Grant, developed a systematic technique of testing Portland Cement, a new material, and made it a reliable and predictable material for the construction industry at large.[50]

The issue of public versus private ownership of the water-supply industry was the cause of lengthy debate in the press and elsewhere. Civil engineers, notably Sir Joseph Bazalgette, sought long and hard to gain control of the whole water-cycle, namely, water-supply and main drainage, but this was not to be achieved in the United Kingdom until the 1970s.

The main underlying historical trend which led to all this activity was the rapid growth of population in urban settings with the consequent problems of poor housing, overcrowding in ill-ventilated and undrained alleys and courts. As we have seen, the advent of cholera in the 1830s was the springboard for vociferous lobbying. But in London at least, in the mid-nineteenth century, it could be said that the state of the Thames itself at Westminster was probably the most effective lobbyer of Parliament.

Public health works also provided a superb opportunity for architects. The need to house massive beam engines or triple-expansion steam plants has left us with a rich heritage of magnificent pumping-stations, which, although strictly functional in character, were frequently beautifully decorated. The Victorians would lavish as much attention to detail on these, often remote, structures as they would on their town halls.

But undoubtedly the most important outcome of all this lobbying, argument, legislation, and civil engineering was not only the dramatic reduction in mortality, but the improvement in the quality of life.

In conclusion, it might be appropriate sum up the contents of this volume on an aesthetic and optimistic note by quoting from George Parker Bidder's address to the 1868 meeting of the British Association for the Advancement of Science when he said:

> The entire question of water supply is one which has already assumed considerable proportions. It deals with the supply of water to the large cities and towns of this Kingdom, with the utilization of that water for a variety of purposes ... with the preservation of the beauty of our rivers, and the prevention of their pollution by the drainage of our towns ... Indeed a more beautiful and interesting subject can hardly engage the attention of engineers; and it may be

50 Grant read two papers on the subject to the Institution of Civil Engineers, which were combined in a book. See John Grant, *Experiments of the Strength of Cement, chiefly in reference to the Portland Cement used in the Southern Main Drainage Works* (London, 1875). See also A.W. Skempton, 'Portland Cements, 1843–1887', *Transactions of the Newcomen Society*, 35 (1962–63), 117–52.

said that it constitutes, in point of fact, the sensational literature of our profession.[51]

Bibliography

Acheson, M., 'Edwin Chadwick: a review of two centuries of public health', *Water and Environmental Management*, 4 (1990), 474–83.

Anderson, L., 'Hard choices: supplying water to New England towns', *Journal of Interdisciplinary History*, 15, no.2 (Autumn 1984), 211–34.

Ashcroft, Anthony, 'John Snow – Victorian physician, medicine in Northumbria', *Studies in the History of Medicine in the North East of England*, The Pybus Society for the History and Bibliography of Medicine (Newcastle upon Tyne, 1993), 246–60.

Baker, M.N., *The Quest for Pure Water* (American Water Works Association, Inc., 1948).

Bateman, J.F., *Metropolis Water Supply from the Sources of the River Severn* (Westminster, 1865).

—, *History and Description of the Manchester Waterworks* (London, 1884).

Bazalgette, Sir Joseph W., 'Presidential Address', *Minutes of Proceedings of the Institution of Civil Engineers*, 76 (1884), 2–69.

Benet, Juan, *Views of the Works of Canal de Isabel II by Clifford* (Madrid, 1989).

Binnie, A.R., *Rainfall, Reservoirs and Water Supply* (London, 1913).

Binnie, G.M., *Early Victorian Water Engineers* (London, 1981).

Blake, Nelson M., *Water for the Cities: A History of the Urban Water-Supply in the United States* (Syracuse, 1956).

Bodson, M., E.Detrienne, and F. Leclercq, *Le Barrage de la Gileppe* (Liège, 1877).

Bolton, F., *London Water Supply* (London, 1888).

Cain, Louis P., 'Unfouling the public's nest: Chicago's sanitary diversion of Lake Michigan, *Technology and Culture*, 15, no.4 (October 1974), 594–613.

Cecen, Kazim, *Sinan's Water Supply System in Istanbul* (Istanbul, 1992).

Chadwick, E., *The Sanitary Condition of the Labouring Population of Gt.Britain* (London, 1842; republished 1964 by Edinburgh University Press).

Clark, E.F., *George Parker Bidder* (Bedford, 1983). See particularly Chapter XVII: 'Sanitary Engineering', 214–29.

[51] Report of the 38th Annual Meeting, at Norwich, of *The British Association for the Advancement of Science* (August 1868), 180–88. G.P. Bidder was addressing the *Mechanics* Section of the meeting.

D'Acre, R., *The Art of Water-Drawing* (London, 1659; reprinted by The Newcomen Society, 1930).

Davidson, I., 'George Deacon (1843–1909) and the Vrynwy works', *Transactions of the Newcomen Society,* 59 (1987–88), 81–95, reprinted in volume 4 in this series, *Dams*, ed. Donald C. Jackson.

Deacon, G.F., 'The Vyrnwy works for the water-supply of Liverpool', *Minutes of Proceedings of the Institution of Civil Engineers*, 126 (1895–6), 26–127.

Dickinson, H.W., *The Water Supply of Greater London*, limited edition, (Newcomen Society, London, 1954).

Dickinson, H.W., *Sir Samuel Morland, Diplomat and Inventor, 1625–1695* (London, 1970).

Duffy, John, 'History of public health and sanitation in the west since 1700' in *The Cambridge World History of Human Disease*, ed. Kenneth F. Kiple (Cambridge, 1993), 200–06.

Dumont, Aristide, *Les Eaux de Lyon et de Paris* (Paris, 1862).

Evans, Mary B., *Water Distribution in Ancient Rome: The Evidence of Frontinus* (Michigan, 1994).

Finer, S.E., *The Life and Times of Sir Edwin Chadwick* (London, 1952).

Fitzsimons, Neal, ed., *The Reminiscences of John B. Jervis: Engineer of the Old Croton* (Syracuse, 1972).

Frazer, W.M., *A History of English Public Health, 1834–1939* (London, 1950).

Frontinus, Julius Sextus, *Stratagems and Aqueducts of Rome*, translated by Charles E. Bennett (Harvard, Mass., 1997). [Loeb Classical Library]

Gibson, W.G., 'The water supply of Cambridge', *Transactions of the Newcomen Society,* 28 (1951–53), 277–8.

Gough, J.W., *Sir Hugh Myddelton* (Oxford, 1964).

Hamlin, Christopher, 'William Dibdin and the idea of biological sewage treatment', *Technology and Culture*, 29, no.2 (1988), 189–218.

Hamlin, Christopher, *Public Health in the Age of Chadwick: Britain, 1800–1854* (Cambridge, 1998).

Hardy, Anne, 'Parish pump to private pipes: London's water supply in the nineteenth century', *Medical History,* Supplement No.11, (1991), 76–93.

Hodge, A. Trevor, ed., *Future Currents in Aqueduct Studies* (Leeds, 1991).

—, *Roman Aqueducts and Water Supply* (London, 1992).

Hoffecker, Carol, 'Water and sewage works in Willmington, Delaware, 1810–1910', *Essays in Public Works History,* 11 (1981), 1–15.

Jansen, Gemma C.M., 'Water systems and sanitation in the houses of Herculaneum', *Medelingen van het Nederlands Institute te Rome,* 50 (1991).

Jardine, R.W., 'James Jardine and the Edinburgh Water Company', *Transactions of the Newcomen Society*, 64 (1992–93), 121–30.

Jephson, Henry, *The Sanitary Evolution of London* (London, 1907).

Jones, F.C., *The Bristol Waterworks Company, 1846–1946* (Bristol, 1946).

Lewis, R.A., *Edwin Chadwick and the Public Health Movement, 1832–1854* (London, 1952).

Longmate, N., *King Cholera* (London, 1966).

Matthews, William, *Hydraulia* (London, 1835).

[Metropolitan Water Board] *London's Water Supply, 1903–1953: A Review of the Work of the Metropolitan Water Board* (London, 1953).

Norrie, Charles Matthew, *Bridging the Years: A Short History of British Civil Engineering* (London, 1956). Good on water-supply and the contractors who specialised in it.

O'Connell, James, 'Chicago's quest for pure water', *Essays in Public Works History*, 1 (June 1976), 1–19.

Owen, David, *The Government of Victorian London, 1855–1889* (Cambridge, Mass., 1982).

Owens, E.J., 'The Kremna aqueduct and water supply in Roman cities', *Greece & Rome,* 38, no.1 (April 1991), 41–58.

Porter, Elizabeth, *Water Management in England and Wales* (Cambridge, 1978).

Porter, Roy, *The Greatest Benefit to Mankind* (London, 1997), especially 'Public health', 405–15.

Porter, Roy, 'Cleaning up the great wen: public health in eighteenth-century London', *Medical History,* Supplement no.11 (1991), 61–75.

Reddaway, T.F., 'London Bridge waterworks after the Great Fire of London', *Transactions of the Newcomen Society*, 19 (1938–39), 217–19.

Regan, C.J., 'Early developments in London's sewage disposal with particular reference to the work of W.J. Dibdin', *Journal and Proceedings of the Institute of Sewage Purification* (1951), 338–47.

Reti, Ladislao, *El Artificio de Juanelo Turriano: Su Historia y su tecnica* (Toledo, 1967).

Robins, F.W., *The Story of Water Supply* (Oxford, 1946).

Rudden, Bernard, *The New River: A Legal History* (Oxford, 1985).

Russell, P., 'John Frederick La Trobe Bateman: water engineer', *Transactions of the Newcomen Society,* 52 (1980–81), 119–38.

Saul, G.M., 'Isambard Kingdom Brunel and the 19th century water industry', *Brunel Bulletin* (Brunel University, July 1981), 2–4.

Saul, G.M., 'John Rennie (1761–1821): one of his contributions to waterworks technology', *Transactions of the Newcomen Society*, 59 (1987–88), 3–13.

Schoenberg, Bruce, S., et al., 'Snow on the water of London', *Mayo Clinical Proceedings*, 49 (1974), 680–84.

Seeger, R.C., 'The history of Melbourne's water supply', *The Melbourne and Metropolitan Board of Works Federation Journal* (1948).

Shadwell, A., *The London Water Supply* (London, 1899).

Simon, Sir J., *English Sanitary Institutions* (London, 1897).

Sisley, Richard, *The London Water Supply* (London, 1899).

Skempton, A.W., ed., *John Smeaton FRS* (London, 1981).

Smith, N.A.F., 'The Roman Dams of Subiaco', *Technology and Culture*, 11, no.1, (1970), 58–68.

—, *A History of Dams* (London, 1971).

—, *Man and Water: A History of Hydro-Technology* (London, 1975).

—, 'The Pont du Gard and the aqueduct of Nîmes', *Transactions of the Newcomen Society*, 62 (1990–91), 53–80.

Snow, Dr John, *On the Mode of Communication of Cholera* (London, 1849).

Sprague De Camp, L., *The Ancient Engineers* (Doubleday & Co. Inc., USA 1960; British edition, London, 1963).

Storrs, F.C., 'Lavoisier's technical reports: 1768–1794. Pt. II (1): The Paris Water Supply; Pt. II (2): "Corrupt waters"', *Annals of Science*, 24 (1968), 179–97.

Szreter, Simon, 'The importance of social intervention in Britain's mortality decline', *c.*1850–1914: A re-interpretation of the role of public health', *Social History of Medicine*, 1, no.1 (1988), 1–37.

Tarr, Joel A., 'Water and wastes: a retrospective assessment of wastewater technology in the United States, 1800–1932', *Technology and Culture*, 25, no.2 (April 1984), 226–63.

Terry, William, *History and Description of the Sheffield Water Works* (Sheffield, 1908).

Vitruvius, Marcus, *The Ten Books on Architecture*, translated by Morris Hickey Morgan (New York, 1960).

Wegman, Edward, *The Water Supply of the City of New York, 1658–1895* (New York, 1896).

White, K.D., *Greek and Roman Technology* (London, 1984).

Wilson. P.N., 'Kendal reservoirs', *Transactions of the Cumberland and Westmorland Antiquarian and Archaeological Society*, 73 (1973), 329–31

1

Attitudes to Roman engineering and the question of the inverted siphon

Norman A.F. Smith

I. A general introduction

Because much has been written on Roman technology and because what has been written bears an air of great authority it is tempting to suppose that the subject has been thoroughly researched and is well understood. In discussions of many aspects of Roman technology the temptation to follow an established pattern has proved irresistible. Continual repetition of the same facts and figures and a sustained commitment to certain ideas and interpretations have consolidated a stereotyped view of history. Not infrequently the original authority for what has become a firmly held and rarely questioned position is disturbingly venerable.

Many of the views that are rigidly adhered to were first advanced in the nineteenth century at a time when the study of classics reigned supreme, architectural history was flourishing and, most influential of all, archaeology was uncovering new material at an unprecedented rate. Such was the enthusiasm to make use of archaeological findings that conclusions were prematurely drawn by people who in many cases were not well qualified to assess the work and capabilities of ancient engineers. Even less excusable, on occasions, were the pronouncements of engineers who, inspired to dabble in ancient history and archaeology for a variety of motives, were prone to expound some manifestly extravagant views, not least in the area of hydraulic technology. Regrettably the fashion is by no means dead even now and the legends continue to be given an occasional airing.

It would be foolish and unjust to deny the value of a large part of the earlier work done by classicists, architectural historians and archaeologists. As a source of basic information awaiting the scrutiny and appraisal of modern historians of technology theirs is important material. This is especially true of the archaeological evidence which in many cases cannot now be examined, so thoroughly have modern developments and re-developments obliterated the traces of Roman engineering over, on and under the ground. On the other hand new archaeological studies continue to add to our knowledge; they range from the more modern applications of underwater archaeology and aerial photography[1] to definitive treatments of Roman remains in geographical areas previously only casually explored.[2] Without doubt, however, there is still much to be done, or repeated, if the object is to understand the Romans as engineers rather than as architects, town planners, farmers, soldiers and so on.

The translation and annotation of a variety of classical texts has made available a wide range of material which is of importance to historians of engineering and technology. To the established and definitive works of E. A. Thompson[3] on military technology, A. G. Drachmann[4] on Roman (and Greek) mechanical technology and L. A. Moritz[5] on grain-milling have recently been added H. Plommer's[6] work on Roman building manuals, O. A. W. Dilke's[7] study of Roman surveying and K. D. White's[8] extensive research into Roman agriculture.

The body of information accumulated by classicists and archaeologists is impressive and of substantial interest. Nevertheless, a very real question remains. Do we yet have a full appreciation of the nature and purpose of Roman engineering which does justice to the material on which such a study could be based? The answer is no, not by a long way. Historians of engineering have as yet failed to compose anything like a coherent and complete account of Roman engineering. Worse, and in many ways surprising, is their failure to address themselves, with a few exceptions, even to specific areas of Roman technology. There is after all every reason to suppose that the engineering historian, looking in a fresh way at the evidence and asking new and more pertinent questions of it, may arrive at interpretations and conclusions different from those of linguists and archaeologists.[9]

The study of ancient engineering generally and Roman engineering in particular is beset with difficulties. Professor M. I. Finley has aptly remarked:[10]

> Paradoxically, there was both more and less technical progress in the ancient world than the standard picture reveals. There was more, provided we avoid the mistake of hunting solely for great radical inventions and we also look at developments within the limits of the traditional techniques. There was less — far less — if we avoid the reverse mistake and look not merely for the appearance of an invention, but also for the extent of its employment.

There is more than a hint here that a major obstacle to comprehending ancient technology is the difficulty of coming to terms with attitudes to technological change and purpose which are unfamiliar to ourselves. Modern technology is characterised, amongst other features, by rapid change and development, by the presence of a profit motive, by an immense reliance on labour saving techniques and by the massive production of consumer goods and services. In antiquity these characteristics were, for the most part, conspicuous by their absence.

Ancient technology evolved in a different social, economic and cultural climate from our own and it reflects this fact. What today we would classify as civil engineering was immensely important throughout a wide repertoire from irrigation and land reclamation to road- and bridge-building, water-supply works and the construction of docks and harbours. The building and agricultural arts also flourished but, comparatively, the mechanical ones did not. Machines characteristically fell into the province of the military engineer; they played only an ancillary rôle in the civilian techniques of building, irrigation (for water-raising), mining and to a small extent farming. The case of water-power is instructive.

Vitruvius' description[11] of the vertical water-wheel is so lucid and unambiguous that there is no way of denying the Romans' familiarity with at least the idea of water-power at an early date, about 25 B.C. Three and more centuries later evidence for a handful of working installations has been unearthed by archaeology.[12]

Modern writers have favoured two interpretations of these facts. Convinced that once the water-wheel was known it *must* have been widely and enthusiastically adopted, some historians turn a blind eye to the slightness of the written and archaeological evidence. At the other extreme the infrequent use of water-power is accepted but surprise is expressed at Roman society's failure to capitalise on what *ought* to have been a revolutionary new idea (as the Middle Ages did). In reality we must face the fact that although the Romans knew how to use water-power they chose not to, at least not to any extent and then only at a late date. The real question is why not, and the answer, ultimately, is that we do not know.[13] What *is* clear, however, is that an attitude to mechanical power prevailed which was fundamentally different from that adopted in later periods, and the fact is important.

Generalisations about the nature of technology — its purpose, the way it progresses and the influences which bear upon it (especially science) — are commonplace. Such generalizations are almost always too sweeping and too simple; as a result they are misleading and often nothing less than erroneous. Two important points have to be borne in mind. Technology's rôle and man's expectations of it have not been the same in different periods; and the fundamental natures of various technologies are intrinsically different. To what extent these considerations help to account for the relative importance of different branches of technology in antiquity it is difficult to say.

Basic needs — irrigation, transport, water-supply, mining and so on — *were* met partly because they were inescapable but partly also because what was required in these cases was more obvious. The building and agricultural arts embodied, to a large extent, essentially simple technical propositions (and components which were static). They also required massive organization and abundant labour for their execution on a grand scale and time, in other words the accumulation of experience, for their technological development.

By comparison the mechanical arts involve concepts which are most difficult to grasp and their application requires labour which is skilful rather than plentiful. Quite what promotes and cultivates mechanical ingenuity — the capacity, in short, to visualize combinations and couplings of moving parts which will perform a rather precise function — is a most fundamental question. Some peoples have demonstrated mechanical ingenuity, medieval European and Chinese for instance; others have not. The Roman Empire's failure to mechanize is perhaps attributable to a measure of mechanical 'blindness' in addition to all the other factors which have been proposed.

To quote Professor Finley again:

> There was more, provided we avoid the mistake of hunting solely for great radical inventions and we also look at developments within the limits of the traditional techniques.

This notion is borne out convincingly by Roman civil engineering. There is virtually no aspect of Roman structural and hydraulic engineering which had not been practised, to some extent, by earlier societies. And in fact Roman engineering generally drew heavily on the accomplishments of captive peoples: the Carthaginians (mining, ship-building and harbour construction); the Etruscans (irrigation, land drainage and reclamation, and arcuate construction); the Greeks (a host of techniques amongst which surveying, hydraulic cement, water-supply, harbour works and ship-building were particularly significant); the Egyptians (irrigation and canal-building); and the Nabataeans (desert irrigation using dams).

Even though the Romans contributed nothing that was fundamentally new to the civil engineering repertoire they did nevertheless apply this repertoire on an unprecedentedly grand scale and achieve an impressive result. Moreover, Roman engineering emphasized three significant developments which had never been prominent before: a widespread application of the semicircular arch, the extensive use of concrete and the concept of public works. In these important respects the Roman civil engineer was the peer of his predecessors.

The catalogue of Roman civil engineering attainments is impressively long. Regrettably many histories of the topic are little more than a chronological ordering of that catalogue. The number of engineering historians who have addressed themselves to a critical assessment of techniques, methods, organization, personnel, costs and so on is conspicuously small. Perhaps the means to probe such issues do not exist, at least not sufficiently to reach substantial conclusions or a complete insight. But until the effort is made and the questions are asked of the evidence there is no knowing what might be revealed.

At present we know virtually nothing about Roman engineers. How many can be named? Far fewer than must have been involved in so many projects. Who carried out the work? Evidently it was very frequently corps of military personnel[14] in which case it would be most interesting to know how many men were involved, to what extent they specialized, how they were organized, where and when they were employed and what techniques they used. Professor Dilke's work[15] on Roman surveying is disappointingly uninformative about the problems of setting out large construction jobs and levelling and aligning routes for roads and aqueducts, not agrimensorial work strictly speaking but unquestionably closely related.

Especially interesting is the nature of the improvements and progress achieved by Roman civil engineers. It is evident that structural evolution, a process which is typical of civil engineering's development in other periods, did occur; it is manifest in aqueduct bridge-building, in dam-construction and in road bridges. Changing shapes and increasing size are probably related to improved methods and new materials, notably concrete. But the nature of the relationship is anything but clear. The need (and ambition) to build on a larger scale perhaps prompted the development of concrete construction; alternatively, the development of concrete and a growing awareness of its strength and capabilities could have been the basis for confident steps towards bigger structures and structural units, a culmination being the amazing roof of the Pantheon.

There is another important issue, and a central one: Roman technical writings, or rather the lack of them. In relation to the immense amount of engineering work carried out the surviving texts — Vitruvius, Frontinus, and a handful of later works[16] — are exceedingly few and not especially relevant. It might be argued of course that another era of great building, the Middle Ages, has not bequeathed to us a very large literature either. But there is a difference. Medieval notebooks and drawings are strictly relevant. There is no escaping the purpose and practicality of Villard de Honnecourt's 'Sketchbook' or Mathias Roriczer's *On the Ordination of Pinnacles*.[17] That surviving drawings of cathedral facades and ground plans really were the ones used is evident from comparison with the existing structures.

What can be said of the status or indeed the purpose of Vitruvius' *De Architectura?* In the first place it must be remembered that *De Architectura* is a unique survival and for that reason alone a considerable importance is attached to it. But whether or not the 'Ten Books' comprise either an accurate or a full picture of Roman building is another matter. Some deficiencies in Vitruvius as a source are self-evident. The date of his treatise, c. 25 B.C., is early in relation to the bulk of important Roman engineering. Vitruvius leaned so heavily on Greek writers and material that his work is concerned as much with Greek technology as Roman. There are many aspects of Roman civil engineering which we know were skilfully and frequently practised which find no place in *De Architectura* at all.

What was the status of the 'Ten Books' in the Roman period? The truth is that we know of no substantial evidence that Vitruvius' book was valued as a building manual by his contemporaries and successors. He was quoted by Pliny the Elder and there is a fleeting reference to him in Frontinus (in connection with the sizes of water pipes). Further, H. Plommer has shown the extent to which the later Roman tracts on building by Cetius Faventinus (4th century) and Palladius (5th century) were based on Vitruvius. Nevertheless, in relation to the bulk of Roman civil engineering as it was practised in the first and second centuries Vitruvius was probably of limited use. As Sir Reginald Blomfield once remarked: 'Of the true vitality and creative power that was latent in Roman architecture I doubt if any glimpse is to be caught in Vitruvius' treatise'. It is an intriguing thought that the status and importance of Vitruvius was very likely far greater during the Renaissance than ever it had been under the Romans. This same exaggerated importance attaching to a sole survival has worked its deceptive influence in modern times.

Frontinus' *De Aquis Urbis Romae* presents problems of a different sort. Unlike Vitruvius, Frontinus was a man of good family and education; he was at one time (A.D. 74–78) the governor of Britain; and he wrote on various aspects of land surveying together with a treatise on warfare.[18] His study of Rome's aqueduct system was based on some seven years experience as the city's water commissioner during which time he worked resolutely to bring some order and efficiency to the operation of a public utility which had been woefully neglected and chronically mismanaged for years. There is an immediacy and sense of personal involvement in *De Aquis* which is necessarily absent in *De Architectura*. The two works differ too in that whereas Vitruvius is wide ranging and ostensibly technical, Frontinus is focused sharply on one specific issue and is not primarily a technical work. Frontinus was above all a civil servant, not an engineer. As a source for engineering history *De Aquis* needs to be used carefully and more critically than has usually been the case.[19]

In essence the state of affairs is thus. Archaeological evidence from all over Europe, North Africa and the Middle East demonstrates the unprecedentedly huge scale on which civil engineering flourished in the hands of Roman engineers. At the same time, however, we know disproportionately little about the practice of Roman civil engineering — the way it was organized, financed, constructed, maintained and surveyed; how it improved and developed; who were the engineers and their work forces. Two important surviving treatises (plus a handful of lesser, and later, ones) are of limited value in throwing light on these matters.

Public water-supply was the most significant aspect of Roman civil engineering. Nothing like its scale had been achieved before, nor was to be again until the nineteenth century. A crucial problem facing European and American cities a century or so ago was the need to supply water on the Roman scale; and, as a public utility, in the Roman manner as well. Essentially it was a wholesale resurrection of Roman engineering which solved the nineteenth century problem. Thus it is no coincidence that one hundred years ago civil engineers cultivated an interest in Roman archaeology and history.

What follows in the second part of this paper is a new look at one particular aspect of Roman water-supply technology — the inverted siphon.

II. Roman inverted siphons

'Inverted siphon' is, in some ways, an unfortunate term for what is in fact a pressurised pipe-line carrying water across a valley or depression. However, the term is firmly established in the technical vocabulary and this is no occasion for trying to eject it. Roman inverted siphons will bear investigation for three reasons. In the first place the subject is of some technological interest in itself and raises significant questions about the materials and hydraulic skill employed. Secondly it is a topic which has rarely been considered (and not at all recently) and even when it has, a number of misconceptions and inaccuracies have been expressed so convincingly that they have gained general currency. And thirdly the siphon issue affords an opportunity to make one or two general points of significance in a broader context.

Historians of technology have rarely noticed, and never emphasized, a significant feature of water-supply in antiquity which is that, with two exceptions, all the essential technological ingredients and variations had been identified, and utilized, before Roman engineers came on the scene.

As early as 700 B.C. Assyrian engineers had laid out a

water-supply system of notable technical completeness at Nineveh,[20] and Greek engineers regularly supplied drinking water over long distances and arranged its distribution.[21]

The two omissions in the pre-Roman repertoire were, firstly, the use of large impounding reservoirs at the sources of aqueducts[22] and, secondly, the application of hydraulic concrete for both structural and hydraulic purposes.

The preoccupation among engineering historians with Roman water-supply is understandable in view of the novel scale on which the Romans worked, the imposing remains and the survival of Frontinus' unique manuscript.

But bearing in mind that the Romans learned much of their engineering from the Greeks, that they regularly (and perhaps frequently) employed Greek engineers, and that Vitruvius relied extensively on Greek sources it is important to notice that a number of Greek aqueducts featured siphons. Among examples which have been identified are those at Pergamum, Patara, Mylasa, Methymna, Catania, Selinus and Syracuse. Lead was evidently a material used to fabricate the pipes and wood, earthenware and stone have been suggested as well.

From many and various opinions expressed by modern writers, a brief selection will indicate what a variety of confused and contradictory views comprise the current position. It was R. J. Forbes who wrote:[23]

> Siphons and bold tunelling are typical for the water-supply systems built by the Greek tyrants and the Hellenistic kings from Sicily and southern Italy to Asia Minor. In Roman times the siphon was used only in some few cases, probably because of leakages and the relatively poor materials available for high pressure.

F. Klemm has taken a different view with emphasis on a non-technical argument:[24]

> Engineering in the Roman Empire was principally State engineering, which accomplished much with the traditional technical means, especially in organizing the making of streets, bridges, aqueducts, war machines and lofty buildings. There were in the Empire more than 186,000 miles of good roads. Ten aqueducts provided Rome daily with some 220 million gallons of water (Plate 1). And the dome of the Pantheon, built in Rome in the second century of our era had a span of over 142 feet 6 inches.
>
> Though the Romans usually constructed gravitational water supplies which necessitated extremely costly aqueducts, yet it must by no means be concluded that they were ignorant of the fundamental principle of communicating pipes for the supply of water by pressure as constructed by the Greeks

before them. Perhaps indeed the difficulties of firm control of a water supply system worked by pressure may have partly determined the choice of gravitation. But it was undoubtedly also a question of different styles of building technique. The Roman wished his technical works to express the political power of the Empire. Here was a reason for the gravity supply system, borne on high through the countryside on mighty arches, whereas the pressure system merely adheres closely to the natural contour of the land. Similarly the arrangement of Roman State buildings was on an axial and symmetrical plan which was an expression of the Idea of Power. This was alien to the thought of classical Greece, who erected her buildings in freedom and without constraint.

In a lengthy and confused discussion Sir Thomas Ashby, whose central point is that Roman engineers were fearful of high pressure and always anxious to avoid it, contributed these views:[25]

> Just as some ridges, however, demanded a tunnel [for an open-channel] and precluded circumvention, so certain valleys, too long to skirt and too deep to bridge, had to be crossed by a siphon. The principle that water thus conveyed rose to its own level was well known, and striking examples of its employment exist. . . . It should be noted that Vitruvius specifies that the lowest point in a siphon should run level for some distance, and should have vents to reduce pressure. Thus, every attempt was made to avoid the natural creation of a pressure-supply, with which these siphons have been confused. The siphons were very expensive to build, since the pipes were small ones of lead, arranged in series to reduce pressure; . . .
>
> It might be asked why the Romans did not take advantage of their excellent skill in cement manufacture to construct the conduits for a siphon in this material, and so to arrive at the theory of a pressure-supply; for the crushing resistance of their hydraulic cement exceeds the figure now accepted as the safe standard.

Some of the points referred to by Ashby we shall look at in more detail shortly but here and now one piece of nonsense can be disposed of. The strength of pressurized pipe-lines has nothing to do with 'crushing resistance'; the strength of an internally pressurized tube is dependant in fact on the material's *tensile* strength.

Finally, the most recent view. In 1974 H.R.H. The Duke of Edinburgh observed that:[26]

> The ancient Romans succeeded in building magnificent unsupported domes like the Pantheon in Rome and those

superb aqueducts like the Pont du Gard in France. But those very aqueducts would not have been necessary if the engineers had realized that water in a pipe will always come back to its original level.

The above extracts are at odds with the evidence and with each other. And above all if Greek engineering, with which the Romans were so familiar, could make successful use of siphons why not Roman engineering? The fact is that inverted siphons *were* a component of Roman water-supply technology and they were regularly adopted and rationally utilized. The high water-pressures induced *were* contained by pipes whose wall-strength was adequate; there is no evidence that leakage was such a serious problem as to be a disincentive; and evidently the use of inverted siphons was not avoided simply because tall aqueduct bridges looked more spectacular.

Already known Roman siphons outnumber the Greek examples. There may well be more. From an archaeological point of view it is important to bear in mind that whereas an aqueduct bridge may survive, if only in fragments, a pipe-line, especially of lead, is very vulnerable, an easy and attractive target for subsequent demolition. In time little trace of it will remain.[27]

A list of the more important Roman siphons which have been located is as follows:

Country	*Place*	*No. of siphons*	*Note*
France	Lyon	9	28
	Rodez	1	29
	Arles	1	30
Spain	Almuñecar	1	31
	Cadiz	1	32
Italy	Rome	1	33
	Angitia	1	34
	Alatri	1	35
	Termini Imerese	1	36
Britain	Lincoln	1	37
Turkey	Aspendos	1	38

By far the most substantial remains are to be found at Aspendos and among the set of nine siphons at Lyon. The most thoroughly documented specimens are those at Lyon.[39]

As for written evidence there is only one source and that is Vitruvius.[40] Here we begin to get into difficulties. Vitruvius on siphons is puzzling, obscure and inaccurate although here and there, as will emerge, he is plausible and straightforward. Unfortunately modern writers, with rare exceptions, have swallowed Vitruvius more or less whole on the subject of siphons and as a result they have offered some strange interpretations of how

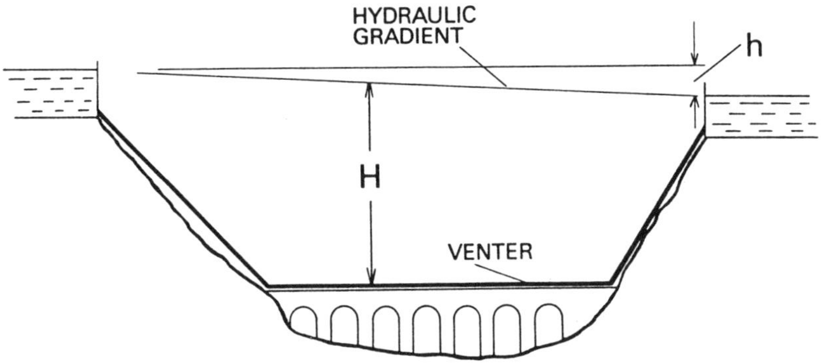

Figure 1. The elements of the inverted siphon.

Roman siphons worked, and even stranger ideas in some cases for their construction.

Figure 1 shows the essential features of an inverted siphon carrying water across a valley from a header tank to a receiving tank. The difference in water-level between the tanks, h feet, is the head lost in giving the water velocity and overcoming frictional resistance in the pipe. The head H feet is a measure of the static pressure in the siphon at any point. The maximum pressure will almost certainly be at the first bend provided that the deepest portion of the pipe-line is substantially horizontal.

In conjunction with Figure 1 we can examine the relevant portions of Vitruvius' account and analyze their contents in order to see how the Vitruvian material has been misunderstood and in what ways it has so successfully misdirected previous enquiries. Everything is taken from Chapter 6, Book VIII of *De Architectura*.

To begin with Vitruvius writes:[41]

> If the distance round such depressions is not great, the water may be carried round circuitously; but if the valleys are extensive, the course [of the siphon] will be directed down their slope. On reaching the bottom, a low substructure is built so that the level there may continue as long as possible. This will form the 'venter', termed KOILIA by the Greeks.

This piece is clear and corresponds satisfactorily with Figure 1. Notice that Vitruvius gives the Greek word for the venter, the lowest section of the siphon which was carried often, but not invariably, on a bridge.

Vitruvius continues by stating:

> Then, on reaching the hill on the opposite side, the length of the venter makes the water slow in swelling up to rise to the top of the hill.

This makes very little sense. If it is an effort by Vitruvius to describe the hydraulics of the situation, and the mechanism which makes the water climb to the receiving tank, it is less than adequate or comprehensible. Previous commentators rarely question or analyze this sentence and existing translations are as various as they are numerous; none make much sense.

The next sentence is acceptable. Vitruvius says:

> But if there is no such venter made in the valleys, nor any substructure built on a level, but merely an elbow, the water will break out, and burst the joints of the pipes.

Sharp bends in pressurized pipe-lines are indeed undesirable. In reality the archaeological evidence suggests that V-shaped siphons were never built, and in fact never had to be. Precisely how the sloping portions of a siphon joined the venter and the extent to which the bend was reinforced is not certain. However, at another point in his account, when discussing the use of earthenware pipes, as opposed to those of lead, Vitruvius says:

> Their joints must be coated with quicklime mixed with oil, and at the angles of the level of the venter a piece of red tufa stone, with a hole bored through it, must be placed right at the elbow, so that the last length of pipe used in the descent is jointed into the stone, and also the first length of the level of the venter.

Presumably this procedure was as applicable to lead pipes as to earthenware ones. Certainly there is no need to go so far as did the French engineer Belgrand who, in 1875, became so worried about elbows that he dispensed with them altogether, and credited the Romans with siphons laid in a continuous curve entirely free of any sharp bends.[42] Nothing suggests that this was ever done; the archaeological evidence is for the shape shown in Figure 1 and Vitruvius is not at odds with it.

Precisely what Vitruvius says next, or is trying to say, remains extremely obscure. The oldest extant versions of Vitruvius' manuscript — from the eighth, tenth and eleventh centuries — and many later versions based on these, are not in agreement and no one rendering makes any better sense than the others. Modern versions vary; for instance Morgan reads:

> And in the venter water cushions must be constructed to relieve the air pressure.

Granger puts it like this:[43]

> Further, stand-pipes are to be made in the bend [Granger really means venter] by which the force of the air may be released.

Others suggest neither water-cushions nor stand-pipes but 'air-valves'.

On the other hand one point is clear. Vitruvius is proposing some sort of device to relieve air pressure in the siphon. In fact he is advising on a non-existent problem. In a siphon carrying water under pressure the problem of air-locking cannot arise simply because the pressure *is* high. There is no air pressure to relieve when the siphon is in flow.

The conviction that Vitruvius must be writing sense at this point, and offering practical and useful advice, has led a sequence of modern writers to some odd conclusions so unscientific and impractical as to be extraordinary. Regrettably there is one body of opinion that can find in this part of Vitruvius an idea for reducing not the *air* pressure in the siphon tube, but the *water* pressure. This is impossible. The pressure in the venter is determined solely by the head above it and it cannot be relieved by valves, stand-pipes, water-cushions — whatever they are — or anything else.

Of all the recent interpretations which have been given some practical form, the prize for incompetence must go to the eminent French architectural historian Auguste Choisy. His extraordinary layout is shown in Figure 2.[44] Quite apart from failing to mention what the Romans used for sky hooks, the arrangement achieves nothing. Choisy is merely substituting two siphons for one, he is adding to the flow length, and achieving not one iota of pressure reduction or anything else. And in any case these idiotic elaborations do not even match the Vitruvian text.

Nineteenth century translators and annotators of Vitruvius, whose work laid the basis for twentieth century interpretations and whose ideas are embedded in these later works, were guilty of two defects in their approach. In the first place they had, more often than not, pre-conceived ideas as to how a Roman siphon *ought* to work, Vitruvius himself being less than explicit, and once they had finalized a design in their own minds and according to their own experience, they then succeeded in twisting and shaping Vitruvius' words to fit their own conclusions.

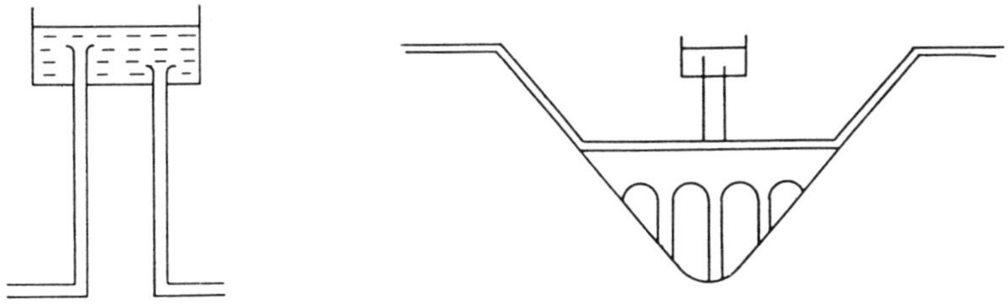

Figure 2. The Vitruvian siphon according to Choisy.

And secondly it is implicit in nineteenth century, and later, writers that Vitruvius is ultimately a reliable source. Any failure to comprehend him is ajudged to be a failure of modern translation and intelligence. This attitude overlooks two possibilities. Conceivably the extant versions of Vitruvius' manuscript are not, in fact, a faithful rendition of the original[45] which necessarily raises obstacles to understanding not likely to be overcome easily, or with any degree of certainty. On the other hand, even if they are faithful renderings, it is possible that Vitruvius did not fully understand his material himself. And if Vitruvius did not know what he was talking about then it is hardly surprising that we cannot find out.

How much of an experienced and practised engineer Vitruvius was is difficult to say. He claims very little personal involvement in engineering work, and later on the only mention of Vitruvius by Frontinus is in connection with plumbing and pipe sizes. Nor is Vitruvius to be judged as a particularly well educated man if the concensus view of his written Latin is a fair one. Indeed it is so bad that there used to be a vogue for attributing him to the third, or an even later, century.[46]

And what of his ability to read Greek, Greek sources being so frequently quoted in the 'Ten Books'? It is conceivable that Vitruvius' imperfect reading of some Greek work on siphons had caused him to confuse the inverted siphon with the true siphon, that is to say a pipeline which at some point carries the liquid flow above the hydraulic gradient. In this situation air-locking can be a problem because the pressure is less than atmospheric; it is a very serious issue in modern aqueducts and was already identified in some early Renaissance works.[47]

Vitruvius' apparent concern over air pressure in his siphons *might* be accounted for by a misreading of one of his Greek sources or by a lack of familiarity with the realities of conventional practice or both.

However, if one reads on to a later statement in Chapter 6 of Book VIII much appears to be clarified. In dealing with siphon tubes made of earthenware Vitruvius says:

> The level of the pipes being thus adjusted, they will not be sprung out of place by the force generated at the descent and at the rising. For a strong current of air is generated in an aqueduct which bursts its way even through stones unless the water is let in slowly and sparingly from the source at first.

The use of a valve, or valves, to facilitate the release of air when a siphon was started up from dry, potentially a very damaging operation, is not only plausible, but desirable. It is interesting too that those venters on the Lyon siphons which are sufficiently complete to allow measurement are not *quite* horizontal. They slope upwards in the direction of flow and this could be construed as an air release device also.

Air valves for the above purpose would serve in another capacity as well. Vitruvius says:

> It is also not ineffectual to build reservoirs at intervals of 24,000 feet, so that if a break occurs anywhere it will not completely ruin the whole work . . . such reservoirs should not be built at a descent, nor in the plane of a venter, nor at risings, nor anywhere in valleys, but only where there is an unbroken level.

However, in the interests of repairs and maintenance some means to drain a siphon tube would be needed.

The essentially straight forward position we have arrived at then is this. A Roman siphon comprised a run of parallel lead tubes connecting a header tank and a receiving tank each at the end of the conventional open channel. Valves in the venter were used to release air when the flow was started up, or they could be opened to drain the siphon when necessary; otherwise they were permanently closed. This specification will not explain all of Vitruvius but nowhere is it seriously at odds with him, and it does conform to the extant archaeological evidence.

Let us now look more closely at some of this evidence. And in view of what is to follow a small digression is in order.

There is a need to be exceedingly cautious when applying modern analytical techniques to ancient, and for that matter not so ancient, technical problems, situations, machines and so on. There is potentially a great danger that the outcome will be no more than a rather futile proof of the obvious. For example, the application of modern stress and stability analyses to historic and surviving dams will often show that these structures are stable and not over-stressed. And sure enough they are still standing. And even if they were not, or even if they were when they should not be — by no means an unusual result[48] — one is not really any further forward. Nothing is learned of the ideas, attitudes and concepts in the minds of the original builders or the design methods they used. A recourse to elaborate modern analysis to demonstrate much that is irrelevant and nothing that is particularly relevant, is rather pointless.

On the other hand the process is sometimes of value. If one calculates the specific speeds of early Fourneyron turbines, it is found that the values are so low as to indicate that it is not correct to regard those otherwise important machines as pressure turbines. And indeed this can be confirmed by a close inspection of the machines' settings. To some extent then, a modern treatment is useful as a basis for rejecting the conventional view that Fourneyron brought the fully developed modern pressure turbine

Aqueduct	*Date*	*Siphon*	*Length (feet)*	*Depth (feet)*	*Fall (feet)*	*Pipe diameter (external) inches*	*Number of pipes*
Mont d'Or	Augustus	Cotte-Chally	1,640	110—120	6½	?	?
		d'Ecully	6,560	210	65½	?	?
Craponne	Augustus	Tupinier	4,260	82	16½	?	?
		Tourillons	15,740	375	115	?	?
Brévenne	Claudius	Grange-Blanche	9,200	290	59	?	?
Gier	Hadrian	St. Genis	2,920	270	19½	10	9
		Soucieu	3,950	304	29	10	9
		Beaunant	8,560	405	30½	10?	±10?
		St. Irénée	1,993	155	5¼	10	7 or 8?

Siphons at Lyon
(compiled from C. G. de Montauzan and A. Grenier, see Note 28, and B. Buffet and R. Evrard, *L'Eau Potable a Travers les Ages*, Liège 1950, pp. 80—85)

into existence single-handed and overnight, so to speak. It is undoubtedly not so simple. In the field of medieval cathedral construction some interesting work has been done recently by means of photo-elastic stress analysis and some valid conclusions have been drawn.[49]

Modern analysis then must be used judiciously and the results evaluated cautiously. Useful conclusions need to be carefully separated from worthless ones and, in addition, basic assumptions and the accuracy of the values of known, or supposedly known, parameters must be assessed carefully.

There are decided limitations upon the use of modern hydraulic theory to investigate Roman siphons. Nevertheless, it will prove at least interesting to try it in the case of a few of the Lyon siphons, the only ones for which we have sufficient information (see table opposite) even to make a start.

The best preserved remains (see Figures 3 and 4) are of the four siphons of the Gier aqueduct, the last of the quartet of aqueducts to be built, probably in the reign of Hadrian. The aqueduct drew its supply from the river Gier near modern St. Chamond and ran a winding course of 75 kms to Lyon. The second of the four piped crossings, the so-called Soucieu siphon, had the following specification:[50]

External pipe diameter	10 inches (0.26 mts)
Number of pipes	9
Length of siphon	3,950 feet (1204 mts)
Fall	29 feet (8.844 mts)
Maximum depth	304 feet (92.822 mts)

Previous enquirers have been obsessed with the problem of siphon strength, some going so far as to deny (as we have seen) that Roman engineers ever used them because the pressures would have been uncontainable. The preoccupation with trying to find in Vitruvius the means to reduce water-pressure is probably associated with this attitude. In fact strength is not a problem, at least not necessarily.

The tensile strength of lead is about 2,000 pounds per square inch and when Belgrand tested[51] some replica Roman pipes in the 1870s they failed at stresses in the range 2,100–2,200 pounds per square inch, equivalent to internal pressures of 18 atmospheres which is something in excess of 600 feet of water.

Let us assume for the sake of being cautious a stress factor of 2, i.e. an ultimate stress of 1,000 pounds per square inch, and an operating head of 400 feet of water, adequate for any of the Lyon siphons. For these conditions a 10 inch pipe would contain the pressure if its wall thickness was ¾ inch. In other words the internal diameter would be 8½ inches. Without doubt this represents a very hefty pipe, bigger by far than any of the standard

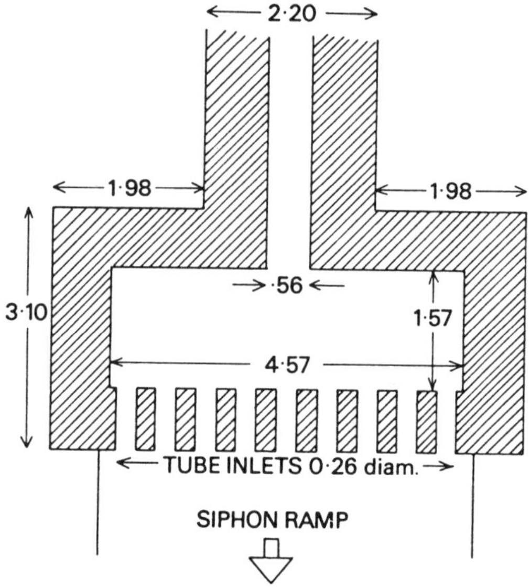

Figure 3. Plan of the header tank on the Soucieu siphon (dimensions in metres).

pipe sizes quoted by Vitruvius or Frontinus. However such pipes must have been made and an extraordinary commitment to large scale soldering was accepted.

Such pipe sizes also represent a very considerable *weight* of lead. One can show that the Soucieu siphon must have weighed, in total, of the order of 1,700 tons and Montauzan's estimate that the nine Lyon siphons between them weighed in at between 10,000 and 15,000 tons is not far-fetched. If the Romans *were* disinclined to build siphons then the difficulties of mining, processing, transporting, fabricating and laying these prodigious weights of lead are the best reasons of all.

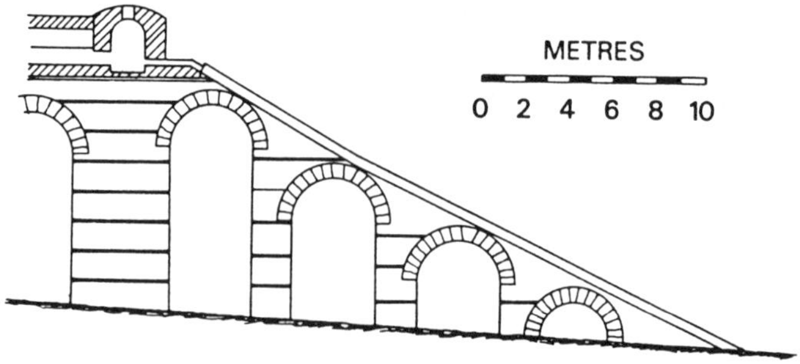

Figure 4. Header tank and siphon ramp on the Beaunant siphon.

Many engineers have attempted to calculate the quantities of water which Roman aqueducts supplied to towns. More often than not this is a hopeless task because of various imponderables not the least of which is the fact that the depths of the covered conduits were determined to allow the movement of workmen rather than the flow of water. And leakage, as Frontinus complains, was frequently very serious. However, compared with an open-channel, a siphon is rather precisely specified and an opportunity exists to carry out a more objective estimate. The standard pipe resistance formula is

$$h = \frac{4\ f.\ L.V^2}{2gd}$$

where h = fall
L = length
V = pipe velocity
d = internal diameter
f = pipe friction coefficient

Applying this formula to one of the Soucieu tubes — and without reproducing the intricacies of the calculation — one can show that the pipe friction coefficient is 0.0038 at a pipe Reynolds number of 2.7×10^5 which signifies, not surprisingly, turbulent flow. The pipe velocity is 4.69 ft/second and the flow of the whole siphon works out at 16.7 cubic feet per second.

Now it is reasonable to suppose that the other three siphons carried the same flow and we can test therefore the performance of the St. Genis siphon, the first of the four siphon crossings on the Gier aqueduct. It used the same size tubes, the same number of tubes and was 2,920 feet long. The pipe friction formula gives a value of h for the St. Genis siphon of 21½ feet in order to provide a sufficient flow capacity. In fact the measured fall across the St. Genis siphon is 19½ feet. The degree of agreement is encouraging and suggests that a standard layout of pipe sizes and number of pipes was used in conjunction with a fixed ratio between length and fall. The L/h factor is 150 for the St. Genis siphon and 137 for Soucieu.

Unfortunately this proposition will not work. The L/h ratio is 281 in the case of the Beaunant siphon, the third on the Gier aqueduct, and higher still, 380, for the St. Irénée siphon which is the fourth and last. In other words the four siphons on the Gier aqueduct were not hydraulically similar and in particular if the Beaunant and St. Irénée siphons discharged the same quantities as those of St. Genis and Soucieu they must have been fitted with more pipes or bigger pipes or both. Conceivably the problem is made even more complicated by another factor.

Over 200 years ago it was first noticed that the venter bridges on the Lyon system were significantly wider than the sloping ramps which carried the siphon tubes down and up the valley sides. Some writers have taken this to indicate that in the depths of the siphon more tubes or bigger tubes were used as compared with the higher levels. It has been supposed that at the lower levels each tube was branched into a pair for the purpose of developing greater pipe strength (the old spectre), one 10 inch tube of ¾ inch wall thickness being divided, for instance, into a pair of 6 inch tubes of ½ inch wall thickness.

Quite what sort of an improvement this represents and whether or not it would have been worth the effort is difficult to say. In any case a simpler interpretation suggests itself. Access to the ramps for maintenance was relatively simple, everything being more or less at ground level. Across the venter bridges however the case for additional width to accommodate workmen, materials and transport is a strong one. Perhaps no further explanation is needed.

As emphasized already, historians have been hitherto much concerned with the question whether or not Roman siphons could have been made sufficiently strong. By comparison hydraulic problems have not been given sufficient consideration and the issue of overriding importance has hardly been broached at all. It is of course the difficult question of how Roman engineers matched the flow of an open channel to the flow capacity of a multi-piped siphon. Flow calculations of the type used above were not developed until the nineteenth century and if Frontinus is a reliable witness (which in this matter may very well not be the case) the Romans had not even deduced that quantity of flow is a function not only of cross-sectional area but of velocity as well.[52]

However, experience and experiment would reveal, presumably, that a given length of aqueduct required a much greater fall per unit length by siphon than by open channel, the frictional resistance being so much higher in the case of a siphon. Given the knowledge that a siphon needed to fall about 1 foot for every 100 feet of its length, we can conjecture that the configuration of a siphon was finalized by judicious trial and error and adjustment, and by simply ensuring that there were sufficient tubes to cope. Using a multi-tube system this approach is substantially possible. It is of interest in this context that the header tank on the St. Genis siphon was originally fitted with ten tube inlets. One of these has been blanked off which may suggest that the siphon in its original form was over-designed. On the other hand in some header tanks, the presence of a high-level side-spillway makes it very clear what happened to an excessive inflow from the open channel. It was simply tipped over the side, a wasteful but effective expedient.

It was Germain de Montauzan's view[53] that the Romans fixed the size of siphons by referring to tables which related such parameters as velocity, fall, length, pipe diameter and so on. Such tables were compiled, so Montauzon supposed, from the observed performance of existing pipe-lines. This idea is not really tenable. It ascribes to the Roman engineer an appreciation of experimental analysis and a capacity to extract useful results from such an approach for which there is no evidence whatsoever. And in any case the number of siphons built was probably not sufficient to make the method viable in the first place.

So far the essence of the construction and operation of the typical Roman siphon has been outlined and while many of the details are obscure the basic proposition is clear. The calculation carried out for the Soucieu siphon at Lyon gave a flow of 16.7 cubic feet per second. It would be foolish to claim a high order of precision for this figure but in suggesting the order of magnitude of the siphon's capacity it is clearly satisfactory. If anything the quantity calculated is too large which means that the effect of pipe-friction has been underestimated.

The next problem is to consider the rôle of siphons in relation to the alternative, that is to say an aqueduct bridge carrying water in an open channel along the hydraulic gradient. Aqueduct bridges were, after all, the conventional and by far the most frequently adopted solution if it was not possible, or expedient, to build an aqueduct at or below ground level. Sometimes, across the Roman Campagna for instance, aqueduct bridges were miles in length. Under what conditions then did Roman water-supply engineers choose to construct a siphon?

The highest Roman bridges, for roads and aqueducts, are listed below.

Pont du Gard	(France)	160 feet
Alcantará	(Spain)	155 feet
Narni	(Italy)	120 feet
Cherchel	(Algeria)	115 feet
Segovia	(Spain)	100 feet
Mérida	(Spain)	85 feet

Only two examples surpass 150 feet and only three more reach three figures. For the siphons, however, at least in those cases for which figures are available, the depths reached are greater (with a few exceptions):[54]

St. Irénée	(Lyon)	155 feet
Ecully	(Lyon)	210 feet
St. Genis	(Lyon)	270 feet
Grange-Blanche	(Lyon)	290 feet
Soucieu	(Lyon)	304 feet

Tourillons	(Lyon)	375 feet
Beaunant	(Lyon)	405 feet
Rodez	(France)	300 feet
Alatri	(Italy)	340 feet

Comparing the two tables the conclusion is inescapable: depth of crossing was the criterion. Something over 150 feet was evidently regarded as the limit for a bridge. Beyond that figure the task of constructing a bridge of masonry and/or brick and concrete was judged to become increasingly impossible as the height rose. Even if the required quantities of construction material could have been made available, the provision of falsework, scaffolding and mechanical lifting equipment would have been prohibitive. And beyond that there was the question of the superimposed weight on the foundations of the bridge's piers at heights approaching, or above, 300 feet. Problems of sheer weight might well have been a powerful disincentive in themselves.[55]

It would be interesting if the depth (or height, depending how one looks at it) criterion could be tested in some context and in fact one opportunity exists. The case in point is possibly significant in view of a development at present under way.

Toledo was not a major city in Roman times but it did possess a water-supply system of some interest.[56] Not the least significant feature of the Roman aqueduct at Toledo — its date is the 2nd century A.D. — was the Alcantarilla dam which impounded water at the channel's head.[57] The aqueduct itself is not particularly notable and its length, 24 miles, was not exceptional. The source of Toledo's water was to the south of the city and therefore, in its final stage of delivery, the aqueduct had to cross the Tagus Gorge, a formidable obstacle.

The remains of the conventional open-channel bringing water from the Alcantarilla dam have been located across the gorge from Toledo and their elevation shows that the depth of crossing involved was of the order of 100 metres. In the 1930s a Spanish engineer, Alphonso Rey Pastor, encouraged by remnants of a bridge's foundations which had been located on the sides of the Tagus Gorge, proposed a rather splendid reconstruction of the one-time Roman aqueduct bridge.[58] It is shown in Figure 5. For his model Rey Pastor took the Pont du Gard although for reasons of geography and chronology the Alcantará bridge would have been more suitable.[59]

Pastor's three-tier bridge is massive by any standards, and not just those of the Roman period if the 19th century example at Roquefavour is anything to go by.[60] The Toledo bridge, if bridge there was, was twice as high as the next highest, the Pont du Gard or the Alcantará bridge. The difference is too great for Pastor's

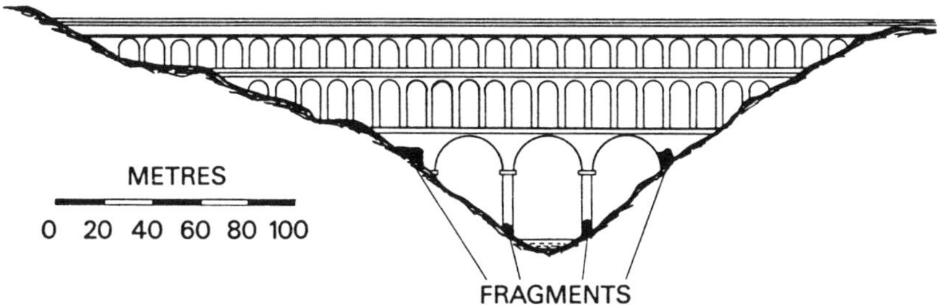

Figure 5. The Toledo aqueduct bridge according to Alphonso Rey Pastor: view looking upstream.

reconstruction to be convincing. The highest, the longest or the biggest specimens of civil engineering works are usually just a *little* in excess of their closest rival. For the highest Roman bridge to have been no less than *twice* as high as any other takes some believing.

Then there is the question of the bridge's fate. Where is it? What has happened to the huge quantity of material it must have contained?[61] Why have only the merest fragments survived until modern times (and indeed until some centuries ago to judge from old prints and engravings of Toledo)? The Pont du Gard and the bridges at Alcantará, Narni, Segovia, Mérida, Metz and a host of other places have all survived, substantially complete in many cases.

The extant fragments of the Toledo bridge are probably very revealing. The four lumps of unquestionably Roman masonry correspond *only* to the *lowest* tier of the bridge, of the order of 50 metres in height and a very high piece of construction according to Roman standards by itself. Indeed 50 metres is directly comparable with Alcantará and the Pont du Gard and at the limit of what Roman engineers regarded as viable.

The inference is obvious. Only one tier of Rey Pastor's bridge in reality ever existed and it was the venter bridge of a siphon some 50 metres (164 feet) deep. Indeed but for the problem of building bridge piers in a very torrential river at a difficult site, the venter bridge might have been even lower.[62]

Plans are advanced in Toledo to modify the Tagus régime, to rebuild Juanelo Turriano's 'Artificio' and its machinery and to house within the building a museum, and to resurrect on the sides of the Tagus gorge some substantial pieces of a 100 metre high aqueduct bridge of reproduction Roman masonry. In fact replica Roman lead pipe would appear to be much more appropriate.

Five conclusions

1. The Romans' use of siphons was calculated and rational. Bridges were not favoured because they were more impressive or

because they presented opportunities for ostentatious displays of structural expertise and grandeur.

So far as was possible Roman engineers went to some lengths, quite literally, to dodge valleys. Up to depths of 150 feet or so aqueducts were carried on bridges. Beyond that siphons were resorted to; it was not necessary very often.

2. Vitruvius' discussion of siphons is very obscure, parts are probably inaccurate, and there is no reason to suppose that our failure to make sense of much of what he says is a failure of modern intelligence. Nor are we necessarily deprived of essential understanding by ignoring or rejecting Vitruvian material. Vitruvius was not necessarily the last word on Roman engineering and very likely, here and there, he is not even the first word.

In the absence of a single text against which *De Architectura* can be compared we simply do not know how good or reliable a source Vitruvius is. Conceivably he is rather inferior and presents a less than full or typical picture of contemporary practice.

3. A technological study of a part of engineering history which has been most frequently presented by classicists, archaeologists and architectural historians suggests a shift of emphasis and the need to reject certain false notions. Siphon strength is not such an important issue as hydraulic behaviour; how Roman engineers produced the desired performance in siphons of differing specification is a mystery.

4. The use of modern analysis to evaluate ancient (and not so ancient) technical problems is of very limited application because of the nature and number of the assumptions which have to be made. The chances of reaching an unhistorical, irrelevant, untypical, inaccurate or silly conclusion are high.

5. Anything like a full understanding of any part of ancient engineering requires a certain level of archaeological and/or written evidence. If the evidence does not reach the requisite threshold of reliable facts, figures and other details then there are distinct limits to the firm conclusions which can be drawn. This is true of the siphon issue and it is undoubtedly true of other aspects of ancient technology. However let it not be thought that the study of ancient technology is either impossible or worthless. There may be obstacles but the field is far too interesting and, if the history of technology is to be seen in its full perspective, too important to be overlooked.

Notes

1. See for example J. W. Shaw: 'Greek and Roman harbourworks' in *A history of seafaring based on underwater archaeology*, edited by G. F. Bass, London 1972, pp. 87—112.

2. For North African dams and irrigation for instance there is C. Vita-Finzi: 'Roman dams in Tripolitania' in *Antiquity*, Vol. XXXV, 1961, pp. 14—20 and plates II—IV and also C. Vita-Finzi and O. Brogan: 'Roman

dams on the Wadi Megenin' in *Libya Antiqua*, Vol. II, 1965, pp. 65—71 and plates XXI—XXVII. For water-supply in Spain the standard work is now C. F. Casado, *Acueductos Romanos en España*, Madrid 1972.

3. Thompson, E. A. *A Roman reformer and inventor* (being a new text of the treatise *De Rebus Bellicis*), Oxford 1952.

4. Drachmann, A. G. *The mechanical technology of Greek and Roman antiquity*, Copenhagen 1963.

5. Moritz, L. A. *Grain-mills and flour in classical antiquity*, Oxford 1958.

6. Plommer, H. *Vitruvius and later Roman building manuals*, Cambridge 1973.

7. Dilke, O. A. W. *The Roman land surveyors. An introduction to the Agrimensores*, Newton Abbot 1971.

8. White, K. D. *Roman farming*, Thames and Hudson 1970.

9. An excellent example of the sort of thing which can be done, although in this case the subject is ancient Egyptian technology rather than Roman, is K. Mendelssohn, *The riddle of the pyramids*, Thames and Hudson 1974.

10. Finley, M. I. 'Technical innovation and economic progress in the Ancient World' in *The Economic History Review*, 2nd Series, Vol. XVIII, No. 1, August 1965, p. 29.

11. Two modern translations of Vitruvius are: *De Architectura*, trans. by F. Granger, Loeb Classical Library, 2 Vols., London 1962; and *Vitruvius. The ten books on architecture*, trans. by M. H. Morgan, Dover Publications, New York 1960. The description of the water-wheel is in Chapter V of Book X.

12. For these installations and further references to them see R. J. Forbes, *Studies in ancient technology*, Vol. II, Leiden 1955, pp. 90—93. The Roman water-mill at Barbegal is described by C. L. Sagui: 'La Meunerie de Barbegal' in *Isis*, Vol. 38, Parts 3 and 4, February 1948, pp. 225—31.

13. On this point see M. I. Finley, Note 10, pp. 35—7.

14. In connection with the construction of an aqueduct tunnel in A.D. 152 at Saldae in Algeria, the engineer, Nonius Datus, refers specifically to gangs of 'Alpine troops' and a 'detachment of marine infantry'; see L. Sprague de Camp, *The ancient engineers*, London 1963, p. 195.

15. As Note 7.

16. For Vitruvius see Note 11 and also H. Plommer, Note 6. A modern edition of Frontinus is *The stratagems and aqueducts of Ancient Rome*, Loeb Classical Library, London 1961. There is also C. Herschel, *Frontinus and the water supply of the City of Rome*, Boston 1899.

17. See J. Gimpel, *The cathedral builders*, London 1961, pp. 107—45.

18. For Frontinus on surveying see O. A. W. Dilke, Note 7. Frontinus on warfare is as Note 16.

19. It is generally accepted, for instance, that Roman engineers were unable to calculate water flows in open channels because they were under the impression that quantity depended only on cross-sectional area of flow and not on velocity. In fact the only authority for this conclusion is Frontinus and it is decidedly problematical whether he can be trusted on a technical matter such as this. Would it not be extraordinary if Roman engineers, who saw so much water flowing down so many channels, should never have deduced that velocity of flow governed quantity just as much as cross-sectional area? For the origins of the current belief in Frontinus' authority see C. Herschel, Note 16, Chapter V.

20. For this scheme and references to it see R. J. Forbes, *Studies in ancient technology*, Vol. 1, Leiden 1955, pp. 155—9.

21. See for example A. Burns: 'Ancient Greek water supply and city planning: a study of Syracuse and Acragas' in *Technology and Culture*, Vol. 15, No. 3, pp. 389—412.

22. For a more detailed treatment of Roman dams see N. Smith, *A history of dams*, London 1971, Chapter 2.

23. As Note 20, p. 161.
24. Klemm, F. *A history of Western technology*, London 1959, p. 48.
25. Ashby, Sir T. *The aqueducts of Rome*, Oxford 1935, pp. 34—7.
26. H.R.H. The Prince Philip, Duke of Edinburgh: 'Research and prediction' in *Notes and Records of the Royal Society of London*, Vol. 29, No. 1, October 1974, p. 18.
27. This point is especially significant in relation to the problem of the aqueduct at Toledo.
28. The standard work on the Roman aqueducts at Lyon is C. G. de Montauzan, *Les aqueducs antiques de Lyon*, Paris 1908. A more modern but briefer treatment is A. Grenier, *Manuel d'archéologie Gallo-Romaine*, Vol. 4, Part 1, 'Aqueducs', Paris 1960.
29. Grenier, Note 28, p. 153 says that the Rodez siphon crossed the valley of the Aveyron in lead tubes at a maximum depth of about 100 metres.
30. The Arles siphon is especially interesting because it was laid, apparently, across the bed of the Rhone. Remnants of the lead siphon tubes which have been dredged up are stamped with their makers' names.
31. See Casado, Note 2.
32. ibid.
33. A branch of the Aqua Claudia was carried by a siphon to the Palatine. See Marion E. Blake, *Roman construction in Italy from Tiberius through the Flavians*, Washington, 1959, p. 123; also E. M. Winslow, *A libation to the gods*, London 1963, pp. 81—3.
34. See Marion Blake, Note 33, p. 82.
35. The Alatri siphon is supposed to have been constructed in 134 B.C., see C. G. de Montauzan, Note 28, pp. 194—7.
36. This siphon, near Palermo, is mentioned by G. Giovannoni: 'Building and engineering' in *The legacy of Rome*, Oxford 1957. p. 467. As yet no further details have been located.
37. The source of Lincoln's water-supply, a spring called Roaring Meg, was lower by 70 feet than the delivery point. It seems most unlikely that the required uphill flow could have been pumped by a machine of the Ctesibian type. It is much more reasonable to suppose that the water was lifted 70 feet at its source, by a chain-of-pots for example, and then 'siphoned' to the town. In her unpublished Ph.D. thesis entitled *Municipal and military water supply and drainage in Roman Britain*, Inst. of Archaeology, Univ. of London, 1970, pp. 46—51, Julie Hanson presents the archaeological evidence thoroughly but her understanding of the technology is less than adequate. In particular her rejection of the idea of a siphon is technically unsound. See also F. H. Thompson: 'The Roman aqueduct at Lincoln' in *The Archaeological Journal*, Vol. CXI, July 1955, pp. 106 ff.
38. This was evidently a big siphon which has left substantial remains of its venter bridge. As yet very little information has been forthcoming; see K. D. Matthews: 'Roman Aqueducts: Technical Aspects of their Construction' in *Expedition*, The Bulletin of the University Museum of the University of Pennsylvania, Vol. 13, No. 1, Fall, 1970, pp. 2—16.
39. See Note 28.
40. C. G. De Montauzan, Note 28, pp. 178—92 and H. Plommer, Note 6, pp. 25—30 offer commentaries on the Vitruvian siphon material. Another effort is P. Grimal: 'Vitruve et la technique des aqueducs' in *Revue de Philologie*, Vol. XIX, 1945, pp. 162—74.
41. Morgan's translation, see Note 11, has been used throughout unless otherwise stated.
42. M. Belgrand, *Les aqueducs Romains*, Paris 1875, p. 80.
43. F. Granger, see Note 11, Vol. 2, p. 185.
44. A. Choisy, *Histoire de l'Architecture*, 2 Vols., Paris 1899, Vol. 1, p. 582.

45. And in fact examples are not hard to find. Some of the manuscripts suggest a slope for open channels (see Bk. VIII, Ch. 6, para. 1) of 6 inches (*semipede*) per 100 feet; others give ¼ inch (*sicilico*) per 100 feet. One of these *must* be a mistranslation. Another example is the choice between *colliquiaria* and *collivaria* in Bk. VIII, Ch. 6, para. 6; see H. Plommer, Note 6, p. 28.

46. Stahl, W. H. *Roman Science*, Madison 1962, p. 92.

47. See for example F. di Giorgio Martini, *Trattati di Architettura Ingegneria e Arte Militari*, 2 vols., Milan 1967, Vol. 1, pp. 160—70.

48. See for example the calculations done on some early Mexican dams by J. Hinds: '200-year-old masonry dams in use in Mexico' in *Engineering News-Record*, 1 September 1932, p. 252.

49. R. Mark: 'The structural analysis of Gothic cathedrals' in *Scientific American*, November 1972, pp. 90—9.

50. Taken from C. G. de Montauzan, Note 28, pp. 197—9 and 217.

51. M. Belgrand, as Note 42, p. 71.

52. See Note 19.

53. C. G. de Montauzan, Note 28, p. 217.

54. Some of the Lyon siphons do not reach great depths, Tupinier about 80 feet and Cotte-Chally probably less than 120 feet. However, on aqueducts where siphons had to be built in any case it would have been expedient to take care of lesser depressions with a shallow siphon.

55. And it is interesting to note that the highest aqueduct bridge ever built, the Roquefavour bridge near Marseilles, reaches just 270 feet and that was erected in 1846.

56. J. P. Martin-Cleto, *El Abastecimiento Romano de Aguas a Toledo*, Toledo, 1970.

57. As Note 56, pp. 4—5 and also N. A. F. Smith, *The heritage of Spanish dams*, Madrid 1970.

58. The reconstruction is illustrated in J. A. Garcia-Diego: 'Restoration of Technological Monuments in Spain' in *Technology and Culture*, Vol. 13, No. 3, p. 427.

59. The Pont du Gard is in France and was built sometime in the reign of Augustus, conventionally 19 B.C. The Alcantará bridge spans the Tagus about 150 miles west of Toledo and is Trajanic, about A.D. 105. Therefore Alcantará is a more appropriate model for a second century bridge in Toledo.

60. See Note 55.

61. On this point and in relation to the conclusion about to be reached see Note 27.

62. Remember that Roman engineers only constructed full-centred semi-circular arches and hence heights were fixed at half the dimension of the span.

2

Our debt to Roman engineering: the water supply of Lincoln to the present day

M.J.T. Lewis

Summary: *Lincoln was one of the four colonia of Roman Britain and therefore intended to serve as an example of cultured urban life to the surrounding region. A good water supply was essential in Roman towns but Lincoln, stand above the 200ft contour, could not receive a normal gravity fed system. This paper examines the archaeological evidence for the water supply to Roman Lincoln and looks at how the problem was solved in later generations. Dr Lewis argues that a comparison of the solutions to engineering problems of even the distant past and the present is an important aspect of industrial archaeology.*

If the subject of this year's Rolt Memorial Lecture looks back to an earlier period than usual, it is for a reason. Tom Rolt was a great man. His qualities were many; and even if he had faults, the last thing he could be accused of was narrow-mindedness. Had his interests and his enthusiasms not been so wide-ranging, so unusually catholic for his day, he would have inspired far fewer with a concern for the industrial past. Yet there is a tendency among industrial archaeologists to be somewhat narrow in their vision: narrow in terms of area, of industry, or of period. While it seems to be generally accepted dogma that industrial archaeology should encompass the whole gamut of industry from Grimes Graves to the computer, how many of us pay more than lip service to this creed? Limitations of time and facilities are one valid reason why this happens, and specialising does have its virtues. But to cast off the blinkers, to take a broader interest in industry before the industrial revolution, to be able to compare the problems and solutions of earlier and later days, does generate a deeper understanding.

Now Tom Rolt may not have been especially concerned with ancient industry but in his broad-mindedness he would have been intrigued, I venture to think, by the enigma of Lincoln's water supply in Roman times. Let us first set the scene and briefly describe the position in medieval and modern times.

Lincoln exists because of its site. The great Jurassic ridge, followed by a major highway since time immemorial, has here been broken through by the River Witham, leaving a truncated spur to north and south. The river crossing on this highway, the first above the fenland, is dominated on the north by the end of the ridge. Here is perched the upper town, the site of the original Roman fortress and incorporating the medieval castle and cathedral (Fig.1). It lies on the limestone, wholly above the 200ft contour, with a maximum height now of about 225ft, or in Roman times perhaps ten feet less. The spring line, running in an arc around the spur where the limestone meets the lias and the upper estuarine clay below, passes through the lower town. Thus the water table is never less than 40ft below the surface of the upper town, where any wells have to have at least that depth. For many miles around there is no significant surface water available at a height greater than that of the upper town.

In the Middle Ages and until Victorian times, while the springs were undoubtedly tapped and rainwater was caught in cisterns, most of Lincoln's water came from numerous wells, public or private. Although some were no doubt contaminated by nearby cesspits, there is no reason to suppose that the supply was much worse or better than was normal for the age. The poor often had to carry their water long distances and by the nineteenth century could be charged 2d a day for the use of private wells. The only exceptions to this basic pattern were low-level conduits fed by springs to the east of the city. In 1260 the Black Friars had a conduit to their house, and by 1535 the Grey Friars had a lead pipe on a similar route from the Monks'

58 INDUSTRIAL ARCHAEOLOGY REVIEW, VII, 1, AUTUMN 1984

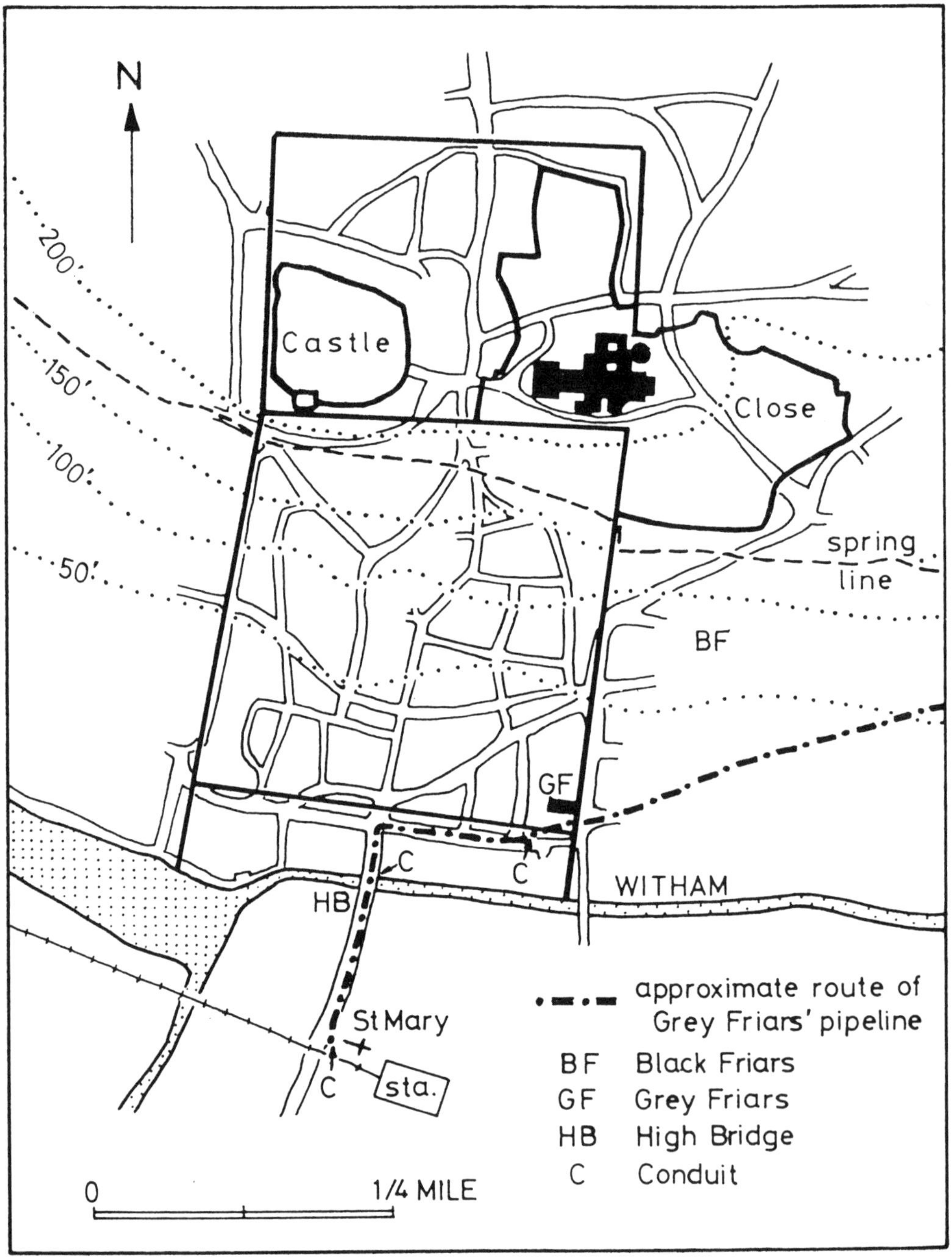

Fig.1 Lincoln: contours and conduits.

Lands. On the dissolution in 1540 the Common Council bought this last supply and extended it via a conduit on High Bridge to the South Ward, where another conduit was set up at St Mary-le-Wigford, which still exists. In sum, the supply was typically meagre; but from its medieval greatness, Lincoln slipped in the sixteenth century into a decline of two centuries and more, and there was little incentive to improve matters.[1]

It was not until the industrial and social revival of the early nineteenth century that reforming spirits began to press for something better. The end result was an Act of Parliament in 1846, establishing the Lincoln Water Company. It was decided not to tap the hard water emanating from the limestone but the softer and supposedly purer water on the southern fringe of the city, where the Prial Brook rose in, and flowed through, the gravel (Fig.2) A 23 acre reservoir, holding 20 million gallons, was created on the brook at Hartsholme, feeding down to filter beds at the new Boultham Water Works. From here two steam engines, of 10 and 45hp, pumped up to a reservoir of 777,000 gallon capacity in the upper town north of the castle, to supply the higher parts, and later to another reservoir holding 1,650,000 gallons at Bracebridge Heath away to the south-east, to supply the lower-lying regions. By 1848 the system was at work, being gradually extended thereafter.[2]

The population of Lincoln, some 14,000 when the water works were set up, grew by the turn of the century to almost 50,000. To meet the increasing demand the Corporation, which had bought the works in 1871, began abstracting from the Upper Witham and from the Pike Drain and Catchwater Drain nearby. This led to disaster. The untreated sewage of the suburbs of Boultham and Bracebridge was discharged straight into the Witham, whence much of it found its way into the water supply. The land traversed by the smaller streams was regularly manured with night soil. Despite warnings of pollution as early as the 1870s, nemesis struck in 1904 and within a year there were over a thousand cases of typhoid, and 131 deaths. As a temporary remedy, water was transported from Newark by rail, and in quest of a permanent solution a borehole was sunk at Boultham to 2200ft, a record depth in this country at the time. Plenty of water was found, but too impregnated with mineral salts to be of use.[3]

So the Corporation was forced to look further afield for a purer supply, and in 1908 it obtained an Act to establish water works at Elkesley in Nottinghamshire, 22 miles to the west. Here it erected a grand pumping station to house two triple-expansion inverted engines by Ashton & Frost, each of 450hp, sited over four boreholes 570ft deep in the Bunter sandstone beds. Each engine could deliver three million gallons a day at normal speed into a 21in rising main which crossed the Trent on an aqueduct beside Dunham Bridge and ended at the new Westgate water tower in the upper town, which was 110ft high and of 300,000 gallon capacity. When it was full, the surplus passed through a trunk main, serving the lower districts en route, 3½ miles to the Bracebridge Heath reservoir whose capacity was enlarged to six million gallons. The tap was turned on at Lincoln in 1911. Since then, the pattern has remained essentially the same. The population has increased, the area served has expanded, and new sources have been tapped. But Elkesley is still the main supply, though without the steam engines which were replaced by electric after one blew a cylinder in 1974.[4]

In essence, then, the problems confronting the modern water engineers of Lincoln have been those of finding an adequate supply of good water and of pumping it to a reservoir or a water tower in the upper town against a head of roughly 200ft and for a distance of 1¾ miles from Boultham or 22 miles from Elkesley.

Precisely the same problems faced the original Roman engineers too. The first occupant of the site of Lincoln was a legionary fortress, set up in about 47 AD and evacuated about 77 when the legion moved north to York. In its place, probably in the early 90s, was established the *colonia* of Lindum: a chartered town, originally a settlement of veterans, and one of only four such in Britain, which enjoyed certain rights of self-government and was intended to serve as a centre of Romanisation.[5] The early *colonia* occupied the upper town, covering the same area — 41 acres — as the previous fortress; the lower town was a later extension of 55 acres spreading down the steep slope to the Witham below, and not enclosed within ramparts until probably the late second century. While the lower town could be supplied at least in part with water from springs, the upper town could not (Fig.4).

60 INDUSTRIAL ARCHAEOLOGY REVIEW, VII, 1, AUTUMN 1984

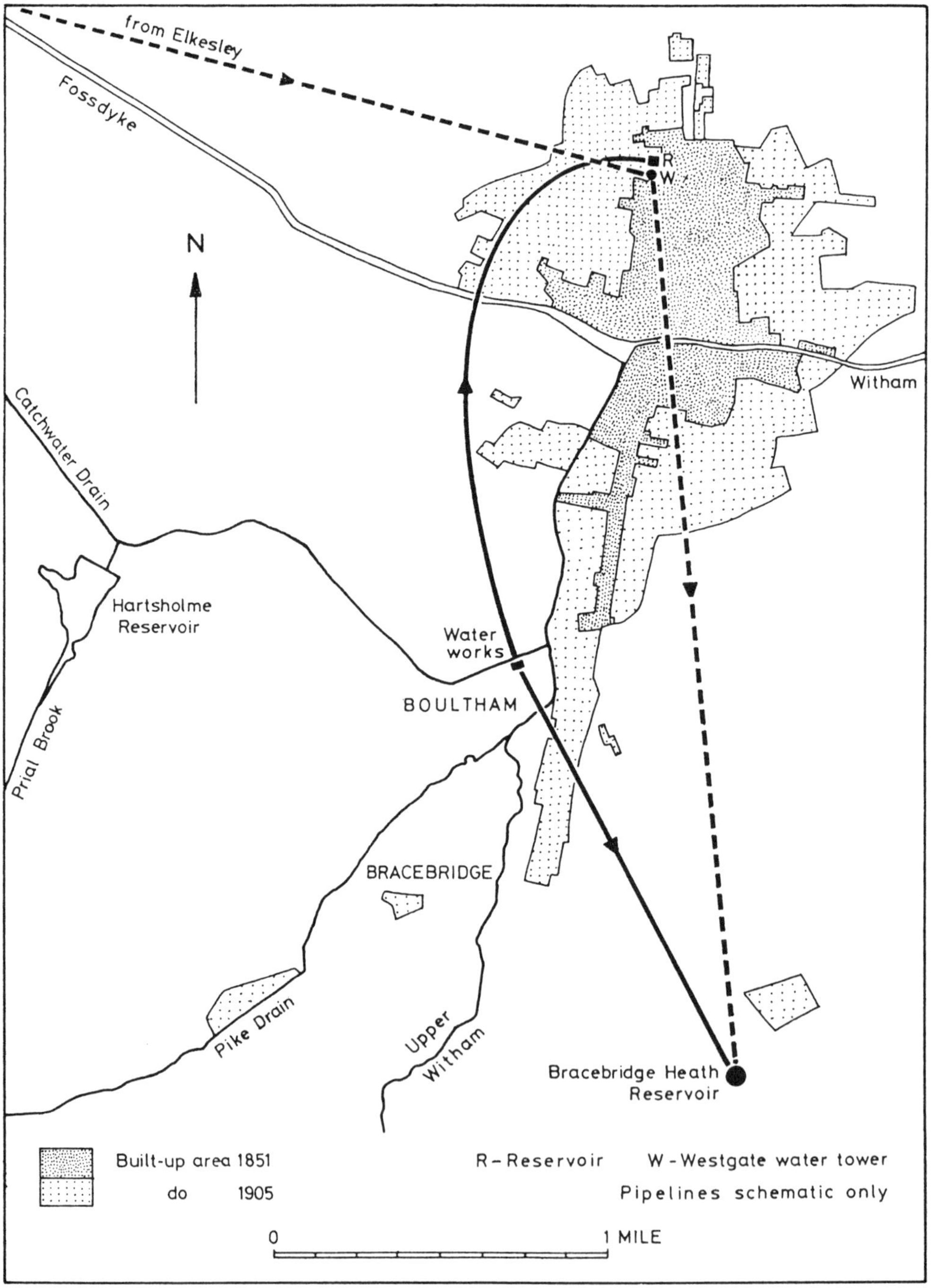

Fig.2 Lincoln: nineteenth and twentieth century supplies.

Yet a copious supply of good water was almost a *sine qua non* for a self-respecting Roman town. To achieve it, little expense would be spared. There might be a dam at the source to impound the water. The aqueduct itself, often lengthy, was not uncommonly in tunnel to avoid pollution or the risk of enemy disruption; otherwise it ran on or near the ground surface and only quite rarely was it perched on that best-known but expensive form of aqueduct, the elevated arcade such as the Pont du Gard, the Segovia aqueduct, or the towering lines of arches that stride across the Campagna to Rome itself. Once at the town, the water was delivered into a *castellum*, a water tower or cistern, sometimes of cathedral-like dimensions, whence it was distributed by pipe to the baths, the public fountains, and to private houses and industry. The domestic plumbing was unrivalled until the nineteenth century. In short, the Roman approach to water supply was very sophisticated and, generally, effective, being assisted by a high standard of surveying which took in its stride such complexities as curved tunnels or long gradients at a constant 1 in 2000.

Wherever possible — which was nearly always — and for obvious reasons, the whole water system would be designed to operate by straightforward gravity flow. The scale of an aqueduct would depend on the topography, on the population served and on the self-esteem of the town in question. Rome itself, for example, boasted at its peak 219 miles of aqueduct, of which only 29 miles were above ground. Lyons had 125 miles, Cologne 48 miles, and so on. Of the supply for the towns of Roman Britain, mostly

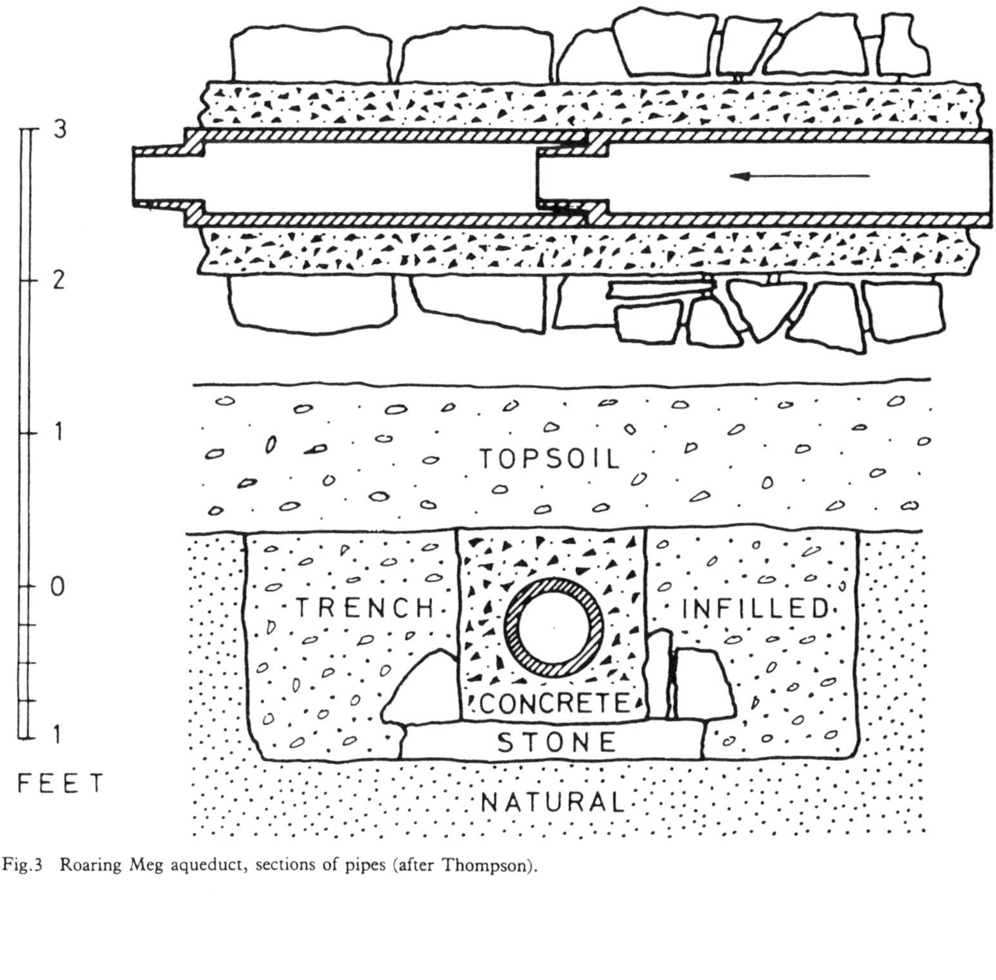

Fig.3 Roaring Meg aqueduct, sections of pipes (after Thompson).

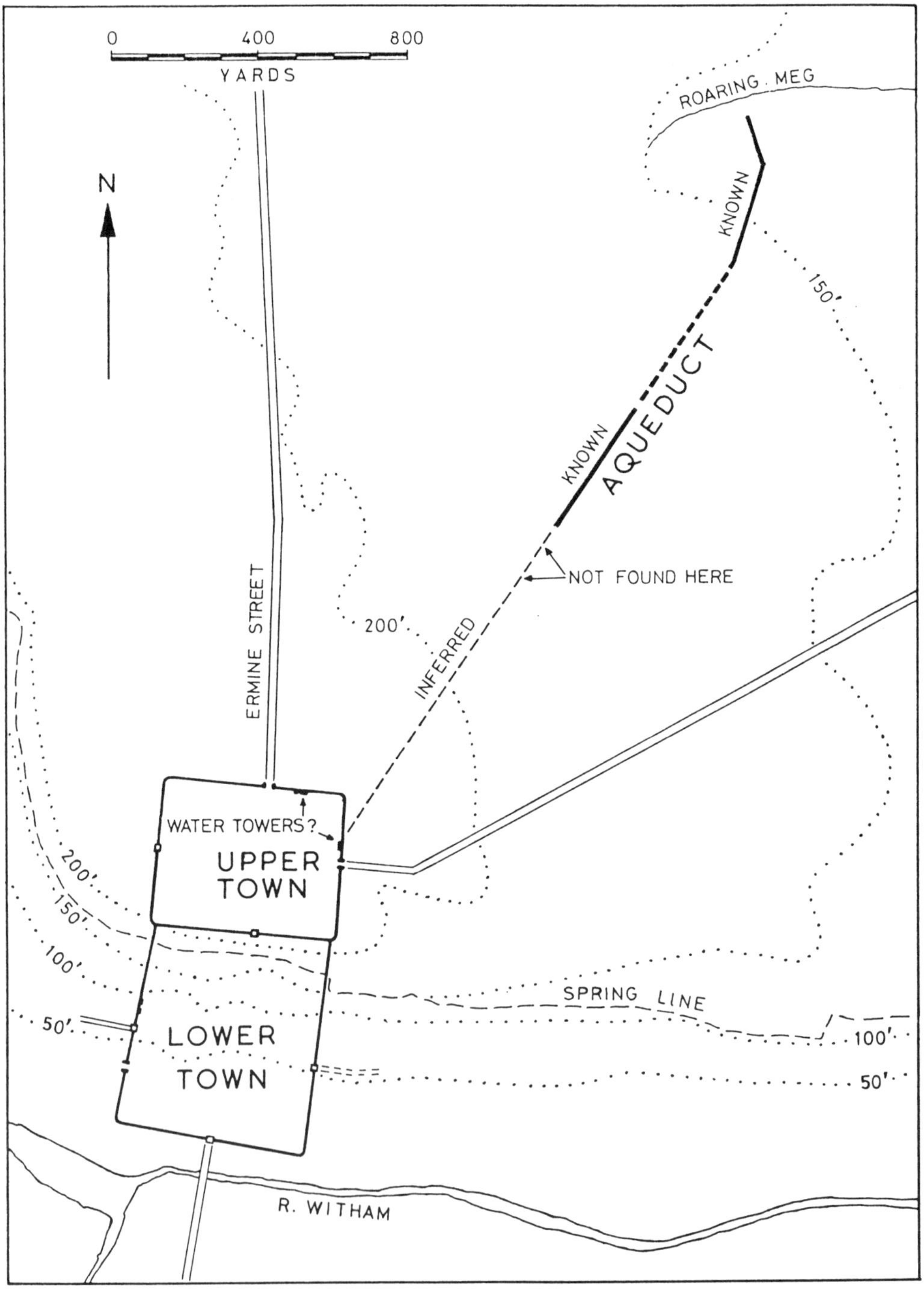

Fig.4 Roaring Meg aqueduct (after Thompson and Wood).

more humble than these great centres, remarkably little is known. The only case where we have a more or less complete knowledge is Dorchester, fed by an 11 mile open leat from the River Frome, meandering around the contours.[6] Similar provision was made at a number of military forts in the north, although the army's demand for water was not as great as that of the civil population.

Lincoln, however, does not adhere to the norm because a plain gravity supply to the upper town was out of the question: there was nowhere for it to come from. Instead, the inhabitants had recourse to a pressure main. It consisted of earthenware pipes, hand-thrown, 3ft long and of 5½in bore except at the connecting spigots where the bore is reduced to 3½in (Fig.3). Thus, unlike ordinary Roman pipes which were of constant bore, the profile is hydraulically terrible, with the constrictions generating much turbulence. The main was laid by digging a trench about 2ft 6in below the modern ground surface, laying a floor of stone slabs, pouring a layer of concrete in shuttering, placing the pipes on it and sealing their joints with — probably — an expanding compound, pouring more concrete until the pipes were totally enclosed in a sheath some 15in square, and finally filling in with soil. The pipeline supplied one of two possible water towers built against the back of the town wall. One, in East Bight, was lined with waterproof cement and measured 55ft long by an unknown width; the other, less certain, lay a little north-east of Eastgate. Both are rather over 200ft OD. From near the north-east corner of the town the pipeline has been traced as various points along the present Nettleham Road as far as a small stream known (optimistically nowadays) as Roaring Meg, 1¼ miles away (SK 988739). Since this point is at 131ft OD, the difference in level is a good 70ft, and the gradient, except perhaps at the input end, was constant at around 1 in 120 up to the town.[7]

Excavations by Hugh Thompson in 1951-2 revealed that towards the input end the pipeline, hitherto aiming north-east, swung to west of north and, as the side of the Roaring Meg valley dipped more steeply down, left the underground trench, to be carried on a raised but solid substructure for 60 yards. Finally it ran for 40 yards more on top of a series of eight piers, presumably arched (Fig.5). Only the foundations of the piers were found, quite slightly built and averaging nearly 5ft long by 5ft 9in wide, except that the last pier was much more massive — 9 to 10ft wide by 16ft long. Although this last pier was still 40 yards from the stream itself, silt deposits showed that in Roman times the water spread out in a pool, natural or artificial, which lapped against the foot of the pier. Excavation in various directions found no trace of any other foundations. From the slight evidence, it looks as if the pipeline was installed in the early second century AD, quite soon after the foundation of the *colonia*.

How then was the water forced up the pipeline

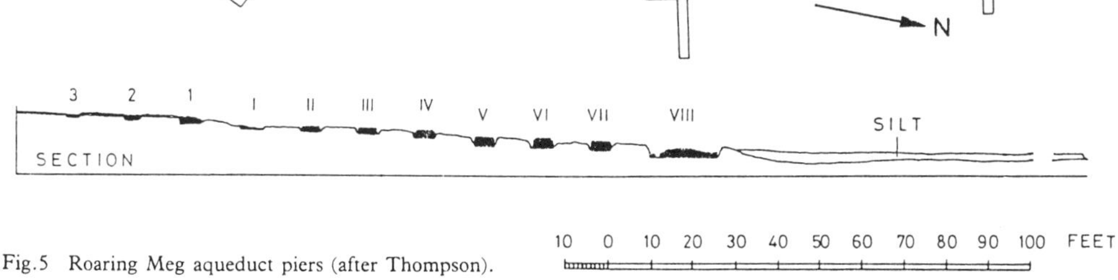

Fig.5 Roaring Meg aqueduct piers (after Thompson).

against a head of 70ft? This is the burning question, to which no certain answer is possible. All we can do is consider the pros and cons of various possibilities. On the face of it, the simplest solution is to postulate a tower over 70ft high on the last pier before Roaring Meg, to which water was lifted by machine and from which it flowed via an inverted siphon to the town. The Romans employed a variety of machines for water-lifting, especially in irrigation: the hand-operated *shaduf* or swape; the foot-powered Archimedean screw; the *noria* or bucket-wheel and its more sophisticated version the *tympanum,* turned by water, by treadmill, or by animals and right-angled gearing. All of these, however, had a very or a fairly small lift and where a high lift as required — as in mine drainage — were operated in series. There is no sign at Roaring Meg of the supports which such a series would need.

The only possible lifting device was the *sakkieh* or chain of buckets, which could lift to much greater heights; at the Stabian baths in Pompeii one such raised water from a well 66ft deep at a rate of perhaps 350 gallons per hour if worked by two men on a treadmill. In Egypt, where they were most common, such machines were usually operated, through gearing, by oxen.[8] The record lift, from a medieval well, is 297ft in two stages. Normally the *sakkieh* is powered from above, but if powered from below it could fit the bill at Roaring Meg, lifting water to a small header tank atop a 70ft tower on the last pier (Fig.6). At this point, however, the theory founders. For such a tower, the foundations at 10ft by 16ft were parlously small and slight. This applies whether the pipe sloped downwards towards Lincoln over the whole 100 yards of the surface section, where the first pier in the series would be some 38ft high, or whether the final pier was freestanding and the pipe dropped vertically down it from the header tank. The Romans were perfectly capable of building structures of this height with entirely adequate foundations. At Roaring Meg they clearly did not, and an elevated tank is ruled out.

Thompson suggested that the piers were there to maintain the regular gradient of 1 in 120 down from Lincoln and to avoid a steepening at the end. If the general gradient is projected, it makes the last pier about ten feet high. On it, he postulated a pump lifting water 10ft and forcing it up the remaining 60ft of height to the town. The Romans indeed used force pumps, and extensively. But on the basis of the only example found in Britain, at Silchester, Thompson calculated that at 50% efficiency and at 20 strokes a minute, operated by hand, it would deliver a mere five gallons a minute along the pipeline or, if worked 16 hours a day, 5000 gallons a day. A number of problems here present themselves. What was the population of the upper town of Lincoln? An informed guess is 3000 to 6000, probably less rather than more.[9] If for the sake of argument we assume a population of 3000, then each inhabitant would get less than two gallons a day. What was the normal supply that town dwellers might reasonably expect? No real answer is possible; but it is estimated, despite many imponderables, that the aqueducts supplying Rome delivered something like 200 million gallons a day,[10] for the benefit of something like a million souls. Two hundred gallons per head per day is generous — the modern norm approaches 40 — but it must be remembered that much of that quantity would flow to waste from the public fountains, unused except perhaps for flushing the sewers. We could not expect anything like this provision for a town such as Lincoln; but two gallons a day is derisory. Besides, on Thompson's figures, it would take nearly an eighth of a horsepower to deliver this miserly quantity. He assumed, too, a piston stroke of 10in; but since the working load on the piston is 262lb, a considerable leverage is needed on the pump handle. If the mechanical advantage were 6:1, reducing the effort required to 44lbs, the handle has to be swung through a 5ft arc, which is hardly possible.

One may quibble with Thompson's assumptions. The stroke was probably only 5¾in, which with a 6:1 lever gives a better swing for the handle of only 34in. The pump efficiency could well have been 75%. But even so, one cannot push the figure for delivery, at 20 strokes, above 4¼ gallons a minute; and the effort would still be an eighth of a horsepower, divided presumably between two men. This is more practicable mechanically. But even assuming, optimistically, that a man could keep up such punishing work — that he could move a 44lb load through 34in twenty times a minute — for four hours on end, and assuming that pumping continued for 16 hours a day, then a total of eight men would be required. It has even

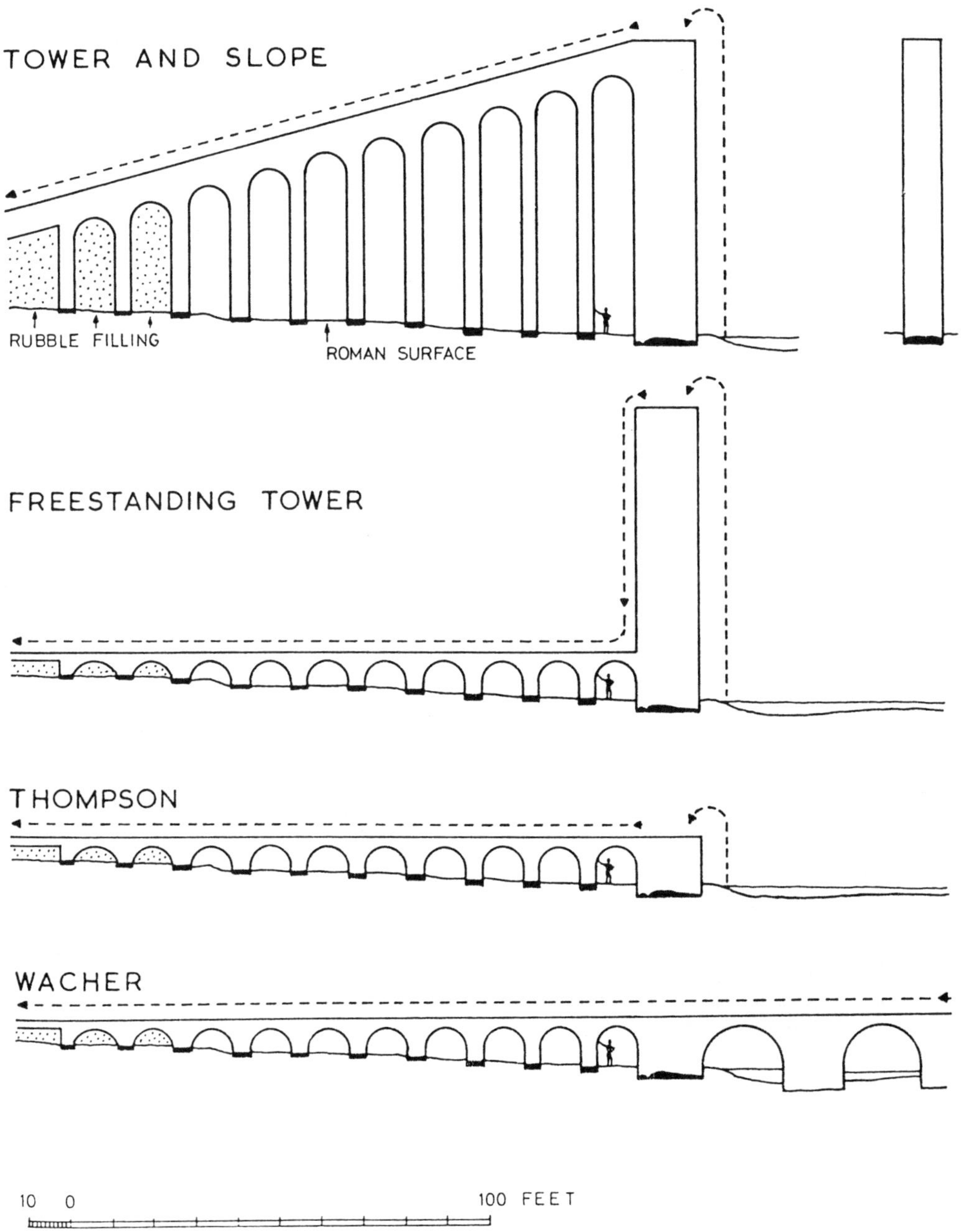

Fig.6 Alternative reconstructions of aqueduct piers.

been suggested that the same quantity could be delivered with no more difficulty or inefficiency by a small fleet of waggons carrying barrels by road.[11]

The next question is whether the Silchester pump was typical. The pump was invented by Ctesibius of Alexandria around 270 BC, and from literary sources we would deduce that its major function was in fire-fighting. No survivals are known from Greek times, but twenty-two Roman pumps have been found, many of them like the Silchester specimen for lifting from quite shallow wells, a few possibly for pumping bilges in ships. They can be very simple affairs of wood, with barrels sometimes lead-lined, leather piston rings, and usually leather flap valves; or they can be extremely precise fabrications of cast bronze, with turned pistons and cylinder bores and on occasion very modern-looking mushroom valves (Fig.7).[12] Their efficiency was surprisingly high. A replica of the very small wooden pump from Belginum in Germany, operated by a single man at 44 complete strokes a minute against a head of 20 metres, was 85% efficient. An actual bronze pump from a Mediterranean wreck was in good enough condition to test, and showed an efficiency of over 95% against an apparently low head. But almost all are very small, few being over 3in bore; and having twin barrels, are

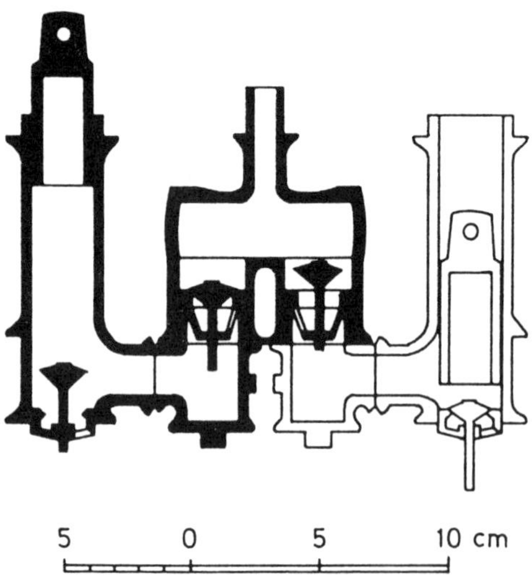

Fig.7 Bolsena 2 bronze pump.

clearly intended for hand operation by rocking lever just like the traditional fire engine.

Only two instances are known of what might be called large pumps, both from a mysterious installation on the foreshore at St Malo whose purpose, and whose power source, is unknown. One of them was a three-cylinder wooden block, the other no less than an eight cylinder block (Fig.8). Both had a bore of almost 6in and a stroke at the most of 8in.[13] Working out some specimen figures for the performance of such a pump, if hand powered, at Roaring Meg, we get the following. At 15 strokes a minute and 75% pump efficiency, one such cylinder would deliver 8¾ gallons a minute or, if worked non-stop, some 12,500 gallons a day. This is barely four gallons per head; little better, and is it practicable? Adding a 16ft friction head for the long pipeline, we find the pressure on the piston is 37 psi, or a load of nearly half a ton. To move that by hand would need enormous leverage, and the limit to the sweep of the handle for manual operation brings the stroke down to a ridicously small figure. Even if it could be done, it would involve a quarter of a horsepower for one cylinder. We are talking of shifts of something like four men each. At least sixteen men's labour to supply at most four gallons a head? It all seems very unlikely.

Is there any way round the problem? One could visualise intermediate pumping stations along the pipeline, although no trace has been found; but they would not reduce the overall workload, only spread it. A pump powered by a waterwheel seems to be ruled out by the distance of the last pier from Roaring Meg, and by the stream's modest flow. Animal-powered pumps, however, might be a better bet. The difficulty is that the obvious transmission of power from a gin circle to a pump is via right-angled gearing and a horizontal crankshaft. But whether the Romans knew of the crank — or more importantly whether they used it — is a highly vexed question. The evidence, in fact, strongly suggests that it was almost entirely foreign to their technology. The next best means of transmission, the cam or trip, they did know, even if they applied it but rarely.[14] It would be technically feasible to construct a horizontal camshaft driven by ox power through gearing, with the cams pushing down the pistons which were then lifted by counterweights.

The eight-cylinder pump from St Malo, with

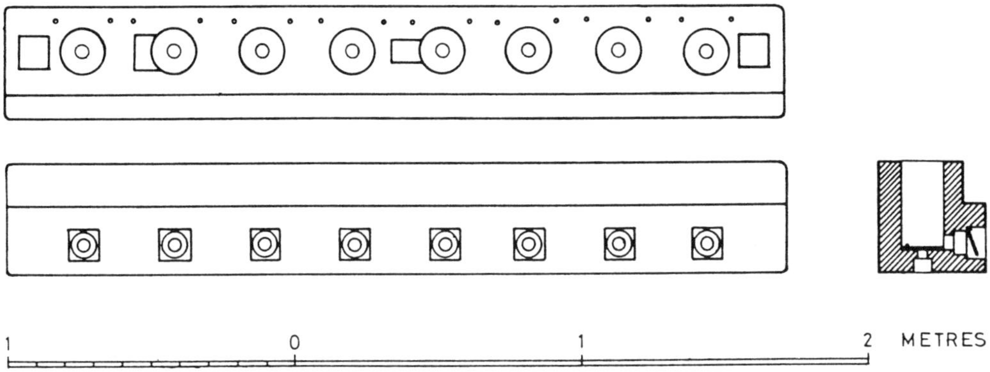

Fig.8 St Malo eight-cylinder pump block; missing leather flap-valves restored from the three-cylinder pump (after Langeout and Meury).

its combined capacity of 6¼ gallons, might be considered as an argument in favour of animal power, permitting far more water to be pumped than by any reasonable use of man power. With eight cylinders, St Malo fashion, and the same rate and efficiency as in the earlier estimate, it would take two horsepower to deliver 70 gallons a minute or 100,000 a day, which equals 33 gallons a head. This is much more desirable a quantity; but there are two drawbacks. Although animals walking in circles were widely used, especially in Egypt and the Middle East, for lifting water, and in the form of the so-called donkey mill, without gearing, for grinding grain, they were not used for anything else. We know of no case of animal-powered pumps, nor of such devices, common in later times, as mining whims. It is a general truth that the ancient world did not mechanise — did not employ other forms of power than human muscle except for transport — to any significant degree, because until the late Roman Empire there was plenty of cheap labour available. Secondly, the mechanics of such an animal-powered pump are awkward. The load of nearly half a ton on each piston would require massive cams; and as far as we know the Romans went no further than crude trips, quite inadequate for the purpose. The St Malo machine, alluring as it is, is probably a red herring as far as Lincoln goes. It lifted, almost certainly, against quite a small head, which means that although its cylinders were large it needed much less energy to work it: within the ability, perhaps, of a small gang of men. At Lincoln with its big pressure head, if animal power were used, a large battery of small pumps would be better than a small battery of large ones.

If, then, pumping by hand was hardly worth the effort and pumping by animal is historically hypothetical and mechanically dubious, what remains? John Wacher argues that the piers at Roaring Meg, while they might smooth the gradient of the pipeline, would not significantly increase its effectiveness: it would have worked just as well with the pipes on or just below the ground the whole way.[15] What function then did the piers have? He suggests that they were part of a conventional arcade which in reality extended right across the Roaring Meg valley and carried an inverted siphon bring water from a higher and much more distant source (Fig.6). There is no technical problem here. If the ancient world could not provide a water supply in an open, non-pressurised, channel, it could and did revert to pressurised pipes, i.e. inverted siphons, across valleys. At least eight Greek and nineteen Roman examples are known, some of them of astonishing conception.[16] The water was carried in solid masonry culverts, in bored stone blocks, or much more commonly in lead pipes. One of the oldest siphons, and the deepest, was at Pergamon in Asia Minor where in the second century BC a 28 mile aqueduct was built, of which the last two miles were in a lead siphon pipe of a foot diameter and perhaps two inches thick. At the deepest point this siphon was about 620ft below the pipe intake, and the greatest pressure was therefore about 270 psi. Its capacity was of the order of 600 gallons a minute.[17]

The best Roman examples are on the four

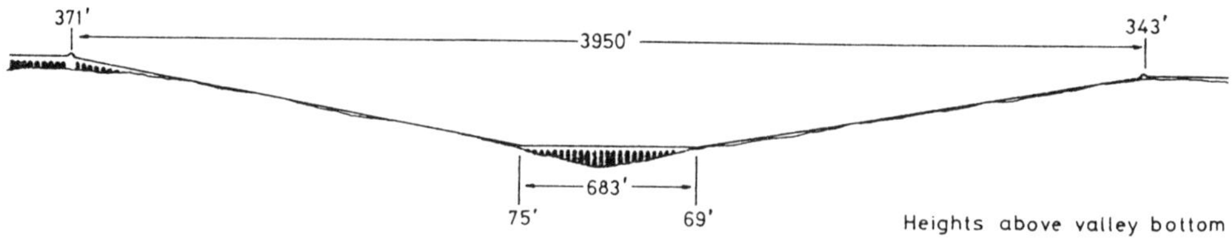

Fig.9 Diagram of Soucieu siphon near Lyons.

aqueducts, totalling 125 miles, which fed the city of Lyons. They included at least nine siphons across the valleys, the longest being almost three miles and the deepest 404ft (176 psi). The Soucieu siphon, the best preserved of them, had a depth of 302ft, not to the bottom of the valley but to a raised arcade standing high above the river and thereby reducing the maximum pressure (Fig.9). For ease of construction and repair, the water was carried in nine parallel lead pipes of 10½in outer diameter. Their combined capacity was of the order of 6000 gallons a minute. This figure gives some idea of the importance of a good water supply to Roman life. So does the weight of lead — about 1700 tons — used on this one siphon, or the 10 to 15,000 tons on all the siphons on all the Lyons aqueducts.[18] It was undoubtedly a costly provision, but presumably was considered more economic than lengthy detours around the valleys. In deciding where to use siphons, where elevated arcades, and where detours the Roman engineer faced questions similar to those faced by the later canal and railway engineers. The choice of route depended on the cost of construction and maintenance. The great Pont du Gard was built as a very tall arcade presumably because the alternative route round the valley would have been more expensive, and because a siphon would have been more expensive and troublesome to maintain.[19]

How does this apply to Lincoln? Wacher points out that the nearest source of water which is higher than the upper town is on the chalk wolds of Market Rasen, some 20 miles away (Fig.10). If the water were tapped somewhere near Otby Top — where coincidentally or otherwise there is a modern reservoir — at a good 400ft, it would give a differential head of a good 200ft. The direction taken by the pipeline when last seen at Roaring Meg — west of north rather than north-east — might be, Wacher argues, because it was aiming to follow the higher ground across the watershed between the Ancholme to the north and the Witham to the south. The highest possible level for the bottom of the siphon would be about 40ft OD near Toft Newton, say 360ft below the intake or 160ft below the delivery point. Tests on the earthenware pipe in its concrete sheath show that its tensile strength is twice that required to withstand the pressure involved. According to Wacher, the pipe could carry about 21 gallons a head, better than by hand pump, but not great.[20] Technically, then, such a siphon would be feasible, though seven times longer than the longest known elsewhere.

Here too, however, there are serious snags. Wacher suggests that the route ran north for some distance from Roaring Meg before swinging a little north of east to take advantage of the Ancholme-Witham watershed. It might be fair enough to opt for extra distance — even at the cost of extra friction in the pipe — rather than for a shorter route at higher pressures. But if it were keeping to high ground on its journey north, why did it not follow Lincoln Cliff — the Jurassic ridge itself? Instead of marching off north from the town for several miles at over 200ft, it starts off north-east and drops to 130ft at Roaring Meg, only to undulate thereafter, as Wacher has it, on either side of the 150ft contour before turning east. It does not smack at all of the careful planning and surveying typical of Roman hydraulic works. Even more tellingly, while the pipeline has been found at plenty of sites on the 1¼ miles between Lincoln and Roaring Meg, not a single trace has appeared along the 20 miles beyond.[21] In particular, Wacher's reconstruction has the piers continuing across Roaring Meg where Thompson's careful search found absolutely no sign of them at all. Technically, the siphon solution is preferable to the pump solution; archaeologically, as far as our present knowledge goes, the evidence favours the pump.

In no pipe yet seen has there been any sediment or any lime deposit. While it is

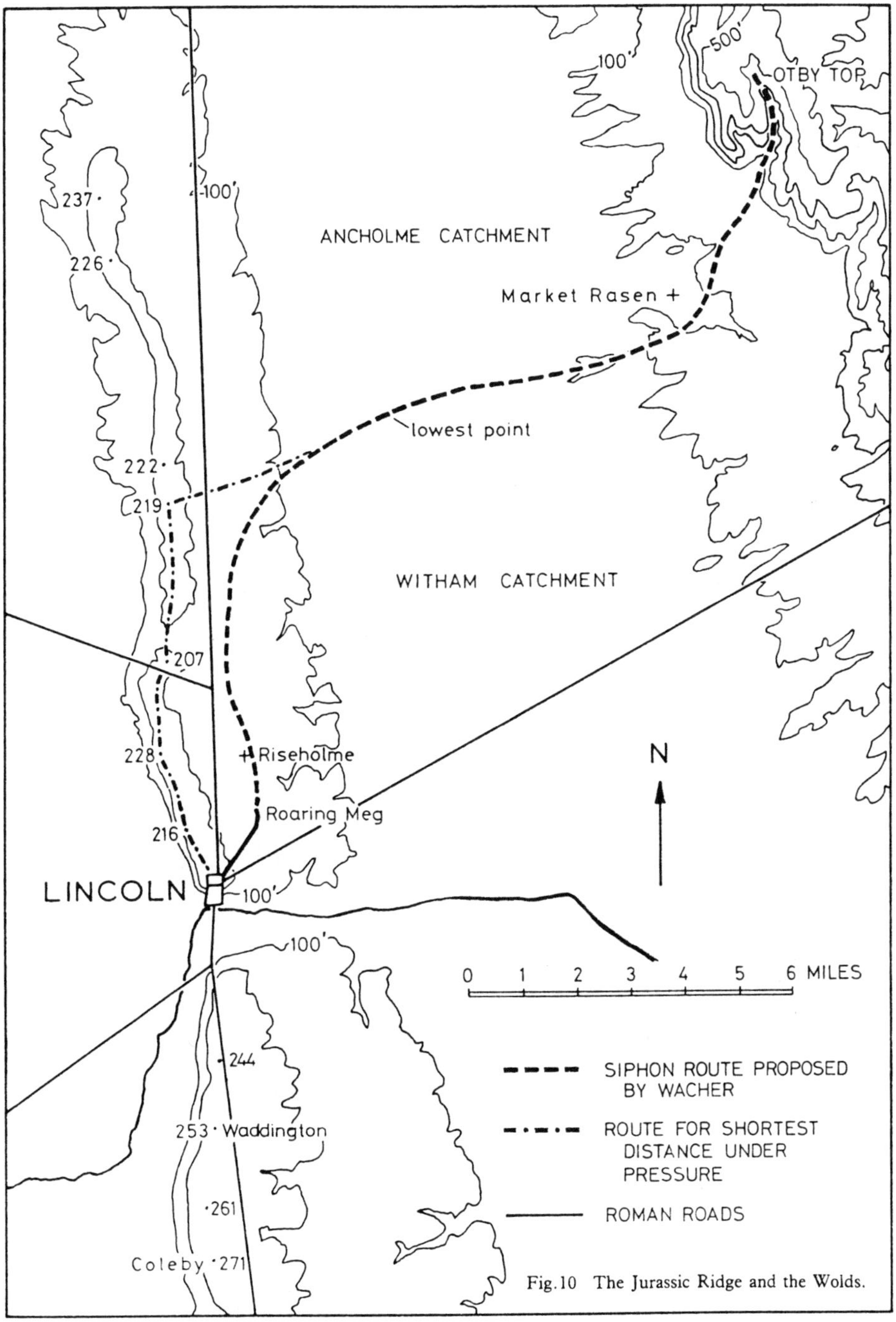

Fig.10 The Jurassic Ridge and the Wolds.

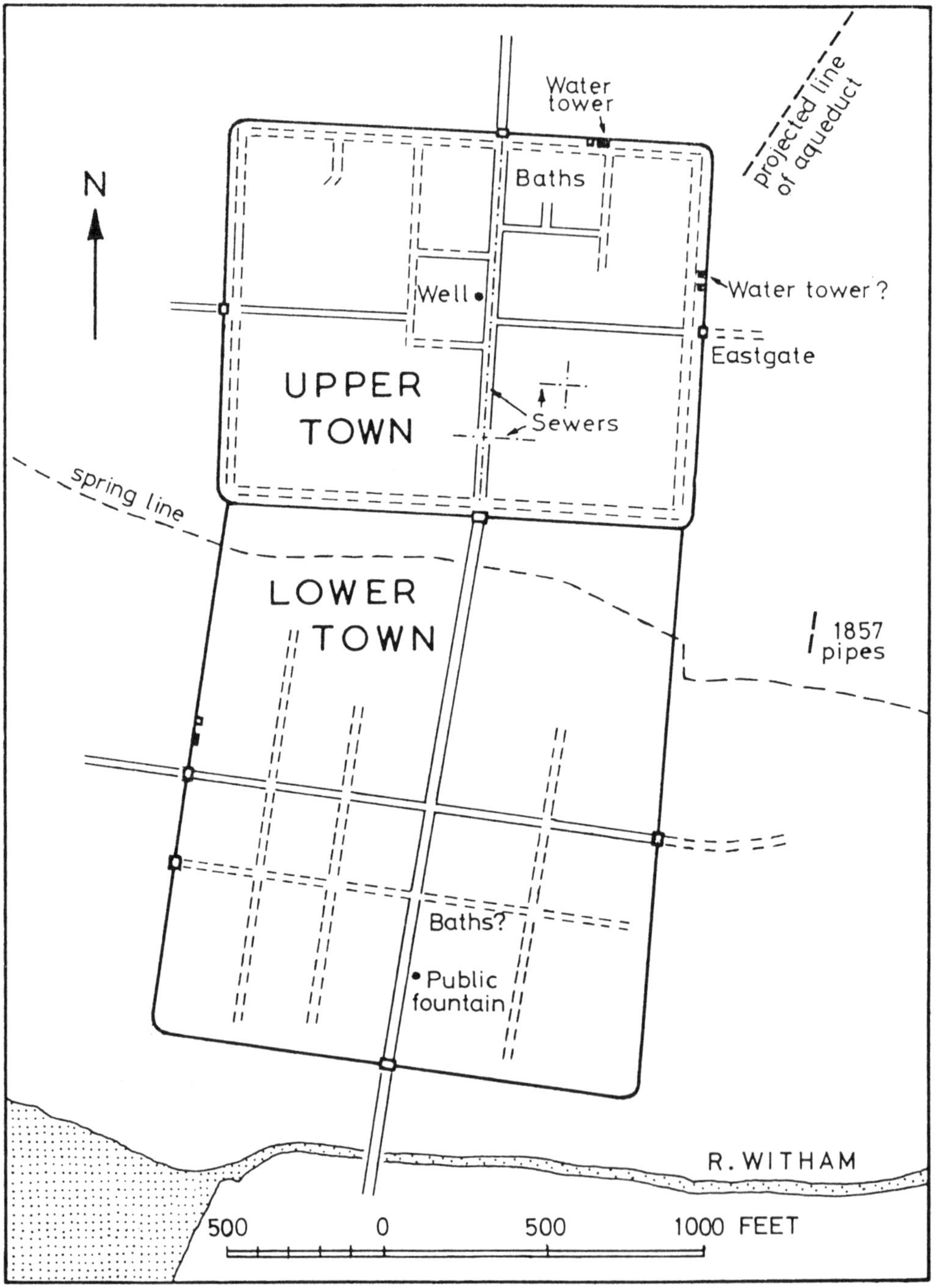

Fig.11 Roman Lincoln (after Wacher).

possible that the water went through a settling tank before entering the pipeline, the absence of lime is suspicious. From several sites it is known that the pipeline went out of use and was replaced by a road at some undetermined but still Roman date. One therefore asks oneself whether the aqueduct, however fed, just did not work. Roman engineers, like engineers of any period, were not infallible. One can think of several instances: the tunnel headings of an Algerian aqueduct not meeting; a dam in a wadi in North Africa, built no doubt in the dry season, facing the wrong way; the 3½ mile tunnel draining the Fucine Lake near Rome, where the miscalculations were such that on the first occasion that the emperor Claudius ceremonially opened the sluices, nothing happened; and the second time, after frantic alterations, the banquet laid on for the imperial party was all but washed away by the unexpected outflow.[22] The Lincoln aqueduct might take its place alongside these fiascos. Was the road laid over the pipeline, one might even ask, to accommodate those road tankers postulated earlier, in lieu of a hydraulic scheme that failed?

Again, it is possible that the Roaring Meg pipeline did not stand alone, and that there were other aqueducts to Lincoln from other sources. A stone culvert that could well be Roman was briefly observed below the lake at Riseholm due north of Lincoln, although this is about the same height as Roaring Meg and a similar pumping problem would arise.[23] Another pressure supply is not impossible from the Jurassic ridge beyond the Witham, where the ground is higher than on the north: around Waddington and Coleby it is 250 to 270ft, and the spring line is at about 200ft. While the head would not be enough to operate an inverted siphon down to the Witham and up to the upper town, the necessary lift at the source would be much less than at Roaring Meg. What might be evidence for such as aqueduct was found in 1857 beside Greestone Stairs on the east of the defences, in the form of a section of pipeline, sheathed in concrete and similar to the Roaring Meg pipe (Fig.11).[24] It is usually interpreted as a branch from the Roaring Meg aqueduct, running downhill to supply the lower town. But since this would not be under pressure, why encase it in such heavy sheathing? The pressure of the concrete cladding strongly suggests a pressure main from the south; either from beyond the river, or even from a local reservoir fed by a source on the spring line just below, on the 150ft contour. In further support of additional aqueducts are other pipes found in Lincoln — it is not known precisely where — of a smooth bore which is hydraulically preferable to the Roaring Meg design.[25] There are also the two candidates for water towers behind the walls. And there is a considerable system of Roman sewers on a grid pattern under the upper town, the biggest measuring 5ft high by 4ft wide.[26] Their scale, unduly large merely for carrying off storm water or merely for admitting sewermen, almost presupposes a plentiful water supply.

Even if Lincoln had no aqueducts that worked, it still had its wells. The finest known is that on the St Paul-in-the-Bail site, enclosed in a room in the east range of the forum. It is very large, almost 10ft square but reduced by arches to 3ft square at the head. The water table is now something like 45ft down; and one might well ask why the Romans chose, on the face of it, to pump a good 70ft up from Roaring Meg over a mile away, rather then 45ft up in the town centre. The answer may be that they began with the well alone, but the aquifer under the town not proving capacious enough to provide the desired quantities, the Roaring Meg aqueduct was added. This might explain the large size of the forum well, to accommodate a pump on a platform some way down. Unfortunately it proved too dangerous to excavate the fill to any depth.[27] In the last resort, the upper town could certainly survive, minus its baths, on wells alone, just as it did in the Middle Ages, although this would surely have involved loss of face in failing to keep up with other Roman municipal Joneses.

We cannot evaluate the Lincoln aqueduct until we understand it, and the solution to the conundrum is still to be found. The moral, however, is closer to hand. Ancient industrial archaeology is every bit as interesting, as exercising and as important as latter day industrial archaeology. And Roman engineers were, within the limits of their technology, every bit as accomplished and even heroic as their later counterparts. Their achievements were not an isolated flash in the pan; many of them survived to exert a direct influence on later ages — not merely water supply systems, but watermills, irrigation machinery, bridges, concrete and building design, to name but a few aspects. Many eighteenth and nineteenth century engineers studied with profit the works of their Roman predecessors, to see what lessons they could learn. So too might we.

Postscript

Since this paper was read, some new information has come to hand.[28] The East Bight water tower proves to have measured about 56 by 28ft, with perhaps a 3ft depth of water; the capacity, at some 2650 gallons, was large by comparison with other known examples. The floor of the tank is at 229.5ft OD, higher than expected, which makes the difference between it and Roaring Meg 98.5ft; problems of pumping would thus be increased. Finally, Wacher's two cropmarks on the suggested inverted siphon route (note 21) have been investigated. One could not be traced on the ground; the other, on trenching, proved to be a modern land drain.

Notes and References

1. J.W.F. Hill, *Mediaeval Lincoln* (Cambridge, 1948), 150, 342; *Tudor and Stuart Lincoln* (Cambridge, 1956), 63-4; *Victorian Lincoln* (Cambridge, 1974), 160; W. White, *History, Gazetteer and Directory of Lincolnshire* (Sheffield, 1856), 104-5.
2. Hill, *Victorian Lincoln*, 161-2; White, 104-5; Lincolnshire Local History Society, *Industrial Archaeology Group Newsletter*, 2.3 (1967).
3. Hill, *Victorian Lincoln*, 160-70, 231-6; *Lincolnshire Chronicle*, 10 March 1905; *Lincolnshire Echo*, 29 March 1906.
4. Hill, *Victorian Lincoln*, 236; *Inauguration of the New Water Supply, 3-4 October 1911* (pamphlet in Lincoln Central Library).
5. For Roman Lincoln in general, I.A. Richmond, 'The Roman City of Lincoln', *Archaeological Journal*, ciii (1947), 26-56; J.B. Whitwell, *Roman Lincolnshire* (Lincoln, 1970), 17-43; J.S. Wacher, *The Towns of Roman Britain* (London, 1974), 120-137.
6. Wacher, 319-20.
7. For the archaeological evidence, F.H. Thompson, 'The Roman Aqueduct at Lincoln', *Archaeological Journal*, cxi (1954), 106-128; K.F. Wood, 'The Roman Aqueduct at Lincoln: recent investigations', *Antiquaries Journal*, lxi (1981), 107-110; brief reports on the East Bight *castellum* in *Lincolnshire History and Archaeology*, 10 (1975), 58 and 11 (1976), 59. Examples of the pipe and its sheathing may be seen in Lincoln City and County Museum.
8. For water lifting, T. Schioler, *Roman and Islamic Water-lifting Wheels* (Odense, 1973); J.G. Landels, *Engineering in the Ancient World* (London, 1978), 58-83. The accounts of the Commissioners of the Arsinoe Water Board (in the Fayyum in Egypt) for 113 AD, preserved in a papyrus, offer an intriguing insight into the operation of a Roman water works, complete with *shadufs, sakkiehs, tympana*, a pump and perhaps Archimedean screws: ed. F.G. Kenyon and H.I. Bell, *Greek Papyri in the British Museum*, 3 (London, 1907), 180-90, partly translated in ed. A.S. Hunt and C.C. Edgar, *Select Papyri*, 2 (Loeb Classical Library, 1934), 536-41. The only known instance in Britain where a town's water may have been lifted by such means is Leicester, where it is plausibly suggested that the Raw Dykes brought the water at a low level to the town wall, where it was lifted to a higher-level water tower: the reverse of the apparent procedure at Lincoln (Wacher, 345-6).
9. Richmond, 66.
10. T. Ashby, *The Aqueducts of Ancient Rome* (Oxford, 1935), 30 estimated 222 million gallons a day. On different grounds, D.R. Blackman, 'The Volume of Water delivered by the Four Great Aqueducts of Rome', *Papers of the British School at Rome*, 46 (1978), 52-72, calculated a smaller quantity for the four biggest aqueducts which, if extended to all ot them, would produce a total of about 177 million gallons.
11. J.S. Wacher, in P.C.G. Isaac, 'Roman Public Health Engineering' (discussion), *Proc. Instn. Civ. Engrs.*, part 1, 70 (1981), 298. Six waggons each carrying 100 gallons or nearly half a ton, the maximum reasonable with ancient vehicles, and with a turn-round time of two hours including filling and emptying, would carry 600 gallons in two hours as against the 540 gallons for a pump. But if the wages of two men were saved, twelve oxen would need feeding. For much technical advice on the hydraulic aspects of the Lincoln aqueduct, I am indebted to Mr D.G. Morgan.
12. The major ancient sources are Philo of Byzantium, *Mechanics*, 4.77.15ff; Hero, *Pneumatics*, 1.28; Vitruvius, 10.7.1-3. For wooden pumps, A. Neyses, 'Eine römische Doppelkolben-Druckpumpe aus dem Vicus Belginum', *Trierer Zeitschrift*, 35 (1972), 109-121; for bronze, T. Schioler, 'Bronze Roman Piston Pumps', *History of Technology*, 1980, 17-38, and especially G. Rouanet, 'Étude de quatre pompes à eau romaines', *Cahiers d'Archéologie Subaquatique*, iii (1974), 49-79.
13. L. Langouet and J-L. Meury, 'La machinerie en bois du haut empire retrouvée à Alet', *Annales de Bretagne*, 70 (1973), 163-182; H. de Saint-Blanquet, 'Une station de pompage gallo-romaine à St-Malo', *Sciences et Avenir*, 1975, 732-6.
14. E.g. an air pump for blowing an organ, operated by trips on a shaft rotated by a windmill, designed by Hero of Alexandria; a *shaduf* likewise powered by trips on a shaft turned by animals, designed by Philo of Byzantium; the strong possibility of a water-powered grain pounder working just like a trip hammer.
15. Wacher, *Towns of Roman Britain*, 126-131. He is probably wrong, however, in arguing that all Roman pumps had to be immersed, and that to have a pump on a platform on the last pier, lifting from 10ft below, would defeat the whole scheme. There is strong evidence from Belginum of a suction pipe below the pump; Neyses, 114-7.
16. The basic ancient source is Vitruvius, 8.6.5-9. For a general survey, see N.A.F. Smith, 'Attitudes to Roman Engineering and the question of the Inverted Siphon', *History of Technology*, 1976, 45-71.
17. G. Garbrecht, 'Die Madradağ-Wasserleitung von Pergamon', *Antike Welt*, 4 (1978), 40-49.
18. The most convenient survey of the Lyons aqueducts is A. Grenier, *Manuel d'archéologie gall-romaine*, 4.1 (Paris, 1960), 118-136.

19. J.B. Ward-Perkins, 'The Aqueduct of Aspendos', *Papers of the British School at Rome,* 23, (1955), 117.
20. Wacher in Isaac, 299-300.
21. Wacher in Isaac, 300, marks on his map two possible features along his route. They are cropmarks observed from the air, and have not been investigated. While they might be left by the pipeline, they could equally represent anything else of any period.
22. *Corpus Inscriptionum Latinarium,* viii, 2728; C. Vita-Finzi, 'Roman Dams in Tripolitania', *Antiquity,* 35 (1961), 18; Tacitus, *Annals,* 12.56-7.
23. Information from Mr F.T. Baker.
24. *Lincoln Rutland and Stamford Mercury,* 6 November 1857.
25. Camden's *Britannia* (ed. Gough, 1806), ii, 366 and figs. iii-iv.
26. Richmond, 36.
27. B. Gilmour and M.J. Jones, 'Lincoln, St Paul-in-the-Bail', *Lincolnshire History and Archaeology,* 15, (1980), 73-76.
28. J.T. Wilford, 'The Roman Aqueduct at Lincoln', unpublished thesis, Hull, 1982.

PORTRAIT OF SIR HUGH MYDDELTON (1560–1631), ENGRAVED BY GEORGE VERTUE IN 1722 FROM THE PAINTING BY CORNELIUS JANSEN, THEN IN THE POSSESSION OF THE EARL OF OXFORD

3

Sir Hugh Myddelton and the New River

G.C. Berry

By the year 1600, the population of the City of London and the liberties which had been increasing rapidly during the past century had probably exceeded 150,000 and London was breaking out of its bounds in spite of fruitless attempts to check its growth.[1] This expansion was only to be expected for London, the centre of an expanding commercial and colonial empire, was attracting people, not only from all over England, but also from abroad.

The City's water supply system which had evolved slowly during the past four hundred years consisted of fountain heads, loosely called conduits, established in certain parts of the City and supplied with water brought in leaden pipes from springs in the suburbs. These were not only becoming inadequate, but were being threatened by building operations. An ineffectual attempt had already been made, through an Act in 1543, to bring water from Hampstead.[2] From 1568 onwards, however, all the new conduits were supplied with water from the Thames. These latter conduits were the Dowgate conduit, supplied by a horse machine at the bottom of Cosen Lane, the conduits at the churches of St. Mary Magdalen and St. Nicholas Cole Abbey (1583) near Old Fish Street, supplied from Peter Morris's waterwheel at London Bridge, and the Aldersgate conduit, the last of the conduits, which was constructed in 1610 by a certain Thomas Hayes, and was supplied with Thames water through wood and stone pipes.[3]

When, in 1581, the City granted a 500-year lease of the first arch of London Bridge to Peter Morris for the accommodation of his waterwheel, important departures were made from the old principles of the conduit system.[4] The City now empowered a private individual to build waterworks of his own with the principal object of supplying individual houses through pipes to be laid in the streets as business directed. The idea of supplying houses on a general scale was new, and in accordance with the reasoning of the time the risk of the new venture was taken by an individual.

[1] For estimates of London's population in the sixteenth and seventeenth centuries see *The Growth of Stuart London* by N. G. Brett-James, London and Middlesex Archaelogical Society, 1935, chapter xx.

[2] 35 Henry VIII, c. 10.

[3] Stow's *Survey of London, Continuation*, 1633 edition, p. 12.

[4] City Records Repertories of the Court of Aldermen 20/201.

Morris was followed in 1594 by Bevis Bulmer, a silver and gold mining engineer, who erected machinery at Broken Wharf to supply the western part of the City.[5] About the year 1600, Edward Wright, whom we shall meet later, endeavoured to raise water from Botolph Wharf, below London Bridge, to Leadenhall to supply houses at the eastern end.[6] Apart from a clash with Morris's successor at London Bridge, which brought about what must have been the first boundary agreement between water undertakings serving London, the fate of this venture is obscure.[7] There was also John Darge, the first water undertaker to be absorbed by the New River (in 1614). He supplied water from Fogwell Pond by leaden pipes and derived a revenue of £60 a year from it.[8]

Although they had granted to Morris and others authority to establish private water works, the Corporation, as their records show, did not divest themselves of the general responsibility for ensuring the supply of water to the City, and the increasing risks of fire and pestilence resulting from a rising and congested population must have added considerably to their anxiety in this field. For although the Thames supply might be considered unlimited, it suffered from two defects : it had to be pumped, and pumping was in its very early stages of development, and it could not have been fresh to those who were formerly accustomed to be served with spring water from the conduits.

The City already possessed an Act for bringing the water of the River Lee to the City (Lee for statutory bodies, Lea for practically everybody else). This Act made no mention of water supply, but authorized the construction of a canal with a maximum width of 80 feet, including the banks, for navigation[9].

The water supply scheme that was to have the greatest influence on the City was devised by Edmund Colthurst of Bath, who had served in Ireland and was sometimes referred to as Captain Colthurst. He was perhaps the Colthurst who presented the ruined abbey to Bath. About the year 1600, he proposed to bring water from springs in Hertfordshire and Middlesex to London, two-thirds of which were to be devoted to scouring the ditches and the remaining third to supply water to particular houses in the Cities of London and Westminster. Queen Elizabeth was made aware of these proposals and referred them to the Lord Treasurer, Lord Buckhurst, for consideration. He in his turn consulted the City Corporation, in 1602, who "did approve the said worke and signified their willinge allowance thereof as a matter very beneficial and no waie p(re)judiciall to them". The queen also directed the sheriffs

[5] For Bevis Bulmer, see *Notes and Queries*, Series 11, IV, p. 401.
[6] City Records Journals of the Court of Common Council 25/180b.
[7] City Records Repertories 26–1/40.
[8] Public Record Office, LR2/28, 2nd book, 2 July 1614.
[9] 13 Elizabeth, c. 18.

of Hertfordshire and Middlesex to inquire whether the proposed channel would injure the inhabitants of the two counties, whether it would diminish the flow of any navigable river, and to what breadth it should be limited. They answered that the inhabitants would not suffer, provided compensation were paid to those whose lands were to be cut and sufficient bridges were constructed and maintained. They also limited the width to 6 feet. The queen died before she was able to grant a licence and Colthurst then applied to King James, who by Letters Patent dated 18 April 1604, gave him the necessary authority.[10] Under this grant, Colthurst was to pay £20 per annum to the king when the River should begin to show a profit. The time allowed for the work was seven years. It would appear from the records that no further steps were taken that year to implement the scheme, but in February 1605 the Court of Aldermen considered a letter from the Privy Council on Colthurst's behalf alleging that the work had already been brought forward three miles.[11]

But other parties were coming forward and committees were appointed by the Common Council to consider their schemes[12] with the result that a Bill was prepared for the November session of 1605, for the purpose of obtaining statutory powers.

This Bill was read for the first time on 30 January 1606, and is recorded in the Commons Journals as "The Bill for the Bringing in of a fresh Stream of running Water from the River Lee, or Uxbridge, to the North Parts of the City of London", which suggests that the City then intended to tap either the River Colne or the Lee. John Strype, Stow's eighteenth-century editor, records that a certain Russell presented a scheme for obtaining a gravity supply from the Colne to Lord Burleigh about the year 1580.

The second reading took place the next day when the Bill was referred to a large committee and Colthurst was invited to attend. The scheme, as it emerged from this committee, was very much on the lines of Colthurst's patent. Among the members of this committee were the following who were to become Adventurers in the New River: Mr. Lawrence Hyde, Mr. Bacchus (Backhouse), Mr. Recorder of London (Sir Henry Montagu), Mr. Middleton (presumably the memorable Sir Hugh). There was also Mr. Robert Middleton (Sir Hugh's brother). The third reading took place on 20 May when, the House being divided, there appeared, for the Noe, 49; with the Yea, 60, and the Bill was passed. At this third reading Mr. Recorder promised the House, on behalf of the City, that they would submit themselves to such orders for recompense to be given to Captain Colthurst as the Lord Chancellor should set down.

[10] 2 James I, part 25.
[11] City Records Repertories 26–2/515b; and H. M. C. Salisbury MSS, part XVI, p. 417.
[12] City Records Journals 26/359, 392.

The Bill then went to the Lords, where Colthurst's recompense was further discussed, the House resolving in the end that stay should be made by the Lord Chancellor for granting any commission under the Great Seal, as required by the Bill, until such recompense should be made to Colthurst for his travail and pain as his Lordship should think reasonable. The Bill was eventually passed with the title: "An Act for the bringing in of a fresh streame of running water to the North part of the City of London". (3 James I c. 18).

The recital to the Act consists of a very brief statement to the effect that it had been found very convenient and necessary to have a fresh stream brought to the north parts of the City from the springs of Chadwell and Amwell, and others in Hertford not far distant from them, and that it had been found, upon view, very feasible and like to be profitable to many. The Act empowered the City at any time to lay out limits of ground for making the river a breadth of 10 feet and no more, to dig the trench and maintain it, to lay earth on either side and to have free access to the cut with men, horses, carts, and carriages. The City on their part were to pay compensation to the landowners and occupiers and all persons injured in their mills standing upon any of the streams from which the water would be taken. In default of agreement, compensation was to be regulated by commissioners appointed for that purpose by the Lord Chancellor under the Great Seal. Four commissioners were to be appointed from each of the counties of Hertfordshire, Middlesex, and Essex, and four more were to be appointed from the City. Damages for breaches, inundations, etc., were to be recoverable by action on the case. The City were also obliged to construct and maintain bridges "in places meet and convenient". An important provision was that until full agreement had been reached with the owners and occupiers of the premises, or the necessary orders agreed upon by a majority of the commissioners, it would not be lawful for the City to dig the new cut nor take any ground for the passage of the water. Once completed, the river would be subjected to the Commission of Sewers and any fines imposed by that Commission for annoyances to the hurt of the River would enure to the benefit of the City.

Soon after the passing of the Act, a survey of the springs and of the proposed course of the river was undertaken on behalf of the City by Richard Staper, clothworker, who was reimbursed £5 6s. 0d. expenses.[13]

The next two years were spent in negotiations with various parties including Colthurst and William Ingelbert. Ingelbert was a land drainage engineer who had assisted the Corporation, possibly as an expert witness, in obtaining their Act. He intended to bring the river from the springs in a trunk or vault "for the sweet keeping thereof".[14]

[13] City Records Repertories 27/265.

[14] City Records Repertories 27/263; Journals 27/89.

The City appears to have been much taken with the idea and obtained a second Act in 1607 for that purpose.[15] Negotiations with Ingelbert, however, seem to have fallen through, for after the passing of this Act he is heard of no more. John Aubrey in his *Brief Lives* states that he was seen later in a poor rug gown like an almsman, sitting by an applewoman at Parliament stairs.[16] He also states, quite wrongly, that Ingelbert was the first inventor of the project, and infers that Sir Hugh Myddelton, whom he calls an Alderman, took the credit for something that was not his, which again was not the case, as we shall see.

In October 1608 Colthurst petitioned the Court of Aldermen, stating that he, his friends, and partners were now ready to construct a river under his patent of 1604 at their own cost, but required to be reassured that the City would not hinder them, by virtue of their two Acts, from bringing water from the Chadwell and Amwell springs.[17] The Court assented, but stipulated that the work was to be executed within two years of the following Lady Day and that no cost should fall to the City. There followed an interval of several months and then on 7 March 1609 Colthurst came forward with a fresh request,[18] this time to have the benefits of the two Acts assigned to his partners on condition that the stream was brought to Islington within two years. This is the first time Islington is recorded, and when one looks at the contour lines one is struck with the aptness of the choice. New River Head, the eventual terminus of what became the New River, is on a southern slope dominating the City, and is in fact the nearest point over 80 feet above high water mark. After bringing the water to Islington, Colthurst offered to allot the City two-thirds of the stream for the public use if they would contribute £2,400 towards the cost of the river. If the City did not wish to contribute, the partners would bear the entire cost and would take all profits. Could it be that this £2,400 represented two-thirds of the original estimated cost of the work for bringing the water to London? If so, it is not surprising that financial difficulties were encountered, as recorded.

The usual committee having been appointed to consider this latest request, they reported a week later in favour of the transfer, but that the partners should execute the work entirely at their own costs.[19]

Then, from the records, a dramatic change takes place, for on 28 March 1609 the Common Council, having considered this last report, passed an Act to constitute not Edmund Colthurst, but

[15] 4 James I, c. 12.
[16] *Brief Lives*, John Aubrey, II, p. 1, edited by Andrew Clark (1898).
[17] City Records Repertories 28/288.
[18] City Records Repertories 28/355.
[19] City Records Repertories 29/3.

Hugh Myddelton, citizen and goldsmith of London, their lawful deputy to exercise their statutory powers.[20]

Now I shall not enter into a life history of Sir Hugh. That is a subject worthy of serious research on its own. I am confining myself to recreating the river as Myddelton knew it. Of his life up till now, I shall merely say that he was born about the year 1560 at Galch Hill, in the parish of Henllan, the sixth son of Richard Myddelton, the governor of Denbigh castle, that his first engineering exercise was, as he has written himself, "seeking of coals for the town of Denbighe".[21] Early in life he went to London and became a goldsmith, but, unfortunately, there is no record of his apprenticeship. He represented Denbigh in Parliament and was the first alderman of that borough. He also sold jewellery to James I.[22] He was, I should say, a great family man.

Myddelton's first introduction to the undertaking that was to mean so much to him was possibly when he and his brother, Robert,[23] were appointed members of the Parliamentary Committee in 1606 to examine the City's first Bill for bringing the waters of Chadwell and Amwell to London. His name does not appear in the list of members who considered the second Bill in 1607. The next reference to him in connection with the New River I find is in his own handwriting. It is the heading at the beginning of the first book of disbursements on the New River, preserved at the Public Record Office, about which I shall have more to say.[24] This heading reads "mon(ie)s paid and Dysbursed for the bringinge of watter from the springs of Chadwell and Amwell to the Citty of london begonne this 20th of Februarie 1608 (i.e. 1609) as folow(e)th". The first two entries read: "To Mr. Stonne for his counsell in my booke from the Citty at twis," £1; and "to Mr. Coulthurst to geve the oficer for warnynge the Comitty to meate in that busenes," 2/6d. These show quite clearly, I think, that Myddelton was Colthurst's partner whilst the latter was negotiating for the transfer of the City's statutory powers. In fact, he was probably one of the "friends and partners" Colthurst had taken unto him when he petitioned the Court of Aldermen to be allowed to proceed with his own scheme in October 1608. If that is so, it would appear that the partners, after considering their powers under his Letters Patent, decided that it would be more satisfactory

[20] City Records Journals 27/377b.

[21] Letter to Sir John Wynn, of Gwydir, 2 September 1625, printed in full in Thomas Pennant's *Tours in Wales* (1810), II, p. 363.

[22] See extracts from Pell Office Records by Frederick Devon, 1836, p. 19; see also W. M. Myddelton, *The Myddeltons of Gwayninog* (1910), p. 18, footnote.

[23] Gery Milner-Gibson-Cullum, F.S.A., *Pedigree of Middelton or Myddelton of Chirk Castle, Denbigh* (1897).

[24] P.R.O.—LR2/27—Records of the Exchequer—Office of the Auditors of Land Revenue—New River account books, 1609–32.

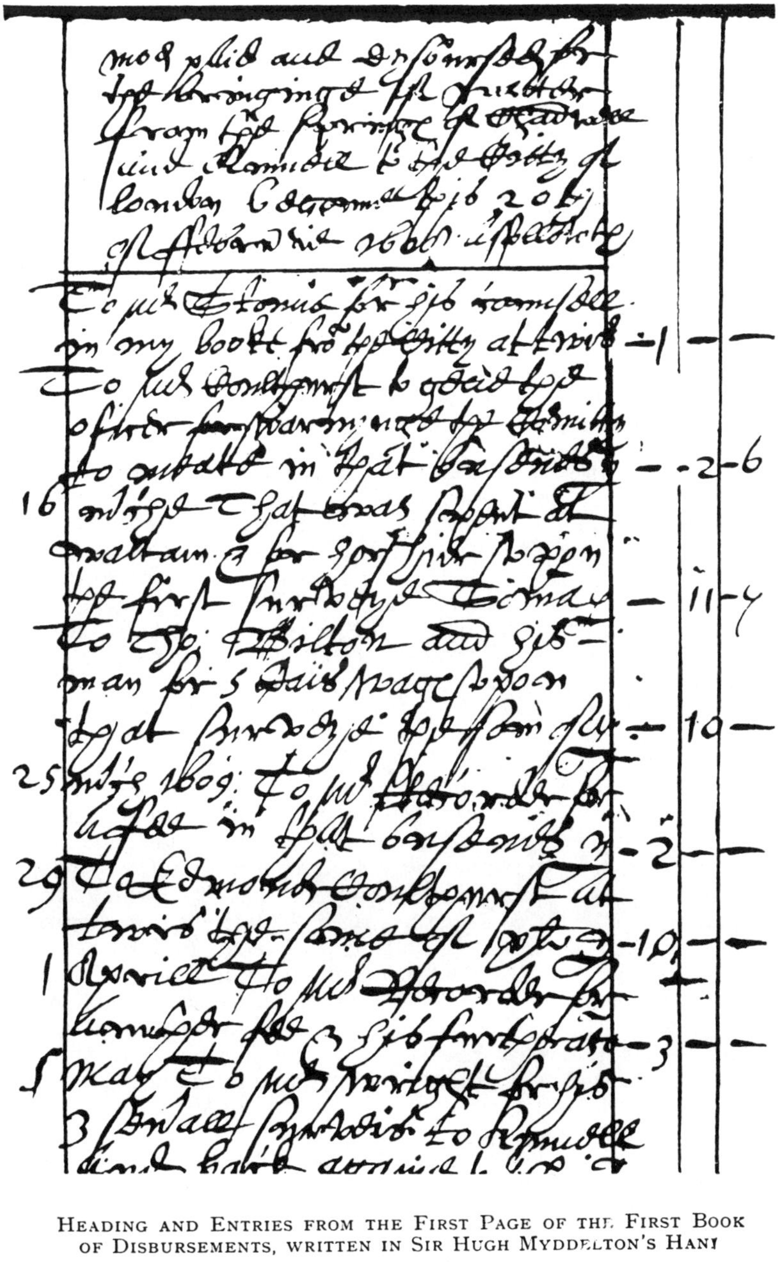

Heading and Entries from the First Page of the First Book of Disbursements, written in Sir Hugh Myddelton's Hand

THE MEMORIAL TO SIR HUGH MYDDELTON ERECTED BY ROBERT MYLNE IN 1800 ON AN ISLAND IN THE NEW RIVER AT AMWELL

to obtain the City's statutory powers. The City's first Act in particular allowed a greater width to the river and also set down a procedure for compensating landowners and others. When it was that Myddelton became acquainted with Colthurst we have yet to learn. He might have met him for the first time when the City's first Bill was being discussed, but it could have been earlier. This is not the first time that there had been a partnership between a Myddelton and a Colthurst. A relation of Sir Hugh's, Sir Henry Myddelton, and a certain Christopher Colthurst, were General and Lieutenant-General of the East India Company's second expedition to the East Indies (the Moluccas) in 1604.[25] However, that might be, there is no doubt that the City preferred to convey the powers and responsibilities of their Acts of Parliament to a relatively wealthy man like Myddelton, rather than to Edmund Colthurst, who they knew had not the means to carry out the work. That the part played by Colthurst in the venture has become obscured was in no way the fault of Sir Hugh who, as we shall see, fully recognized the services he had rendered.

In appointing Myddelton their deputy, the City stipulated that he should enter on the work within two months and complete it within four years.[26] On 1 April 1609 they granted him their letter of attorney,[27] and on the twenty-first of the same month a formal agreement, drawn up by the Recorder, Sir Henry Montagu, was duly completed. Sir Henry Montagu, later first Earl of Manchester, became Lord Chief Justice after the dismissal of Sir Edward Coke. He was also the brother of Bishop Montagu and, what seems more than a coincidence in view of Colthurst's connexion with Bath, he gave Bath Abbey its west door.[28]

But I must revert to the first four pages of Myddelton's accounts, which are in his own hand and record expenditure up to 12 August 1609. He advances small sums to Colthurst for the payment of wages, the first of which, dated 5 May 1609, is some indication of when the work started. These might have included an allowance to Colthurst, for under 2 August there is the first specific entry of 14s. 0d. to Mr. Colthurst for his weekly allowance. With the exception of one gap, which I shall come to later, this weekly payment was made regularly to Colthurst until August 1616, when he probably died.

The first record of expenditure on the construction of the river, apart from the inevitable legal charges, is 11s. 7d. spent on 16 March for horse hire at Waltham for the first survey. On 5 May there is an entry "To Mr. Wright for his three sev(er)all surveis to

[25] See *The Voyage of Sir Henry Middleton to the Moluccas*, Hakluyt Society, Series II, 88, (1943).
[26] City Records Journals 27/377b.
[27] M.W.B.—M.B./230.
[28] Revd. Richard Warner, *The History of Bath* (1801), p. 244, footnote.

Amwell and back againe to Islingt(on)", £20 3s. 0d. This Mr. Wright would be none other than Edward Wright of Garveston, Norfolk, who a few years before had tried to raise water from the Thames at Botolph's Wharf for the supply of the City. A mathematician, the discoverer of the principle underlying the Mercator projection, the friend of Lord Napier of logarithm fame, he was also an astronomer of some renown. It is doubtless to him that the Rev. Wilhelm Bedwell, the vicar of Tottenham, refers in his Description of Tottenham (1631) when he writes of the New River, "He who first chalked out the way, we know was a Minister of London, our English Tycho.[29] A week later, Myddelton pays a further £40 to Wright "before hande to be my Arts man", and from 9 September he is paid a weekly allowance of £2. On the 5 May again there appears another principal character, one who was to be associated with the river for many years. He is William Parnell, the principal supplier of timber for the river works. On this day he was paid 12s. 6d. "for 5 Dais laboure and travell in the last survey with Mr. Wright", and £2 18s. 0d. for 600 "of bords for the springe headd and for the caredge thereof". Other payments recorded are, on 20 July "To Mr. Corbet, Clearke of the Counsell, for the Lords l(ett)res to the Justices of Peace and for thear open plackard in my behalf to all officers, the som of" £2. On 2 August, there is an entry "For a Comyssion owt of the Channsserye to Compounde with men for ther lande the som of" £2 10s. 0d. This commission was an essential item in the procedure laid down by the Act for settling landowners' compensation.

From 11 August the accounts are kept weekly and in detail by Myddelton's first clerk, Edward Hughes. Who this Edward Hughes was I am not sure, but he might be a descendant of Edward Hughes of Holt Castle on the Denbighshire-Cheshire border, the Receiver-General to Queen Elizabeth of the revenues of North Wales and Cheshire.[30] The work is now properly under way and by the end of August over fifty labourers appear by name, and by the end of September over 130 are engaged. To the end of the year the number of named labourers never falls below forty-seven and is several times above 100. In December the position is somewhat obscured by such entries as "To John Gilderson and his compa(nie) in parte for worke done by the pole", £3 10s. 0d., without any reference to the number in the company. The same evidence of activity subsists until the week ending 27 January 1610, when the number falls to seventeen. From then on until November 1611, very little progress is made. The obstruction and difficulties encountered by Myddelton have been mentioned time and again whenever the story of the river has been told but here, in these accounts, the result of the obstruction stands out clearly. The work

[29] Tycho Brahe (1546–1601) famous Danish astronomer.

[30] See *Notes and Queries*, CXCV (1950), "Links with Shakespeare—VII", pp. 385–6.

was not merely slowed down, it was virtually stopped for about twenty-two months. After 10 February 1610, Wright's allowance ceases, though Colthurst continues to be paid up until 11 June. Throughout the whole of the period, the clerk's wages of 12s. 0d. a week remain undisturbed. In fact, for certain weeks his and the molecatcher's wages (3s. 4d.) are the only items recorded. Work on the river was limited to maintenance, but there was no intention of giving up. John Allen, Robert Essex, Thomas Stone, and Edward Toaler, the first of the walksmen to be paid quarterly, received their first quarter's wages at midsummer 1610. As they only received £1 altogether, it can only be assumed that their duties were part-time.

It has often been repeated that the river had been brought as far as Enfield when the work was stopped, and that King James became interested in the undertaking while watching the excavations in Theobalds Park.[31] The books, however, reveal that the cut had not reached the northern limits of Cheshunt when the great hold-up occurred. It would seem that work came to a standstill while the first of the great loops, that at Wormley, was being negotiated. We gather that the obstruction was caused solely by human opposition, although the details are not clear. No insuperable engineering problem had been encountered.

As they had promised, the City came to Myddelton's assistance, and in April 1610, an appeal was made to the Lords of the Privy Council to be "media(t)ors to the Kings most excellent Ma(ies)ty that all lette and hindraunce may be removed and way given to Mr. Midleton". The opposition, however, went to the House of Commons, then mainly composed of country gentlemen, who had just completed six years of constitutional skirmishing with His Majesty. There was every reason to believe that the House would have changed its mind about the New River. After all, relatively few members had voted for and against it. The river, too, was now becoming a reality and not some vague, stupendous scheme that might never materialize. By the beginning of May, a certain Mr. Beaulieu had found the to-do over the matter in the House sufficiently remarkable to write to his friend (Mr. Trumbull) abroad about it (see Winwood's *Memorials*, III, p. 160). On 18 May, a Bill for the repeal of the City's two Acts was read for the first time. This led the Court of Aldermen a week later to appoint a deputation to attend Sir Julius Caesar, the Chancellor of the Exchequer, Sir John Herbert, Secretary of State, and Sir Thomas Parry, Chancellor of the Duchy, all members of the Privy Council and of the House or Commons, and "earnestlie move and intreate theire lawfull favours for the hindring" of the Bill as the stream since the passing of the Acts "hath ben already brought onwards about tenn miles at the Charges of Mr. Hugh Midleton . . . amounting already

[31] See Samuel Smiles, *Lives of the Engineers*, (1861), I p. 115.

to the some of three thowsand pownds. . . .",[32] a pious exaggeration, as we shall see. The Bill was read a second time, nevertheless, and was referred to a committee on 20 June. The Commons Journal for 30 June mentions a "petition from Gentlemen of the Country that attend", and an order to meet in the afternoon to appoint ten members to view the river and certify the allegations contained in the petition. The names of the ten members were :—

> Mr. Alford, Sr. Wm. Strowd, Sir Geo. Rivers, Sir Roger Owen, Sir Dudley Diggs, Sir Tho. Holcroft, Sir Edw. Montague, Sir Tho. Beaumont, Mr. Talbot Bowes, and Sir Warwick Heale.

As the end of the session was at hand, however, they were instructed not to go on the survey until the next session.

What were presumably the objections of the supporters of the Bill and the answers thereto are given in an undated document among the State Papers.[33] They are ranged in two groups, (*a*) against the diversion of the springs ; (*b*) against bringing the river through many men's lands. The diversions of the springs would lower the level of the Lee, thus hindering navigation, reducing the power of the water mills and causing the meadows near the river to become parched. Twelve objections are listed under the second heading, but I will content myself with four : (1) By bringing the river through so many lands, the nature of the soil in many places would be altered and made worse as meadows into bogs and quagmires and arable land into "squallie" ground. The answer to this was that it appeared that the soil adjoining was the better and not the worse for the river. (2) The mangling of their lands lying in common fields into "quilletts and small peeces whereby the peeces cut off being mingled among other mens are made unfit for that use." Answer : "It anie man's lands be devided or cut the lawe provideth that he shalbe recompensed". (3) That the Church shall be wronged in their tythes without remedy. Answer : Doctor Atkins will quarrel at a trifle for where the cut is a mile in length in his parish it amounteth not to one acre of ground the tythe whereof is not worth above 2s 6d a year. (4) Lastly, that the Mayor etc. have not done anything in this business for themselves for the bringing thereof to London but have by an Act of Common Council irrevocably granted their whole interest . . . to Mr. Middleton who doth the same for his own private benefit. Answer : If the Mayor and Citizens would not adventure upon so uncertain a work, Mr. Myddelton deserveth the greater commendation in adventuring his money and travail for the good of the City and if the City find good and the country no hurt though Mr. Myddelton gaineth by it, he deserveth it well.

In the next session, the survey was apparently forgotten, for after waiting till nearly the end of November, the City sent the Commons a humble petition asking them to appoint members to

[32] City Records Repertories 29/231.
[33] P.R.O.—S.P. 14 lxxviii, no. 106.

view the work and report to the intent "that unles the sugestions prove true . . . a worke of so publique use having allready receaved probacon of two Sessions of p(ar)liament may be no longer hindred by causeless opposicon".[34] But this does not appear to have had any effect on the House. In the following February,[35] King James, resentful at personal attacks made by certain members, dissolved Parliament and decided to do without one for a time.

This put Myddelton in a quandary. Parliament, which was to decide the fate of the river, had been dismissed, leaving his opponents free to obstruct. Statutory powers, after an undecided Parliament had been dissolved, must have felt somewhat insubstantial to the undertaker of what was then an unprecedented and controversial public work. Not knowing how long it would be before he should obtain liberty to proceed from Parliament, Myddelton immediately petitioned the City for an extension of time, and at the end of the month the Common Council, "finding no fault in him", agreed that he should be allowed seven years from the following 21 April.[36] The old agreement was surrendered and a new one was completed on 28 March 1611.

But the work remained at a standstill for many months after. Then, sometime during the summer, negotiations began between Myddelton and Robert, Earl of Salisbury, the Lord Treasurer, on behalf of King James about the future of the river. The river had already been driven through the Lord Treasurer's lands, e.g. at the Ryefield, Hoddesdon.[37] Meetings probably took place as well between Myddelton, the pioneer of pipe smoking (according to Malcolm's tradition[38]) and the royal author of the "Counterblaste to Tobacco". But perhaps Myddelton had given up smoking by then. Anyhow, King James was in a strong position either to defeat the river entirely or to encourage it. At that time he was anxious to find new sources of income to raise his precarious revenue. The negotiations with the Commons over the great Contract, whereby he was to receive £200,000 per annum in exchange for his feudal dues, had only recently failed. Although it has often been repeated that Myddelton applied to King James for financial assistance, it is possible that James, animated by the prospects which the New River might offer when completed, took advantage of Myddelton's predicament to become a partner in the venture. That seems to be the view taken by John Fuller when he wrote "But oh what an injury it was unto him (i.e. Myddelton) that a potent and idle spectator should strike in (Reader I could heartily wish it were falsehood what I report) and by his greatness possess a moiety of the Profit which the unwearied endeavours of the

[34] City Records. Letter book DD/223.
[35] 9 February 1611. *H. of Lords Journal*, p. 684.
[36] City Records Journals 28/176b.
[37] M.W.B.—N.R.Compensation receipt no. 5.
[38] J. P. Malcolm, *Manners and Customs of London* (1811), I, p. 217.

aforesaid knight had purchased himself". But Fuller is a little harsh on the king, for he did have something very valuable to offer and that was power to overcome opposition after some eighteen months of Parliamentary indecision. Some years later, Myddelton advised his friend, Sir John Wynne, who was trying to interest him in a large engineering project he had in mind: "first of all his Majesty's interest must be got".[39]

But to return to the negotiations. On behalf of His Majesty, Salisbury agreed to remove all unjust and unlawful impediments which would hinder the bringing of the waters to the City and the continuance of the current or river after it had been brought there; to allow Myddelton to cut the river through the royal lands "without compounding or paying anything therefore"; and to grant the necessary commissions as prescribed by the Act. On 24 August, Salisbury appointed a certain Miles Whitacres to be informed on behalf of His Majesty of all expenditure on the river to be made thereafter. In accordance with the terms of the agreement, Whitacres and Myddelton were to enter all disbursements in books to be signed by both. The books thus signed were to be a sufficient warrant to the Lord Treasurer to authorize repayment to Myddelton of half the expenditure incurred. For all disbursements made before 24 August, Myddelton was to deliver an account to be sworn to by him on oath. In consideration of this powerful and financial support, King James was to receive for ever "the one halfe of the benefitt p(ro)fitt and comoditie of the saide water and of the conveyinge of the same to and from all places. . . ."[40]

Here, then, is the origin of the disbursement books I have referred to, which came to the Public Record Office *via* the Land Revenue Office in the last century. This bargain, between King James and Hugh Myddelton, produced a minute and fascinating description not only of the construction of this amazing work but also of its maintenance for many years after. The story, in fact, is taken down to the end of the year 1631, when the Crown ceased to be directly interested in the undertaking.

The first book, which ends at 24 August 1611, was kept by Edward Hughes, except for the first four pages which are all in Myddelton's unmistakeable handwriting. Every page bears Myddelton's signature and at the end there is an endorsement to the effect that the total sum of the book amounting to £1,139 15s. 11½d, had been sworn to by Hugh Myddelton, Esq., before Sir George Snygge, one of the Barons of the Exchequer. The other seven books are signed by Myddelton and Whitacres and later by Myddelton and Edward Ball (Whitacres's successor). They are all kept by William Lewyn, Myddelton's second clerk. I am not sure why William Lewyn replaced Edward Hughes at this juncture, but it appears, from one

[39] See (21) above.
[40] Patent Roll, 10 James I, part 10, no. 19.

of the first disbursements recorded by Lewyn, that the sum of 16s. 7d. was paid to Myddelton for three days' expenses at Hoddesdon when he went to take the accounts of Mr. Hughes,[41] the clerk, who was sick. Hughes may therefore have died or become incapacitated. William Lewyn was probably the father of Sir Justinian Lewyn (1613–1673), for there is a reference, in the charred remains of the New River Company's minutes, to negotiations in 1637 between Sir Justinian and the Company in a dispute between that body and William Lewyn.[42]

The formal agreement between the Lord Treasurer and Hugh Myddelton was eventually completed on 1 November 1611. One provision bears witness to Myddelton's disposition. It allows him at his own expense to derive a quill out of the main pipe that was to supply the City, in order to serve "the poore people gratis aboute Ste. Johns Streete and Aldersgate Streete which are not able nor fitt to paie for their water. . . ."[43]

The agreement with the Lord Treasurer was superseded by an agreement with King James, dated 2 May 1612.[44] It is more or less a repetition of the former agreement, but there is one provision of the greatest importance. His Majesty commands all his subjects not to "moleste inquiett trouble or hinder the said Hughe Middleton nor attempte use or putt in use anie unlawfull courses by sleightes clamor or otherwise to the hinderance . . . of bringing the saide springes . . . to the cittie of London or to the p(re)servacon of the continuall passage . . . of the said river upon paine of his maiesties highe displeasure and uppon such other paine and penaltie as by anie laws or statutes of this realme can or maie bee inflicted upon them, or anie of them, for theire disobedience and contempte in that behalfe. . . ."

Having described how King James became possessed of one half of the expected profits of the New River, it might be as well here to see how the interests of the other partners or Adventurers were administered. As we have seen, Myddelton and his assigns were appointed the City's deputies for their own benefit with power to exercise all the functions authorized by the City's two Acts. At an unspecified date Myddelton, in whom the undertaking was vested, divided it into 36 shares which he assigned to his partners, the "Adventurers", in proportion to their contribution. According to the New River Company's Charter, this division and allocation would have taken place after the work had begun, Myddelton having found the charge of the work "greater and heavier than at first was expected, the success thereof doubtfull and the opposicon

[41] P.R.O.—LR. 2/27, 2nd book, 1 September 1611.
[42] See also *Notes and Queries*, CXCVI (1951), p. 508, "Family connections of Bishop Brian Duppa", by Sir Gyles Isham.
[43] See Patent Roll, 10 James I, part 10, no. 19.
[44] Ibid.

made against it very strong". By a Deed dated 8 May 1612, that is a few days after the agreement with King James, two of these shares were conveyed to Sir Henry Nevill, of Billingsbear Park, Berkshire, who "is and hath been an adventurer in the said worke".[45] This was certainly not the first assignment for there is a clear reference in the deed to "others formerlye made". Sir Henry's consideration was £200 paid "long before the sealing . . . of theis presents" and a promise to pay one-sixteenth of all charges when expenditure shall have exceeded £3,200. It would be a mistake, therefore, to say that a share in the New River cost only £100. The reason why Sir Henry was to pay one-sixteenth towards the work (excluding the king's half of the expenditure, and receive only two thirty-sixths or one-eighteenth of the profits (excluding the king's half) was that it had been arranged that Edmund Colthurst should receive a life interest in four of the shares whilst contributing nothing to the cost of constructing the river, "in consideracon of great labour and endeavour by him bestowed. . . ." After Colthurst's death, two of the shares were to go to Colthurst's heirs and the other two to Myddelton in respect of his "great industrie travell and paynes takinge . . . in and about . . . the . . . finishing . . . of the saide worke. . . ."

The change of clerk I mentioned just now was not the only alteration which took place at this time. A tough policy was adopted towards opposition. Thus at the end of August, a warrant is obtained, for 2s. 6d. "to bringe a seditious fellowe . . . who p(er)suaded ye water bearers of London to petition to his Ma(ies)tie against ye newe Ryver" before the Recorder (Sir Henry Montagu).

Although a working agreement with the Lord Treasurer was reached by September 1611, work on the New River was not resumed in earnest until the end of November. The first sign of the revival was an advance of £60 to William Parnell, in October, for bridges for the river.[46]

With the resumption of work, further changes took place. First we have the appearance of Edward Pond on the scene. Pond, who died at Peterborough in 1629, was a mathematician and the author of a well known almanac which survived into the eighteenth century. According to the accounts, he was consulted at the end of November 1611, with a certain "Mr. Blagrave" about taking the plott (presumably marking out the ground) between Cheshunt and Theobalds. I had hoped to identify "Mr. Blagrave" with John Blagrave of Reading, a distinguished mathematician, but the mass of evidence so far consulted asserts that he died a few months earlier, in August. There is no further mention in the accounts of Mr. Blagrave after 7 December, but Edward Pond remains with Myddelton until just after the triumphal inauguration at New River

[45] M.W.B.—M.B. 9/79.
[46] P.R.O.—LR. 2/27, 2nd book, 12 October 1611.

Head. There is no doubt that Pond took the place of Edward Wright as Myddelton's "Arts Man". Wright had reappeared only once, in September 1611, when he was paid £1 for trying the level in Cheshunt Park at the north end of Cheshunt. Wright died in 1615 and an obituary of him recorded at Gonville and Caius College, Cambridge, his old college, states that he was "the first undertaker of that difficult but useful work . . . but by the tricks of others" was prevented from finishing it.[47] Bedwell also wrote that before the work was altogether finished, he was put by it, and others employed to make an end of it.[48]

Now although Wright was probably the first to make a detailed survey of the course of the New River and directed the mathematical part of the work in the early days, there is no doubt that the responsibility for the undertaking fell on Myddelton. What led to the substitution of Pond for Wright I do not know. It is possible that Myddelton, who was not entirely a free agent now, was prevailed upon to employ Pond by his royal partner. Pond himself was fully aware of his share in the work, for there is a note in his almanac for the year 1627 (it probably appeared in earlier editions) in the usual list of principal events in the history of the world, which runs :—

> Since the river from Ware to Lond begun by me Ed. Pond 15 (years)
> Since it was finished by me (the sole Artist) Ed. Pond 14[49]

Very different from Sir Hugh, who in the letter to Sir John Wynne previously quoted, wrote of his achievements : "Few are the things done by me ; for which I give God the glory".

Edward Pond was paid £2 6s. 8d. a week regularly for his entertainment, until the last week of his employment. He then was the expert who guided the river on its wandering course. He owned a level, the details of which are unfortunately missing, and a little shelter was made for it out of 4 ells of canvas at 16d. the ell. "Davie Gryffethe", the trumpeter, and "Rees Uphughes" carried it about for a week in February 1612, and were paid a shilling a day each. He was doubtless responsible for the couplet, dear to compilers of year books and calendars, which appeared in the early editions of his almanac :—

> If faults escape in Printing give each man his right ;
> A letter overslipt, is but an oversight.[50]

Another important change, affecting Myddelton personally, was that he too was now receiving a weekly allowance. The first entry (25 January 1612) runs :—

> To Mr. Hughe Midleton for a fortenightes expences, for hymselfe his man & tooe geldeinges . . . att xlvjs viijd the weeke, £4 13s. 4d.

[47] Charles Hutton, *Mathematical Tables* (1785), p. 28, footnote.
[48] Wilhelm Bedwell, *A Brief Description of the towne of Tottenham Highcrosse* (1631).
[49] B.M.—P.P. 2465, 1627.
[50] B.M.—P.P. 2465, 1604.

This allowance continued to be paid to him regularly until 1620, when a different arrangement was made.[51]

I must also mention "Howell Joahnes". He is brought in by Myddelton right at the beginning of the accounts, but only incidentally as "my man Howell". He is first employed on the river in the first week of September 1611 as a ganger at 7s. 0d. a week, and remains with the undertaking until his death in 1637. In 1622 he was paid 100 marks for "his great paynes taken in the workes" and Sir Hugh left him £20 in his will. After the completion of the river, he supervised the labourers in the streets and was later in charge at New River Head, where he lived. Whilst at New River Head, there is every reason to believe that he made the most of his opportunities, for shortly after his death the New River Company passed the following resolution which survives as a charred fragment :—

> (At) this nexte Courte the Company will make Choice of one whom they will putt in (the place of) Howell Jones Deceased, for looking to the walke & water house at Islington And it is ordered that none shalbe there suffered hereafter to sell drink or . . . or to use victualling in any sorte whatsoever.

The year 1612 dawned full of hope and as "a token of the new yeare" £15 7s. 0d. was distributed to such "as have given greate furtherannce tooe the workes and have taken greate paynes & have done manye kyndenes theirein (as ap(er)ethe by a p(ar)ticular byll)", unfortunately missing.[52]

Once the work got under way, rapid progress was made and in under twenty-two months the river was brought from the northern boundary of Cheshunt to Clerkenwell by a devious route, as one can see, over thirty miles long.

For it was in the afternoon of Michaelmas Day 1613 that the Lord Mayor, Sir John Swinnerton, accompanied by two Adventurers in the New River, Sir Thomas Myddelton (the Lord Mayor elect and Sir Hugh's brother), and Sir Henry Montagu (the Recorder) witnessed the first issuing of the river into the Round Pond at New River Head.[53] The opening of the flood gates was preceded by a colourful pageant and the delivery of a speech in verse composed by Thomas Middleton, the dramatist. I shall spare you a rendering of it, but here are the names of some of the artists mentioned in the speech. First the overseer, "an ancient soldier and an Artisan", Captain Edmund Colthurst, the clerk, William Lewyn, the mathematician, Edward Pond, the master of the timber works, William Parnell, the borer, Richard Parkes, and lastly, the paviour, Thomas Horne.

[51] P.R.O.—LR. 2/33.
[52] P.R.O.—LR. 2/27, 2nd book, 4 January 1612.
[53] Stow's *Survey of London, Continuation*, 1633 edition.

I will now trace the progress of the advance party from the first mention of place names in the accounts. Naturally, this does not mean that the channel had been completed at those places. The trench often had to be deepened or bottomed and the banks had to be raised and spread. I assume from the evidence of leakages that the river was completed in sections and controlled by sluices and floodgates.

Spital Brook was reached by 2 December 1609, and by 16 December work was going on in the vicinity of Broxbourne Church, for at that date the sexton kindly obliged with his ropes to pull down a tree on payment of 3d. There is no specific reference to Amwell spring at this time, just an entry at 23 December about a pin and a padlock for the sluice there. Again, there is no specific reference to the highway bridges at Ware and Broxbourne, though the one at Broxbourne does not seem to have been built until November 1611. No place-names are mentioned in the first book after Broxbourne, but we come to Mr. Purvey's land, which appears to have been in the Wormley loop, and the land held by Mr. Warde, His Majesty's trumpeter, of Mr. Kindlemarshe.

When work is resumed after the hold-up, the first names to appear are Windmill Field (Wormley) and Brookefield, Cheshunt (week ended 11 January 1612). Cheshunt Park (opposite Brookfield) appears a week after. Then comes Killsmores (Cheshunt), 11 February 1612; Aldbury (Cheshunt), 7 March; Theobalds and Northe Field (Enfield), 11 April; Hatchmore Grove, 25 April; Churchbury Field (Enfield), 16 May; Butts Farm (Edmonton), 13 June; Mynceing Wood (Southgate), 11 July; Wood Green, 25 July. Then there is a long interval while consolidation is taking place. We resume with Hornsey Highway, 6 March, 1613; Newington the following week; Holloway (Islington), 20 March; finally, the Mantels (Clerkenwell), the site of the Round Pond, 10 April 1613. The fact that the inauguration did not take place until five months later shows that much remained to be done. Moreover, the first cut did not always succeed and had to be filled in again. A new cut would then have to be made. There is abundant evidence of this.

Having traced the course, I should mention a few of the landowners whose lands were cut by the New River. First, the worthies who allowed free passage for the river, being moved thereto (in the words of their acquittances) "by the great love they bore to the City of London". At Amwell there is the Rev. Thomas Hassall, the Vicar, a true friend and supporter of Myddelton, assisting him on more than one occasion to judge from the books. Myddelton not only gave donations to the poor of his parish and remembered them in his will,[54] but also, in October 1612, presented

[54] For the Rev. Thos. Hassall's care for the poor of his parish, see *History To-Day*, V, p. 697, "Annals of the poor", by C. M. Matthews.

the church with a Bible "in good speed of the river", a copy, I suppose, of the recently published Authorized Version. I understand that it has not survived. The other worthies were William Frankelonde of the Rye in the parish of Stanstead, John Bailey, John Saring, and Robert Michel of Hoddesdon.

Among those who preferred some compensation were :—

The Lord Treasurer (Salisbury) and his tenants, Thomas and William Smith	Land in the Ryefield.
John Thorogood	Land at Hoddesdon.
Eme Parrett	Land at Spital Brook.
George Fordham and his landlord, Mr. Kindlemarshe	Land at Wormley.
Thomas Westwood	Land at Cheshunt.
Sir Henry Baker of Sissinghurst, Kent	Land at Enfield.
Sir Christopher Martyne of London	Land at Butt's Farm, Edmonton, the Bush Hill Frame was built on his Land.
Joseph Jackson	Land at Edmonton. The Jacksons owned Broomfield Park.
George Huxley and John Wylde	Land at Edmonton.
The Countess of Dorset	Land at Tottenham.
Dr. Peter Lilly	Land at Hornsey.
Edmond Nowell, Lessee of Robert Earl of Salisbury	Land at Edmonton.
Alderman Barkham	Land at Edmonton and Tottenham.
Sir Francis Popham	Land at Newington.
Brey Norris	Land at Holloway, Islington.
Sr. William Compton, Lord Compton, of Canonbury House	Land at Islington.
Dame Alice Owen, a friend of Myddelton's, for her tenant Mr. Ashe	The Hermitage Field, Clerkenwell.
Samuel Backhouse (one of the Adventurers)	The Mantells, Clerkenwell.

"The Mantells" or "Commandry Mantells" formerly belonged to the Knights of St. John of Jerusalem.[55] Sir William Compton's compensation, £75, was the highest single sum paid out, but Samuel Backhouse and his tenant, Henry Ward, obtained £100 between them.

The course of the New River, as originally constructed from Chadwell Spring to New River Head, was about 40 miles long. The total fall between these two points is about 18 feet, which gives an average fall of about 5 inches per mile. But the course was intercepted by sluices, e.g. at the spring head, at Amwell, Broxbourne, and Theobalds, which produced drops of varying depths so that the fall between sluices was very much less, between 2 and 3 inches a mile. The maintenance of this near level course accounts for its many twists and turns. This was no Roman aqueduct ; it was more like Chesterton's rolling English road.

[55] W. J. Pinks, *The History of Clerkenwell*, 2nd edition (1881), p. 491.

One can see how faithfully it followed the 100-foot contour. To cross the streams, the river was carried in small wooden troughs near the point where the level of the river and the level of the stream coincided. Sometimes the intervening stream was carried over the river by what was called a flash, the stream being confined in a trough. It must have been a very wet business, as the water level of the stream would have to be raised to the level of the trough. It was possible by means of the flash to tap the stream, if necessary. There were flashes at Wormley, in the Chase (Enfield), in "Mincing Wood" and other places. There were other devices for carrying off the land water :—

> for a pysher att Woodgreene 24 foote longe & a foote square toe Carrie ye Land water – 15 –[56]

Another advantage of pursuing this wandering course was that large aqueducts were avoided. When the river was first made, only one of these was constructed, at Bush Hill, and this is explained by the 100-foot contour. The gap between this contour on either side of the Salmon Brook valley is narrow, while the detour formed by the line itself is very great in proportion.

The original construction of this aqueduct, the Bush Hill Frame as it was called, is not described in detail as the whole of the work was entrusted to the ubiquitous Parnell for lump sum payments. Happily for the record, repairs were soon required,[57] and the following articles were bought : 400 boards to line the bottom of the frame to keep the caulking in at 9s. 0d. the 100 ; 3 barrels of tar at 20s. 0d. the barrel ; 900 of Rosin at 8s. 0d. the 100 ; 700 of pitch at 10s. 0d. the 100 ; 500 of oakum at 9s. 0d. the 100, and 100 lbs. of hard tallow to mix with the pitch tar and rosin. A load of tileshards was also bought to underpin the frame. At the same time, Alice Crew, a regular supplier, was paid £5 3s. 0d. for half crown nails, two shilling nails, double tens, single tens, and sixpenny nails.[58] Shipwrights at 2s. 0d. a day were employed to caulk the frame. No lead is mentioned and no plumbers are employed. Parnell was no plumber. This is of interest, for later in the century this same aqueduct was lined with lead. It also appears that the frame was originally supported along part of its length by timber arches as described in Howes's "Chronicles".[59]

When they first started excavating for the river up at Ware, a mysterious machine called "The Ingen" was employed. From the following extracts there is no doubt that it was an imposing contrivance :—

[56] P.R.O.—LR. 2/28, 1st book, 15 May 1613.
[57] P.R.O.—LR. 2/28, 2nd book, 22 January 1614.
[58] P.R.O.—LR. 2/28, 2nd book, 19 February 1614.
[59] Edmund Howes, *Chronicles* (1632), p. 1015.

Mr. Hellam for xiij horses and his men two daies to drawe the Ingen, £2 3s. 4d.	(14/10/1609)
for the rudder post, for the yoke and timber used about the Ingen, 4s. 4d.	(28/10/1609)
to Essex for bread & drinke for the Carters when the Ingen went, 8d.	(28/10/1609)
to Harvey & Saringe for watching the waights in the Ingen, 8d.	(28/10/1609)
to Richard Williams & Thos. Levesey for watching the Tumbrell, 8d.	(28/10/1609)
to Lupton for his cattel to draw the Ingen, 8s. 0d.	(11/11/1609)
A new shaft for the Ingen, 2s. 4d.	(11/11/1609)

In O. L. Dick's edition of Aubrey's *Brief Lives* (1949) there is a note of Aubrey's about a country man who undertook to save the Adventurers a large sum of money by employing a very large plough instead of digging up the earth. He had this special plough made which was drawn by seventeen horses and is said to have saved them a great deal of money.[60] This sounds like the "Ingen". It does not appear in Clark's edition of Aubrey. I am sorry to say the Ingen did not last the course, for in 1610, during the great hold-up, it was taken home to Mr. Thorogood's at Hoddesdon and was never heard of again.

Middlesex and Hertfordshire family and place-names abound in the books. It appears also that a number of Welshmen were inspired to follow Sir Hugh. Thus we find "Rees Uphughes", Thomas Davies, Morgan and Griffeth Evans, John Hughes, William Jones, Davie Roberts, Morgan and Hugh Williams, Thomas Edwards, Thomas Griffeths, John Lewis, Davie Price, Robert Howell, etc. There are one or two exotic beauties like Polydor Guarderope, and the euphonious Averey Lacey. Only one death is recorded, that of Thomas Trenson, "the Scottishman", in December 1609. Two shillings was contributed towards his funeral. There is no evidence at all of Dutch experts being employed.

General labourers, numbering well over 200 in some weeks, are paid a wage of 10d. a day. Working in the water sometimes earned an extra 2d. a day, but not always. A good many of them would follow the excavators who worked either individually or in gangs at piece work rates. Piece work by excavators was paid at so much the pole or rod, according to depth, although the price for a certain depth would vary according to the soil, presumably. Here is a fairly average scale of rates :—

	s.	d.		s.	d.
½ foot ..	0	4	5 feet ..	2	8
1 foot ..	0	10	6 feet ..	3	4
1½ feet ..	1	0	7 feet ..	4	6
2 feet ..	1	4	8 feet ..	5	10
3 feet ..	1	5	9 feet ..	6	2
4 feet ..	2	0	10 feet ..	9	6

13 feet .. 12s. 0d.

[60] John Aubrey, *Brief Lives*, O. L. Dick's edition (1949), p. 198.

Greater depths were excavated than this : (1/2/1612).

> Cleament Lambard for carrieing the bottom in howsemores, 10 foote, 9 foote & 8 foote deep(er) after 6 foote deepe conteyneinge 4 rodd (after 4s. 4½d. the rodd), 17s. 6d.

In December 1609, gauges were introduced for measuring the work. "Single gauge" was paid for at 1s. 2d. the rod ; "double gauge" at 2s. 4d. the rod ; and "treble gauge" at 3s. 6d. The system did not last, although it was revived for a short time in January 1612, when the work on the river was resumed.

Overwork (overtime ?) is mentioned :—

> To William Smith, mould taker, his wages and for over worke done, 5s. 0d.

A large drum costing 14s. 0d. was used to call the workmen together.[61]

A great number of carpenters were employed, usually at 1s. 4d. per day. They appear all the way along the course, wharfing the banks and erecting bridges. Wooden bridges were cheap. What could be more reasonable than :—

> to Jeffreye Bennett for settinge upp xxj Carte Bridges in Mr. Hawkins sev(er)all groundes att xl/s (40s. 0d.) a Bridge, £42.[62]

The following entry relating to brick bridges reveals an old English custom which I believe has not been known to have existed at Cheshunt before :—

> For three thowsand of brickes toe make upp the brick bridge att the Cucckinge stoole in Chesthunt feilde att xijs the thowsand, £1 16s. 0d.[63]

On 1 May 1612, £40 was delivered to Thomas Springfield and Robte Nottingham :—

> toe make p(ro)vision of Brick & lyme for bridges and other necessarie uses for the Ryver, for wch theye have entred into bondes to his Ma(ies)tie, for the makeinge of soe manye bridges or other works as shall amounte to the some of fortie poundes.

The bricklayer then earned 18d. a day.

The river also had its consultants :—

> Toe Mr. Browne a Master Worke man of London for his paynes tooe come and viewe the bryck Brydges wch Springfield & Nottingham made, 5s. 0d. (2 November 1612).

> To Mr. Burgen for his paynes & charges toe surveye all the workes & for his Directions for ye furtherance of the worke, £4 (27 February 1613).

As one can well imagine, King James took more than a superficial interest in the river and gave his orders. Thus in August 1613, £30 is spent in making banks on the upper side of Bromley Grove, in Theobalds Park, to keep the river from flooding the Grove, according to His Majesty's express commandment. Again, in

[61] P.R.O.—LR. 2/27, 2nd book, 21 March 1612.
[62] P.R.O.—LR. 2/28, 1st book, 20 February 1613.
[63] P.R.O.—LR. 2/28, 2nd book, 2 July 1614.

November that year, the cut in John Gyrton's Park, at Enfield, had to be altered as His Majesty found fault with it.

The main preoccupation was to maintain the left bank of the river for the land fell away on that side most of the way. Up at Ware, Amwell, Hoddesdon, and Broxbourne particularly, there are continual references to raising the banks and strengthening them with clay and piles.[64] Leakages occurred, as was only to be expected.[65] Water clay and lime were used for repairs.

Just before Christmas 1614 the New River overflowed about Brookfield, in Cheshunt, as a result of a great land flood, and compensation costing £16 14s. 4d. was paid out.

At Ware, many loads of chalk were laid down to raise the highway above the level of the river in July 1612 and again in March 1614. There are many references to "Casteing upp a Calfe", i.e. clearing the river of a mound of earth formed by the banks caving in.

Trees, of course, got in the way and had to be felled or uprooted (stocked). To fell three great oaks cost 3s. 6d. To stock a great oak cost 2s. 0d. Just "a tree" cost 1s. 0d., and un unspecified number of hawthorns, 3s. 0d. Most of the trees were encountered in April and May 1612, at Enfield.[66]

Cutting weeds in the river is a time honoured occupation going back to the days when the river was still being constructed. Thus at 24 October 1612—"For an Iron Hooke for Morgan Williams tooe cutt wedes", 1s. 4d. Weed cutting in the river between Chadwell Head and Amwell Spring occupied a fortnight in August 1615 and cost 14s. 0d.[67]

Most animals had to be kept under control but moles, the great enemy, had to be extirpated. Mole-catchers were employed and mole traps were bought to protect the banks. They appear to have been more active in the earlier period of construction. Gratings were erected to keep out hogs and geese, and at New River Head the tampins, or plugs, had to be lined to keep the frogs out of the distribution system, but they got in.

The carriage of materials was an important item :—

> To John Adams for Carrieinge 50 Loades of Chalke to rayse ye bankes at Amwell, 4d. p(er) loade, 16s. 8d.

Money was heavy, but had to be carried round for paying wages, etc. :—

> For the portadge of viij C li, 2s. 6d. the hundred, £1.

[64] E.g. P.R.O.—LR. 2/27, 2nd book, 5 September 1612; LR. 2/28, 2nd book, 19 June 1613.

[65] E.g. P.R.O.—LR. 2/27, 2nd book, 23 November 1611; 13 June 1612.

[66] E.g. P.R.O.—LR. 2/27, 2nd book, 23 May 1612; 30 May 1612.

[67] LR. 2/27, 2nd book; LR. 2/29.

I must now turn to the distribution of the water from New River Head. Myddelton's Round Pond was constructed in 1613–14. It was wharfed with oak boards; 4,800 were ordered at 7s. 6d. the 100, and was surrounded by a brick wall, probably the same wall which appears in Hollar's etchings of 1665. The builder was Stephen Boone, another regular contractor and bridge builder. Boone was also responsible for the brickwork of the water house and the chimneys.

The construction of this house, which was built on the south side of the Pond, is given in some detail.

Iron bars and casements for the windows, etc., and keys, hooks, and hinges, spikes, broad nails and other iron work for the house were supplied by Ferdinando Jefford at a cost of £27. J. Yates provided 220 feet of glass for the windows at 5d. the foot. The building contained a counting house, the Middle Room, garrets, and a staircase.

In the second week of November 1613, Richard Sellbye, carpenter, is paid £1 6s. 0d. for making the gallery within the cistern house to stand upon to turn the stopcocks. The following week, 2s. 6d. is spent on hinges and nails for the trap door going down into the cistern house. The cistern itself received the water from the Round Pond through a grating and distributed it to the wooden pipes which were let into its south side.

This brings us to wooden pipes and main-laying. As early as the middle of April 1612, John Bartlemew, pump maker of Ware, was paid 10s. 0d. travelling expenses for going into Berkshire to make choice of elms, elm being the only timber employed for pipe making throughout the period of the accounts. He probably looked over Billingsbear Park, near Wokingham, the seat of Sir Henry Nevill, whom we have already met as one of the Adventurers, for Sir Henry was paid £67 in March 1613 for elms. Berkshire, however, was not the only county to be searched. Middlesex, Essex, Buckinghamshire, and Kent also added their quota. Among particular places mentioned are North Fleet, Dartford, Staines, Maidenhead, and Walthamstow. The record consignment appears to be one from Sir John Garrard (Lord Mayor in 1601): 420 loads, at 22s. 0d. the load, from his estate near Faversham, Kent. But this was exceptional. Most consignments were under 100 loads. The trunks were generally brought up or down the Thames and landed at the Bridgehouse yard, where there was pipe-boring machinery on a wharf known as "Pipe-borers' Wharf".[68] Elms were also landed at Bear Quay, near the Tower, and stored at the City's store in the Green Yard, Leadenhall.

The elm trunks were laid on movable trestles and were bored through by being pressed hard against long augers turned by horse machinery. There is no mention of the machinery in the books;

[68] See Aytoun Ellis, *Three Hundred Years of London River* (1951).

elm trees come into the Bridgehouse yard and pipes come out. Lump sum payments on account are entered and only rarely does the length bored and the diameter appear. From an entry made in March 1623 it appears that boring and jointing 162 yards of 3-inch pipe cost 7½d. a yard.

The diameters of pipes recorded in the accounts are 3 inches, 4 inches, and 6 inches. When bored, these pipes were sharpened like pencils at one end (the top) and bored out at the other end (the butt). To effect a joint, the top end of one pipe was rammed into the butt end of another.

The following entry, which appears under the week-ending 4 September 1613, is of some interest: "Toe Thomas Cock for Elmes & Hoopes for the pypes . . .", £15 10s. 0d. This is the earliest reference I have found to the iron hoops which were driven into the wooden part of the butts in order to strengthen them. Hoops for this purpose varied between 5d. and 7d. each.

The principal pipe borers were Richard Parkes, John Bartlemew, and later, Avery Lacey. For carrying the pipes through the streets, George Hubberd, car-man, would charge between 1s. 6d. and 2s. 2d. the load.

Digging the pipe trench and paving worked out at between 3½d. and 4½d. a yard. 1,012 yards between Pye Corner and Ludgate, and between Newgate and the Cross in Cheapside cost 4¼d. a yard. Richard Taylor, pipe-layer, earned 2s. 0d. a day, while his assistants, William Price and Morgan Evans, earned 1s. 0d. a day. Much of the work was done at night, at the same rate, but night watching was only worth 8d. a night. There is a continuous demand for candles at varying prices, often 4d. a pound, sometimes more, and for links at 2s. 0d. a dozen.

The first line of pipes was laid from New River Head across the fields to St. John Street, thence to West Smithfield, Pie Corner, and Newgate, where one branch proceeded to Cheapside and the other went down the Old Bailey to Ludgate. This work was completed in January 1614. Soon after, an extension was made from Newgate to the conduit at Holborn (i.e. Snow Hill) and beyond to Holborn Bars, and from Ludgate up the hill towards Temple Bar which was reached by October 1614. From these pipes others were laid to serve the side streets. A second line of pipes was driven down Goswell Street in 1617. Foster Lane was too narrow for laying a wooden pipe, so a lead pipe was put down instead.

No attempt was made to compete with the London Bridge Water Works. Very few pipes were laid south of St. Paul's, Watling Street, and Candlewick Street.

The surplus rubbish had to be disposed of: "To Richard Pressgrave (a regular contractor) for carrying a hundrethe nynetee and tooe Loades of rubbish out of the streetes att xd. p(er) Loade", £8.

To control the water in the pipes, wooden and sometimes brass stopcocks were used. The brass stopcocks were mainly supplied by Edward Aspinall, whose charges ranged from £2 10s. 0d. for a 2-inch stopcock to £10 for a 6-inch one. Two large wooden stopcocks laid in the Old Bailey cost £2 2s. 0d. in carpenters' wages alone. Iron was only used for tyres (i.e. lining) and for keys to turn the cocks. Keys were expensive and there were light-fingered individuals then, as there are to-day :—

> For a newe Iron key toe tourne the woodden Stopp Cockes weigheing fiftie poundes at 4d. the pounde the old keye being stolen awaye from the workemen in the nighte, 16s. 8d.

Lead, at 15s. 6d. the hundred, was extensively used in repairs and solder, at 8d. the pound, was much in demand for joining the brass ferrules to the tenants' lead pipes. Old bootlegs were bought when obtainable, for placing underneath the pipes to catch the solder. Tanned sheepskins ("Bazells", a term still used in book-binding) were used for repairs. There are also occasional references to caulking chisels which must have come in handy.

The pipes were hauled through the streets on pairs of coach wheels, and iron chains and tarred ropes were used for lifting them and letting them down into the trenches. Clay was sometimes rammed around the pipes to stop leakage.

In spite of all this activity, the profits were not forthcoming, and in 1614 Myddelton applied to the City for a loan of £3,000 to keep going, which was granted to him.[69] Up to the end of November 1614, i.e. the fourth book, £18,525 0s. 1d. had been spent on the undertaking, and King James, in accordance with the agreement, paid half of this—£9,262 9s. 6d. .*sic*.. Not immediately, though, for Myddelton had to wait until 1617 before he obtained the last £915 4s. 7½d.[70] I have split this £18,500 into years and added a small amount to bring the expenditure down to the end of 1614. This gives us some idea of the tempo of the work. Thus :—

20 February to 30 December 1609	..	£707 was spent.
6 January to 29 December 1610	..	£305 was spent.
5 January to 28 December 1611	..	£361 was spent.
4 January to 26 December 1612	..	£4,424 was spent.
2 January to 24 December 1613	..	£6,473 was spent.
1 January to 3 December 1614	..	£6,407 was spent.

The great rise in expenditure in the years 1613 and 1614 is accounted for by the purchase and boring of elms and the laying of pipes. So much of a water undertaking's capital is buried

[69] City Records Repertories, 31–2/396.

[70] P.R.O.—LR. 2/34. Myddelton's statement. It contains an error of 6½d. in the amount due from the King.

underground. £9,262 9s. 6d. was the total amount Myddelton received from the Exchequer. All subsequent contributions on James's part were left to be paid out of the profits when they should arise.

There was no rush to take the water and the number of tenants rose too slowly for any quick return. Here is the result of the first year's collection. It begins with the summer quarter of 1614 :[71]

Number of paying tenants		*Rents received*	*Arrears received*	*Fines received*
		£ s. d.	£ s. d.	£ s. d.
1614 Midsummer	37	21 4 4	12 18 8	385 17 4
1614 Michaelmas	175	81 10 2		
1614 Christmas ..	351	137 17 6	6 18 0	193 11 0
1615 Lady Day..	384	155 12 8		

The usual rates were 5s. 0d. and 6s. 8d. a quarter, which is what John Milton's father paid. Sir Hugh himself paid 10s. 0d. a quarter for his house in Wood Street.[72]

King James now began to show signs of impatience. In December 1616, a letter was sent by the Council to the Lord Mayor stating that His Majesty had been informed "that but few, in respect of the generallytie and such for whome it was principally intended doe take that water" and that it was not to be thought that so much had been done "to noe use or purpose other then the hinderance & prejudice of such as were by yow deputed to undertake the same", and accordingly the Lord Mayor was required to order and provide that "the said water may be taken into all such houses within that Citty and the liberties as either oute of necessity or convenience may make use of the same. . . ."[73] Discussion took place on the means to be adopted, but although a rise appears in the amount of fines, or laying-on charges, collected, there is no startling rise in the water-rate collection. The situation is confused by the great number of arrears and the growing number of bad debts.

In 1618, another letter was addressed to the Lord Mayor, requesting him to turn down a proposal made by certain brewers to take over Dowgate conduit and its pumping machinery.[74]

As the New River had not, in the words of the Acts of the Privy Council, "sorted to that effect as was expected either in pointe of proffitt or otherwise",[75] it was decided to discuss with Myddelton whether it would be best for His Majesty's profit that the works should be taken either entirely into His Majesty's or into the Adventurers' hands. The outcome of these discussions was the incorporation of the 29 Adventurers by Charter, dated 21 June 1619, under the name of "The Governor and Company of the

[71] P.R.O.—LR. 2/35.
[72] P.R.O.—LR. 2/36.
[73] Acts of the Privy Council, XXXV, 1616–17, p. 99.
[74] Act of the Privy Council, XXXVI, 1618–19, p. 54.
[75] Acts of the Privy Council, XXXVI, 1618–19, p. 157.

New River brought from Chadwell and Amwell to London", with Myddelton as the first governor, Robert Bateman, deputy governor, and Rowland Backhouse, treasurer. The other Adventurers were Myddelton's son and heir, Hugh, his brother, Sir Thomas, the Lord Mayor, and his nephews, William and Richard ; Henry Myddelton, son of David Myddelton, deceased, Timothy Myddelton, Esquire, and William Myddelton, draper ; Sir William "Burlacie" the elder, and Sir William "Burlacie" the younger ; Sir Lawrence Hyde and Nicholas Hyde ; Samuel Backhouse and his son John ; William Bateman, Sir Henry Montagu, Sir Robert Killigrew, Sir Henry Nevill, John Packer, Humphrey Hall, Peter Vanlore, Marmaduke Rawdon, Henry Vincent, John Farrer, Edward Pritchard, James Bearblock, and Gabriel Newman.[76] The first meeting took place in Serjeants' Inn, Fleet Street, in the chambers of Sir Henry Montagu. Later meetings were held at Myddelton's house. At his death, Myddelton held 13 of the shares, some of the Adventurers being his nominees.[77]

Meanwhile, important new works were in progress. These were the boarded river at Highbury and the tapping of the River Lee near Ware. The Highbury Frame was constructed in order to shorten the river and carry it over the Hackney Brook Valley. The original course reached westwards as far as the Holloway Road and Ring Cross, a forgotten district in Holloway, Islington. Details of construction of this work are given extensively. In the first place, an imposing bank was thrown up which took from the middle of January 1618 to the beginning of November to complete. The trough itself was put together by carpenters during the months of August and September 1619, thus allowing a certain amount of time for the bank to settle. This frame was also caulked with pitch and tar and no lead was used.

The number of paying tenants had passed the 1,000 mark by Michaelmas 1618,[78] and now it became necessary to take further steps to ensure the supply. A Royal Commission was therefore appointed to enquire into the possibilities of taking water from the Lee. Their report must have been favourable, for by the middle of June 1620, work is proceeding to that end. Soon the bargemen began to object and the dam which had been put up at Ware end was pulled down, but was presumably set up again.

In 1621, Myddelton undertook to manage the whole of the undertaking for £800 a year, but not to pay for extensions to the system. The allowance proved insufficient, and in the following year it was increased to £1,000. Few details of the work are now recorded.

[76] Patent Roll. 17 James I, part 16, no. 67.
[77] Wills from Doctors' Commons (Camden Society), LXXXIII, p. 92.
[78] P.R.O.—LR. 2/36.

The river was surveyed from time to time by parties of Adventurers :—

> For the Charges for dyett & horse Meate for those that were appointed to Survey the state of the Ryver from London to Ware theire beinge tenne in Nomber, & tenne horses, they being three dayes uppon the same Surveye, £8 17s. 0d.[79]

Argument continued over the administration of the works, and King James and his advisers tried to get the Adventurers to surrender their shares, but they hung out. They informed Sir William Pitt, one of the negotiators on His Majesty's behalf, "that their shares for the present did arise but to 28 li. a man. That if his Maiestie by regall power and proclamacon would assist them they might then bee raised to 100 li. a peece. . . ." Unfortunately for His Majesty, "by the condicons obtained in their Incorporacon they have gotten this advantage that his Mats. moytie being undevided from theirs cannot be improved wthout improveing theirs". Sir William dared not advise the use of proclamations by which "it may be feared that the benefit must be raised more by fines and vexacon then by reforming therby or improving the workes".[80]

The year 1622 was momentous in many ways, although it did start badly for King James, who in January narrowly escaped drowning in the frozen New River at Theobalds after his horse had stumbled.[81] But the profits were in sight at last, and at Michaelmas £433 2s. 7d. became due to the king after £370 17s. 10d. had been deducted for unpaid contributions.[82] On 19 October, Myddelton was created the 193rd baronet in recognition of his engineering exploits, the first engineer to be honoured.[83] The three feats cited in the patent were firstly the New River, secondly the silver-mining venture in Wales, and thirdly the reclamation of Brading Harbour in the Isle of Wight. These last two certainly stand out as evidence of the great energy and resilience displayed by this undaunted man, for they were undertaken during the disappointing period after the river was completed. The Welsh mines by all accounts were successful and probably helped to calm the king's impatience. It is a wonder how Myddelton held all these concerns together with communications as they were. The mines were in the wilds of Cardigan, and an open warrant was sent out from the Council on Myddelton's behalf in 1618.[84]

In 1623, it was the turn of the City to recognize "the great and extraordinary benefitt and service" they had received from the water. In the repertories of the Court of Aldermen for 15 November of that year, special mention is made of its great use at

[79] P.R.O.—LR. 2/34, 26 March 1625.
[80] B.M. Add. MSS. 29, 974.
[81] John Nichols, *The Progresses . . . of King James the First* (1828), IV, p. 749.
[82] LR. 2/33, 30 September 1623 ; LR. 2/34.
[83] B.M. Harleian MS. 1507, no. 40 (Alteration of arms).
[84] Act of the Privy Council XXXVI, 1618–19, p. 100.

"many great fires happened within this Citty and chiefly last night at a verie terrible and fearefull fire which might have greatly endangered the Cittie had that needefull water binn wanting".[85] This occurred at the house of Sir William Cockayn, Alderman, in Broad Street.[86] The token of recognition took the form of a gold chain and a jewel. They both appear in the portrait of Sir Hugh by Cornelius Jansen in the Goldsmiths' Hall.

In 1625, King James died and was succeeded by King Charles. In the same year a violent plague attacked the City. A contemporary account, for what it is worth, gives 35,417 as the number of deaths.[87] Whatever the figure, the plague had a definite effect on the water rents. The number of tenants decreased and the arrears increased. The profits dwindled and only recovered slowly.

About the year 1630, King Charles decided to give up his share of the undertaking called "The Royal Moiety" and to receive instead a yearly rent of £500. Charles was also considering another scheme, the Hoddesdon scheme, which must inevitably have become a rival of the New River, but which would be a far more profitable venture as far as he was concerned, for by merely granting patents and authorities, he was to receive £4,000 a year.[88] It never came to anything and disappeared in the Civil War. The Royal Moiety in the New River was reconveyed to Sir Hugh by indenture, dated 18 November 1631,[89] which also created the £500 annual rent known as the Crown Clog. At the same time, King Charles remitted his share of the unpaid arrears amounting to £1,290 0s. 4d.[90] The first payment of £500 actually became due at Michaelmas 1631.

But what of Sir Hugh, who had borne the strains and stresses of the undertaking since the beginning in 1609? Although we cannot, without further evidence, gauge the financial embarrassment suffered by him at any time when bringing the river to London, the anxiety must have been great indeed, not only during the period of construction, but also during the lean years that followed. He was now over 71 and a very sick man. Three days after the completion of the agreement with King Charles, he made his will, and within a few weeks, on 10 December, this great Welshman died, whether of grief at being deserted by his royal partner or happily with the knowledge that at last the New River undertaking, though far from prosperous, was free from what must have been an uneasy partnership, we have yet to learn.

[85] City Records. Repertories 38/12a.
[86] Edmund Howes, *Chronicles*, p. 1035.
[87] Ibid., pp. 1041–2.
[88] C.S.P. Dom., 1627–28, p. 114 ; 1628–31, p. 550.
[89] Patent Roll. 7 Charles I, part 20, no. 159.
[90] P.R.O.—S.P. 39/30, no. 61.

He was buried at the Church of St. Matthew, Friday Street, where he had been a churchwarden in his day. This church, which was burnt down in the Great Fire, was rebuilt by Wren in 1685. It was finally demolished in 1883, when the New River Company took the opportunity of enquiring about the site of his tomb, but after considerable search the churchwarden, Mr. I. S. Rowsell, informed the Company that no trace of it had been found.

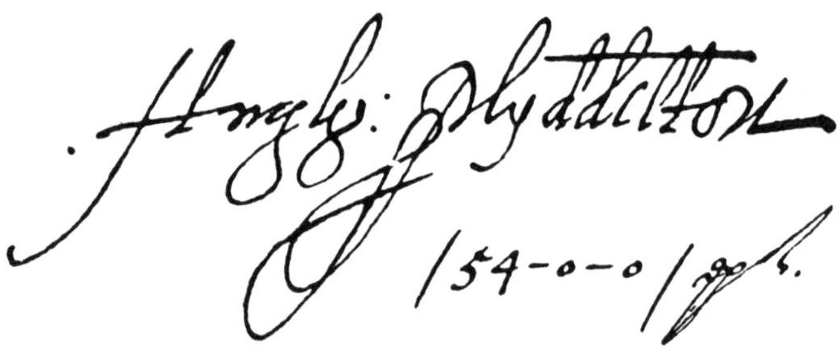

Sir Hugh Myddelton's Signature
from the First Book of Disbursements

4

George Sorocold of Derby: a pioneer of water supply

F. Williamson

IT is remarkable that a man of such high ability as George Sorocold is not mentioned in the *Dictionary of National Biography*, or, indeed in any other biographical dictionary, yet he was the first man outside of London to install 'house to house' water supply, and in his time had a great reputation as an engineer.

The family of Sorocold was widely spread in Lancashire at an early date, but information is not available as to where precisely the family originated, or as to the derivation of the name. The latter appears in various forms which need not be enumerated here. If the name is divided into two syllables, the first may appear as Sar, Ser, Sor or Sur, and the 'r' may be doubled, the second is -cole, -coale, -call, -cale, -cold, etc., with a connective, o, a, e or i. Experts on surnames and place-names whom I have consulted can throw no light on the etymology of the name. It may possibly be connected with some such place-name as *Saurescales*, mentioned by Ekwall in his *Place-names of Lancashire*, but the middle 's' does not occur in any form of the surname so far discovered.

The family was evidently fairly well-to-do, possibly of yeoman stock, some of them being landowners and armigerous. In 1633 Thomas Sorocole contributed his pedigree at the Herald's Visitation of London 1633-4,[1] he was a son of Ralph Sorocole, of Manchester; and Thomas Sorocold, of Barton, near Manchester, registered his pedigree at the Herald's Visitation of Lancashire in 1664.[2]

[1] *Harleian Soc.*, vol. 17, 1883. [2] *Chetham Soc.*, vol. 88, p. 176.

From Thomas Sorocole, of London, mercer, two lines descended whose pedigrees are given in Burke's *Landed Gentry*, 1937, under Serocold and Skeels. Unfortunately with regard to George Sorocold we have no information whatever as to his parentage, or the date and place of his birth, and we can only surmise that he may have been born in 1668, the son of James Sorocold, of Ashton, Lancs, who died in 1675. This James, at the same time as his brother George, was admitted scholar at Gonville and Caius, Cambridge, 9 May, 1645, when they were aged respectively 18 and 17. Their father was George Sorocold, gent, of Ashton in Makerfield, Lancs. 1668 was not too late a date for the birth of our subject as we shall see later.

George Sorocold is first encountered on his marriage with Mary Francis at All Saints', Derby on 7 Dec., 1684, and I think he must have been a stranger to Derby at the time of his marriage as the parish clerk made a hopeless mess of his name, entering it in the register as "Souerould," whereas in later entries we find normal spellings. Whatever credence we attach to the statement of Ralph Thoresby, the Yorkshire historian, who was personally acquainted with Sorocold, it is safe to say that George was very young at the time of his marriage, and if Thoresby is correct he cannot have been more than sixteen. This is the extraordinary statement he makes: "our ingenious hydrographer, Mr. George Sorocold has already had thirteen children (of whom eight were living at the same time) all single births, and all nursed by his wife, before he was twenty-eight years of age."[1] I leave my medical readers to discuss the probabilities of this, which I have not been able to verify by the corroboration of parish registers.

It seems obvious that at the time of his marriage he was not living in All Saint's parish, although he may have been in one of the other four Derby parishes, because the

[1] Thoresby, *Ducatus Leodiensis*, 1715.

first baptism of one of his children at All Saint's was on 20 April, 1691 whereas if there were any semblance of truth in Thoresby's statement, this should be Sorocold's sixth child at least. There were other children of his baptized and buried at All Saint's, and his wife was buried at the same church, on 13th April, 1728. The date of Sorocold's death is a complete blank, and we cannot be sure whether he survived his wife or not. In the registers she is simply described as Mrs. Mary Sorocold, neither wife or widow. Sorocold's wife was the daughter of Henry Francis, apothecary, one of the most respected men in Derby. His house has just been pulled down, 1936. That something untoward happened to Sorocold is clear from the letter dated 2 March, 1717[1] written by the Mayor of Derby to Baron Parker, with reference to the navigation of the Derwent, where he refers to Sorocold as "the ingenious, unfortunate, mathematician." Defoe[2] tells a story of Sorocold falling under the big waterwheel at the Silk Mill but concludes by saying he suffered no hurt at all. Perhaps at the time Defoe wrote this may have been the case, yet there may have been unfortunate after effects of which Defoe was ignorant. He gives no dates to assist in arriving at an opinion.

By the testimony of his contemporaries best able to judge Sorocold was looked upon as an extraordinarily good engineer. His competitor, Thomas Savory, engineer and inventor said that "in composing such sort of engines [for supplying towns with water] I think no person has excelled the ingenious Mr. George Sorocold."[3] Beighton said that the London Bridge works were better than the famous installation at Marly, France. Sir Thomas Johnson, of Liverpool, introduced him to the Corporation of Liverpool, as a very ingenious man, to advise them on the construction of their first dock. Sir Godfrey Copley,

[1] See below the account of the Derwent Navigation.

[2] See below the account of the Silk Mills, and of the Sheffield Water supply.

[3] Savory, *Miner's Friend*, 1702.

founder of the Copley Medal, said that Sorocold's engine at Marchant's London Waterworks was the best piece of work he had seen. We are told that Sorocold was one of the two millwrights of his time who never failed in what they undertook, because they considered the perfection and success of their work first and then profit afterwards. This is easy to believe as several of Sorocold's machines continued to work continuously for very long periods, in some cases for 150 years, until at last steam-engines ousted waterwheels.

Sorocold devoted most of his time to public water supplies, but this was by no means his only line. His first recorded job was the hanging of the bells of All Saint's, Derby (now the Cathedral) in 1687. He was consulting engineer for the first Liverpool Dock in 1708, and probably for the first great wet dock at Rotherhithe. His millwork at the Derby Silk Mills was one of the wonders of the century. It was more than fifty years before any other factory was built equal to the latter either in England or abroad. He was also engineer for making the Derwent navigable.

It will be seen that Sorocold has a just right to more consideration than has hitherto been accorded him and as he was generally known as George Sorocold, of Derby, during his working life-time, this seems to be an appropriate place to record, what is known of his work.

The only previous attempts at anything like an account of Sorocold are by Mr. Henry Peet,[1] and Mr. Rhys Jenkins,[2] to both of whom I am indebted.

Seeing how closely Sorocold was associated with hydraulics, it is strange that the only patent granted to him was for something entirely unconnected with water.

[1] Peet, Henry, *Thomas Steers*, engineer of Liverpool's First Dock, with an appendix on George Sorocold, in *Trans. Hist. Soc., Lancs. & Ches.*, vol. 82, 1930.

[2] Jenkins, Rhys, *George Sorocold*, in the *Engineer*, 18 Oct., 1918. Reprinted in *Collected Papers of Rhys Jenkins.*

He received letters patent, 1st Jan., 1704 (not 1703 as given in the reprint of 1857), for " a new machine for cutting and sawing all sorts of boards, timber and stone, and twisting all kinds of ropes, cords and cables by the strength of horses or water."[1] There is no specification filed with the application so that details are not known, but the patent may have had some connection with the machinery for boring elm trees for use as water-pipes, which he worked from his water-engine at Derby in 1692.

WATER SUPPLY. The most important part of Sorocold's work lay in the installation of public water supplies, in which branch of engineering he was a practical pioneer. Sorocold designed sets of pumps worked by water wheels, which forced water up to cisterns on the highest level available, from which sufficient pressure was obtained to give a supply of water direct to the houses of those who were prepared to pay for the service.

It is true that Peter Morris installed a water supply in London on similar lines, as early as 1582, but Sorocold seems to have been the first to introduce the same principle to provincial towns, beginning with Derby in 1692. Moreover, Sorocold's wheels and pumps were much superior to those of Morris, which they superseded at the London Bridge Works. The water wheels were constructed to rise and fall in accordance with the level of the stream which turned them, a principle patented by John Hadley, 3rd March, 1693,[2] exactly a year later than the lease granted to Sorocold to empower him to construct the Derby water works. There is no specification filed with Hadley's patent, which by the way claimed several different inventions, so we cannot be sure that it was exactly the same device as used by Sorocold, although it probably was, for the two men are found in collaboration on the construction of Marchant's London Water Works.

[1] *Patent of George Sorocold*, 1703, No. 369, Patent Office, 1857.

[2] *Abridgments of Specifications relating to Hydraulics*, 1617-1866.

* Readers should note that for the purposes of this reprinting only pp. 43–47 and pp. 65–93 have been reproduced from the original.

Whoever was the real inventor of the rising and falling water wheel as will be seen later that Sorocold used this innovation in his first work at Derby.

Local historians generally have neglected the history of water supply, and many have scoffed at the efforts of Sorocold and his associates, but there could be little wrong with pumping systems which continued to work satisfactorily from about 1700 until 1850, when the growth of population rather than inherent defects rendered other methods necessary. It should also be remembered that by 1850 untreated river-water in industrial areas had become too impure for human consumption.

In the following pages an attempt will be made to describe the various water supplies which are known or reasonably assumed to be the work of Sorocold, and details of his methods will come out in the various accounts.

Our knowledge of the towns where Sorocold put in water supplies is due largely to Ralph Thoresby, the Yorkshire historian, who was personally acquainted with the engineer. In his *Ducatus Leodinensis*, 1715, after describing the water engine erected by Sorocold at Leeds, goes on to say that he has done the like at Macclesfield, Wirksworth, Yarmouth, Portsmouth, Norwich, King's Lynn, London Bridge, Deal, Bridgnorth, Islington New Works and Bristol. To these should, of course, be added Derby, and Exeter, (mentioned by Celia Fiennes as being similar to what she had seen at Derby and Islington), and also Sheffield where Defoe says the water-engine was constructed by Sorocold.

It will be seen that in several instances Thoresby's statement is corroborated by other evidence, but in other cases Sorocold's name does not appear in local records. Usually Sorocold did not retain any financial interest in his works once they were constructed, and leases were granted direct to those who intended to carry on as proprietors, and although there can be little doubt that Sorocold was the man really responsible for designing and carrying out the work, his name may not appear in the records. In some instances no records remain of what actually happened beyond the bare fact that some sort of water supply was installed.

In the absence of other records I am prepared to accept Thoresby's list as entirely reliable.

DERBY, 1692. Derby seems to be the first place at which Sorocold installed a public water supply, which is what we should expect seeing that he was living in the town at this date.

It is fortunate that the terms on which Sorocold began his undertaking is preserved in abstract[1] although the original document is lost, and from it we get a clear idea of the procedure.

" Deed whereby the Mayor and Burgesses of Derby gave to George Sorocold, of Derby, gent., the mill commonly called Gunpower Mill, near St. Michael's Mills, and two sluices adjoining and the little Byflatt [island] whereon the mill stands, with free liberty to erect a water-house, a water-wheel and other engines, laying pipes for conveying water into the streets, lanes, and passages within the borough, to hold for a term of 99 years at an annual rent of £3 with certain conditions as to opening the streets, digging trenches, and repairs, etc., the said George Sorocold to begin work within three months next ensuing, and to lay pipes through the King's Streete, Iron Gate, Market Place, Rotten Roe, and the Corne Market, and to the Gaole Bridge in the said Burrough within three years next following.'

On the back of the document is a memorandum to the effect that the said George on his completion of the said waterworks may have liberty to surrender the lease, leaving the waterhouse, water-wheels, sluices, bridges, pipes, pavements, etc. in good repair.

[1] *Cal. of Ancient Records of the Borough of Derby*, 1904, p. 48, Deed No. 135.

Signed. Samuel Fletcher, Mayor, and by George Sorocold and witnessed by John Bagnold, Town Clerk, Thomas Bourne and Joshua Bourne. Dated, 5 March, 1691/2."

It is certain that Sorocold set to work immediately he got his authority, for John Houghton[1] in his *Collection* No. 38, of 21 April, 1693 says, " one Mr. George Sorocold has set up a waterhouse to convey water-pipes to all the houses in the town [Derby] that desire it, and it is likely it will be much used."

A few years later Celia Fiennes[2] visited the town and gives an account of the water-engine, which shows it could be raised and lowered like the later one at London Bridge.

" The River Derwent runs by the town and turns many mills, and the water engine which turns the water into the pipes that serves the town, the same wheels grind also, but they do it for a halfpenny a strike, which is the same measure as our bushill. At this engine they can grind if it is ever so high a flood, which hinders all the other from working at the flood, they are so choaked up, but this they can set higher or lower just as the water is. There are bays which they make with stones to keep the water to run to the mill and thence it falls againe into the Derwent; there is also a fine stone conduit in the Market Place, which is very spacious, well pitch'd, a good Market Cross." (*op. cit.* p. 140).

A fuller description is given by William Wolley in his manuscript history of Derby, printed in Simpson's History, p. 176, and written in 1712.

" St. Michael's Mills belong to the Corporation, but there is also adjoining thereto a Silk throwing mill built by Mr. Cocket [Thomas Cotchet], of great curiosity [this

[1] John Houghton, *Collection for Improvement of Husbandry and Trade.* A weekly news-letter.

[2] Celia Fiennes. *Through England on a Side Saddle in the time of William and Mary*, 1888, p. 140.

preceded Lombe's mill], and on the hither side the water-engine, invented by Mr. Sorocold with a great deal of art, which at the same time with one wheel throws up water to a cistern joining St. Michael's Church, which from thence is conducted in pipes, and supplies all parts of the town at a cheap rate, and turns a malt mill and bores elm trees for pipes, all at the same time and all managed by one man. It pays to the land tax £50. 12s. 4d. per annum."

It seems from the foregoing that the cistern or reservoir was adjoining the church, but Hutton[1] says it was "at the top of the church, ... Perhaps this is the must useful church in Derby, though preached in but once a month." Hutton is probably the most unreliable writer who ever professed to be an historian, and in this instance, he is directly contradicted by two references which carry more weight than his facetious statement. Simpson quotes an extract from the churchwarden's accounts of St. Michaels, "paid every Michaelmas by Mr. Sorocold, 5s. for *the ground* on which the cistern stands" (p. 298), although undated this is clearly during Sorocold's lifetime, and again he quotes from a terrier of lands, etc. belonging to St. Michael's delivered into the Bishop's Consistory Office in 1786 (*ibid.*, p. 391), "Item—Twenty shillings yearly to be paid to the vicar or minister as a rent for the cistern *standing in the churchyard*, to be paid by the owners of the water-works." It is certain from this that at the time Hutton was writing the cistern was in the churchyard and not *on* the church itself.

Although Sorocold carried out the installation, he appears to have been financed by the Corporation, at least to some extent, and they probably took over the management of the waterworks afterwards. At a meeting of the Derby Company of Mercers held at the house of Mr. Howel, in Derby 1st March 1691/2, it was resolved that Mr. William Franceys, apothecary should pay over to the

[1] William Hutton, *Hist. of Derby*, 1791, p. 14.

Mayor and Burgesses, the sum of £40 for carrying on the waterworks for the town, according to agreement between the Mayor and Burgesses and Mr. George Surrocold, to be lent for three years without interest.[1]

The Mercers had a great difficulty in getting their money back, and after twenty years ten pounds of the loan were still owing.

From a deed of 1738 it appears that the profits of the waterworks went toward lighting the streets.

"Counterpart of a lease made by consent of John Gisborne, mayor, from Thomas Borrow, esq., and other Trustees appointed to administer the rents, etc., arising from the Borough Waterworks, and the interest from a sum of money given by Margaret Chambers, for the support and maintenance of 80 lamps or lights for the use of the Borough, to Benjamin Grainger, Stocking frame-knitter, of Derby, of the gunpowder Mill, now used as an [water] engine house, and the malt mill, with sluices thereto belonging. Term 40 years, rent £23. 6s. 8d. Witnesses: William Bateman, Richard Whitby, 11 Apr., 1738."[2]

Before Sorocold's time there had been no direct supply of water to houses, and it had to be carried from the two rivers, various wells and conduits. The Friars had a conduit before the suppression of the lesser monasteries, there was one in the market place in 1610, and another Becket Well erected in 1652, the last being still in existence. Sorocold's system supplied only that part of the borough which lies between the River Derwent and Markeaton Brook, then called Odde Brook, at least that is the inference from the terms of the lease, and also from the topographical features of the town.

In 1829[3] the water engine was still the main source of

[1] *Derby Company of Mercers*, ed. by H. H. Arnold Bemrose, in D.A.J., vol. 15. Original minute book in Derby Pub. Lib.

[2] *Cal. of Anc. Records*, No. 285.

[3] Glover, II, p. 417.

supply to the town, but when Glover says the reservoir was on the top of St. Michael's Church, he is merely quoting Hutton's words without acknowledging the quotation.

Bagshaw[1] in 1846 adds the information that there were four miles of pipes carrying water to the consumers.

The frontispiece of Simpson's *Walks through Derby*, 1827, taken from an old picture shows Sorocold's wheel and engine house before the silk mills were erected.

BRIDGNORTH, 1706. Mentioned by Ralph Thoresby as being one of the places where Sorocold put in a water supply, and although his name has not been discovered in local records, yet the character of the waterworks indicates the hand of Sorocold, and the reference to it being the work of those who erected the new water work at London Bridge is definite proof.

The earliest description of the Bridgnorth works is in Cox's *Magna Britannia*, vol. 5, 1727, p. 676, which thus describes it.

" The water here is very good and wholesome, with which the town is plentifully supplied partly by several conduits to which it is conveyed in leaden pipes from a spring half a mile from the town. This was the noble gift of the great benefactor to this place, Sir William Whitmore Bart., created 17 Jas. I, partly by a water engine, which throws up such a quantity of water of the Severn (as is sufficient for the whole town) to the top of the Castle Hill, which is sixty yards higher than the river. This wondrous piece of art was performed by those that erected the new waterworks at London Bridge."

The water was carried through a four inch main to a cistern having a capacity of about 6,500 gallons.[2]

The wheel and pump house is figured in Buck's view of Bridgnorth, published in 1732, but there is a better picture of it in R. Godfrey's engraving after Paul Sandby,

[1] Bagshaw, S., *Hist. etc. of Derbyshire*, 1846, p. 81.
[2] Bagshaw, S., *Hist. etc. of Shropshire*, 1851, p. 618.

R.A., c. 1778, which through the kindness of Dr. W. Watkins-Pitchford, of Bridgnorth is reproduced to illustrate this paper. It shows the water-wheel, and the 'engine-house' adjoining, in which were housed the pumps, this arrangement being usual with Sorocold's installations.

The wheel and pumps continued to be the source of supply until 1857, a continuous life of 150 years, when a steam-engine and reservoir were constructed on the other side of the river.

BRISTOL, 1696. One of the towns where Thoresby tells us that Sorocold put in one of his installations. In 1696 an Act of Parliament[1] was obtained for supplying the City with fresh water, and a fair account is given by Barrett, in his *History of Bristol.* Although Barrett does not mention Sorocold that is no reason for doubting Thoresby's word, if Sorocold were only interested as engineer. It is curious to note a Richard Berry as one of the undertakers, who might be the Richard Barry who was associated with Sorocold in other undertakings.

" Besides these conveniences [conduits and pumps] for the supply of water, that no part of the City should be without this useful element, it was brought in elm pipes from a large pond or reservoir, a mile from Lawford's Gate, to the remote parts of the city that stand most in need of it for their families and businesses that require a great supply. This reservoir is filled by a large wheel engine, erected at a place on the bank of the Avon two miles above Bristol bridge. An Act of Parliament passed for this purpose [An Act for the better supplying the City of Bristol with Freshwater, received the Royal Assent, 10 April, 1696]; and in August 1696, Daniel Small, of London, draper, Christopher Fowler and Richard Goddard, of London, merchants, and Richard Berry and Samuel Sandford, citizens of Bristol, on behalf of themselves and

[1] *House of Commons Journals.*

PLATE VII.

Waterwheel and Engine-House at Bridgnorth constructed by George Sorocold, 1706.
From an engraving by R. Godfrey after Paul Sandby, R.A., *c.* 1778.
This is typical of many of Sorocold's installations.

Plate VIII.

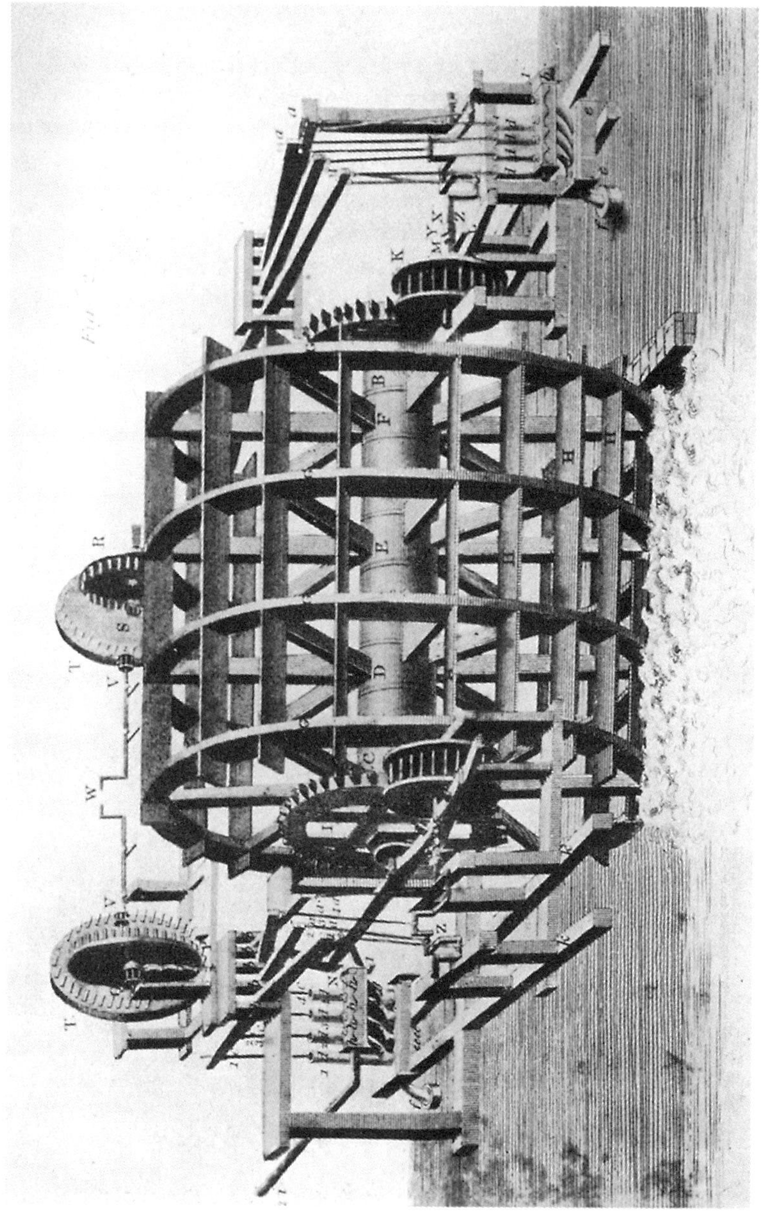

One of the Water Engines at London Bridge constructed by George Sorocold, 1701-1703. Reproduced from a drawing by Henry Beighton, 1731, in Philosophical Trans. of the Royal Society; abridged edition. Beighton shows only eight of the sixteen pumps for the sake of clearness.

others, contracted with the mayor, etc. for supplying the inhabitants with fresh water at reasonable rates."

They were allowed to bring their pipes through any person's lands (except houses, gardens and orchards) from Hanham Mills and other places, in aqueducts or pipes. "The undertaking was perfected at great expence, the whole being divided into 95 shares at £65 a share, and water brought in at the easy rate of 40s. a year to each family who reserved it."[1]

Celia Fiennes[2] who visited Bristol about 1698, describes nothing but conduits, which is rather remarkable as she seems to have been interested in water supplies.

In 1903 a number of elm-pipes were excavated, some of which had 2¼ in. plug-holes in the sides, evidently meant for service connections.[3]

Deal. This was one of the towns where, according to Thoresby, Sorocold installed one of his water-engines, and here we discover the remarkable fact that two rival supplies were being installed at the same time and apparently both completed.

In 1688, James II granted letters patent, dated 21st July, to William Rider, and Edward Burdett,[4] giving them the sole liberty of building a conduit head, wharf, and other necessary buildings on the beach at Deal

What these two actually accomplished is not known, but 23 Jan., 1699/1700, William Warner presented a petition to the House of Commons setting forth that the town of Deal had always been short of water, and that he had found out a method "which had already cost him several hundred pounds and will do more" which the inhabitants were satisfied would effectually supply their wants, and prayed to be allowed to bring in a Bill.

[1] Barrett, W., Hist. etc. of Bristol, 1789, p 101.
[2] Op. cit.
[3] *Trans. Bristol & Glos. Arch. Soc.*, Vol. 27, p. 335.
[4] *House of Commons Journal*, 6 Feb., 1699/1700.

Leave was granted and Mr. Brewer brought in a Bill on 1st Feb. which was then read a first time.

Five days later (6 Feb. 1699/1700) William Rider presented a petition to the House, reminding the House of the Letters Patents granted to himself and Edward Burdett, and going on to say that the Patentees had been at great charges in bringing engineers from London to Deal to view the premises, and to direct what was necessary to be done (this is no doubt, where Sorocold came in), " But, the Petitioner is informed that one William Warner is endeavouring to advantage himself by the Petitioner's discovery, and for that end hath *begun* a work for bringing freshwater to Deal," prays to be heard by counsel before Warner's bill is passed.

Commons ordered the petition to lie on the table until Warner's bill was read a second time. On 14 Feb. the Bill and petition were read and the Bill negatived.

William Warner being for the moment cleared out of the way, Rider petitions again on the 16th inst., and to his preamble about his Letters Patents adds:—" But the Petitioner, who purposeth to supply water from a River called the North Stream, it being the opinion of able engineers, that it cannot be done by well-water [this sounds like Sorocold again], cannot effectually perform this undertaking without breaking up some small part of the soil of some persons, lying between the town and the river, and praying leave to bring in a Bill."

Leave was granted and Sir Robert Davers brought in a Bill on 21 Feb., 1699/1700, which was read a first time on 26 Feb., a second time on 29 Feb., and committed.

All seemed plain sailing for Mr. Rider, but Warner was by no means beaten. On the 16 Mar., 1699/1700, the Mayor, Jurats and Inhabitants of Deal presented a petition to the House in his favour, setting forth " that Mr. William Warner, about a year since, offered to erect a waterwork for supplying them with good water, and begun the same at great expense, and hath made considerable progress therein, and the Petitioners being informed that Mr. Rider hath presented a Bill to this House to supplant Mr. Warner in his undertaking, which will be a great detriment to the Corporation, and praying that they may be heard by their Council against the Bill before the same do pass."

On the 13 Mar. the Committee having examined the Bill and heard Council, returned the Bill with some amendments, but when the question was put that the Bill be ingrossed—it passed in the negative.

A year later William Rider [spelled Ryder this time] petitioned the House again, 4 Mar., 1700/1, after introducing his Letters Patents goes on to say, " The interest of which Patent is also, by mortgage, vested in the Petitioner," he must have bought out Edward Burdett, and asks leave to bring in a Bill. This time the Marquis of Hartington and Sir Robert Davers were authorized to prepare and bring it in. It was duly presented and read a second time and committed on 20 Mar., 1700/1.

Even now Warner was not defeated and he presented another petition on 27 Mar., 1701, which provides a little more information. He states, " that having liberty from the Archbishop of Canterbury, and a lease for 99 years from the town of New Deale, to enable him to erect water works for the use, etc., hath expended about £1600 in the undertaking, which may be effectually completed within two months time; but William Ryder, esquire, hath obtained leave to bring in a Bill for supplying Deal with freshwater, notwithstanding that his Bill for that purpose the last session was rejected upon a petition of the inhabitants of the town against it, and praying that the Petitioner may be heard by his Counsel against the now Bill in this House."

The Committee examined the Bill and heard Mr. Warner, and made amendments to the Bill, which are not recorded, and 8 May, 1701 it passed and was sent to the Lords. The Lords made amendments, striking out a paragraph referring to Rider and Burdetts letters patents, with which it passed, and on 12 June, 1701 the Act received the Royal Assent.

It might be assumed that Warner was prevented from going on with his scheme, but this was not so, as his authority did not rest on the consent of Parliament, and we learn from the *London Gazette*, 16-20 April, 1702, that " Mr. Warner and partners have actually brought very excellent water into the town of Deale."

Warner's scheme was to supply water from a well, while Rider's was to supply it from a river, and both appear to have been carried into effect and to have continued in use side by side until after the passing of the Deal Water Act in 1840.

EXETER, 1695. Exeter is not one of the towns where Sorocold is known to have acted as engineer, nor is he mentioned as one of the undertakers, but the system is so like others which he was known to have designed, that no apology is needed for mentioning it here. There is also the direct statement of Celia Fiennes that the Exeter plant was similar to those she had seen at Islington and Derby, which we know were Sorocold's.

" Behind this building [i.e. the Guild Hall which is still standing as she described it] there is a vast cistern which holds upwards of 600 hogsheads [31,500 gallons] of water which supplies by pipes the whole city. This cistern is replenished from the river, which is on purpose turned into a little channell by itself to turn the mill, and fills the engine that casts the water into the trucks which convey it to this cistern. The Water Engine is like those at Islington and at Darby as I have seen, and is what now they make use of in diverse places either to supply them with water or to draine a marsh or overplus of water."[1]

[1]Fiennes, *Op. cit.*, p. 209.

According to local historians an Act of Parliament was obtained in 1694 [there is no record of this in the House of Commons Journal], to convey water from the Exe. An engine on a very ingenious model was erected at the head of the New Leat, which notwithstanding the elevated situation of the city plentifully supplied, by wooden-pipes, such of the inhabitants, who on payment of an annual rent, are desirous of being furnished therewith.[1]

On 10 Nov., 1694, Jonathan Pyrke and Richard Lowbridge contracted with the Mayor, etc., concerning waterworks, and on 12 Feb., 1695 the Mayor, etc., made a grant to them of the waterworks of the city and several parcels of land for a term of 200 years.[2]

Oliver[3] mentions the second grant and elaborates it a little, giving the undertakers as Jonathan Pyrke, of Stowbridge, co. Worcester, gent., Richard Lowbridge and Andrew Crowley, of the same place, ironmongers, and Daniel Dennett, of Gloucester, carpenter, who agreed at their own expense to erect and maintain a water engine.

Celia Fiennes probably visited Exeter about the end of 1698, but she gives no definite dates, however, the plant was then quite newly constructed as can be inferred from letters in the City archives.

On 19 Feb., 1700 the Chamber was ordered by the House of Commons to furnish not later than 8 a.m. on 2 Mar., 1700, an account of the value of the Estate of which the City stands seized and possessed. In reply the Clerk had to say that the information could not be given as the accounts were not up to date, as " before that time expired the waterworks were taken in hand, and everyone was soe much busied thereabout that nothing could be done."[4]

[1] Jenkins, *Hist. of Exeter*, 1806, p. 193.
[2] H.M.C. *Records of Exeter*, p. 285.
[3] Oliver, *Hist. of Exeter*, 1861 ed. p. 149.
[4] H.M.C., *Records of Exeter*, p. 226.

Although, as I have said, Sorocold's name does not occur in any available record, I am certain that in 1694 there could not have been local engineers capable of designing a pumping installation similar to Sorocold's without his direct assistance.

ISLINGTON NEW WORKS. Thoresby says that Sorocold put in one of his installations at Islington New Works and Celia Fiennes tells us that the water-engine at Exeter was similar to those she had seen at Islington and Derby. Yet in all the accounts of the New River Works which I have seen, there is no mention of this. The New River was an artificial river originally 38¾ miles long (now 28), opened in 1613, and designed by Sir Hugh Myddelton to bring water from the river Lea; it is still an important source of supply for London. At various points along its length there are 13 deep wells from which water is pumped into the river, and Sorocold may have been responsible for one of these. In 1889 a founder's share was sold for £122,800.

KING'S LYNN, c. 1698. King's Lynn is one of the towns where Ralph Thoresby tells us that Sorocold put in one of his installations, and although local records do not confirm this it will be seen that a pumping system was introduced at King's Lynn about the time of Sorocold, so that there seems no reason to doubt Thoresby's statement.

Henry J. Hillen in his *History of King's Lynn*, vol. 2, p. 793 gives a fairly long and detailed account of the early water supply of the town. In early times, as was usual, water was carried in pipes to conduits, purely by gravitation, and there is no need to quote details. "In 1682 the common dike was recast and a wind-mill erected, whereby the water was raised, by means of tubs or barrels, from the river to a reservoir or 'condytt or receipt,' a wooden tank in a round tower of the town wall."

This of course was a more primitive arrangement than

Sorocold's, but " During the 18th century the water was raised thirty or forty feet either by horse-power, the wind, or a force derived from an artificial waterfall which worked a machine termed a water-engine. Horses, however, were seldom employed, unless there was an exceptionally dry season, or in case the other machinery was out of gear. The old water-engine was found to be in a hopeless condition, and as the ingenuity of the Lynn machinists was of no avail, a new engine was obtained (1780). Although the new machine, similar in type, but three times as powerful as the old one, was quite adequate to the requirements of the town, the Assembly suddenly decide to have nothing less than a Steam Engine."

A Newcomen engine was installed, but it proved so expensive in coal consumed that it was discarded and the old water-engine brought back into use.

An illuminating sentence in Hillen's book suggest that the water-engine must have been put in about 1698 or just a little earlier. " From 1686 to 1698 the expenses of maintaining the water-works exceeded the income by £288. 13s. 6d., the income being £1,338. 14s. 2d. and the disbursements £1627. 7s. 8d."

The large income seems to indicate a house to house service, although Hillen does not say so. If later accounts were available they would probably show a good profit after 1698.

LEEDS, 1694. In this undertaking Sorocold was associated with Henry Gilbert, of Netherseal (then in Leicestershire, but now in Derbyshire having been transferred in 1897), and there is a fair amount of information as to the sequence of events. Henry Gilbert may have been a surveyor, and I find that Mr. Mundy of Markeaton paid him the sum of £113. 16s. in 1686, but the entry does not say what it was for.[1]

The Corporation Minute Book of Leeds,[2] records the

[1] Mundy MSS. in Derby Pub. Lib. [2] Recently printed by the Thoresby Soc.

agreement made between the Corporation and the two engineers which is as follows.

" THE COURT of Major, Aldermen and Assistants holden for the said Burrough the third day of March in the sixth yeare of the Raigne of our Sov'aigne Lord and Lady William and Mary, by the grace of god over England, &c., King and Queene. [1694]

Whereas att a Meeting of the Major, Aldermen and Comon Councell Henry Gillert of Nether Soale in the County of Leicester, Esq., and George Sorocold of the towne of Darby, gent., have proposed and doe designe to lay an engine to convey water from the River of Aire through the streets to the sev'all houses within the towne of Leeds aforesaid or to soe many of them as shall purchase the same of them. This Corte therefore takeing the same into consideracon doe judge that it wilbe a worke of publique benefitt and deserves great encouragement, for the encouragement therefore of the said undertaking and in consideracon of the some of Forty pounds to be paid by the said Mr. Sorocold to the Treasurer of this Corporacon for the use and benefitt thereof doe order And it is hereby ordered that the said Mr. Sorocold, his heires, executors, Admi'strators and Assignes shall forever hereafter be exempted by reason of the profitts and rents which he or they shall make of the said water workes of and from all and all manner of taxes, layes and assessements to and for the poore Constable and highwayes, the Mill by which the said Mr. Sorocold conveyes the water and workes his engines onely excepted And that to be taxed and assessed att the onely usuall rate that it has bene taxed and assessed for seaven years last past."

Ralph Thoresby's diary has several entries which show that Sorocold lost no time in getting to work.

1694, July 12th. Sent for by Mr. Mayor, Mr. Recorder, and Mr. Sorocold, about waterworks.[1]

[1] Thoresby's *Diary*, vol. 1 p. 258.

1694, Aug. 22nd. Was several times with Mr. Sorocold's workmen, who this day first began in Kirk Gate to lay the lead pipes to convey water to each family.[1]

On the 1st October, 1694 the licence for laying pipes in the sheets was signed,[2] and under this date Thoresby duly records that he " dined with the Lords [of the manor] and with them subscribed Mr. Sorocold's lease for the new waterworks," although as we have seen the work was already begun. On the 17th of the same month Thoresby was " at the New Waterworks, a most ingenious contrivance." In his *Ducatus Leodiensis*, 1715, he makes further reference to the waterworks, " The engine . . . at Pitfall, is for conveying the river-water by lead-pipes to the several parts of the town, which was performed Anno, 1695, by the ingenious Mr. George Sorocold, the Great English Engineer." Further, in his museum, Thoresby had several rings of the different lead pipes used in Leeds, the gift of Sorocold.

The water-wheel was situated on the River Aire at Pitfall near the north-east end of Leeds Bridge and the reservoir was near St. John's Church at the top of Briggate, giving sufficient pressure to supply the houses. It will be seen that the Leeds installation was similar to the plants he installed at several other towns.

London Marchant's Waterworks, 1696. In 1696, Sorocold was engaged on a water supply scheme in London in association with John Hadley, the engineer who patented the rising and falling waterwheel in 1693. This we know from a letter[3] dated 3 Mar., 1696, written by Sir Godfrey Copley, Bart., of Sprotborough, Yorkshire, and M.P. for Thirsk:—

" To Thomas Kirke. I have been this day, and am to meet tomorrow, Mr. Saracole and Mr. Hadley. I have

[1] *Ibid.*, p. 261.
[2] *Manour Booke of Leeds*, Thoresby Soc., vol. 9, pt. 3, 1899, p. 281.
[3] *Cat. of Stowe* MSS., 747, f. 63.

seen his engine consisting of three mill wheeles with small cranks at each end of the axletree, which raises Tems water, and all are carried by one stream of kennel water . . . one wheel being under another, I do think the best piece of work I have seen."

This is very high praise from a man of Sir Godfrey's standing, as he was a Fellow of the Royal Society, and the Copley Medal, founded in 1736 under the terms of his will, is still the premier award of the society.

Copley was clearly referring to the installation known as Marchant's Water Works at Tom's Coffee-house in St. Martin's Lane. Hugh Marchant and others obtained letters patent in 1694 giving them the right to use all such waters as run down the common sewers within the Bill of Mortality—except those within the City and Liberties of London, and those invested in the City of London by Act of Parliament—for driving mills for raising Thames and other water for the supply of the public. Another patent in 1696 authorised them to erect over-shot wheels, lay pipes, etc., in the river, and to take Thames water for public purposes. Stephen Switzer, *Hydrostaticks and Hydraulicks*, 1729, describes the St. Martin's Lane plant as " the work of one, Mr. Sorocold, a very good engineer," and gives an illustration of the water-wheel, etc. An over-shot wheel was geared to a three-throw crank shaft, and each crank was coupled by a link to an intermediate point in an overhead lever pivoted at one end; the passage rods, three in number, were hung from the outer ends of the levers. Switzer does not mention the use of three water wheels in tandem as described by Copley.[1]

LONDON BRIDGE WORKS, 1701. The early water supplies of London were from wells and conduits from which water was carried in buckets and carts to the consumers. Bridge water-works were developed in

[1] From a paper on Sorocold in the *Collected Papers of* Rhys Jenkins, where Switzer's illustration is reproduced.

Germany early in the 16th century, and in 1581 the Corporation of London granted a lease to Peter Morice (variously spelled Morryce, Moryce, Moris and Morris) said to have been a Dutchman or Fleming, but who none the less was a free citizen of London, to erect an engine within the first arch of London Bridge. This engine pumped the water up to a standard at the north-east corner of Leadenhall, and from thence supplied the eastern part of the City. Two years later a second arch was similarly leased. This was undoubtedly the first installation of pumping machinery in England for the supply of water for domestic purposes, and there does not seem to have been any extension of this system to the provinces until Sorocold's time. The Great Fire of 1666 seems to have quickly destroyed Morice's machinery, and in 1667 authority was given to Thomas Morris, a grandson of Peter, to rebuild his water-house adjoining London Bridge. The property remained with the Morris family until 1701, when Thomas and John Morris sold their rights to Richard Soame citizen and goldsmith, and others, for £38,000. Soame also obtained the lease of a third arch, actually the fourth arch in order from the river bank. In this arch a new machine was built to the design of George Sorocold, who was also responsible for reconstructing all the machinery.

Edward Hatton in his *New View of London*, 1708 says, "that besides the old work erected by Mr. Morris, the new, placed in the fourth arch of the Bridge, consists of two wheels with seven engines, set up about the year 1702, so there are in all thirteen engines. They are the contrivance of that great English engineer, Mr. Sorocold, whereby Thames water is raised from the north end of the Bridge to a very great altitude, by which means many parts of the city, etc., are served with Thames water. The flux and reflux of the water working the engine."

In the British Museum is a contemporary manuscript which also records Sorocold's connection with these works.

"There was an engine set up at the Bridge foot in Queen Elizabeth's time, for raising water to serve the City, and since improved. Lately they have taken in another arch of the Bridge, and the ingenious Mr. Seracoll hath contrived a most admirable machine there."[1]

Soame's company at the same time leased the City conduits for £700 a year and the company was divided into three hundred shares at £500 each, (£150,000 capital). When the company was dissolved in 1822 the shares had been increased to 1500.[2] The bridge was rebuilt at this time and the waterworks removed.

A detailed description of one section of the new works was given by Henry Beighton in the *Philosophical Transactions* 1731 together with an illustration which is reproduced with this paper (Plate VIII). Beighton's description has been briefly summarized as follows.

The waterwheel was 20 ft. diameter with floats 14 ft. long and 1.5 ft. deep. The wheel was coupled by gearing giving an increased speed in the ratio of 2.2 to 1 to two four-throw crankshafts, one on each side the waterwheel. Connecting rods from these crankshafts oscillated overhead beams, to the ends of which the pump rods were fixed, so that the one waterwheel worked 16 pumps. The pumps had bores of 7 in. and strokes of 2.5 ft. with the waterwheel turning at 6 r.p.m. the pump displacement was about 880 gals. per minute against a head of 120 feet equivalent to a h.p. of 32. The waterwheel could be raised and lowered to suit the state of the tide, a mechanism patented by John Hadley in 1693. This was found to be of little use and was seldom used.[3]

[1] Harleian MSS. 5000. Printed in the *Gentleman's Mag.*, 1807, p. 419.

[2] Foord, A. S. *Springs, streams and spas of London*, 1910.

[3] *Pumping Machinery*, vol. 2, p. 42, Science Museum Handbook; and *Collected papers of Rhys Jenkins*.

The complete waterworks consisted of four waterwheels driving 52 similar pumps and it is said that 11,724 tons of water were raised per day. This is 1820 gals. per minute, and equals a water h.p. of 66.

Beighton says the machines at London Bridge were far superior to those so much famed at Marly in France which pumped water from the Seine for use at the Palace Gardens of Versailles, as the latter are very ill contrived in the cranks and some other parts.[1]

The timber-work was all admirably well executed; and the composition and contrivance, both for strength and usefulness, not exceeded by any he had seen.

Henry Beighton was in practice as a civil engineer, and a respectable member of the Royal Society, so that his account of the London Bridge Works is one to be respected.

MACCLESFIELD. One of the towns at which Sorocold installed one of his water-engines according to Ralph Thoresby, whose word we have no reason to doubt, but in this instance there is no comfirmatory evidence.

The only information available of any water supply at Macclesfield has been kindly supplied by Mr. Arthur Smith, of Gawsworth, who is engaged on a work on the Common Lands of Cheshire. He points out that Charles II in 1685 granted a charter to Macclesfield authorizing the Mayor, Aldermen and Burgesses to 'convey water in and through pipes or otherwise to the Borough of Macclesfield aforesaid from the springs in the commons or waste grounds near the borough.'

Mr. Smith says the water works established under this charter were gravitational and did not involve pumping, the commons being several feet above the level of the town. The supply thus conveyed sufficed until 1849 when it was augmented from a neighbouring valley.

[1] *Philosophical Trans. of the Royal Soc.*, abridged, vol. 7, from 1724 to 1734. London, 1809. Gives a key to the lettering on the illustration reproduced with this paper, and full extracts from Beighton's paper.

NEWCASTLE-UPON-TYNE, 1698. The work of Sorocold in relation to water supply must have aroused wide-spread interest and in 1694 we find Dr. Jabez Cay, of Newcastle-upon-Tyne writing to Ralph Thoresby, of Leeds,[1] for particulars of the water supply there.

Newcastle, 6th Nov., 1694.

Dear Sir,

Yesterday I received your kind letter enclosed in one from Whitehaven, I would beg a favour of you. I hear there is a man come to Leeds with a design to furnish the town with water, after a manner somewhat extraordinary; they say he forceth the water almost half a mile upon an ascent (the ascent being about four yards perpendicular in twenty), from the river to his conservatory [reservoir]. Now I would gladly be informed of the truth of this; and what sort of an engine it is by which he does all this, that I may know whether it be applicable to our coal works or no? and what the true height is, to which he raiseth his water? and whether his encouragement from the town be like to answer the expense of so great a work? . . . .

Sir, Your Servant,

Jabez Cay.

I have no information as to what resulted from Dr. Cay's enquiries, but on 14 Feb., 1698 a bill was promoted in Parliament "for better supplying the town of Newcastle with fresh water," and this passed the Lords, 10 June, 1698. The House of Commons Journal gives no information about the details of this scheme, but on 25 Feb., 1701, a petition was sent to Parliament by William Yarnold, gent., on behalf of himself and other proprietors and undertakers, setting forth that in pursuance of an Act passed in the 10th year of His Majesty's reign, and by a lease made by the Mayor and Burgesses, the petitioner has at great expense brought sufficient fresh water to serve

[1] *Letters of Ralph Thoresby*, vol. 1, p. 178.

the town, from a place called The Zeme, in the county of Durham, about four miles from Newcastle, and the High Sheriff had held three inquisitions to enquire what damage the owners of lands through which the water is conveyed, had sustained thereby, as the Act directs. And the petitioner asked leave to bring in a Bill to confirm the inquisitions and lease, etc.

NORWICH, 1694. One of the towns mentioned by Ralph Thoresby as having one of Sorocold's water-engines, and here we have direct confirmation of it.

As at other places the town relied on wells, ponds, streams and conduits until Sorocold came along with a direct supply to the houses.

In Sept., 1694 an agreement was made between the City of Norwich, and Richard Barry, of Westminster and George Sorocold, of Derby, gentlemen, that in consideration of £200 and a yearly rent of £25, they or their executors for 99 years should have the power to supply water and arrange for distribution " for the general use of such of the inhabitants as should at reasonable rates and rents purchase a share in the same, for the service of their respective houses or families, or for other conveniences."

Barry and Sorocold agreed to commence within nine months of the signing of the deed, and within six years to have lead, or wooden-pipes in the principal streets of the City. In case of fire the City retained the right to cut any of the pipes to obtain water to extinguish the fire.

By another agreement between the City and Richard Soame, merchant and George Sorocold they acquired a lease of the three water mills known as New Mills.[1]

Ordinarily the leases and licences granted by the City would have been sufficient, but for some reason unknown, on 14 Feb., 1700, a petition was presented to Parliament by " the Mayor, Sheriffs, Citizens, and Commonalty of the

[1] *Abstract of the several deeds . . . relating to the Water works and New Mills*, Norwich, 1786.

City of Norwich," setting forth that they had granted to Richard Barry, esquire, and George Sorocold, gentleman, several licences and privileges upon their undertaking to supply the inhabitants of Norwich with river water, and another agreement made between the petitioners and Richard Soame, merchant and George Sorocold for the same purpose.[1]

They thought it "absolutely necessary" that these indentures should be confirmed by Act of Parliament.

Leave to bring in a Bill was granted, which at the committee stage had a clause added for setting up lights within the City, and on the 21st Mar., 1700 it passed the Lords as "An Act for confirming a Lease and certain Indentures between the City of Norwich, and Richard Barry, esquire, George Sorocold, gent., and Richard Soame, merchant; and for enlightening the streets of the said City."

Apparently Sorocold got to work before the act was passed, for when Celia Fiennes[2] visited Norwich about 1698 the work was in progress. She describes "A little building on which they were at work, design'd for a water house to supply the town by pipes into their houses with water."

NOTTINGHAM, 1696. There may have been other engineers constructing waterworks on the same principle as Sorocold, but if so we know less about their work than we do of Sorocold's. In the circumstances it may be excusable just to mention what little is known of Nottingham as the water engine there was on similar lines to the one at Derby.

The old Waterworks Company obtained a lease from the Corporation in 1696 and had a water engine on the River Leen at the bottom of Finkhill Street, and "the main wheel of the hydraulic machine set in motion a number of crank levers, and itself moved, like the great

[1] *House of Commons Journal.* [2] Op. cit., p. 120.

waterwork wheels at London Bridge, by the power of water. . . . leaden-pipes supply some of the lower parts of the town with water, while the higher parts are supplied from a cistern behind the General Hospital.[1]

This was the main source of supply until after the passing of an Act in 1822 for supplying water from the River Trent, which, of course, provided a much greater quantity of water, but this Act (p. 57) definitely prevented any interference with the old waterworks on the banks of the River Leen.

PORTSMOUTH, 1694. Another of the towns where Thoresby tells us Sorocold had erected one of his water-engines.

As usual the town relied on wells and conduits, but " there was, however, an abortive attempt to provide something better in 1694, when a lease was granted to Richard Barrey, of London, and George Surracolde, of Derby, empowering them to dig up the streets, lay pipes, and supply the town with good and wholesome spring water for a term of 99 years. There is however no record of anything being carried out, and presumably the powers lapsed."[2]

Sorocold was associated with Richard Barry at Norwich, and in view of the reliability of Thoresby as a recorder, and of Sorocold as an engineer, I am inclined to think that lack of supporting evidence does not cancel Thoresby's positive statement.

SHEFFIELD, 1697. Sheffield is not mentioned by Thoresby as one of the places where Sorocold installed a water-engine, nevertheless we have positive evidence that he did so, which out-weighs any negative assumption of local historians. Daniel Defoe in his " Tour," 1727,[3] speaking of Sheffield, says " there is a fine Engine or Mill also for raising water to supply the town, which was done

[1] Blackner, *Hist. of Nottingham*, 1815, p. 25.

[2] *Guide to Postsmouth*, prepared for the British Association meeting held there in 1911.

[3] Everyman's ed. vol. 2, p. 184.

by Mr. Seracoal, the same who fell into the River at the throwing-mill[1] at Derby."[2]

The fullest account of water supply at Sheffield is that of Mr. William Terrey,[3] formerly manager of the Corporation waterworks, unfortunately he does not give any references so his authorities cannot be followed up.

According to Mr. Terrey the first distinctly artificlal water supply in Sheffield was the construction of a reservoir by a man named Barker from whom it got the name by which it is still known, Barker Pool. The water was carried from the reservoir in pails and barrels. In 1793 it was done away with, the name only remaining.

In 1697, Peter Whalley, an engineer of Nottingham, obtained a lease from the Duke of Norfolk (Lord of the Manor) of a piece of land just above Lady's Bridge, where the exchange Brewery now (1908) stands. This piece of land was known as " The Isle of Wight " or the Island as a mill goyt surrounded it. He also obtained the lease of another piece of land near Barker Pool, with license to break up the streets for laying pipes to supply the town with water.

It was Whalley's intention to erect an engine near the river, by which he could force up the river water to the reservoir at Barker Pool, whence it was to be distributed in pipes over the town. He had the support of the inhabitants, but died when he had only made small progress.

However the scheme was taken up by five public spirited citizens who seem to have taken over Whalley's leases. Although Mr. Terry appears to think this scheme was never carried out, the direct evidence of Defoe is in my view quite conclusive that the work was completed and by Sorocold who may have been called in after Peter Whalley's death, or he may have been associated with him in the first place.

[1] Throwing-mill = silk throwing or spinning mill.

[2] See ante, p. 57. [3] *History and Desc. of the Sheffleld water works*, 1908.

Mr. Terrey says,[1] "it is almost certain that the repeated failures to pump water from the river into the town were due to the fact that the only means of conveying water was by wooden pipes. . . These pipes, though strong in the body, would be excessively weak and unreliable at the joints, and, though perfectly efficacious for conveying water downwards by gravity, would not be suitable for pumping water from a lower to a higher level."

Mr. Terrey is wrong in both arguments, for lead-pipes were used by Sorocold at Leeds in 1694, and wooden pipes proved strong enough to stand the pressure at many places throughout the country.

Prior to the Sheffield Act of 1830 all the water pipes in Sheffield were of wood, and there were thirteen miles of them. When taken up some had been in a century and none for less than seventy years, yet the pipes and even the wool packing in the joints were found to be as perfect as when they were put in.[2] This is a wonderful testimony to the value of these pipes. According to Mr. Terrey the pipes were bored by hand at Sheffield, but as we have seen Sorocold was boring by machinery at Derby in 1692.

SPROTBOROUGH. This was the seat of Sir Godfrey Copley, Bart., who, as we have seen, met Sorocold and Hadley in London in 1696, and it is not surprising to find that he had a water-engine installed on his property.

Ralph Thoresby tells us in his diary, that he was at Sprotborough on 15 Feb., 1703, "assisting Mr. Kirk and Mr. Arthington in taking a level for the new canal, that is now making from the water engine (which is very curious and conveys water to a large cistern upon the roof of the hall, a vast height from the foot of the hill) to the corn, mill, thence he can go to Coningsburgh Castle on one hand, or Doncaster on the other."[3]

It is true Thoresby does not mention Sorocold in connection with the water-engine at Sprotborough, but

[1] Op. cit., p. 14. [2] Op. cit, p., 22. [3] Thoresby's Diary, vol. 1, p. 412

there is a possibility if not a probability that it was his work, and possibly further information may come to light.

WIRKSWORTH. Thoresby tells us that Sorocold put in a water supply, but there is no other evidence available. Wirksworth is only fourteen miles from Derby, and it is quite possible that Sorocold put in a pumping plant at one of the lead-mines which at that time were very active in and around Wirksworth. Some sort of water supply there must have been as in 1827 new cast-iron pipes were laid down to give a better supply of water for the town.[1]

YARMOUTH, GREAT, 1694. This is another of the towns which, according to Ralph Thoresby, was supplied with water by Sorocold.

In 1694 the Corporation entered into an agreement with Richard Barry, of London, and George Sorocold, of Derby, to supply the inhabitants with good and wholesome fresh spring water. They proposed to collect a supply from wells on the Denes into a large reservoir, and distribute the water through the town by pipes for which purpose they had permission to break up the streets. A somewhat similar plan was proposed by Mr. Dodd in 1810, but, being opposed in Parliament, was abandoned.[2]

A later writer[3] describes Barry and Sorocold as " two adventurers from London," so they were in the language of 1694, but not in that of to-day, for ' adventurer ' we should now say ' contractor.' The same writer goes on to say that their attempt was abandoned, as was another in 1753.

In view of Thoresby's definite statement, it seems unreasonable to say Barry and Sorocold failed in their undertaking.

The compilation of a record such as this would not have been possible without the assistance of many who possessed special knowledge of the history of the various

[1] Ince, Thomas, MS. notes on Wirksworth, 1827, in Derby Public Library.
[2] Manship, *Hist. of Gt. Yarmouth*, p. 307.
[3] In *Yarmouth Mercury*, 2 Mar., 1907.

places with which I have had to deal, and I cannot conclude without expressing my sincere thanks to the following:—Messrs. W. A. Atkinson, Knaresborough; Gilbert Berry, Librarian, Macclesfield; The Trustees of the British Museum; Messrs. Percy Clare, Reference Library, Bermondsey; H. W. Dickinson, Purley; A. A. Gomme, Librarian, Patent Office Library; R. J. Gordon, Librarian, Leeds; T. Walter Hall, Sheffield; William J. Harris, Librarian, Islington; George Hayward, Librarian, Norwich; Miss I. F. Jackson, Librarian, Deal; Rhys Jenkins, Hermitage, Berks.; J. P. Lamb, Librarian, Sheffield; Dr. E. Bosdin Leech, Manchester; Mabel Lady Mellor (née Serocold); Messrs. Henry Peet, Birkenhead; James Ross, Librarian, Bristol; H. Sargeant, Librarian, Portsmouth; C. H. Senior, Librarian, King's Lynn; Col. O. P. Serocold, Taplow, Bucks.; Arthur Smith, Gawsworth, near Macclesfield; Clifford Stott, Librarian, Rochdale; H. Tapley-Sopey, Librarian, Exeter; Major W. H. Tapp, Wimbledon; Thomas Thornton, Town Clerk, Leeds; Dr. W. Watkins-Pitchford, Bridgnorth; and R. G. Watlow, Librarian, Yarmouth.

POSTSCRIPT: After the foregoing was in print, Mr. John Hobbs, of Derby, drew my attention to three documents calendared in Moulton's, *Cat. of Palæography, etc.*, 1930, which show a connection between the district of Derby and the George Sorocold, whom I have assumed to be the grandfather of the engineer. The documents are dated respectively 1641, 1645, and 1647, and concern the assignment by George Sorocold, of Ashton in Makerfield, Lanc., gent., and William Whitfield, of Roby, Lanc., yeoman, to Anne Moseley, younger daughter of Rowland Moseley, esq., deceased, and sister of Sir Edward Moseley, of Rolleston, co. Staff. and Collyhurst, co. Lanc., of the manors of Etwall and Hardwick, co. Derby for the sum of £812.

This transaction was not known to Lysons and other local historians, who all make the manor of Etwall pass in 1641 direct from Sir William Gerard to Sir Edward Moseley, whereas it is clear from these deeds that there was a previous transfer.

Etwall lies six miles south-west of Derby, and Hardwick, now Hargate Manor is one mile further away in the same direction.

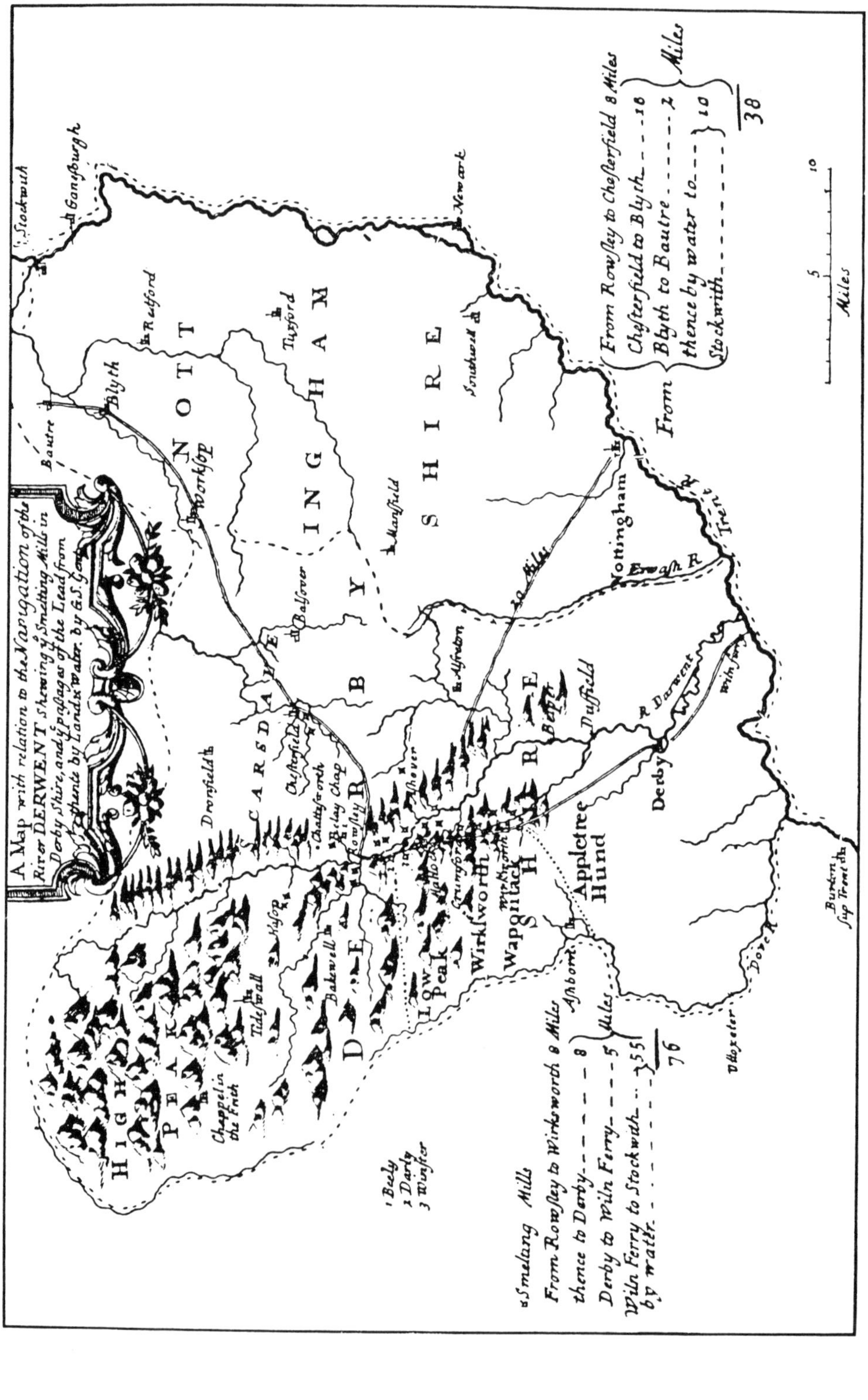
A Map with relation to the Navigation of the River DERWENT Shewing ye Smelting Mills in Derby Shire, and ye passages of the Lead from thence by Land & Water. by G.S.
From Rowsley to Chesterfield 8 Miles
Chesterfield to Blyth - - - 18
Blyth to Bautre - - - - - 2
thence by water to - - Stockwith - - - - - - - 10
From Miles
38
From Rowsley to Wirksworth 8 Miles
thence to Derby - - - - - 8
Derby to Wiln Ferry - - - 5
Wiln Ferry to Stockwith by water - - - - - - - 55
Miles
76
Smelting Mills
1 Beely
2 Darly
3 Winster
Miles
NOTTINGHAM SHIRE
DERBY SHIRE
HIGH PEAK
LOW Peak
SCARSDALE
Wirksworth Wapontack
Appletree Hund
Stockwith
Gansburgh
Newark
Retford
Tuxford
Blyth
Bautre
Worksop
Southwell
Mansfield
Nottingham
Erwash R.
Trent
Balsover
Alfreton
Chesterfield
Dronfield
Derby
R. Darwent
Duffield
Belper
Ashborn
Dove R.
Uttoxeter
Burton sup Trent
Chappel in the Frith
Tideswall
Bakwell
Rowsley
Chatsworth
Wiln Ferry

5
The old water-supply of Seville

George Higgin

THE new Seville Waterworks, now in progress of construction by an English company, will destroy an ancient system of supply which has been in operation for more than seven hundred years.

Constructed in the year A.D. 1172 by the Moor Jucef Abu Jacub, whilst the City of Seville was still under the domination of the Saracens, it still exists in great part in the same original form in which it was first instituted—a standing monument to the conservative influences of the climate and people of the region where it is situated. Within a few short months all the most interesting features of this work of Moorish art will have disappeared.

The ancient Moors were always celebrated for their skill in dealing with water, and their application of it to irrigation, and many examples of their work in this respect still exist in Spain, in the primitive condition in which their constructors left them more than six hundred years ago; indeed it may safely be said that all the existing irrigation-works in Spain, now in active use, were originally designed and constructed by the enterprising Saracens.

The Author is not, however, aware of any existing system of town-distribution constructed by the Moors, so ancient and complete as that of Seville; and as this presents many points of interest to the engineer, a short account of it may be of use both in preserving an historical record of a work which is nowhere else recorded, and also as showing how, seven centuries ago, the then little understood principles of hydraulics were applied.

The old Seville water-supply may be divided into three parts.

First. The collection of the water.

Second. Its conduction to the city.

Third. Its distribution inside the city.

The first portion of the works, the collection of the water, is much older than the two latter portions, inasmuch as it is believed to have been done by the Romans in the early part of the Christian era, whilst Spain was still under their domination, and was used as a supply for the ancient city of Ontivar, a city which has long disappeared, and left only its name to tell of its former existence. The water is collected by a tunnel constructed in a rock of tertiary formation which forms the basin of a small river called the Guadaira. This river rises in the Sierra de Moron, near the town of Moron in the province of Seville, and after a short course falls into the Guadalquivir close to the city of Seville. The valley of the river for some distance above Seville is composed of a very porous argillaceous limestone, resting on a bed of impermeable clay, and at the junction of the two strata numerous springs occur, the principal of these being in the neighbourhood of the town of Alcalà, which is situated on the river Guadaira, at a distance of 13 kilometres from Seville. The head or starting-point of the tunnel referred to is in a depression near the edge of the river and at some 600 metres above the last houses of the town. In this place, and at 8½ metres below the ground, there is a circular chamber, excavated in the rock and communicating by a shaft with the surface, and from the bottom of this chamber numerous strong springs of clear water rise and flow away through a tunnel excavated in the rock. The height of the tunnel varies from about 4 feet to 5 feet 6 inches, it has a mean width of about 3 feet, but it is very irregular in shape and dimensions. The level of the water in this chamber is, in round numbers, about 22 metres above the streets of Seville.

From the first chamber the channel runs under the town of Alcalà, and almost in the centre of the town it is joined by a second tunnel, coming from another direction, and at a much higher level, the water of which is utilized to turn a most curious underground flour-mill, to which descent is made by means of a gallery leading down from the principal street of the town. The combined waters run on in a westerly direction under the town, and on emerging from it follow the course of the river for some little distance, till they arrive at the Arroyo de Zacatin. The tunnel makes a short detour at this point, keeping well under the bed of the stream, and afterwards continues on till it leaves the broken ground, and comes out into the open country at a distance of rather more than 3 kilometres from its origin. The depth of the tunnel below the surface of the ground varies according to the nature of the latter, but it is never less than 40 or 50 feet, and in some cases it is much more. Shafts communicate with the surface at every 50 or 60 metres, these being those probably used for the construction of the work. During the whole course of its progress through the tunnel, the stream receives additions to its volume from the numerous little springs encountered in its route, and it also loses somewhat from filtrations in those points where the tunnel approaches more nearly to the edge of the hills. At the upper part of the depression down which the Arroyo de Zacatin flows, a tunnel has been perforated into the hills, probably at the same time as the original work was made, and a considerable stream so obtained is utilized to actuate five flour-mills. At the point where this stream crossed above the tunnel, arrangements were made by means of shafts and connecting-tunnels by which the waters from the Zacatin tunnel could be turned into the main tunnel in case of shortness of supply. The total volume of water collected by the tunnel in the course of the 3 kilometres of its length, varies according to the nature of the seasons from 4,000 to 6,000 and even 8,000 cubic metres per day, say 880,000 to 1,320,000 and 1,760,000 gallons per day. The water is of excellent quality, a little hard, having about 16 degrees of hardness, but entirely free from organic matter, and containing only a very moderate quantity of salts in solution; it may be described as a first-class water for dietetic purposes. The Zacatin tunnel produces an average supply of about 500,000 gallons per day.

Considering the date of the construction of the tunnel, that the use of powder was not then known, and that for the distance of 3 kilometres the tunnel was excavated in the solid rock at a depth below the surface sometimes exceeding 150 feet, the industry and engineering talent displayed by these early masters of the profession may well be admired.

Conduction to the City.

During the time of the Romans the water after leaving the tunnel was probably led direct to the town of Ontivar, but after the destruction of this latter place and the occupation of the country by the Moors, the water was led by an open conduit down to a place called Torreblanca. Here it received an additional supply from another aqueduct and tunnel, the origin of which is lost, and from hence the united streams flowed on to a point now called the Cruz del Campo, 1,100 metres from Seville, giving power during their course to nine flour-mills.

From the Cruz del Campo the water was conducted by an aqueduct to the dividing house outside the old walls at the gate known as the Puerta de Carmona. Both the aqueduct and the watercourse are said to have been the work of the Moor Jucef Abu Jacub, A.D. 1172.

At a much later period, in order to avoid the pollution of the waters in the open course, the Town Council of Seville abandoned the nine mills which were their property, covered in the upper part of the course, and constructed another aqueduct for the purpose of carrying the water by a shorter channel over a depression in the ground, between Torreblanca and the Cruz del Campo. The level of the water at the Cruz del Campo before entering the old aqueduct is 11½ metres above the lowest part of the city, and at the dividing house outside the Carmona Gate it is nearly 9 metres.

Distribution of the Water within the City.

On the conquest of the City of Seville by the Spanish king, commonly known as Saint Ferdinand, he, in accordance with the custom of those times, took possession of the aqueduct and water, and the people were only allowed by royal favour, whatever water remained after the wants of his palace were satisfied, in the words of the old charters, of the surplus which was left after the "real casa y cocina" (royal house and kitchen) were satisfied. Immediately after the conquest, the king of course made large grants of lands and houses to his faithful followers, and at the same time he allotted to them portions of the water. Grants of water were also made to the numerous convents, which were then and afterwards established in the city. These grants are for the most part still valid, and the descendants of these old warriors, such as the Duke of Medina Coeli, and others, still claim a large share of the waters.

As time advanced, however, and the powers of the aristocracy decreased, whilst that of the townsmen increased, the administration of the water gradually fell into the hands of the town council. Public fountains were erected for the poorer classes, and portions of the water were sold or let off to the inhabitants. By degrees the quantity left for his majesty's palace grew smaller and smaller, and of the portion now claimed, his administrator considers himself lucky if he receives one-fourth part, whilst many of the old grantees are compelled, against their will, to suffer a similar reduction.

The Author will now describe the system, by which the waters were divided and conducted to the house. Before doing so, however, it will be advisable to describe the unit of measure used in the regulations for supply. This is the "paja," the Spanish word for straw, and was doubtless originally meant to be as much water as would flow through a pipe having an internal diameter similar to that of a straw. Even supposing the diameter of a straw to be unvarying, it is evident that the amount of water flowing through it would depend directly on the head, and it can scarcely be conceived that the old Moors were not cognisant of this; but neither tradition nor written evidence has handed down any definition as to the head that was assumed, as that necessary, and though doubtless popular custom assigned some more or less regular head, yet at the present day it is impossible to find any reliable data on which to base it.

As regards the dimensions of the "paja," this was fixed at a comparatively early date, and amongst the archives of the municipality of Seville there is preserved an ancient document on vellum, in which the dimensions of the orifices necessary for the various discharges are shown graphically. On Fig. 1, the Author has reproduced the sizes from $\frac{1}{4}$ paja up to 10 pajas.

The heading of the old document referred to says:—

"The ancient and modern measures used in the distribution of the waters of the Carmona aqueduct of this city of Seville, in the times of our lords the Catholic kings, and of their predecessors, such as our lord the king Don Juan of famous memory.

"By order of my lord Don Juan Francisco Navarette, of the council of H.M. and his judge of the royal tribunal of this city of Seville, and inspector of his royal palaces, and of the waters of the Carmona aqueduct.

"Made in the year, 1657.

SEBASTIAN DE RUESTA."

According to this document the diameter of the orifice for "one paja" is as nearly as possible 6 millimetres, or say $\frac{1}{50}$th of a Spanish foot, and the discharge of the various orifices is shown to vary as the square of the diameter.

There exists also in the municipal archives a brass box, which is used, or was used, for measuring the water supplied to any house, in case of complaint. This box, shown by Fig. 2, is 10·82 inches square and 5·70 inches deep, and contains on three of its sides openings marked respectively from $\frac{1}{8}$ "paja" to 19 "pajas;" on the third side there is a scale, which, starting from the centre of the orifices, marks 2 Spanish inches, each inch being divided into 12

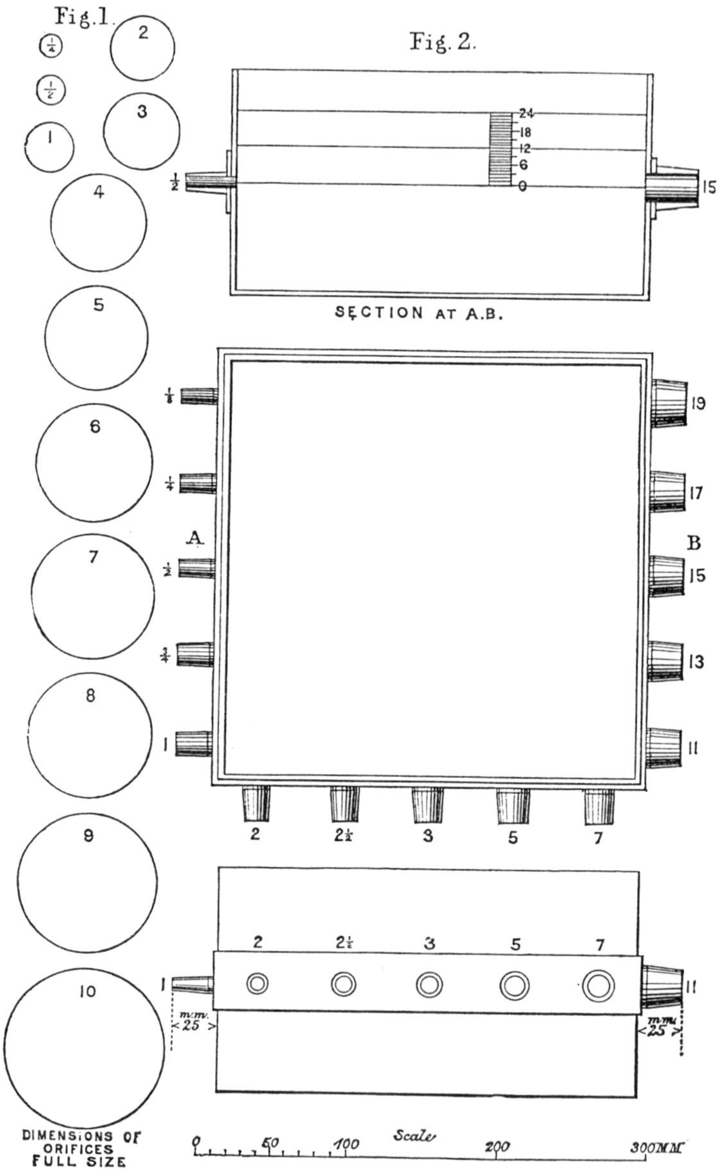

Old Water-Supply of Seville.

Details of Metal Box for Measurement of Water, preserved in the Archives of the Town Council.

FIG. 3.

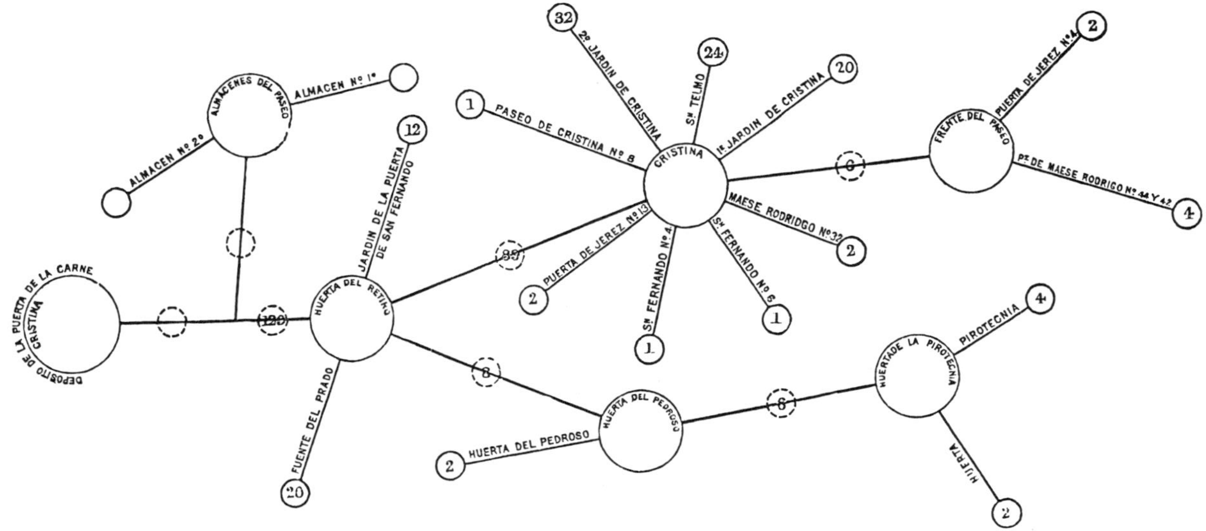

OLD WATER-SUPPLY OF SEVILLE.

RAMO DE CRISTINA.

Numeros.	—	Cuartas.	Numeros.	—	Cuartas.	Numeros.	—	Cuartas.
..	Almacen }	..	..	Jardin de la Puerta de San Fernando	12	13	Puerta de Jerez	2
..	Id. }					4	Id.	2
..	Fuente del Prado	20	..	Id. de Cristina primo	32	..	Pirotecnia	4
4	San Fernando	1	..	Id. Id. segundo	20	8	Paseo de Cristina	1
6	Id.	1				..	Palacio de San Telmo	21
..	Huerta del Pedroso	2	32	Maese Rodrigo	4			
..	Id. de la Pirotecnia	2	42 y 44	Id.	2		Suma	130

lines. It would appear from this box that the proper head should be 1 inch, as it is probable that the gauge was meant to ascertain whether any given pipe was discharging more or less than a "paja," and how much.

A series of very careful measurements made by the Author with this gauge, to determine the value of the "paja," gave with 1 inch head, the following results:—

Dimension of Orifice.	Discharge in 24 hours in Litres.
$\frac{1}{8}$ paja	270·84
$\frac{1}{4}$ "	434·17
$\frac{1}{2}$ "	604·19
$\frac{3}{4}$ "	1152·00
1 "	1661·53

According to these experiments, the value of 1 "paja" would be equal to about 365½ gallons, and this is probably more than most of the old grantees receive, although there is a general idea prevalent in Seville, that the "paja" ought to be nearly double the amount mentioned above. The relation between the various quantities is as close as could be expected under the circumstances, as the constructors do not appear to have understood the influence of the adjutages, and these are all the same length, that for $\frac{1}{8}$ "paja" being the same as that for 19 "pajas."

The Author will now proceed to describe the system of distribution:—

The water on arriving at the city is delivered by the aqueduct into a covered reservoir, which is under the charge of an officer of the municipality, and to which no one has access without permission. The plan and section of this reservoir are shown on Plate 13, Figs. 4 and 5. The wall bounding the reservoir on one side, which is a half octagon, contains eight orifices, corresponding to eight principal mains serving the city. The first of these orifices, which is near the level of the floor, corresponds to the main supplying the royal palaces, and the opening communicates direct with an earthenware pipe. The remaining seven orifices communicate with a second chamber, from the bottom of which the pipes for the supply of each principal main lead off perpendicularly. These orifices are all provided with a bronze frame and diaphragm, pierced with holes of various dimensions. The orifice belonging to the royal palaces is 14 inches lower than the remaining seven, so that His Majesty has a decided advantage over his subjects.

The translation of the original document, preserved in the archives describing these various orifices, is as follows:—

No. 1. Pipe belonging to the royal palaces, which takes all the water of which it is capable. Its dimensions on the upper part is 14 inches, and on the lower part 9½ inches; it does not possess its proper frame.

No. 2. Known as "Barca Rota," in which is placed a bronze frame, having one opening, whose dimensions are equivalent to 23 "pajas" of water.

No. 3. Known as "Colegial del Salvador," in which is found another frame of the same metal, with five openings. The first of 11 pajas; the second of 13; the third of 13; the fourth of 14; and the fifth also of 14, making together 65 "pajas" of water.

No. 4. Known as the City pipe, in which the water travels to the public fountains, the square of San Francisco, Calle Laguna, the Triana Gate, and the Square of La Magdalena, where is found another frame of the same metal as the former ones, having three orifices. The first of 24 "pajas"; the second of 30; and the third also of 30, making up 84 "pajas" of water.

No. 5. Known as "San Pablo," is also of bronze. It consists of two orifices; the first of 18, and the second of 21 "pajas," making together 39 "pajas" of water.

No. 6. Known as "Medina Sidonia," of the same metal as the others, in which are six circular orifices, as are all of this principal main. They consist: the first of 36; the second of 16; the third of 13; the fourth of 13; the fifth of 11; and the sixth of 18; which added together make 107 "pajas" of water.

No. 7. Known as "Baños de San Juan de la Palma" consists of two orifices. The first of 22, and the second of 17, together 39 "pajas" of water.

No. 8. Known as that of the "Duke of Medina Coeli," consists of one sole orifice of 18 "pajas" of water.

It will be noted that whilst many of the openings have a number of orifices of a specified size, the entire body of water delivered by them is carried off by one down pipe in the second chamber. The reason of this is to be sought in the historical antecedents of the system. As fresh grants were made by the king, or at a later period by the municipality, a new hole was bored in the frame, the size of which corresponded to the dimensions shown on the old documents, for the number of pajas granted. The main pipe was always supposed to be, and indeed was, large enough to convey all the water required; the measurement into the main pipe was made at this chamber, and the measurement to the proprietors was made, as will be hereafter described, at the various distributing-boxes in the city.

The head of water over the orifices is that due to the difference of head between the two chambers, and is generally an inch or two. No means were provided for throttling the down pipe, but as the number of pajas admitted at the distributing chamber is, of course, exactly that distributed to the proprietors, the head of water is supposed to be balanced so as only to give the requisite discharge. Were the pipes sound and the distribution accurate, this would be so in practice. There is, however, a good deal of inequality in the distribution, and there is a large loss from leakage; but from long practice the distributors manage to make a pretty fair distribution to the various mains in chamber. The leakage is compensated for by a general discount on all the houses; and if, as often takes place, one proprietor gets more than his share, it is taken off some other proprietor, who has not had the astuteness to satisfy the official distributor. The principle of measurement is sound; it is virtually a discharge through a thin plate under a known head, and is the best and most efficient means of measuring. The openings in this main chamber have no adjutages; from which it is evident that the original constructors of this work knew more of the true principles of hydraulics than the constructors of the gauge-box.

The pipes for the conduction of the water through the streets were originally of earthenware surrounded by brickwork; many of these pipes, put down hundreds of years ago, are still doing service, but in some cases they have been superseded by cast iron. The smaller services are all of lead.

As there is occasionally a head of 25 or 30 feet on the pipes, it is evident that the leakage from the old class of pipes was, and is, considerable. It is wonderful, however, to see how well they have stood the pressure. As a matter of fact, however, out of the 4,000 metres of water per day brought into the city by the old aqueduct, not much more than one-half ever reaches its final destination.

Figs. 7 and 8, Plate 13, will explain how the distribution is made to the houses. Fig. 8 is an exact copy of one of the old plans kept in the municipality, and shows the mode of distribution in one of the small districts. The circles shown on the plan represent the dividing-boxes, the section and plan of which are shown on Fig. 7. The distribution is effected by means of a small metal box, placed in the outside wall of one of the houses, at such a height that the level to which the water will rise at this portion of the city shall about one-half fill the box. A lead pipe about 4 inches in diameter inserted in the bottom of the box communicates with the main. The side of the box is perforated with as many holes as there are participants of the water flowing from this box, and the dimensions of the orifices correspond to the amount of pajas each one should receive. The water flowing from the orifices is caught in a small hollow made of cement, from which a lead pipe leads it down into the street and thence to the house of the owner. Many of these boxes serve for the division of the water from other boxes. In Fig. 8, Plate 13, it will be seen how this is done. The box G, at the corner of the Calle del Carpio, receives its water from a pipe coming down that street, and supplies five private houses and five secondary boxes, three of which are shown on this plan, lettered G′, G″, G‴, and the remaining two are in another district. This box therefore contains ten orifices of varying dimensions.

The large book preserved in the municipality contains plans of every district supplied similar to the one copied here, and also a series of plans or diagrams, of one of which Fig. 3 is a copy, showing the boxes situated on each main, and the houses supplied from each box. Thus the Christina box supplies eight houses, the number and street of each being written on the diagram. At the side of each diagram is a statement showing the street and the number of each house, and the amount of water it is to receive counted in fourth parts of "pajas."

The old boxes were mostly of iron or bronze, the metal being from $\frac{1}{8}$ to $\frac{1}{4}$ inch thick; they are of various shapes, some being oval and some square; as a rule they are about a foot square, and 9 inches deep. They are at varying heights above the street, according to the level of the latter.

It is evident that, theoretically, this system of measuring the water was good, and if a uniform head could be kept, there would be nothing to object to. Practically there were many drawbacks. There was no effectual way of controlling the head; the boxes were fixed in accordance with the height to which it was ascertained by actual experiment the water would rise; but once fixed, many causes combined to render the pressure a constantly varying one. The men in charge of the various districts manipulated the boxes in accordance with their hereditary practical instincts, and a kind of rough-and-ready justice was administered. Thus, in one box where there was too great a flow of water, either from the box having been fixed originally too low or a

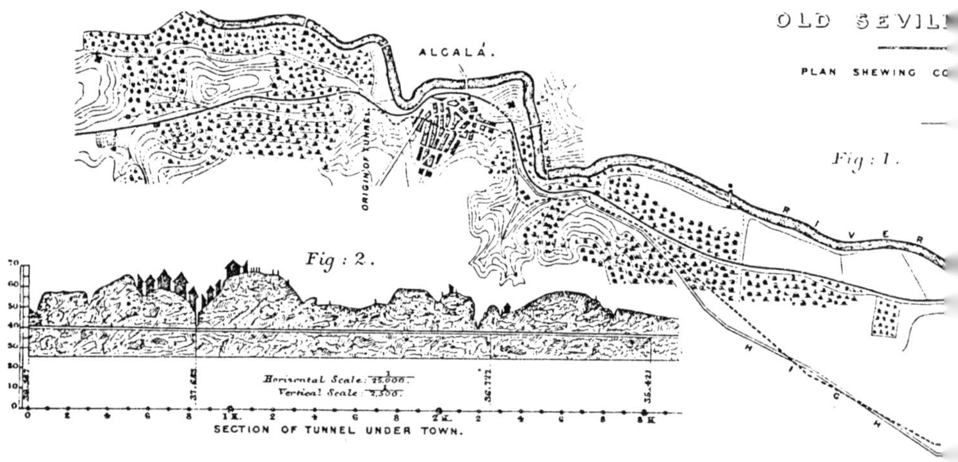
OLD SEVIL
PLAN SHEWING CO
Fig: 1.
ALCALÁ.
ORIGIN OF TUNNEL
RIVER
Fig: 2.
Horizontal Scale
Vertical Scale
SECTION OF TUNNEL UNDER TOWN.

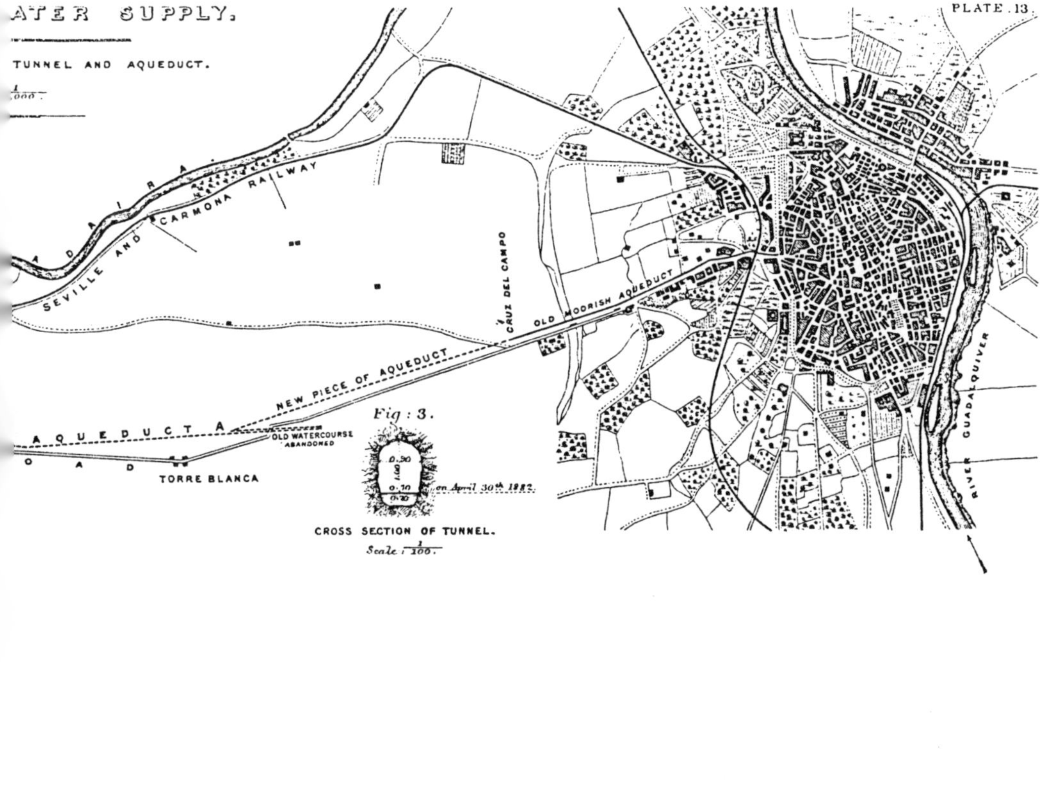
ATER SUPPLY.
TUNNEL AND AQUEDUCT.
PLATE.13.
SEVILLE AND CARMONA RAILWAY
CRUZ DEL CAMPO
OLD MOORISH AQUEDUCT
NEW PIECE OF AQUEDUCT
AQUEDUCT A
OLD WATERCOURSE ABANDONED
TORRE BLANCA
RIVER GUADALQUIVER
Fig: 3.
CROSS SECTION OF TUNNEL.

MAIN DIVIDING CHAMBER AT THE CARMONA GATE.

Fig: 4.

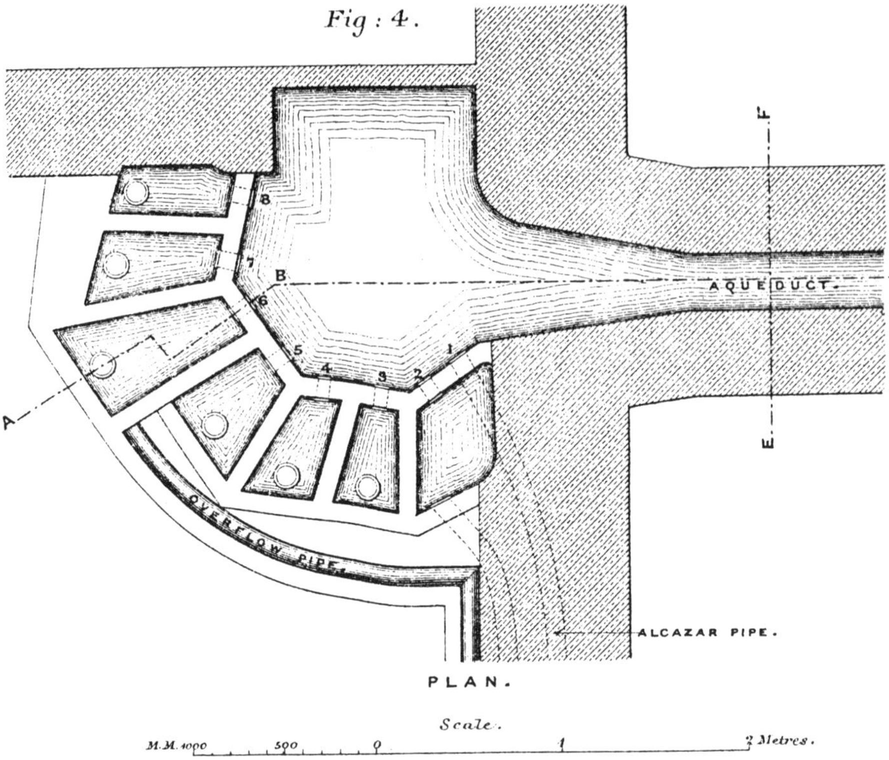

Fig: 5.

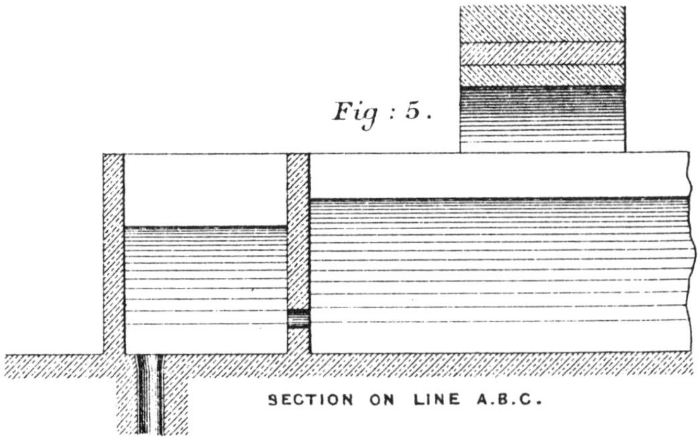

SECTION ON LINE A.B.C.

Fig: 6.

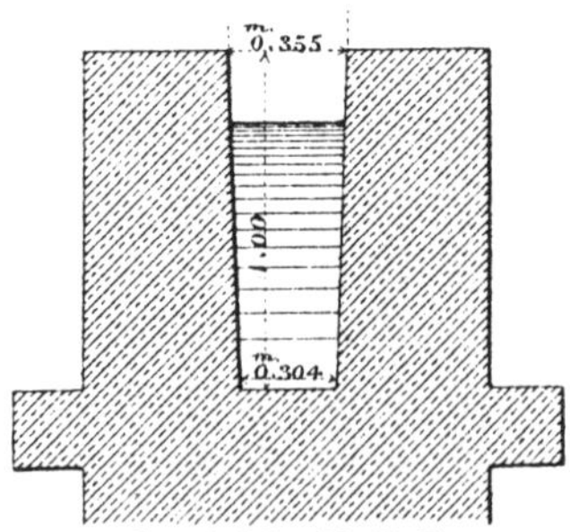

SECTION AT E.F.

Fig: 7.

SUBSIDIARY DIVIDING BOX IN CITY.

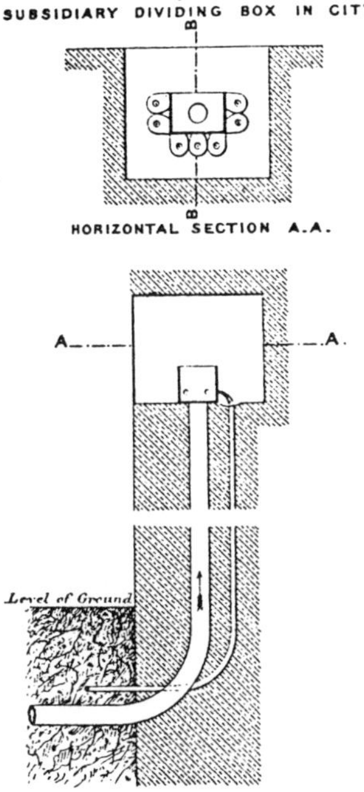

VERTICAL SECTION B.B.

Fig: 8.

Copied from the Register Book in the Municipal Archives of Seville.

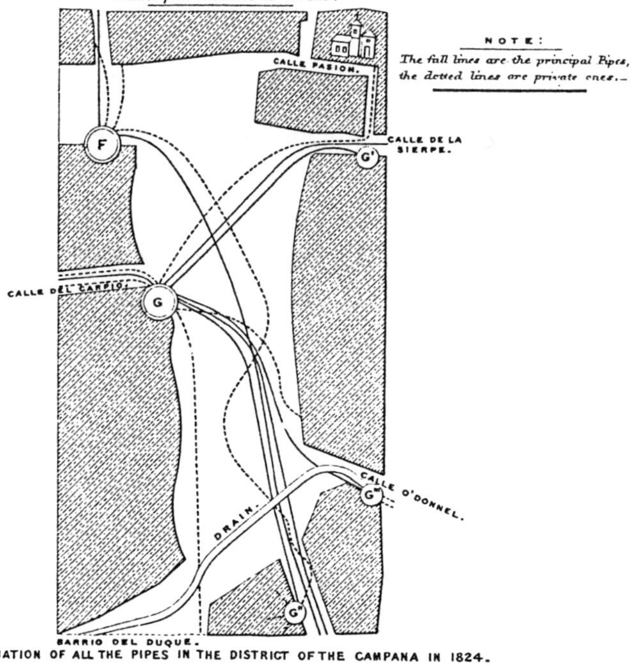

SITUATION OF ALL THE PIPES IN THE DISTRICT OF THE CAMPANA IN 1824.

greater pressure having been given by the laying of new and larger mains, the Author saw that a half brick had been placed on the inlet-pipe, and by this means the head was kept down to about 4 inches. The keys of all the boxes are kept by the head water-bailiff; and it is easy to understand what power this gave to a shrewd practical man who understood, in his own way, how to increase or diminish the supply to each house. His position thus became a lucrative one. To obtain one's full supply it was necessary to be on good terms with him, whilst those who had not thus provided for themselves found their supply daily growing smaller and smaller. At the same time a kind of practical balance was maintained in each district.

It will be easily understood that, after its numerous journeys, and the large amount of friction to which it was subjected, the water had not much power remaining when it arrived at the houses. In few houses, even in streets at the lowest level, did it rise many feet above the level of the floors; almost all, if not all, the houses which have water, collect it in a marble cistern, from whence it is pumped up by hand-power to the upper floors.

The new waterworks, now far advanced towards completion, will sweep away all the old system of distribution. They take the water at the conclusion of the tunnel and pump it up by steam-power to reservoirs constructed near the town of Alcalà, from whence it will be conducted by a 21-inch main to the city and distributed under a mean pressure of 25 metres head, so as to rise to the top floors of the highest houses.

Before all the data for doing so vanish, the Author thought that a description of these ancient works would probably interest the profession, as giving an account of perhaps one of the oldest systems of house-distribution known in connection with the water-supply of towns.

The communication is illustrated by four sheets of small scale drawings, from which Plate 13 and the woodcuts, Figs. 1, 2, and 3, have been prepared.

6

Portsmouth's water supply, 1800–1860

Mary Hallett

INTRODUCTION: WATER SUPPLY ON PORTSEA ISLAND BEFORE 1800

All settlements need access to water for domestic, agricultural, industrial, or public use. Whenever there is an inadequate natural supply of water at hand for any of these needs, it has to be supplemented. Waterworks were among the first and technically most demanding mechanical and civil engineering projects. The earliest ones were for irrigation: water was lifted into canals either by manual devices like the Egyptian 'shaduf', or by mechanical ones such as the Persian 'saquieh' or waterwheel. Ancient Rome, however, was one of a number of contemporary cities with a public water supply. Its water was brought in aqueducts up to sixty miles long from lakes in the surrounding hills to public fountains in the streets.

Until the nineteenth century Portsmouth obtained fresh water from springs on Portsea Island, either within the town walls, or in the adjacent common fields. In the sixteenth century the chief source was from the springs on Four House Green where Henry VII built his naval brewhouses. By the eighteenth century there were also public wells or pumps in all of the main streets, in addition to an unknown number of private ones.

Existing supplies were maintained by keeping wells and pumps in good repair and by regulating the water carters' prices. While all had access to the common pumps, some people were neither willing nor able to collect their water themselves. Even those with their own wells often preferred an alternative supply for drinking. They were supplied by the water carters, whose prices in 1690 were fixed at 2d a half-hogshead. (One hogshead = 54 gallons).

Although these regulations remained the same, they applied to very different situations. Over time the population grew, but in irregular spurts, chiefly during wartime, when the number of jobs in the Dockyard sharply increased. In the sixteenth century, Camden noted of Portsmouth that

> 'in war-time it is much frequented, at other times scarce at all; the inhabitants being more attentive to war and navigation than to trade'.

Increases in the demand for water from the enlarged population were met traditionally by sinking new wells. The first proposal for a new service for the inhabitants came in 1694, when two gentlemen, Richard Barrey of London and George Sorocold of Derby, obtained permission from the Corporation to operate a waterworks. From known springs in White Swan Field, opposite the present inn of that name in Commercial Road, they intended to lay on the town's first piped water supply.[1]

Sorocold has been described as 'the most eminent British engineer of his day'. Certainly he was a pioneer of waterworks technology. He developed a wheel for lifting water into a conduit. Unlike earlier ones, his wheels rose and fell with the level of the stream that turned them. They were installed at Norwich, Leeds and Exeter, all places with flowing water. Nothing seems to have come of his plans for Portsmouth, although he paid the rent for a site until 1709.[2] Judging by the source of the water this is hardly surprising, because White Swan Field boasted below-surface springs rather than streams. There is no evidence that he intended to construct a steam pumping-engine, similar to those developed by Savery and Newcomen.

Yet despite periodic shortages, there was no immediate effort to devise a better supply. A contemporary local writer observed that the inhabitants

> '. . . are badly supplied with water, having none but what partakes of a saline quality; and even this is so scarce that were it not for showers of rain, which is most industriously catched by everyone, people could not possibly subsist. As it is, if their constitutions were not of a very peculiar make, they would long before now have attempted a remedy for so great an evil; but when such a thing is proposed, they wisely answer that their fathers and grandfathers contented themselves with what water the town afforded and why should they irreverently quit the road of their progenitors, not considering what danger they are in of a conflagration, especially at low tide; and yet by such sagacious reasoning they are yet content to pay, one house with another, 40s per annum for so wretched a supply of this commodity'.[3]

Contemporary accounts seem to illustrate the existence of not one but two separate water problems in the town. Most commentators concentrated on the probable absolute shortage of water, when population growth strained existing supplies. This was a temporary problem, usually solved by digging more wells. The pamphlet quoted above hinted at another and less tractable problem—that of annual relative shortages caused by the seasonal fluctuations of well yields. The town was supplied mostly by shallow-wells, sunk into the porous sand or gravel strata and fed by surface drainage. Yields from these wells tended to fluctuate with the rainfall and to go dry during the spring and summer months. If this inconvenience was regarded as an inevitable annual

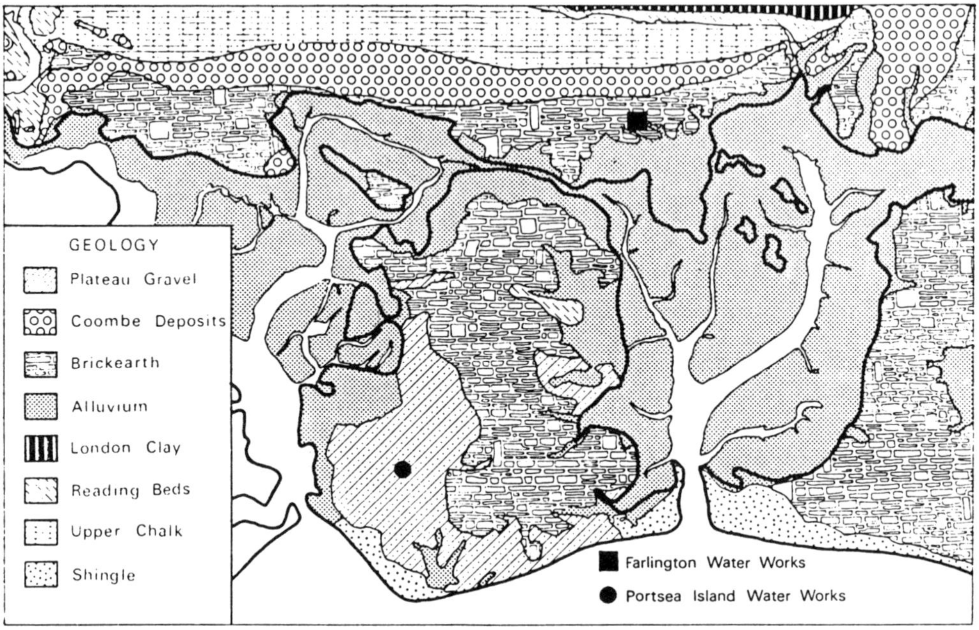

Geology of Portsea Island, showing sites of the two waterworks.
Drawn by the Cartographic Unit, Portsmouth Polytechnic.

occurrence, it may explain the conservative inhabitants' failure to seek a remedy.

Portsmouth continued to be supplied by wells and pumps, nominally regulated by the Corporation, throughout the seventeenth and eighteenth centuries. The sole change was administrative: when the various Paving Acts relating to Portsmouth and Portsea were passed between 1764 and 1792 the Commissioners assumed responsibility for the public water supplies.[4] Although the Commissioners initiated no changes in the methods of water supply, their records contained brief references to an intended Waterworks Bill in January 1795. Unfortunately there is no other information about this petition, which apparently never reached the House of Commons. It was not until 1809 that promoters availed themselves of Acts of Parliament for the construction of waterworks.

ORIGINS OF THE PORTSMOUTH WATERWORKS

The first Act of Parliament to provide a piped-water supply for Portsmouth was passed in 1741.[5] It permitted the use not only of a new means of conveying water, but also of a new source, from springs on the mainland some miles from the town. Although Colonel Smith, Lord of the Manor of Farlington, who had obtained the Act, died before he could carry out his project, its powers passed to Peter Taylor (M.P. for Portsmouth, 1774–7) who began by digging

> 'an immense well in Crookham Copse near Purbrook, and then carrying an archway of brick through Portsdown Hill above the level of Portsmouth, anticipating to meet with springs, the flow of which might readily be

conveyed with pipes, but after an immense expense of mining and brickwork through the whole bed of chalk, till the workmen emerged at an opening behind Farlington Church, not a drop of water was met with, and the speculation failed.'[6]

Shortly afterwards the Dockyard was busier than ever, with work created by the Napoleonic wars. Portsea's accompanying transformation from 'a solitary waste to a busy town' was reflected in the first national censuses:

	Portsmouth	Portsea
Population		
1801	7,839	24,327
1811	7,103	31,365
% Change in population		
1801-1811	—9	+29

Despite the town's population increase, there is no evidence of an absolute shortage of water. Nevertheless it was during this period that waterworks were first constructed in Portsmouth. Though it had long been technically possible to provide a piped-water supply, it had rarely been economically viable before the early nineteenth century, when both reliable pumping-engines and durable, standardised cast-iron pipes became available. Almost as many waterworks were statutorily incorporated in Britain between 1801 and 1810 as had been throughout the eighteenth century. In Portsmouth, the growth in numbers and prosperity, at least of business and professional men, might have been expected to provide both the capital and market for a convenient new service.

By 1808, there was little disagreement about the desirability of a waterworks for Portsmouth—the problem was rather who should provide it. Two groups of promoters emerged, both with apparently good claims to the confidence of the inhabitants. They were each led by men possessing not merely general engineering skill, but also the more particular experience of setting-up waterworks in London. The importance of an engineer's role can hardly be over-emphasised. It was confidence in his judgement that encouraged prospective investors to supply the necessary financial resources for the undertaking. He advised on the source of supply, the site of the works, and the technical problems of distribution. Any mistakes were apt to be expensive, and, as it happened, problems originated in the doubtful quality of the engineering advice received.

A public meeting in September 1808 rejected Ralph and Barrodall Dodd's proposal to pipe water from Farlington for £60,000, in favour of William Nicholson's plan to supply the town from the White Swan Field Springs for £32,000. In addition to its apparent cheapness, another merit of this scheme was its determination to be a predominantly local concern. Its subscription list was closed within twenty-four hours, a fair indication of the initial enthusiasm for the project. Over 75 per cent of its intended proprietors lived in Portsmouth:

PORTSEA ISLAND WATERWORKS COMPANY—SOURCES OF CAPITAL, 1809[7]

PORTSMOUTH		
Gentlemen	13.5%	
Professional Men	19.5%	
Manufacturers	17.3%	
Distributors	27.0%	
Total Portsmouth		77.3%
LONDON		
Miscellaneous		22.7%
		100.0%

William Nicholson, 1753–1815.
Reproduced by courtesy of the British Museum.

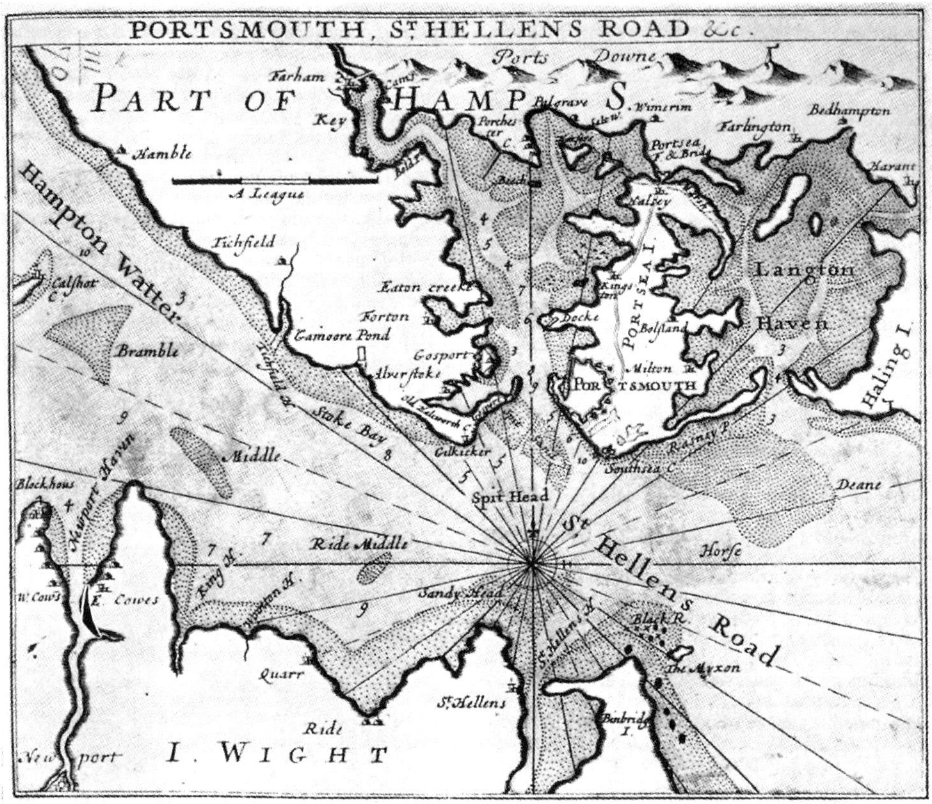

Portsmouth, St. Helen's Road, c. 1765.
Reproduced by courtesy of the Trustees of the British Museum.

The chief economic attraction of the site was its proximity to both Portsmouth and Portsea, obviating the need for heavy expenditure on expensive mains. Its water, however, proved rather less of an asset. Because it was on Portsea Island, this source suffered the same disadvantages as other immediate supplies, namely fluctuating yield and doubtful quality. The proposed well was rather an infiltration pit, about sixty feet across, fed from underlying sand and gravel beds, and liable to contamination by sewage. At this stage there appears to have been no debate about the water's quality.

However Portsmouth was destined to have not one but two waterworks, for the Dodds refused to succumb to the fate ordained for them by the Mayor's meeting. Instead they obtained the assignment of Colonel Smith's Act of 1741, intending to pipe water from springs in Farlington Marshes. The capital for this project came from gentlemen and businessmen in London and the Home Counties, many of whom had recently invested in similar undertakings there. (The ease which which both concerns acquired capital illustrates the ready availability of funds for investment at this time. It may not be very surprising that there was a surplus of such funds in Portsmouth, where there were few other opportunities for the investment of war profits. The influx of capital from London, however, is to be noted as one more example of the role of the metropolis in provincial economic development.)

If there is no evidence of further public discussion about the wisdom of constructing two waterworks for Portsmouth, each group of promoters certainly attempted to discourage the other from continuing, although neither had much room for manoeuvre. Unfortunately for the Farlington group, the 1741 Act expressly allowed other promoters to provide water from other sources. With other objections overruled, the Portsea Island Waterworks Company was duly incorporated in May 1809.[8] Their engineer then published a pamphlet[9] 'proving' that there was insufficient good water at Farlington to supply the town, though his criticisms would have been better applied to his own company's source. Nevertheless it is of interest to note Nicholson's belief that there was no absolute shortage of water:

> 'Portsmouth is at present supplied with good water, carried by carts from wells in the vicinity; some of which are so productive as singly to afford a quantity equal to one-fifth of the whole demand'.

It therefore seems that piped-water was intended to be sold more as a service than as a commodity, with the promoters doubtless hoping it would replace existing supplies. The two companies then raced to be first in supply, and perhaps to corner the piped-water market.

CONSTRUCTION OF THE PORTSMOUTH WATERWORKS, 1808–11

The Farlington Waterworks were designed by the Dodds, who then handed over to the resident engineer, George Buck, appointed to supervise its construction. A 34-horsepower steam engine* costing £2,000 was ordered from Boulton and Watt's Soho factory. It was brought by canal to Paddington where it was loaded on to a ship for Portsmouth. From the Ordnance Wharf there it was transported to Farlington by road. Some months later one of Boulton and Watt's mill-wrights came, as was customary, to supervise the fitting together of the engine.

Having received quotations from manufacturers of cast-iron pipes in Scotland, the North, the Midlands and South Wales, the Farlington Company purchased most of its pipes from the Butterley Iron Works near Derby, which charged the lowest prices delivered to Portsmouth. It has been estimated that the Farlington Company spent over £40,000 on 28 miles of main and service pipe: the Portsea Island Company used only 9½ miles, at an estimated cost of £9,600. The average cost per mile of pipes was thus £1,500 for the Farlington Company, but only £1,000 for their rivals. This difference is accounted for by the higher percentage of main pipes used by the Farlington Company:

	PIPES USED	
Company	*Mains* (7–15 *in. diam.*)	*Service* (2–6 *in. diam.*)
Farlington	34%	66%
Portsea Island	14%	86%

It needed not only more but larger mains to provide a sufficient head of water. On the other hand, the service pipe figures give some indication of the relative sizes of their prospective markets: less than 14,000 yards were laid by the Portsea Works, but 35,000 yards by the Farlington Company.

Bricks, clay, timber, ironmongery, lime and tiles were bought locally. The most expensive item was the bricks needed not only for the engine-house, workshop

* Beam pumping engine; 36-inch diameter cylinder; two boilers; crank motion; jackhead pump 17 inches in diameter, with an 8-foot stroke, and two buckets and clacks mounted.

and outbuildings, but also to line the reservoirs. Over one million were used at a cost of £3,700.

Various groups contracted to build the works, which involved a number of tasks requiring different degrees of skill. The millwrights were the expert mechanics who designed the patterns for the pipe connections and fitted the cocks and fire-plugs. So indispensable were they that they cost the Company not only their wages but also the price of improved accommodation, to induce them to stay at Farlington. While the bricklayers built the engine-house on the Marshes, a gang of some eighty labourers excavated three reservoirs near by. The water was to be pumped from the two collecting basins on the Marshes into the service reservoir on Portsdown Hill, whence it would flow to the town by gravity.

This gang was then employed, but not always fully occupied, on pipe-laying from 1809 to 1813. Absenteeism was one problem. At Christmas there were predictable disagreements about the acceptable duration of the festivities with 'strong beer remaining too long in men's heads which prevented their returning to work as soon as could have (been) wished'. Among the summer's distractions was the Portsdown Fair, all too conveniently sited just up the hill. There were other occasions when the labourers chose to work elsewhere, albeit temporarily, especially at harvest time. The other problem was the difficulty of securing delivery of sufficient supplies of pipes to keep the gang busy. In June 1810, with one of the few references to the current political situation, the Butterley Works apologised for delays, '. . . through the Local Militia having lately been embodied, in which many of our Moulders Serve'. Another time a furnace collapsed, halting production for several weeks. But most delays were caused by the shippers' dilatoriness. Yet by March 1811 Harrison's gang had driven over five miles of mains, from the works along the turnpike road to Portsbridge and down the London road towards the town.

When the Farlington Company began driving* pipes in the towns, they met a further though not unexpected problem, namely the Paving Commissioners' disapproval. The ensuing arguments were essentially disagreements about legal rights, and hinged upon interpretation of the relevant statutes. As the Farlington Company began laying pipes in Portsmouth in August 1810 it was from the Paving Commissioners there that the first threat of trouble came for Buck:

> 'I this morning went down to Portsmouth with a gang of men and set them to work in laying down pipes in the High Street, they had not been at work long, before several of the Commissioners came, and in very angry and peremptory language desired us to desist: but I of course kept the men at their work; however in about an hour afterwards they brought four men with shovels who, as fast as we opened the trench, filled it up again, and put an entire stop to our progress; I immediately went in search of Mr. Spicer [the Mayor] to request his interference, but he refused to act in the business, not knowing how far he was justified in so doing—His answer was, that if Councel [sic] were hard on the subject and it was determined that we have sufficient Authority to go on, he would then afford us his Assistance as chief Magistrate—The Commissioners have done this by virtue of a supposed authority, contained in a clause of the Paving Act, whereby any person is prohibited from making any *alteration* in the pavement of the streets without the consent of seven or more of them—From the nature of the Clause it is evident it does not extend to us as it is not our intention to make any alteration whatever'.

The Commissioners were then shown Colonel Smith's Act, with its clause allowing promoters to lay pipes in the towns provided that they repaired any street damage.

There was no street-fighting with the Portsea trustees, who embarked instead on a protracted legal wrangle about the assignment of the 1741 Act to the current promoters. They were fined the statutory 40/- for pavement-breaking at Winchester Assizes in March 1811, but this verdict was reversed and judgement given for the Farlington Company in the Court of King's Bench in the following year.

The Portsea Island Water Company had far less trouble in establishing its right to lay pipes in the towns, because the Improvement Commissioners had agreed to the relevant clauses when the water company's incorporation bill was drafted. They castigated both companies alike, however, for leaving the roads littered with equipment or poorly repaired. The Farlington Company was censured because workmen '. . . neglect to duly Light, Watch and Guard such Excavations to the great annoyance of the inhabitants'. But a few days earlier the Portsea Island Company had received a terse letter about the 'shameful and disgraceful State of the Road and Pavement in High St.', with similar instructions to make immediate repairs.

Meanwhile the Portsea Island Company was similarly occupied building its works. Unfortunately the records have disappeared, but it is possible to piece together an account from the Farlington Company records and local newspapers. By 1811, the Portsea Island Company directors had called up £25,000 of their nominal capital, early extended to £40,000. Of

* Technical term for *laying* pipes.

9

Map of Portsea Island, 1810, from
1" to 1 mile Ordnance Survey Map of Hampshire.
Photograph by the Portsmouth City Engineer's Dept.

this it is known that they spent £1,900 on land in White Swan Field purchased from the Corporation, £3,000 on fees to Nicholson the consulting engineer, and £3,000 on steam engines, also from Boulton and Watt. As this company intended to supply customers directly from its well, two 20-horsepower engines were needed; one to lift the water, the other to force it through the mains.

Each concern was predictably most interested in the other's progress. The Farlington Company's London directors received vivid accounts from Buck, whose visits to White Swan Field often coincided with some mishap:

> 'On Thursday last a serious accident befell Nicholson's well, about an hour before I was there—while the miners were at work deepening the well, the pressure of water and surrounding earth drove in the Curb until both sides met, and entirely prevented further progress. Since which, the earth surrounding the top of the well and contained within the Walls of the building has sunk downwards about four feet perpendicularly and sideways at the same time, leaving the foundations of the Engine House without support, in consequence of which they have been obliged to prop up the walls—Since the above occurred, no works has been done, but the entrances to the House are nailed up to prevent the damage from being known'.

As a result of other similar setbacks, the local company became rather sensitive about the publicity of its misfortunes—one of the disadvantages of being so close to the town.

Despite these problems the Portsea Island Company was the first to go into service, on 29th April 1811. At Farlington completion was hindered by delays in pipe delivery. Some five months later, on 24th September, they too were ready to supply Portsmouth, although their first attempt to pump water through the mains proved something of an anti-climax:

> '... the water discharged from the waste Water Pipe in the Government Ditch with very great force and numbers of the Inhabitants flocked out to see the Farlington Water and was in hopes of being able to have sent clear water into the town.... But from the short time that the Engine had been at work, it appeared that most of the Joints of the Main between Cosham and the Road leading down to the Engine House were leaky to such a degree so as to prevent any further working of the Engine until the whole of them are repaired which will take several days.... No such thing ought to have happened with lead joints if proper Attention had been paid to the execution of the work'.

In contrast, the local newspaper report was typically prosaic, merely noting the arrival of the water, though it did refer to the company 'throwing their water in the moats' prior to the official supply. The service to Portsea was started on 3rd October. Thus three years after the waterworks had been projected the town had two concerns eager to serve those prepared to pay for piped-water.

According to Nicholson the average daily demand for water per head for all uses including waste was twenty gallons. Multiplying this figure by the decennial population returns shows that the estimated daily demand for water in Portsmouth rose from 660,000 gallons in 1811 to 1,440,000 gallons in 1851. The Farlington Company's 34-horsepower engine worked fourteen strokes per minute, lifting fifty gallons at each stroke. Its maximum daily output was therefore approximately one million gallons. The original wells in White Swan Field were reputed to yield ten or eleven hogsheads per minute, or between 778,000 and 855,000 gallons in twenty-four hours. Taking relative engine capacity into consideration, it is believed that this company's 20-horsepower engine could have raised about 800,000 gallons per day. The Water Companies' total daily maximum possible output was therefore approximately 1,800,000 gallons. Together they had ample capacity to meet all the town's requirements for water until the 1850's:

Comparison between the Estimated Total Demand for water and the Maximum Potential Piped Supply, in Portsmouth, 1811–1861

Year	Population of Portsmouth (000)	Estimated daily town demand for water, at 20 gallons per head (000) gallons)	Maximum Potential Daily Piped Supply (000 gallons)		Possible Excess Piped Supply (000 gallons)
			Farlington Company	Portsea Island Company	
1811	42	840	1,000	800	960
1821	47	940	1,000	800	860
1831	50	1,000	1,000	800	800
1841	53	1,060	1,000	800	740
1851	72	1,440	1,000	800	360
1861	95	1,900	1,000	800	—100

In addition, water supplies from carters and private and public wells continued to be available.

WATERWORKS COMPETITION FOR CUSTOMERS, 1811–1815

Both companies set out to sell the particular virtues of their piped water: its fitness for every purpose and its convenience. In common with most contemporary waterworks, however, the Portsmouth companies offered only an intermittent supply, pumping water into the mains for just a few hours a day. The potential market for piped-water had been physically restricted by the companies' decisions to drive their pipes only in

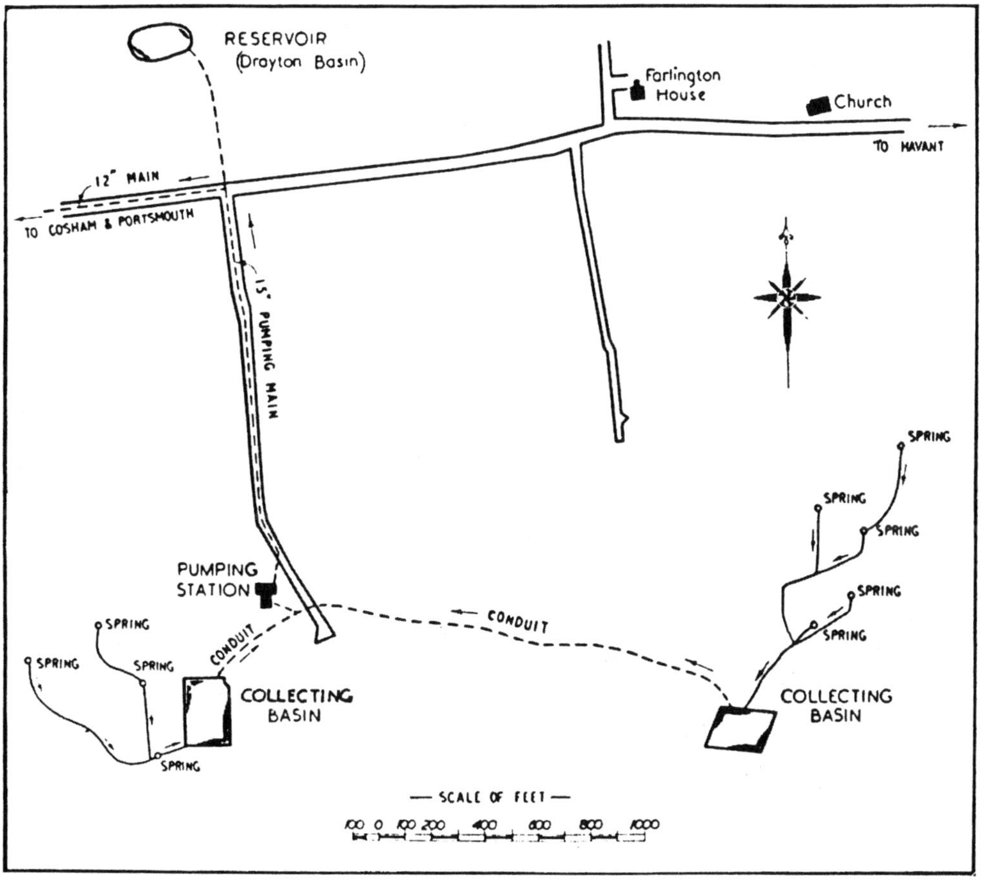

Diagram of site of Farlington Waterworks.

Reproduced from D. Halton Thompson, *History of Portsmouth and Gosport Water Supply*, with permission of the Directors of the Portsmouth Water Company.

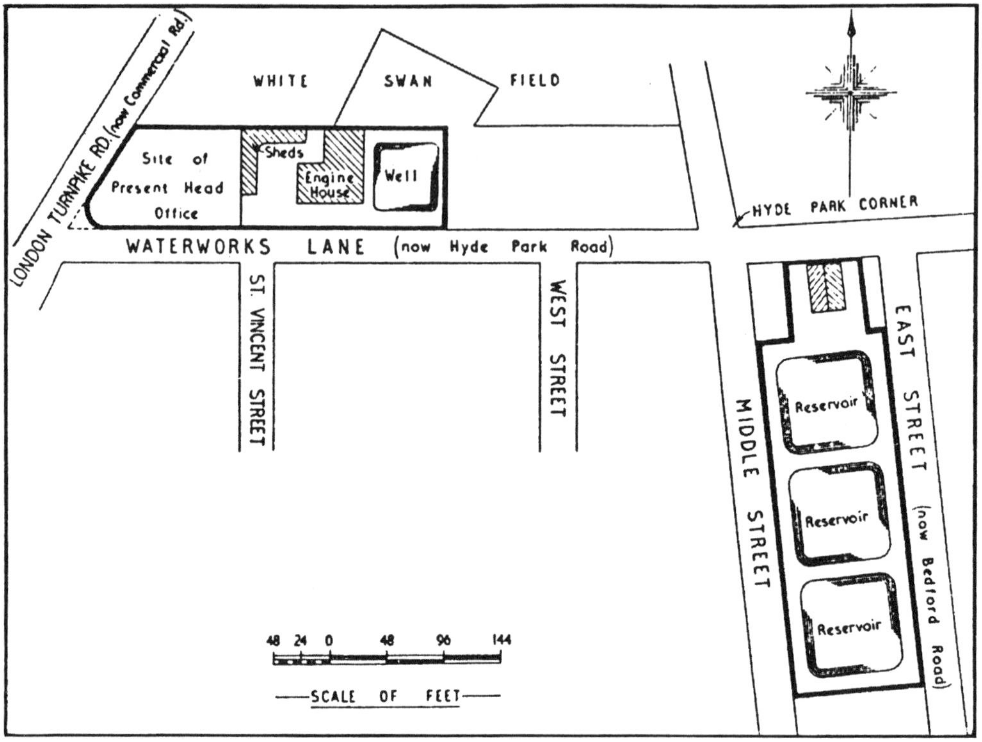

Diagram of site of Portsea Island Waterworks.

Reproduced from D. Halton Thompson, *History of Portsmouth and Gosport Water Supply*, with permission of the Directors of the Portsmouth Water Company.

the same principal streets. Each firm was therefore forced to compete not only with traditional supplies but also with the other company.

As well as providing a free standpipe for the poor, the Portsea Company offered a major concession to prospective consumers: they would bear the cost of the lead pipe which connected the houses to the mains, until March 1812. This was a clever plan because it reduced one of the chief drawbacks to having a piped-supply, the capital cost to the customer. (The price paid for the water depended on the rateable value of the property. It ranged officially from 16 shillings a year for houses rated at £9 and under to 38 shillings for those at £39; charges for property rated at £40 and above were negotiated individually. Thanks to the competition for customers, the actual prices paid were significantly lower.) The Farlington Company unwillingly matched the rival's free lead-pipe offer. Having exhausted the original capital, but with no rent yet received, neither company could afford such generosity. The concession was withdrawn in February 1812.

The situation highlights perhaps the most firmly held conviction of both companies, that customers once obtained would remain with that company. This partly explains the desire of each concern to be first in supply. Yet the Farlington Company's later start was no disadvantage. Though its rivals had had five months to acquire useful experience of running their works, they also had a number of dissatisfied customers awaiting with interest the start of the Farlington supply. Once both companies were operating, the mere provision of piped-water was insufficient to retain customers, who soon learnt to compare quantity and quality. The Portsea Island Company was slow to appreciate the significance of this situation. Lower prices meant little without customers. A defensive letter appeared in the local paper, blaming the company's lack of favour on malicious rumour. The Farlington Company were more inclined to blame its competitors' 'illiberal and unwarrantable conduct' in attracting custom under false pretences. The chief problem was the poor quality of the Portsea Island Company's water. Because of the site of its well and the absence of a service reservoir there was a lot of sand in the water. In December 1811 the engines were stopped, so that over seventeen feet of sand could be removed from the well. After this they decided to invest in settling ponds. Early in 1812 it was announced that another £10,000 was to be raised, 'the expenses of the undertaking having so far exceeded the estimate'. Essential though these reservoirs were, the new shares were not taken up with alacrity. The proprietors' dilemma is easy to understand—£25,000 had already been raised to provide a service which was proving unsatisfactory. Now more funds were needed, with no guarantee that in the meantime the Farlington Company would not make further inroads into their present fickle market. Yet without settling-ponds the Portsea Island Waterworks could not function long. In the event 478 of the 800 shares were taken up, providing £5,975; the shortfall of £4,025 was raised on a mortgage. As some small compensation for this unpopular outlay, however, the Company were soon able to sell forty tons of washed sand to a local builder. The water was still occasionally unfit to use: one customer boiled some and then 'put it into a Clean Tea Pot prepared on purpose and after standing some time it came out of a deep Orange Colour'. Another complained of whitings 'turned red as bull's beef'. Though understandably infuriated by these reports, the Company aggravated matters by the treatment of customers who wanted to change to Farlington water, employing crude threats to ensure loyalty.

Farlington water was not of perfect quality either: it could contain mud or weeds, although one director suggested that the weeds improved the water. In opting for immediate cleansing, the management most sensibly pandered to the public's less scientific whims. Nevertheless other forms of pollution remained long unremedied. The manager noted quite casually in 1815 that, '. . . cattle gets into the Eastern Reservoir and thickens the water very much, it should be fenced off'. This happened when boundary fences were damaged: fencing the reservoirs was intended to double protection. Four years later these reservoirs were still unfenced, as a Captain Shearman found when he fell into the Drayton Basin on Portsdown Hill in endeavouring to extricate his dog. The *Hampshire Telegraph* pointed out that as 'there is no fence or railing round it, or any other means by ropes hanging into the water, to prevent such accidents, or to assist a person to get up the wall, . . . he must inevitably have perished, had not a game-keeper happily seen him fall in'.[10] Despite these incidents, many customers chose to pay the Farlington Company's higher charges, while the Portsea proprietors were forced to incur additional expense in attempts to render the water more palatable.

In contrast, the quantity of piped-water was neither a cause for complaint nor an immediate competitive issue—probably because both companies' capacity exceeded the market's demand. But the Farlington Company noted that its source was the more reliable. In common with other springs on Portsea Island, the local company's fluctuated seasonally. Their rivals

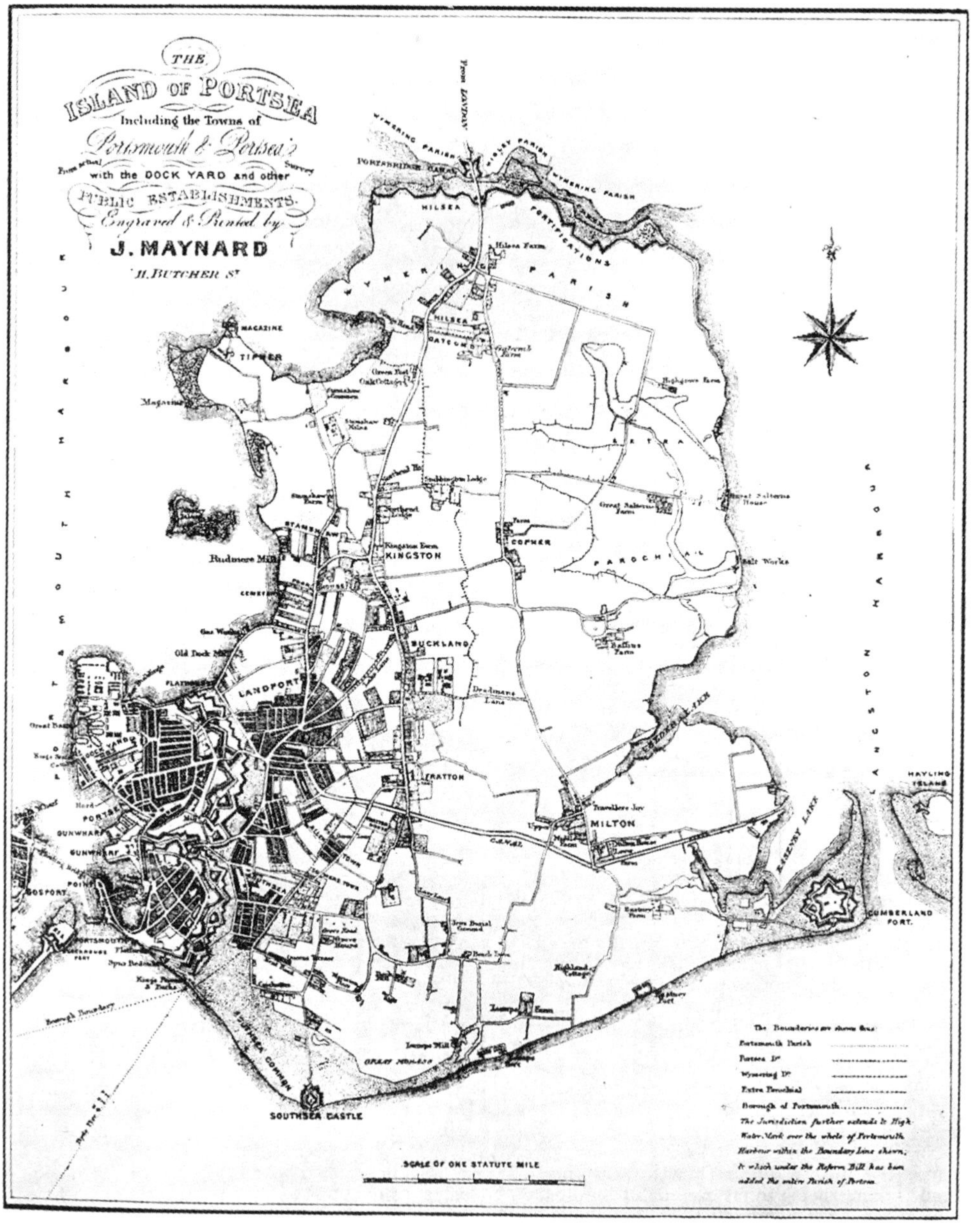

Maynard Map of Portsea Island, 1833.
From the City Record Office, Photograph by the Portsmouth City Engineer's Dept.

could expect to attract customers when, 'as the weather gets warmer, the springs in this Island do not increase in quantity nor improve much in quality'.

Details of how many customers were acquired are conspicuously absent from the records. One hundred and five houses were laid on in the first two months of service. An official report of 1850 suggested that in 1812 the Farlington Company was supplying only 500 out of the total of some 7,000 houses in the town.

The manoeuvres for customers must also be seen against the background of prosperity and population growth, largely engendered in Portsmouth by the Napoleonic Wars. Initially at least the Farlington Company found no real difficulty in attracting customers. It is suspected that there was some social prestige in being among the first consumers to have a piped-water supply. More interesting is the evidence that from the beginning it was not only the professional classes who were eager customers. Though the first rental lists show a surgeon and a solicitor, they are outnumbered by the modest traders, butchers, bakers, drapers, plumbers and carpenters. All these apparently could afford not only the price of the water, but also the capital cost of lead pipe and storage cisterns. These domestic consumers proved to be the most important customers for piped water. Each company had hoped to obtain a lucrative contract to supply the Dockyard, which however was still self-sufficient in the 1850s. The various barracks also had their own supplies, although the Ordnance and the Royal Marines took Farlington water for £30 and £40 per year respectively. Then in 1813 both companies acquired premises in Bath Square, to use as water-houses for supplying commercial shipping. From tanks there the water was siphoned through a leather hose into barrels which were taken aboard. Neither company earned much, for the trade was irregular and payment even more so, while the cost of rental collection by contractors was high.

After domestic consumers, the water companies' most valuable customers were the brewers. At least nine of them—Charles Barton, William Deacon, George Garrett, William Garrett, Edward Goodeve, Kempster Lipscombe, William Matthews, and Robert and William Stroud—invested in Portsea Island Company. Most of them also took that company's water, although they all had their own supplies. Then once again events played into the hands of the Farlington Company. Within days of the start of their supply Messrs. Pike's, the most important Portsea Island Company customer, had a whole brew spoiled by sandy water. The embarrassed management swiftly offered a reward of two hundred guineas for information about the person who had 'lately thrown some noisome or offensive matter into the water',[11] supplied by them. This attempt to lead suspicion away from the company convinced few. Pike's demanded the immediate removal of the local company's supply and reverted to the use of a well near the Halfway Houses. Their example was followed by the other brewers, until by February 1812 the only one still taking White Swan water was Stroud of Portsea.

Meanwhile the Farlington Company was having rather more success. Between 1811 and 1830, they secured the custom of more than half the brewers:

Total number of Brewers listed	16	100%
Breweries supplied by Farlington Water Company	9	56%
Breweries suppiied by Portsea Island Water Company	1	6%
Breweries relying on own water supply	6	38%

The Farlington Company's hard water had the advantage over that from the Island: for while the Portsea Island Company's soft water was more suitable for brewing porter hard water produced better ale. Most Portsmouth brewers were ale-men.

One other group did not purchase piped-water, but nevertheless benefited from the waterworks' existence. The Improvement Commissioners continued to supervise the town's public pumps, but also received a free piped-supply for fire-fighting from each company—though this was necessarily limited to the hours of service. As compensation, however, the Farlington Company manager proposed that the local fire insurance companies should make some contribution, but there is no record of the companies doing this.

By 1813 the Farlington Company's annual income was £2,800, and costs were £1,800. The first 6 per cent dividend was paid in 1814, not on the original capital of £60,000 but on a £15,000 loan raised in 1810 and converted into 300 £50 'New Shares' in 1813. Like its rival, this Company originally underestimated the cost of building its waterworks. But while the Farlington Company distributed only 1.2 per cent on the total capital, there is no record of the Portsea Island Company ever paying any dividend to the shareholders.

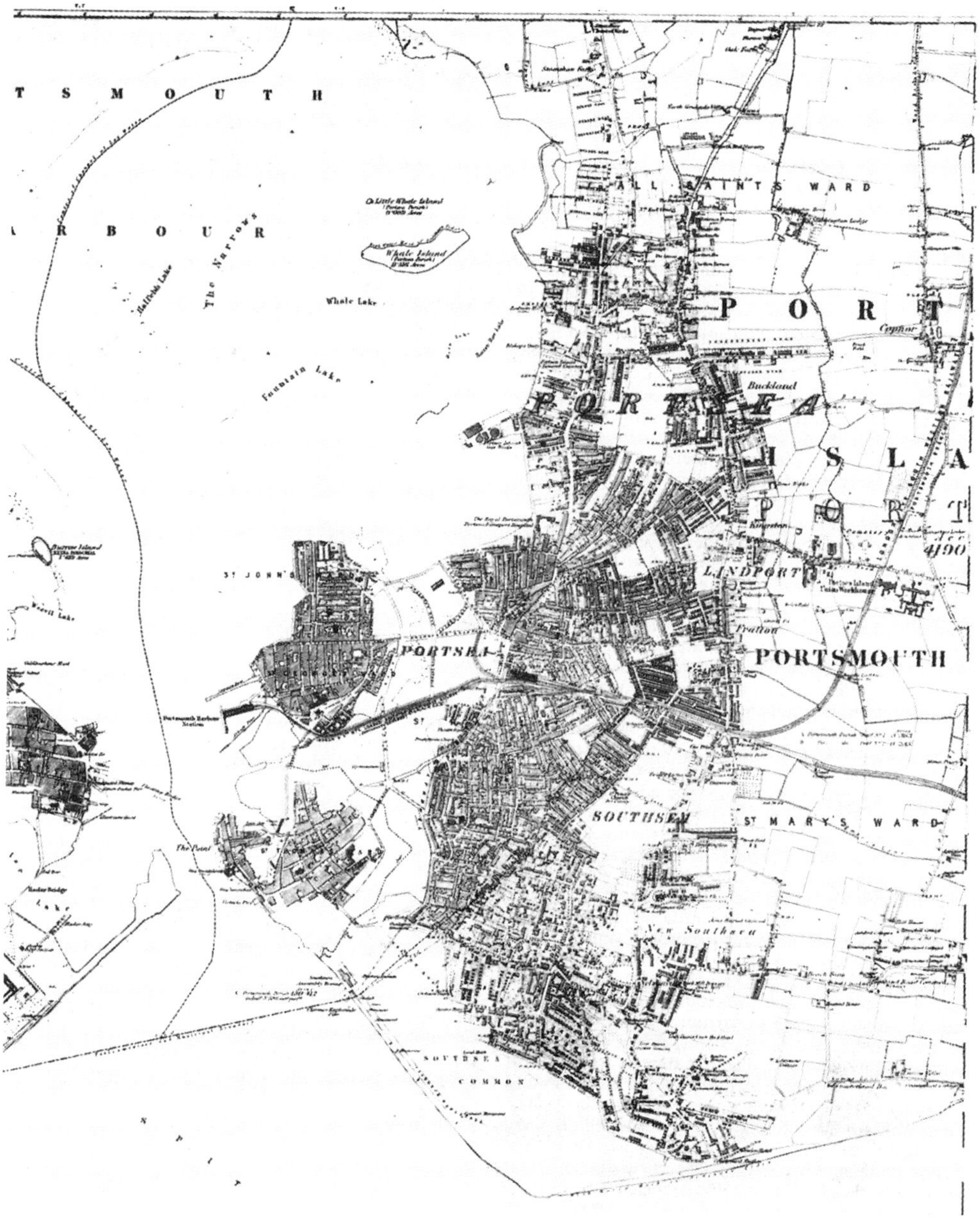

Map of Portsmouth and Portsea, 1870, from 6" to 1 mile Ordnance Survey Map.
Photograph by the Portsmouth City Engineer's Dept.

EFFECTS OF THE POST-WAR SLUMP AFTER 1815

By 1815 the Farlington Company was finding it difficult to secure prompt payment from consumers. This problem had begun even during the war prosperity, possibly because the fascination of the piped supply was diminishing: with no more price-cuts to counter flagging interest, some failed altogether to pay their rental. The Board believed that the main reason for the arrears was the local manager's 'insufficient exertion'. George Plummer soon decided to resign rather than 'accede to their liberal offer' of 10 per cent commission instead of his former £300 salary. That changing managers would not solve the revenue problems was soon apparent. Back in 1815 Plummer had commented on local attitudes to the recent peace: 'I believe many a poor Neighbour here would not inwardly Grieve if Boney should be the means of stirring up Another War on the Continent'. A year later this hope had vanished, and with it the town's recent prosperity. Once again conditions in the Dockyard largely dictated the town's economic fortune. Bellingham, the new manager, now reported that 'times are now so very much altered and the discharges from the Dockyard so great it strikes a gloom over the whole place'. During the war the Dockyard had employed more than 4,000 men, or 36 per cent of the estimated male labour force in a town of forty-two thousand. With peace came redundancies, until only some 2,000 were still employed at the Dockyard. Weekly wages there fell from £6,000 to £3,000.[12] Simultaneously the Admiralty chose to cut the wages of those still employed instead of ordering additional discharges, 'from knowing there is no existing means of employ out of the Dockyards'. The immediate effect was an increase in those receiving poor relief. Even in 1816, 600 people were in Portsea Poor House, with another 3,276 on out-relief, out of a district population of some 33,000.[13] With the contraction of employment and the diversion of a larger proportion of remaining income to relief a period of economic decline began. General loss of confidence was illustrated by the collapse, among other businesses, of Godwin's Bank in November 1818. One of the partners was Thomas Minchin, a proprietor of the Portsea Island Waterworks. A slackening of population growth also reflected the rundown:

> 'Peace had stopped the one industry of Portsmouth and with it the growth of its population, which only increased from 1811 and 1831 . . . an abnormally slow rate for an English town at that time'.[14]

William Cobbett recounted the new distress in a town where

> 'houses which let there for forty or fifty pounds a year each now let for three or four shillings a week each; and thousands perhaps, cannot be let at all to anyone capable of paying rent. There is an absolute tumbling down taking place, where, so lately, there were such "vast improvements" .'[15]

It is suspected that comparatively few of the water companies' customers were among the unemployed: lowly-paid dockyard men being improbable company tenants. But even those customers still employed could well be under pressure: it is easy to see how the volume of trading could fall, while the real and money burden of the poor rate increased, leaving lower net disposable income. Bellingham reported that growing numbers of tenants were applying for reductions in their water-rates, especially where there were now fewer tenants in houses supplied by the Company. The Board was quick to prune costs, ordering 6d per chaldron off one coal bill, because of the fall in price of horses, corn and labour. But it refused any tariff cuts, declaring that the tenants had previously been underrated. It was probably true that because of the two companies' competition many customers had not paid the full cost of the water. Even so, the commercial wisdom of effectively raising prices during a depression may be doubted, particularly when other sources were available.

Predictably, price became a more significant factor than quality. The Farlington Board recognised the necessity of proving that the Company's piped supply was a bargain. An annual rental of thirty shillings cost less than 1d per day, for a supply of thirty gallons. It was alleged that a similar quantity would cost 12d to 15d from water-carriers. The comparison is misleading: the carriers' charges were nearer an eighth of a penny per gallon than ½d; and the directors assumed that instead of a piped supply only one other source would be used. It seems more likely that when money was short water would be obtained from several sources, free from wells for washing and by limited purchases from carters for drinking. Also the piped water cost omits the initial capital cost of lead pipe and cisterns, which should be discounted over time. If the convenience of a piped supply were sacrificed it was possible to reduce the money cost of water.

The effects of the slump are shown in the Farlington Company's Account:

FARLINGTON WATERWORKS COMPANY
Summary of Accounts, 1813–1829[16]
(*All figures to nearest* £10)

Year	Calculated Total Revenue	Calculated Total Costs	Calculated Profits
1813	2,800	1,910	890
1814*	3,000	1,670	1,330
1815*	3,270	1,860	1,410
1816*	2,600	1,950	650
1817*	1,820	1,390	430
1818	1,340	1,100	240
1819	930	1,000	—70
1820	1,300	1,280	20
1821	800	1,140	—340
1822*	1,690	1,050	640
1823*	1,670	1,130	540
1824*	1,770	1,340	430
1825*	1,970	1,140	830
1826*	2,000	1,480	520
1827	2,160	1,700	460
1828*	2,310	1,270	1,040
1829*	2,510	1,570	940

* Dividend paid.

Domestic consumers thought to have been over 70 per cent in arrears by 1818 were largely responsible for the sharp falls in revenue. Even in 1829 income had not recovered to the 1815 peak. Simultaneously 70–80 per cent of the firm's costs were fixed or unavoidable, leaving little room for manoeuvre. Reserves were exhausted by a distribution of dividends.

After the war, therefore, expansion required the use of outside capital. That this was available locally is illustrated by the new enterprises which were floated. The Portsmouth and Arundel Canal Company was incorporated in 1817 with an initial capital of £125,000, of which at least £20,000 was reputed to come from Portsmouth people. It gained further powers to raise additional funds in 1822 and 1829. Also in 1822, the Portsmouth Gas Company was incorporated with a capital of £31,800.

The water company's performance in the post-war slump might help to explain why investors were attracted to fresh enterprises. But the fact that the Farlington Company never approached the market for funds also reflected a general policy of retrenchment. There is evidence, however, that the Company benefited modestly and indirectly from the canal's existence. People living close-by applied for Farlington Water, 'in consequence of the salt water from the canal having found its way into the wells in the Neighbourhood'.

The records imply that the Portsea Island Company was even more seriously affected by the slump. Whereas the Farlington Company chose passively to await the return of its customers, in 1815 and again in 1818 the other company proposed the positive measure of uniting the two concerns, to consolidate remaining revenue. In 1819 these were modified to suggest supplying the whole town with water from Farlington and converting the White Swan site into a gas works. Yet more ideas were discussed in 1825. That the Portsea Company took the initiative each time is perhaps some indication of its problems. Negotiations foundered each time because the Farlington Company refused to accept responsibility for the other's debts. The White Swan Works needed yet more resources to make the concern viable, yet the past record discouraged new investment. Interest on the existing £12,000 mortgage was considerably in arrears. In 1828 a new Act of Parliament[17] was obtained, intended to offer security to any new mortgagee: if interest was not repaid regularly, mortgagees could foreclose and force the sale of the concern. This power was invoked in August 1828, when a decretal Order in Chancery put the firm up for auction. Seeing the opportunity to remove the rival company, the Farlington Company authorised its solicitor to bid up to £4,000 for the works. At the sale he bid up to £5,100, but ironically in an attempt to force up the bidding, three of the mortgagees were themselves saddled with the works for £5,300. From the auction the mortgagees received 8s 8d in the £; the original shareholders nothing. Of £52,000 capital raised, £46,000 was written off in 1828, clear evidence of the cost of failure.

The new owners, banker George Grant, wine-merchant Edward Casher, and innkeeper Andrew Nance, wanted to negotiate with the Farlington Company. However, as the terms proposed by these three were deemed 'totally inexpedient to treat upon', they decided to run the works themselves. Though no water company records survive for the 1830's, the two works appear to have competed undramatically, again taking their cue from the economic state of the town. Census figures suggest that Portsmouth continued to stagnate after a partial recovery from the slump: its population rose from 50,389 in 1831 to 53,032 in 1841, a mere 5.2 per cent increase, smaller even than the 9.6 per cent change between 1821 and 1831.

PROBLEMS OF A UNITED WATER COMPANY AFTER 1840

After thirty years' competition, the two water companies ceased to operate independently on 25th March, 1840, when the United Portsmouth, Portsea and Farlington Waterworks Company was formed. Casher and Nance received £5,125 in cash, plus one-sixth of the new company's shares, for their Portsea Island works. This United Company had a monopoly of piped-water supply in Portsmouth. It was also under new management, for the former provincial agents retired. They were replaced by Henry Bottrell, whose son Richard became Company Secretary. These two proposed changes guaranteed to improve the Company's efficiency. They decided to reduce the amount of arrears, both by threatening to cut off the supply and by exploiting the inconvenience otherwise occasioned by dry weather: Henry Bottrell noted how the most long-standing arrears were those of the rich, adding that there was 'no such poverty as had long been represented'. They also managed to cut the cost of collecting the rentals by inducing the Clerk to accept this additional task for a slightly higher salary.

The new company planned to supply the town wholly from Farlington. More immediately controversial was the decision to work the engine six days a week, but to supply each district only on alternate days. Though this arrangement need not have affected the total quantity of water available, it was argued that the new service was inconvenient. Many customers could not store sufficient stocks to cover the days when the Company would not supply them afresh. It was also doubted whether the water would be fit to drink after several days in an open tank. Given the Companies' past records, allusions to previously splendid service were unconvincing. Nevertheless it was generally agreed that the new company was not providing a satisfactory service, even if this opinion was not aided by public ignorance of the legal powers and responsibilities of the waterworks. Far from granting concessions, the Company raised brewing and military contract prices from January 1841. But to meet the charge that 'not only are the rents raised by the Waterworks Company, but the inhabitants have suffered most cruelly from the want of a sufficient supply of water', the Improvement Commissioners resolved to erect more public pumps. This certainly solved the immediate problem of securing supplies for the increasing population. It was not until later that the dangers to health of drawing water from the midst of increasingly dense settlement became apparent.

By September 1840 the Farlington works were supplying seven-eighths of the total output of piped-water, serving all of Portsea and most of Portsmouth and Southsea. Henry Bottrell planned to 'stop Mr. Casher's Engine for ever in 10–14 days'. Any hopes that these arrangements would provide a service to belie all the criticism were short-lived. On 13th November, 1840 the reclaimed land at Farlington was flooded and the sea-wall gave way. As the springs in the Marshes were inundated with salt water, the waterworks were immediately put out of action. Though winter gales had often caused floods, the sea-wall had always held. Now as Casher said, 'You could have gone in a boat from Portsmouth Harbour to the waterworks'.

A London engineer, William Tonkin, had been commissioned to inspect the United Company's works. His major criticism, weighted by the recent flood, was of the Farlington site:

> 'The Engine was fixed in the wrong place, instead of requiring a Mound it should have been fixed further North, on the hard ground near the turnpike road, at the foot of the [Portsdown] Hill; with a culvert opened parallel to it, there is not a doubt but there would be sufficient water discovered to supply the tenantry. . . . Such a position would have been a security from all inundation of the sea.'

Yet it was surely not surprising that the original promoting engineers should have chosen these known springs in the Marshes, in view of Peter Taylor's failure to find water near Portsdown Hill thirty years before. It is possible, however, to criticise the management for paying insufficient attention to the vulnerability of the site. There was no attempt either to raise the sea-wall or to ensure routine maintenance. Though in 1840 the water flooded in from surrounding land, the only part of the sea-wall to give way belonged to the Company.

The first repairs were unsuccessful, for in March 1841 the sea broke through again. One difficulty was that timber piles could be driven into the hole only at neap tide, which occurred but once a fortnight. Materials were carted in, because barge costs were too high. The immense efforts to seal the breach swiftly are illustrated by the fate of one unfortunate small carter who '. . . loaded his cart so much that in coming down the [Portsdown] Hill, his Horse sunk under the weight and broke its back, of which it died on the spot, the shafts of his Cart were also broken by the fall, and injury done . . . to the amount of over £7'.

Forty or fifty labourers then packed in two thousand tons of chalk, clay and shingle. Meanwhile an emergency service was run from the Portsea Island works, although by June 1841 'there were four stoppages at the Engines at Mr. Casher's works in eight days . . . Friday afternoon boiler "leaked like a sieve" and towards evening two small holes on each side burst into one large one, let out the whole of the water and extinguished the fire . . . On Monday they were again able to start, but by no means satisfactorily'. On 19th August, 1841, however, the town was altogether without piped-water, for the Portsea Island boiler finally disintegrated before the Farlington springs could be used. The service was resumed six days later when for the first time since November 1840 some water was pumped from Farlington.

At midsummer 1841 the United Company was nearly £10,000 in debt:

	£
Amount owed for	
Portsea Island Works	3,125
Sea-wall repairs	5,145
New Engine for Portsea Works	1,400
Miscellaneous fuel and maintenance bills	300
	9,970

£5,700 was borrowed from eight shareholders after the service was resumed. Revenue was then expected to recover and any profit be diverted to repayment, despite Casher's fear that 'If we continue to have so many out goings . . . we shall never have any income and thereby cheat Sir Robert Peel'. Yet by 1844 the average annual surplus was put at £2,300, thanks probably to staff economies and a technical improvement in the new engine which reduced fuel costs per unit of output by 40 per cent. By October 1844, a little more than three years after the flood, all debts were repaid except the £3,125 owed to Casher and Nance—there is no record that they ever received this money.

During these years too, Richard Bottrell was accused of conspiring with Tonkin to share the profits on the purchase of a new engine for the Portsea works, and attempting to defraud the Company by issuing false share certificates. There was insufficient evidence for legal action, but both Bottrells left the Company.

In 1840 the United Company's capital had been reorganised in 840 £50 shares. It first paid a dividend of 4 per cent in 1846. An average of 5 per cent was then paid every year until 1859, when a final distribution of 12 per cent was made before the concern was taken over by the Borough of Portsmouth Water Company.

They were supplying approximately 4,000 houses in 1845, and about 5,000 houses in 1850, out of a total of some 12,800 in 1851. Far from exploiting either a monopoly of piped supply or the widening market, the Company settled for the steady profits from a low level of activity. If this sounds unenterprising, it is not inconsistent with a wish neither to incur further substantial debts nor to risk additional capital in a hitherto precarious concern.

Judging by the difficulty of ensuring a quorum at annual general meetings after 1846, little interest was shown by the proprietors. Thirty-five years after the Company's foundation few of the original shareholders were alive. Their holdings had either been inherited or sold to new proprietors, most of whom lived away from Portsmouth. Any charge of inertia could, however, be countered by the argument that this was irrelevant while profits were being made. Effective control of the Company had passed to the two proprietors with large holdings and homes in Portsmouth, Casher and Nance, the last owners of the Portsea Island Waterworks. Casher's early dominance of the United Company now suggests that his business ability had prompted the generous terms at the Union in 1840.[18]

ECONOMIC EXPANSION IN PORTSMOUTH AND PORTSEA IN THE 1840's

The formation of the United Company's monopoly of piped-water coincided with renewed prosperity in the town. After twenty-five years of relative stagnation the population again began to increase rapidly. Between 1821 and 1841 the total town population had increased by less than 8,000. In the 1840's it grew by almost 20,000, from 53,000 to 72,000, a growth rate of 36 per cent which exceeded that during the Napoleonic wars. During the 1850's it increased by a further 23,000, or 31 per cent, to 95,000. Again the major reason for the town's expansion appears to have been a new increase in employment at the Dockyard, this time during peace. The stimulus to naval shipbuilding came not from outright hostilities, but from the fear of a French invasion.[19] During the 1840's the Dockyard labour force expanded to some 6,000, or 30 per cent of the estimated total male labour force of the town. Between 1843 and 1848 the yard was extended by 15 acres to over 115 acres, in order to accommodate a steam-basin and four new docks. The New Buildings area of Portsea had to be cleared to provide part of the site.

These four acres cost the Admiralty £60,000, of which the Water Company claimed £1,553 11s 8d, in compensation for the loss of revenue from displaced customers!

If economic conditions improved in Portsmouth in the 1840's, however, social ones did not. Although the area of settlement was further extended beyond the limits of the old towns, its density continued to rise. It is clear from censuses that provision of additional houses lagged behind population growth:

Comparison between the rates of increase of population and of the number of inhabited houses, in Portsmouth, 1841–1861

<table>
<tr><th rowspan="2"></th><th colspan="2">Rate of Increase of Population</th><th colspan="2">Rate of Change in number of Inhabited Houses</th></tr>
<tr><th>1841–51</th><th>1851–61</th><th>1841–51</th><th>1851–61</th></tr>
<tr><td></td><td>%</td><td>%</td><td>%</td><td>%</td></tr>
<tr><td>Portsmouth Town</td><td>9.5</td><td>4.5</td><td>— 1.3</td><td>— 4.9</td></tr>
<tr><td>Portsea Town</td><td rowspan="3">42.3</td><td>12.5</td><td rowspan="3">+ 22.8</td><td>+ 1.0</td></tr>
<tr><td>Kingston</td><td>31.0</td><td>+ 25.4</td></tr>
<tr><td>Landport</td><td>55.0</td><td>+ 39.2</td></tr>
</table>

Although these figures could mask improvements in the quality of housing, it is believed that these were more than offset by the lack of other adequate amenities, chiefly paving, lighting—and water supply.

It is estimated that the total daily demand for water in Portsmouth rose from 1,060,000 gallons in 1841 to 1,440,000 gallons in 1851. During the same period, the United Company's output is calculated to have risen only from 130,000 to 170,000 gallons per day:

UNITED COMPANY
Estimated Supply of Water in 1840 and 1850

	1840	1850
Coal Consumption per week	2 tons	2½ tons
Output of Water per ton of coal	472,000 galls	472,000 galls
Output of Water per week	944,000 galls	1,180,000 galls
Average supply of water available per day	133,000 galls	167,500 galls

The Water Company was therefore meeting less than 10 per cent of the estimated total demand in 1850. The quality of service enjoyed by some customers was less good than even this modest level of activity might suggest: while a piped-supply was nominally available six days a week, those with cisterns above ground level obtained water only on the three days when the engine was worked.

With the Water Company catering for only a small percentage of the market, the town's water supply had to be supplemented by sources on Portsea Island. In the short run it was possible for the Improvement Commissioners to obtain a sufficient quantity by maintaining existing public wells and sinking additional ones. What now appears as the most disturbing water problem was the reliance on these sources when their quality was deteriorating. Indeed the very increase in the demand for Portsea Island water seems largely to have caused this deterioration, for there was an increase not only in the population, but also in the density of settlement. However, this settlement was built on porous strata of sand and gravel, and in the absence of proper drainage neighbouring wells were liable to pollution by sewage seeping through the subsoil.

Water on Portsea Island was thus not only uncertain in quantity but also poor in quality. In contrast, while the quantity of water pumped from Farlington was insufficient, its quality was good. The obvious answer was to increase the Water Company's capacity to meet the town's requirements. Yet during the 1840's no resources were ploughed back into the Company because the profits were distributed to the shareholders, nor were any outside funds either sought or proferred.

In contrast with the 1820's and early 1830's when few investment projects were initiated in the locality, a number of enterprises were floated from 1838 onwards, chiefly in transport. Locally sponsored projects included a Floating-Bridge Company (1838), the Camber Docks Improvement undertaken by the Corporation (1839), and the Portsmouth Harbour Pier Company (1846). On a less ambitious scale, the first horse-omnibus service was started in 1840. The first railway to Portsmouth was an extension, opened in 1847, of the Brighton and Chichester Railway. By then the Direct London to Portsmouth Railway Company had been incorporated, although the line was not opened until 1859. It is not suggested that all or even most of the necessary capital was raised in Portsmouth. Yet it seems likely that the bulk of the funds for the Floating Bridge, the Camber Improvement and the Harbour Pier, amounting in total to £70,000, were obtained locally. Another investment opportunity, though one difficult to quantify, which arose directly out of the town's renewed expansion, was in house building. According to censuses, between 1841 and 1851 some 2,900 new houses were finished, a 30 per cent increase in the total stock.

In the 1840's the inhabitants of Portsmouth had access to a variety of new enterprises in which to invest. Although railway investment in particular was highly speculative, this itself obviously appealed to many.

While there is no record of the local water company ever seeking funds, as a mature concern with an erratic dividend record it would have found difficulty in attracting them. Yet given the very existence and the experience of the Water Company there was little incentive for public bodies like the Improvement Commissioners or the Corporation to set up an alternative piped-supply. But this is to explain solely in economic terms why there was no further investment during the 1840's in the waterworks. The other major reason was that it was not yet deemed to be socially necessary, because the relationship between the poor quality of Portsea Island water and the health of the town was not then generally recognised.

CHOLERA AND RAWLINSON'S REPORT, 1848-50

At this time the national debate on urban conditions was becoming more urgent, especially in the new large towns of the north and midlands from which the impetus for reform came, but the authorities in long-settled towns like Portsmouth preferred to discuss and criticise the proposals of others. Thus although the Portsea Commissioners thought that the Health of Towns Bill was a good idea in principle, they opposed the new power structure in May 1847 on the grounds that:

> '1. If passed into law it would materially interfere with all local management, and tend to increase Government patronage at the expense of Rate-Payers.
>
> 2. It would entirely supersede all existing locally elected Representative Boards'.

While there was an obvious element of self-interest in the arguments, the chief reason for the continued opposition was the increase in relative as well as absolute cost which would be incurred by the rate-paying inhabitants. Under the proposed new scheme Government establishments were not to be rated. Nevertheless there is evidence that even in 1848 the Portsea Commissioners expected a local Board of Health to be set up in Portsmouth, for they intended to press for the appointment of their own clerk W. O. Marshall as Clerk to the new Board.

Other measures were taken up with more alacrity. After the Nuisances Removal and Diseases Prevention Act was passed in 1846, the Portsmouth Commissioners appointed a Surveyor and Inspector of Nuisances and a Health Officer. They also contracted with the Water Company for their streets to be flushed for a quarter of an hour per day between Michaelmas Day and Lady Day, for £1 a week. The Portsea Commissioners continued to organise their street-watering from carts, but they followed this Act by ordering their Sanitary Committee to report on the state of Portsea. Their findings in February 1848 were typical of those for most nineteenth-century towns. In addition to describing the filthy condition of courts, the state of the pavements and failings of the scavenger, the committee stressed the deficiencies of the water supply:

> 'The supply of water afforded to the Inhabitants of these localities is exceedingly limited, and altogether insufficient for ordinary Household purposes, much less for cleaning or flushing the Gutters. Several of the public pumps were constantly out of order. Necessity of providing if possible a more ample supply, so that it may not only be used for ordinary purposes but for the general cleansing of the Town'.

From the Portsea Commissioners' minutes it is clear that no effective remedy was introduced before the outbreak of cholera, which caused 152 deaths in July and August 1848. But the fear of a renewed outbreak in September did spur the Sanitary Committee to produce plans for combating the disease.

Though similarly stricken by the epidemic, the inhabitants of Kingston, Landport and Southsea had no Commissioners to carry out such preventive measures. But the newly-formed General Board of Health was empowered to investigate, upon request, areas in England and Wales where the annual mortality rate exceeded 23 per thousand. Even before the cholera, between 1841 and 1847, the average annual death rate in Portsmouth had reached 25.37 per thousand. The wards of St. Mary's, All Saints, and St. Paul's therefore petitioned for an investigation into whether the state of the town warranted a Local Board of Health. Robert Rawlinson, a Civil Engineer and a Superintendent Inspector of the General Board of Health, began hearing evidence in Portsmouth in December 1848. He presented his report in 1850. In his conclusions, Rawlinson found that[20]

> 1. The Borough is not healthy as it might be, because of ill-paved and unclean streets, imperfect privy accommodation, crowded courts and houses with large exposed middens and cesspools; and no adequate power for local govt. exists.
> 2. The excess of disease is distinctly traced to the un-drained and crowded districts, to deficient ventilation, to the absence of a full water supply, and of sewers and drains generally.
> 3. The condition of the inhabitants could be improved, comforts increased and rates reduced by
>   (a) A perfect system of street, court, yard, house drainage.

(b) A constant and cheap supply of water under pressure, laid on to every house and yard to the entire superseding of all local wells and pumps which are proved to be expensive.

(c) The substitution of water closets of soil pan apparatus for the more expensive and noxious privies and cesspools which exist, and by regular and systematic removal of all refuse at short intervals.

(d) Properly paved courts and passages, and by a regular system of washing and cleansing all courts, passages, slaughterhouses, footpaths and surface channels in the borough and by maintaining good roads throughout the district.

The major part of his report was devoted to explaining how the lack of an effective sewage and drainage system endangered public health by polluting local water supplies. A local physician, and founder of the Royal Portsmouth hospital, and an enthusiastic Sanitary reformer, Dr. W. C. Engledue, had given evidence indicating the need for deep drainage:

> 'At present the island of Portsea was one large cesspool, for there could not be less than 16,000 cess-pools daily permitting 30,000 gallons of urine to penetrate into the soil. Just reflect on the character of the well-water of a district which became mixed every year with 365 times 30,000 gallons of urine, to say nothing of a host of other abominations too numerous to be mentioned on the present occasion!'

The fortifications were blamed for intercepting the natural drainage, but this was only one aspect of the problem in a crowded town on a low, flat and damp site. Rawlinson therefore recommended a pumped system of house drainage. He estimated the annual expenditure at £1,470, though this could be offset by the sale of refuse.

Having positively correlated the state of the drainage and the island well-water, Rawlinson proceeded to criticise the piped-water supply:

> 'This evidence is clear and distinct as to the deficiency of the present water supply of the towns, and, indeed, to the whole borough, and this, where there is a company in existence, proves that the water they supply is bad in quality, deficient in quantity, and the price charged excessive; or that the company have not thought it worth their while to accommodate their mode of supply to the convenience of the people so as to induce all parties to use their water'.

He also stressed the convergence of the economic and social interests of the Company and the town:

> '(It) shows how many mistakes may be committed where a full and comprehensive view has not been taken, but where private speculation has alone been the ruling motive. The entire failure of the scheme, even in this respect, ought to be instructive. The pecuniary benefit of individuals or companies is more or less connected and bound up with the beneficial advantage of the community. For ignorance and mismanagement a day of retributive reckoning is sure to come'.

Yet the experience of Portsmouth seemed rather to illustrate the divergence of the interests of the town and the Company. As sanitary conditions in the town worsened, the Company began to make its first consistent profits for years: in these admittedly limited terms, it was then a success.

Rawlinson favoured a waterworks, preferably municipally-controlled, providing a constant service for all the inhabitants. In particular this would reduce the high cost of water to the poor, who had most need and least access, under the existing system. This marked a significant shift in attitude. He implied that every citizen had a right not just to a supply of water, but to good quality and plentiful quantity. He regarded a comprehensive piped-supply in Portsmouth as a necessity, which was in sharp contrast to the views held forty years previously, when piped-water was a luxury available only to those able and willing to pay for it. To counter any suggestion that the town could not afford the cost of improvements, Rawlinson declared that the annual computed cost of the insanitary conditions between 1841 and 1847 was almost £60,000.

DELAYS IN IMPROVING THE TOWN'S WATER SUPPLY, 1850–59

Rawlinson's account was given added point by the return of cholera in July and August 1849, when 676 deaths occurred in six weeks, with a final toll of over 800 victims. Yet the committee appointed by the Portsea Commissioners to examine Rawlinson's report in June 1850 felt unable to recommend its application to Portsea. They believed that the Commissioners already possessed sufficient powers to deal with the situation, and cast doubts on the impartiality of the inspector's judgement:

> 'The Report . . . has the appearance of having been got up with a view to accomplish the ultimate design of that Board—the extension of the principle of Centralisation'.

A battle then ensued between the Commissioners and the Corporation. After a further investigation in 1851 a provisional order for a local Board of Health in Portsmouth was made, with the support of the Corporation, but its confirmation was successfully opposed by the Commissioners. The Council had refused a ratepayers' poll on the Public Health Act, although they should have been consulted. In any case the Commissioners were strongly opposed to vesting control in the Corporation, because this would upset a previously acceptable balance of power. In 1853 the Corporation again voted on the Public Health Act, accepting it by one vote but postponing any action on it indefinitely.

This leisurely wrangling contrasts oddly with the urgency of Rawlinson's report. But the existing statutory authorities were excused from taking immediate action by the disappearance of cholera, despite a scare in 1852. Their failure to take the initiative, however, meant that action was either not taken at all or was both delayed and piecemeal. Nothing was done about drainage: a comprehensive plan submitted in 1862 was rejected. But Landport and Southsea did obtain an Improvement Act in 1857.

It is tentatively suggested that the Water Company itself responded to Rawlinson's report with a threefold increase in output per pumping day between 1850 and 1856. At the same time its revenue increased only some 25 per cent, from £4,000 to £5,000. Together these factors would indicate a sharp fall in the price per gallon. Yet because the number of houses they supplied was put at 5,000 in 1850 and 4,500 in 1856, any such improvement seems to have been restricted to existing customers—hardly in the spirit of Rawlinson's report.

Despite Rawlinson's remarks on the Company's failings, it was because the concern was going well in 1850 that Casher was opposed to its possible sale to the Corporation. While a number of new contracts were signed after 1850 with the railway companies and with the various barracks, the Company still refused to extend its mains to supply new residential areas. Whereupon one enterprising architect, T. E. Owen, laid on a private piped supply from his well in Elm Grove to his housing development in Southsea. At the same time the Company was ready to oppose potential new competitors: when it was rumoured in 1852 that a Portsmouth, Gosport and Southsea Waterworks Company was to be floated, they immediately canvassed the Admiralty support of Sir Francis Baring against the Town Council. This bill, however, never reached Parliament, though it was said to promise to meet the Public Health Act requirements. In 1850 the Corporation had approved the proposal that they should acquire powers to buy out the United Company. The matter was not further discussed till 1855 when a council committee again reported in favour of the municipal control of the water supply. But consideration of the Company's findings was postponed for a year, possibly because the Mayor then was Andrew Nance, a leading Water Company proprietor.

In the absence of a lead from the Council three private citizens, Benjamin Bramble, builder, Dr. W. C. Engledue, founder of the Royal Portsmouth Hospital and an enthusiastic sanitary reformer and R. W. Ford, a young Portsmouth solicitor, took matters into their own hands. They formed themselves into a provisional committee for seeking a new Act to improve the town's water supply, in February 1856. By October 1856 they had received a report from their consulting engineer, Henry St. John Diaper,[21] who recommended the purchase of the United Company, followed by the installation of a larger engine to supply the whole town from Farlington or from springs two or three miles east at Bedhampton and Havant. He put the maximum cost of these improvements at £35,000. The proposals were discussed at a public meeting in December 1856 and received the Corporation's approval in February 1857. Despite the past record, these promoters believed it was possible to provide both a piped-water supply and a return for shareholders. . . . To emphasise this idea of a 'consumers' company', they proposed to raise its capital in £10 shares.

The Borough of Portsmouth Waterworks Act was passed in July 1857, with power to raise £80,000—£42,000 to buy out the United Company at its nominal value and £35,000 for the improvements to the present system.[22] All this money had to be raised before the new company could go into service. It may be that the ensuing delay was caused by the effects of the Crimean War, but the United Company formally signed itself over to the new Company on 21st December, 1858, just fifty years after the original concerns had been promoted. At that time the United Company was supplying approximately one-third of the houses in the town. Within fifteen years the new Borough of Portsmouth Waterworks Company was providing a piped supply for almost all the inhabitants.[23]

NOTES

Reference in the text, unless otherwise stated, are to the Portsmouth Water Company Records in the Portsmouth City Record Office (hereafter referred to as P.C.R.O.), and in particular to the Letterbooks, 1808–1817; 1840–1859, and the Directors' Minute Book, 1815–1830, in Boxes HT/39–40.

[1] P.C.R.O., CE/5, *Book of Constitutions*, 5 Apr., 1694.

[2] *Ibid.*, CF/7.

[3] Anon., *The Boro'—being a faithfull tho' humerous account of one of the strongest garrisons and seaport towns in G.B.*, published by Robert Wilkins (Portsmouth, 1748).

[4] 4 Geo. III, c. 92, *Portsmouth Common Paving Act*, 1764.
8 Geo. III, c. 62, *Portsmouth Paving Act*, 1768.
32 Geo. III, c. 103, *Portsea Paving Act*, 1792.

[5] 14 Geo. II, c. 43, *An Act to enable Thomas Smith, Lord of the Manor of Farlington, to supply the Town of Portsmouth with good and wholesome Water, at his own proper costs and charges.*

[6] Henry Slight, *History of Portsmouth* (Portsmouth, 1828), p. 14.

[7] 49 Geo. III, c. cxviii, *Portsea Island Waterworks Act*, 1809.

[8] *Ibid.*

[9] William Nicholson, *A Letter to the Incorporated Company of Proprietors of the Portsea Island Waterworks* (London, 1809).

[10] *Hampshire Telegraph*, 20 Sept. 1811.

[11] *Ibid.*, 14 Oct. 1811.

[12] W. G. Gates, *Illustrated History of Portsmouth* (Portsmouth, 1900), p. 606.

[13] R. B. Wade, ed., *Story of Eighteenth Century Portsmouth* (Portsmouth, 1957), p. 26 of typescript.

[14] Sir J. H. Clapham, *An Economic History of Modern Britain*, vol. I, pp. 43–44.

[15] William Cobbett, *Rural Rides*, vol. 1, 19–23: Nov. 1822.

[16] Estimated from *Letterbook* 1812–17 and *Directors' Minute Book*, 1815–30.

[17] 7–8 Geo IV, c. cxxxviii, *Portsea Island Waterworks Act*, 1828.

[18] Casher and Nance were also prominent in local public life, Casher being Mayor of Portsmouth in 1843 and 1844, and Nance Mayor in 1854.

[19] C. J. Bartlett, *Great Britain and Sea Power, 1815–1853* (1963), pp. 242 ff.

[20] R. Rawlinson, *Report to the General Board of Health on the Sewage, Drainage and Water Supply of Portsmouth* (1850).

[21] P.C.R.O., *Portsmouth Water Company Records*, Box HT/41.

[22] 20–21 Vic. c. xiv, *Borough of Portsmouth Waterworks Act*, 1857.

[23] D. Halton Thomson, *History of Portsmouth and Gosport Water Supply* (Portsmouth, 1957), p. 99.

ACKNOWLEDGMENTS

For their help in the preparation of this paper the author is much indebted to Prof. A. Temple Patterson, Dr. L. Gittins, Dr. E. P. Hennock, Mr. R. Windle, Mr. M. J. W. Willis-Fear, the Portsmouth City Engineer's Department and Miss J. M. Hallett.

Farlington Company Engine-House on Farlington Marshes, 1808–1928, from the south. Photograph from the Portsmouth Water Company's archives deposited in the City Record Office.

7

The impounding reservoirs of the Newcastle and Gateshead Water Company, 1845–1905

R.W. Rennison

1. INTRODUCTION

The Newcastle and Gateshead Water Company is unusual, if not unique, in having had impounding reservoirs designed by the three greatest 19th century water engineers, Simpson, Bateman and Hawksley. When the Company was formed under the name of the Whittle Dean Water Company in 1845 it was to James Simpson that the promoters turned to design the first reservoirs at Whittle Dean. Initially the consumption of water was at a rate of 0.7 million gallons per day (m.g.d.) but by 1850 this had risen to 1.6 m.g.d. and the Company then constructed two further reservoirs in the same group at Whittle Dean. In 1853 water shortages exacerbated an out-break of cholera which occurred on Tyneside and it was as a result of this epidemic that John Frederic Bateman was first involved in the area's water supply; it was not, however, until 1863 that he was appointed as consulting engineer to the Company, which by this time had changed its name to the present title. As a result of the expansion of the towns on Tyneside consumption had increased to 4.4 m.g.d. by 1861 and it was Bateman's task initially to design and supervise the construction of a new impounding reservoir at Hallington. In spite of this increase in the Company's resources it was still found necessary to construct further reservoirs, an impounding reservoir of considerable size at Colt Crag together with a very much smaller holding reservoir. Later the Hallington group was added to by the construction of the West Hallington reservoir although it had no catchment of its own and relied solely on water transferred from Colt Crag. The two groups of reservoirs, Whittle Dean and Hallington, still did not provide the resources which the Company required and in 1888 it was decided to ask Thomas Hawksley to design a further reservoir at Catcleugh, some 45 miles north west of Newcastle. By this time the daily consumption had risen to 14 m.g.d. and in order to provide supplies for Newcastle's growing industry three Acts of Parliament were obtained, in 1889, 1894 and 1902 respectively. It was not until 1905 that the Catcleugh reservoir was brought into service, by which time consumption had risen to 19 m.g.d. It is with the construction of the three groups of reservoirs and with their respective engineers that this paper is concerned. (Fig. 1, Page 28)

2. WHITTLE DEAN RESERVOIRS

The three principals involved in the formation of the Whittle Dean Water Company were William George Armstrong (1810–1900), a solicitor but later to found a notable engineering company in Newcastle, Richard Grainger (1797–1861) a substantial speculative builder, and Thomas Sopwith (1803–1879) a civil engineer interested additionally in public health matters. The collaboration between these three men, aided by a Provisional Committee, resulted in plans of the proposed works being deposited in Parliament in November 1844 and a prospectus published in February 1845.[1] (Fig. 2, Page 30)

THE IMPOUNDING RESERVOIRS OF THE NEWCASTLE AND GATESHEAD WATER COMPANY

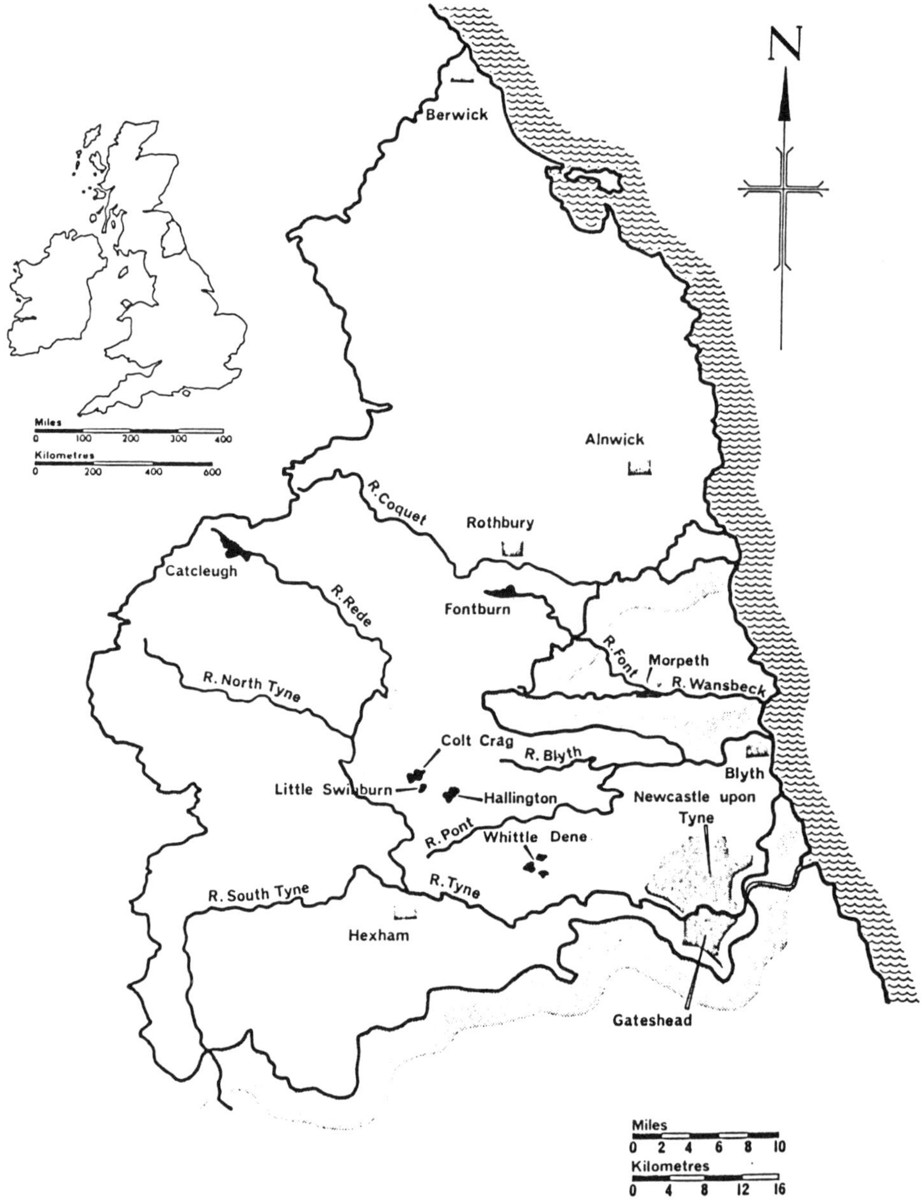

Fig. 1 Reservoir locations in relation to area of supply.

THE IMPOUNDING RESERVOIRS OF THE NEWCASTLE AND GATESHEAD WATER COMPANY

The plans, and the prospectus, showed a single impounding reservoir on the Whittle Burn, some 12 miles west of Newcastle with a 24″ diameter pipe carrying water from it into Newcastle. At no time was the capacity of the reservoir stated other than the remark that it would be "capable of containing about half a year's supply (so enabling consumers to).... be entirely independent of the stream during the most protracted season of drought that is ever likely to occur in this country".[2] The reservoir was to be built at an elevation of 360 ft., thus obviating the need for pumping, on "ground......of a perfectly impervious nature, eminently favourable for the retention of water and the security of the basin against leakage"[3]; the site was free from mining operations.

The deposited plans and the drawings in the prospectus were prepared by Sopwith, in spite of the comment that the design had been prepared "under the advice of Mr. James Simpson (1799–1869)—well known as one of the first engineers for the construction of water works in the kingdom".[4] Simpson had discussed with Sopwith, on 25 January 1845,[5] matters relating to the works and it was under his supervision that the estimates of cost had been made. Subsequent to the initial preparation of plans the Company wrote to the Duke of Northumberland, the principal landowner affected by the proposed new works, to the effect that Grainger (possibly as the promoter with suitable labour immediately available) wished to construct "a small dam—six or eight feet in length and two or three feet high—across the stream...for the purpose of gauging the quantity of water which the runner is capable of supplying";[6] the Company also asked permission to put down boreholes to a depth of some ten feet for site investigation purposes. As a result of the request, gaugings were carried out from 9 January 1845 until April 1846, by which time construction was under way. During the seven month period beginning at the earlier date the average stream flow had amounted to 5 m.g.d., the consumption of water in Newcastle and Gateshead at that time being some 700,000 g.p.d. Cognizant of variations in flow the Company, in its first report, admitted to periods of considerable reduction in flow but, nevertheless, was of the opinion that "the *aggregate* produce,....when *equalised* by means of a reservoir of adequate dimensions, will be redundantly sufficient."[7]

From the time of the preparation of the prospectus, stream flows at Whittle Dean had evidently increased, the prospectus having stated that after a period of dry weather and "according to the testimony of highly respectable persons resident in the neighbourhood, they were as low as they are ever known"[8] but still able to produce some 1.7 m.g.d., twice the quantity required. Calculations were also produced to show that from the 4340 acre catchment, assuming a 24 in p.a. rainfall, the run-off would be 1,181 mill. galls. p.a., four times the then consumption and it was this figure, 3 m.g.d., which the directors considered themselves able to promise consumers; run-off, however, was assumed to be 50%, a high figure.[9] The Bill went before Parliament during the early summer of 1845 and received the Royal Assent on 30 June 1845; on this day Sopwith's association with the Company terminated and Simpson became Engineer, he having appeared with Sopwith and Armstrong before a House of Lords Committee.

After this time, and for no recorded reason, a decision was evidently made by Simpson to construct the necessary storage as a series of reservoirs rather than the single unit proposed initially. Part of the reason was probably the fact that two roads were to be affected by the proposals and their continuance made logical the formation of earth embankments rather than viaducts to carry them; natural sub-division of the reservoirs resulted. Preparation of contract drawings was completed by December 1845, and on 13 March 1846 the contract was awarded to John Ridley and Robert Atkinson, local contractors.[10] (Fig. 3, Page 30) The contract called for the construction of a group of five reservoirs to a much more sophisticated design than the original proposal. From the Lower reservoir, from which water was to be taken into supply, the remaining reservoirs ascended in two directions, north and west, to collect water from the two streams draining the catchment in each direction terminating at the upstream end in a small 'subsiding reservoir', the principal purpose of which was to hold the incoming water when turbid, so allowing settlement prior to its being stored in the lower reservoirs. The Lower reservoir was to be constructed at an elevation of 270 ft., the Northern and Western five feet higher and the subsiding reservoirs some three feet higher again. A by-wash was incorporated into each arm of the group to enable the reservoirs to be by-passed completely in times of flood, the water courses passing under the two roads by means of masonry arches and uniting downstream of the Lower reservoir.

The contract drawn up for the work did not include full quantities but nevertheless specified the works in some detail and included a schedule of prices to be applied in making any adjustments to the lump sum tender; craftsmen were charged at a daily (10 hours) rate of 5/-, labourers at 3/3. The embankments forming the reservoirs were of the type by then generally adopted, earth having a slope of 1 in 3

THE IMPOUNDING RESERVOIRS OF THE NEWCASTLE AND GATESHEAD WATER COMPANY

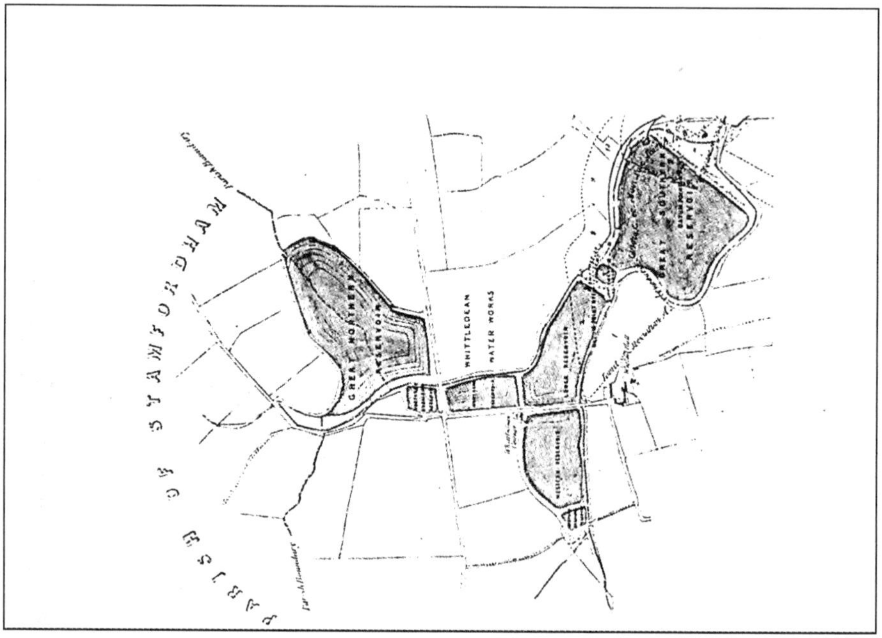

Fig. 3 Whittle Dean reservoirs as built. 1865.

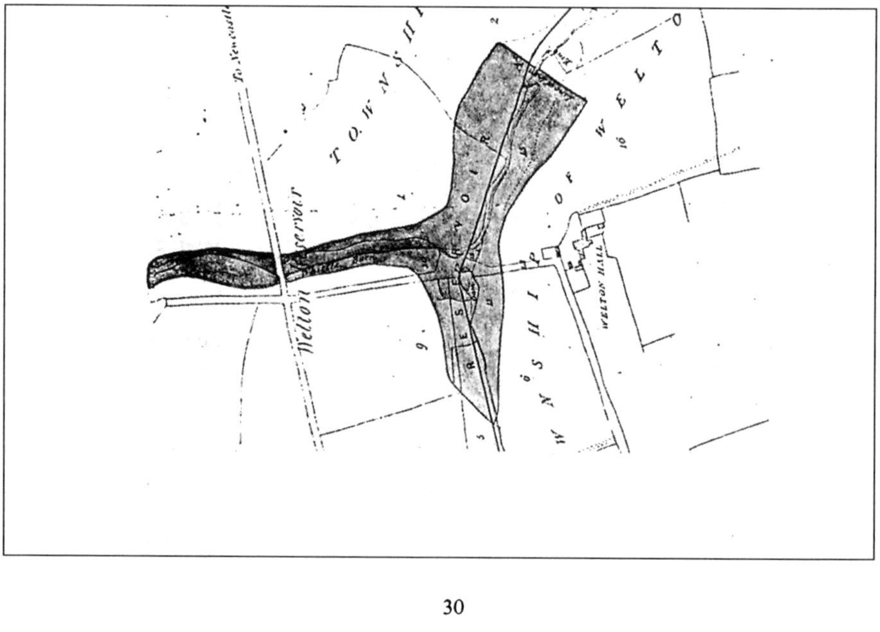

Fig. 2 Whittle Dean reservoirs. 1844.

and incorporating a puddle clay core with a batter of 8 to 1 on each face, the core being carried down at least 3′ 6″ into solid, impermeable ground. The specification—the general clauses are very similar to those of the present day—stated that only "the best and most tenacious portion of the clay (should be) used for puddling and pounding"[11] and the embankments should be brought up in layers not exceeding 9 ins. in depth, rammed solid as work proceeded. The arrangement adopted by Simpson for the reservoir outlets was to form a forebay 6 ft. in width and faced in masonry; to withstand the horizontal thrust imposed by the retained earth cast iron struts or 'sustaining bars' at vertical centres of five feet at the lower levels but reducing to eight at top water level were used, spanning horizontally at 10 ft. centres between the masonry walls, the joints of which were generally of lime mortar although Roman cement was used to secure the iron struts and to make watertight the wall where pipes passed through it.

The outlet arrangements of the various reservoirs differ. The two subsiding reservoirs have masonry forebays with a single outlet pipe leading to the adjoining lower reservoir. These two latter reservoirs, the Northern and the Western, also have forebays but from each lead two outlet pipes, one at the bottom of the reservoir and one at its mid-height, each controlled by a sluice. (Fig. 4, Page 32) The two 24 in. diameter pipes unite immediately downstream of the puddle core and this main discharges into the Lower reservoir. A 24 in. diameter sludge pipe runs from the bottom of each reservoir and under the bed of the Lower, at the discharge point of which the main is joined by the scour from that reservoir. The construction of the Lower reservoir outlet is more complicated in that the forebay abuts on to an outlet chamber, placed immediately upstream of the puddle core, which receives water from the reservoir at three levels by means of pipes controlled by sluices, so enabling the upper layers—thought to be the most pure—to be taken into service no matter what depth of water was in store. From the chamber the main 30 in. diameter outlet—with provision for duplication--runs through a tunnel 6 ft. wide and 8 ft. high. (Fig. 5, Page 32)

Outlets from the other reservoirs were laid on 18″ thick brick walls carried down to solid ground. Iron gratings were incorporated to prevent debris being drawn into the pipes. All ironwork was purchased by the Company for building in by the contractor and sluices were eventually purchased from Simpson's Company in Pimlico. Other ironwork was manufactured by the Ouseburn Ironworks in Newcastle.[12]

Construction of the reservoirs began in April 1846 by which time a Resident Engineer, Robert Nicholson (1808–1855) had been appointed to supervise the work and act as Simpson's representative. He, in turn, employed a superintendent, William Reid, who was resident at Whittle Dean from April 1846 to May 1847, the period in which the major part of the work was carried out. From the diary[13] kept by Reid during the above period it has been ascertained that Simpson, in fact, visited the works only on four occasions, although Nicholson was in contact with events on a much more regular basis. Other than the washing-out of a partially formed embankment due to floods following a thaw, problems were principally due to poor workmanship; masonry walls were not properly backed, clay puddle was placed without the specified gravel content, materials used for backfilling and for forming embankments were laid in excessively thick layers, masonry was laid in earth instead of mortar and stone pitching was incorrectly placed. In addition, problems were caused by the heavy drinking of the 200 men employed on the works—housed in a camp adjacent to the Western reservoir—and by the contractors attempting to carry out earthworks in ground affected by frost. In spite of problems, completion was achieved by November 1847, almost within the eighteen month period initially specified, and the reservoirs were brought into use in April 1848.

At their completion, the reservoirs had cost £29,616 for contract work together with £5,193 for ironwork and pipes, a total of £34,808.[14] The value of the tender submitted in March 1846 had been £24,914 but there are no details extant concerning the cause of the 20% increase incurred; this figure in turn showed a vast increase over the estimated total cost of £14,191 given in the prospectus of 1845, although Simpson sought to justify the great increase by commenting that the reservoirs, with a capacity of 215 mill. galls., were some 30% larger than originally envisaged.

In January 1848 Simpson, in a report to the directors, expressed his confidence in the success of the undertaking and in the year following Nicholson commented on the satisfactory performance of the reservoirs, which gave no cause for complaint until 1850 when, due to a summer of low rainfall, storage fell to less than 40 mill. galls. At this point Nicholson proposed the construction of an additional reservoir at Whittle Dene but before this, however, it had been found necessary to form a small pond, downstream of the Lower reservoir. Of negligible storage capacity and built at small cost this reservoir was inspected

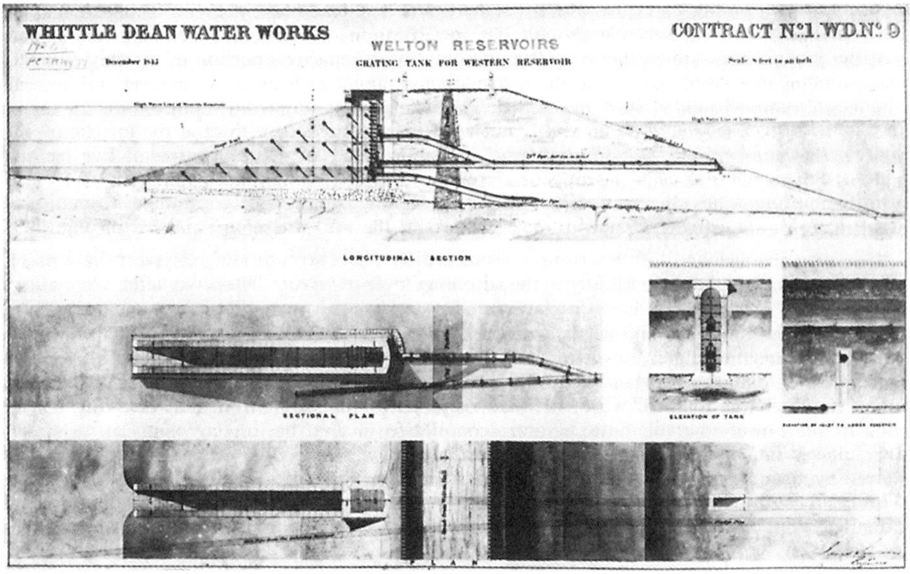

Fig. 4 Whittle Dean: Western reservoir embankment, 1845.

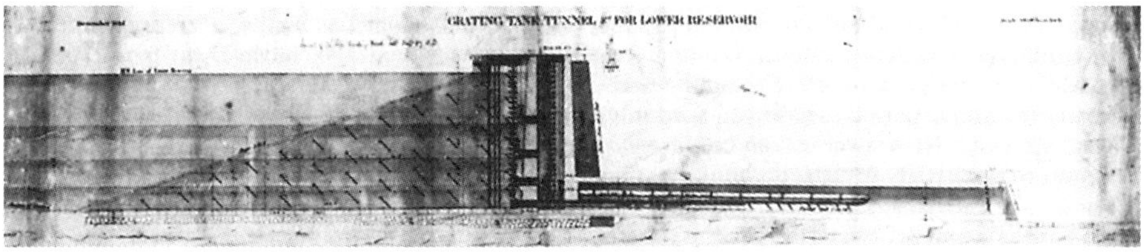

Fig. 5. Whittle Dean: Lower reservoir embankment, 1845.

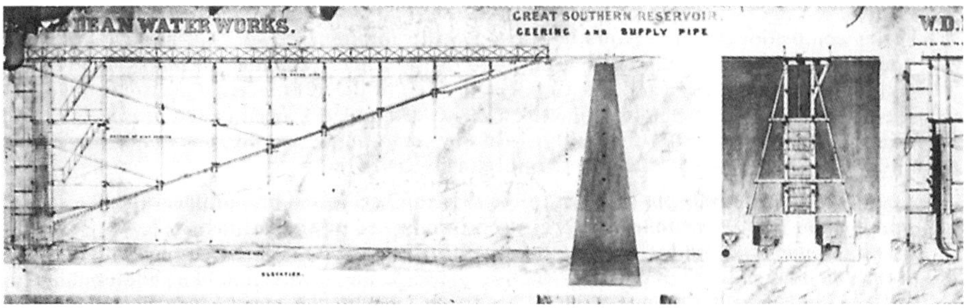

Fig. 6. Whittle Dean: Great Southern reservoir outlet, 1854.

by the directors on 12 July 1850, when it "appeared effectually to gather all the leakages from the large Reservoirs", [15] problems in satisfactorily forming and filling the cut-off trench for the Lower reservoir embankment having been encountered during its construction. The preparation of drawings and contract documents for the new reservoir, the Great Northern, was put in hand by Nicholson and tenders for the work were received on 6 March 1851, the contract being awarded, again to a local contractor, Richard Cail, for the sum of £3,025.[16] Work began immediately and by July earthworks were virtually complete and pitching was proceeding; the final cost was £3,114.

The draw-off arrangements for this reservoir were somewhat different from Simpson's earlier design in that a brick culvert carried the main to a valve well, slightly upstream of the puddle clay core. In the well a sluice controlled flow into the outlet pipe. The new reservoir impounds water, at a level some 20 ft. higher than the others, from a stream flowing from the north eastern area of the catchment. The reservoir itself, of 125 mill. galls. capacity, is similar in the embankment details to Simpson's but it differs in that no by-wash is provided, neither is a subsiding reservoir located at its head; it is notable for the extreme steepness of its embankments, steeper than 1 in 2.

The eighth and last of the group of reservoirs, the Great Southern, was conceived in October 1853, six weeks after the outbreak of cholera on Tyneside. In an effort to increase supplies of upland water, so eliminating the need for occasional pumping from the river Tyne, Simpson and Nicholson were directed to report upon future sources. Their report suggested that an aqueduct be constructed to bring water to Whittle Dean from the north west. The watercourse would intercept streams flowing generally south westwards and convey water by means of a three mile long tunnel to the river Pont and thence to the Whittle Dean group of reservoirs where additional storage would be provided. On 31 July 1854 the contract for the reservoir was awarded to Benjamin Lawton[17] in the sum of £17,498. Again designed by Nicholson this reservoir was very similar to those initially constructed under Simpson's direction. After the drawings were produced, however, a change was made to the draw-off arrangement, eliminating the forebay with its 'sustaining bars'. The revised design comprised a vertical pipe 24 ins. in diameter standing in the reservoir and connecting with the outlet main. The stand-pipe incorporated four flanged joints connected only by locating bolts; by this arrangement each joint could be opened some nine inches. The upper pipe terminated in a domed top, in turn connected to a screwed rod and cross-head, by means of which the four outlet points—at each flanged joint—could be opened progressively. Access to the operating mechanism was by means of a substantial timber 'geering' or gantry, 150 ft. long and 55 ft. high (see Fig. 6), a structure very similar in design to the timber railway viaducts and colliery staithes common in the district. The outlets could thus be operated so as to discharge water from the upper layers of the reservoir, no matter what its level; its effect was thus identical to Simpson's arrangements of pipes with sluices. The design of the pipework has been attributed to Armstrong[18] who by this time had become the Company's Chairman; it is not unreasonable to suppose that his influence induced Nicholson to modify his designs. During the construction of the reservoir Nicholson died and his place was taken by his nephew and assistant, John Furness Tone (1822–1881)[19] under whose direction the reservoir was completed, not without some acrimony as Lawton was obliged to suspend work so as to enable some 120 mill. gallons of water to be stored in the unfinished reservoir for use in the summer of 1856. He was paid the sum of £1000 on account of the suspension and subsequently the reservoir was brought into use, storing 260 mill. gallons during 1857; the final cost was £26,911.

3. THE HALLINGTON GROUP OF RESERVOIRS

3.1 East Hallington. As the Whittle Dean reservoirs were the work of Simpson, so the Hallington group was the responsibility of John Frederic Bateman (1810–1889), appointed as Engineer in 1863.[20] Before this, in 1861, Bateman had visited Hallington to assess the suitability of the area as the site for a reservoir to impound the water from the Hallington North Burn,[21] already utilised by the Company by virtue of the aqueduct proposed by Nicholson. In a report to the directors, Bateman recommended Hallington as being capable of cheaply providing storage of 600 mill. galls.—compared with the Whittle Dean total of 530 mill. galls.—and so providing 3 m.g.d. from a catchment of 6000 acres, assuming a run off of 8 inches per annum, "perhaps more than experience warrants"[22]; the reservoir would cost £20,000. Bateman also asked that rain and stream gauging be instituted, recommending that the former be located at low, mid and high points of the catchment in an attempt to provide reliable figures. Although the science of rainfall had become more widely understood over the past decade Bateman's apparent uncertainty is evident in his remarks that he could not "help thinking that the fall of rain must increase the further you go towards the hills".[23] In a later report Bateman considered his earlier views vindicated in that this was, in fact, the case, the figures for the period October 1861 to February 1862 being 7.78 ins. at Whittle Dean, 9.44 in Hallington Village and 10.18 at Woodford; the mean of the several Hallington readings was 9.91, an increase of 27% over Whittle Dean.

As a result of his gaugings Bateman postulated the dry year rainfall at Hallington as about 24 ins. p.a., comparing this with the 19½ ins. experienced at Whittle Dean in the dry years of 1850 and 1851; deducting 13 ins. for loss, 11 ins. would remain for impounding. The report went on to discuss the capacity of the reservoir proposed, with a catchment of 4000 acres, and concluded that it should be capable of producing 2 m.g.d. in the driest season; to this figure 1.5 m.g.d. should be added for springs. The capacity of the reservoir should be 624 million gallons and the cost was estimated at £25,000. In his later report Bateman also looked ahead to the construction of further reservoirs as "the Swinburne District may be so easily acquired and forms so natural an extension of your collecting ground, that it must necessarily come under consideration"[24] but also he anticipated, correctly, the policy of the Company a century ahead in his view that "the softer water of the Tyne, if it can be satisfactorily deprived of colour, may become an unobjectionable source to the inhabitants"[25].

Bateman's proposals concerning the Hallington reservoir were agreed to by the directors in November 1861 and the requisite Act of Parliament received the Royal Assent in May 1863; although Bateman had recommended an expeditious start it was not until 1869 that work began. Due to the Company having enlarged its Newburn pumping station it had been possible to abstract additional water from the Tyne and so postpone construction of additional reservoirs but the increase in consumption, and high pumping costs, made necessary the increase in upland resources. Bateman was instructed to proceed with the work in September 1868, whereupon he promptly stated that his fee would be £1,750, no doubt an attempt to avoid such misunderstandings as had occurred in the past[26]. After the preparation of the necessary documents, tenders were received in July 1869—the lowest was that of J. B. McGuire, £19,000 [27]—and work began in August 1869.

As constructed, Hallington reservoir has a surface area of 125 acres, a maximum depth of water of 36 ft., a capacity of 685 mill. galls., a top water level of 505.4 ft. A.O.D. and a catchment area of 5450 acres. Due to the open nature of its location the reservoir is contained by three earth embankments 1360, 1770 and 900 ft. in length. All embankments have a slope of 1 in 3 on the upstream face and 1 in 2 downstream and in addition to the puddle core—a maximum of 14 ft. thick at the base and 8 ft. at the top—the contract drawings show an inner section of the dam as being formed of "clayey or most adhesive material—2 feet layers";[28] the drawings also show coarser material placed in layers four feet thick in the outer sections. In common with the Whittle Dean reservoirs all inner slopes are faced with stone pitching and the cut-off trench reaches a maximum depth below ground level of some 20 ft. Where crossed by pipes it was filled with concrete, otherwise with puddle clay.

Compared with Simpson's draw-off arrangements for the Whittle Dean reservoirs, those at Hallington are rudimentary. Bateman's drawings show a combined outlet and scour pipe, 24 ins. in diameter, together with a 12 in. diameter drain, provided to discharge water encountered while forming the cut-off trench for one of the embankments. In spite of doubts having already been entertained as to this arrangement—similar to that used at Dale Dyke where failure had occurred only 5 years before and in the resultant inquiry on which Bateman had been involved—control valves were placed only on the downstream end of the pipes which ran under the embankment at its highest point, albeit in a shallow trench excavated in undisturbed ground and surrounded in concrete. In order to prevent the seepage of water along the line of the pipe a wrought iron plate was placed at the crossing with the puddle core. The overflow from the reservoir comprised an open masonry channel 10 ft. wide and some 5 ft. deep fed by a weir with a crest 50 ft. long. The channel joins the piped compensation discharge in a gauging basin, so permitting the measurement of flow. (Fig. 7)

Within ten weeks of the commencement of work the contractor, McGuire, found himself in financial difficulties. By this time the value of work completed amounted to £2,823 and the Company, due to a guarantee having been arranged at the time of making the contract, was able to take over the work, purchasing materials on site and employing McGuire as its agent but withholding from him £2000

THE IMPOUNDING RESERVOIRS OF THE NEWCASTLE AND GATESHEAD WATER COMPANY

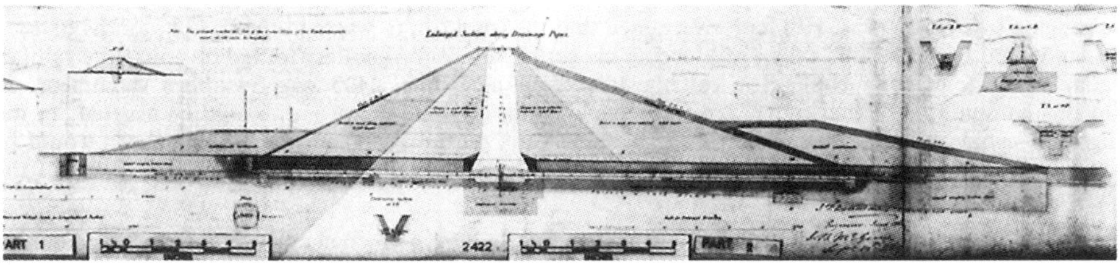

Fig. 7. East Hallington reservoir embankment, 1869.

of the money owed, to be reimbursed on completion; McGuire was paid a salary of £300 p.a. Details of the construction of the reservoir are scant; valves were obtained from Sir W. G. Armstrong & Company, stone was brought by a Company railway—complete with a Black Hawthorn locomotive—from Mootlaw quarry, some two miles distant, and clay was found on site. Progress was good and by December 1870 it was decided that impounding should begin, an event forced upon the Company by the difficulties experienced in maintaining supply. Unusually, it was not Bateman who certified the reservoir as being able to be used but one of the directors, Christian Allhusen, a manufacturing chemist[29]. Work continued, on a reduced scale, during 1871 but in September faults were found in the western embankment. Details are not available but it would seem that slips had occurred, these being countered by forming berms with a slope of 1 in 5 at the foot of each face of the embankment. Like the main embankment, the western, too, had pipes beneath it, provided to drain the springs encountered during construction.

By December 1872 all work had been completed and plant sold, McGuire had been paid his £2000 and had left, Bateman's fees had been settled and the reservoir had been brought into service. The directors were well pleased by the outcome; they estimated that the reservoir had cost only a quarter of what could have been expected for such a capacity and, in addition, it had been found to hold 685 million gallons compared with Bateman's initial estimate of 600. On these two counts, and because of its "soundness of construction [and] the admirable design and execution of the works, this reservoir will add one more to the existing evidences of Mr. Bateman's sound judgement and ability".[30]

3.2 Swinburn and Colt Crag. These sanguine views were soon to be modified, principally by the dry summers experienced in 1873 and 1874 when it was only by pumping water from the river at Newburn that supplies could be maintained. As early as August 1873 the views of Bateman's assistant George Henry Hill (1827–1919) were sought, Bateman at that time being in Buenos Aires. Hill pointed out in his report[31] that the Company's gravitational supply was safe only up to 7 m.g.d. while consumption had recently reached 8.5 m.g.d.; as demand would increase further and as no works could be completed within three years he recommended river pumping as soon as storage began to fall. As records of rainfall had been maintained since Bateman's initial report there were now figures for nine years; from them, Hill inferred that Bateman had, in fact, underestimated the Hallington run-off by two inches p.a. and he pointed out that the respective annual rainfall figures for Whittle Dean and Hallington could now be taken as 24″ p.a. and 28.8″,[32] with dry-year figures taken as having available 2 ins. p.a. less,[33] an over-estimate. Hill calculated an additional available supply of 7 m.g.d., necessitating the construction of two reservoirs, one with a capacity of 355 mill. galls. on the Swinburn and one of 655 mill. galls. at Hallington; water would be carried to Hallington by tunnel, the total cost being some £105,000.

Hill looked upon these proposals as lasting only for four or five years, and recommended that a longer term solution be found. At this period the long-term proposals became the subject of much public discussion in the towns of Newcastle and Gateshead and although details cannot be examined in this paper[34] they did encourage the Company to pursue the problem of increasing its resources. In 1875 an

THE IMPOUNDING RESERVOIRS OF THE NEWCASTLE AND GATESHEAD WATER COMPANY

investigation was requested of Bateman and, due to his being ill, a further report was received from Hill, who considered all possible sources then under public scrutiny and discussed in greater detail his proposals of two years earlier. In this later report[35] the question of rainfall was not greatly enlarged upon whereas the gauging of streams was. Hill had ascertained that the total run to waste in the first 8 months of 1875 had amounted to 1500 mill. galls., reinforcing his earlier desire for greater storage to utilise the rainfall available on the existing Hallington catchment and the additional 4400 acre Swinburn catchment. In droughts comparable to that of 1873/4 a safe yield of an additional 3 m.g.d. would be assured. In the long term—some 12 to 15 years hence—further reservoirs on the Warksburn and Blackaburn would be required. Within three weeks of receiving Hill's report to the directors prepared to submit plans of the new works to Parliament.

In January 1876 trial borings were taken in the Swinburn district and in May the Company's proposals came before a Parliamentary Committee. The drawings of the works, signed by Bateman and Hill, show three reservoirs situated on the Dry Burn, the Upper, Little and Lower Swinburn, with capacities of 417, 34 and 345 mill. galls. and elevations of 687, 553 and 543 ft.; an aqueduct, partly in tunnel, was to carry the water to Hallington. In Committee[36], fierce opposition to the proposals was encountered, principally from the land-owner most affected, and it was at his instigation that an alternative site was suggested in substitution for the Upper Swinburn reservoir. Parliament rejected the lower reservoir but the Act sanctioning the other works became law in July 1876. In August the several reservoir sites were visited by Bateman, accompanied by Thomas John Bewick (1821–1897), who had been responsible for suggesting the construction of the alternative reservoir, and a report submitted to the directors who at that time agreed to construct only the Little Swinburn reservoir and aqueduct—so enabling additional water to be more quickly available—until such time as a decision could be made as to a larger reservoir. Bewick was not convinced that the Little Swinburn reservoir was necessary at all—other than reducing the length of aqueduct, it is not—but the Company would not consent to its abandonment and in April 1877 it was agreed to proceed with the reservoir proposed by Bewick, to be known as the Colt Crag reservoir and having a capacity of 1068 mill. galls. at an elevation of 661 ft.;[37] together with the small reservoir and aqueduct[38]; the requisite Act of Parliament was obtained in July 1877. (Fig 8)

Bewick's proposals had much to recommend them. Excluding Little Swinburn, common to both schemes, the Company's initial intention would have provided storage of 762 mill. galls. compared with Bewick's 1068 mill. galls. at Colt Crag. The Company's scheme, however, would have necessitated the construction of five embankments to impound the two reservoirs while two suffice at Colt Crag, the total earthwork quantities—very approximately—are 350,000 cu. yards for Hill's reservoir and 100,000 for Bewick's. In terms of cost, the substitution of the Little Swinburn embankment by an additional length of aqueduct would have made little difference.

Parliamentary plans for the second scheme are in Bateman's name and the contract drawings are signed only by him. The Colt Crag reservoir comprises a main embankment 620 ft. long with a maximum height of 57 ft.—the reservoir's water depth is 52 ft.—together with a secondary embankment at its western end 630 ft. long and 13 ft. high.[39] The Little Swinburn embankment is 200 ft. long and 28 ft. high. Overflows from the two reservoirs skirt the embankments; that at Colt Crag comprises a channel approximately 450 ft. long and 16 ft. in width fed by a 133 ft. long horse-shoe weir while that at Swinburn consists of a channel 12 ft. wide and some 300 ft. in length with an inlet weir 50 ft. long. The reservoir outlets are much superior in design to that at Hallington. Both are by means of valve towers, (Fig. 9) six feet in diameter internally, constructed in brickwork and set into the embankment upstream of the puddle core; brickwork tunnels eight feet in diameter, constructed in solid ground, pass beneath the embankments and form the principal discharging facilities. Both valve towers have two outlets, at high and low levels, controlled by cast iron rectangular sluices. The Colt Crag outlet and overflow both discharge into the watercourse, and thence to Little Swinburn, while the outlet from the latter is into the aqueduct leading to Hallington. Compensation water, to a maximum of 200,000 g.p.d., is provided from this latter reservoir. (Fig. 10)

Having been instructed by the directors to design Colt Crag to retain the maximum possible quantity of water, Bateman was able by January 1878 to seek tenders for the construction of the several works. In April the contract was awarded to William Rigby of Worksop in the sum of £36,845 [40] and work began immediately. To supervise the contract Bateman nominated William Fox as his assistant and

THE IMPOUNDING RESERVOIRS OF THE NEWCASTLE AND GATESHEAD WATER COMPANY

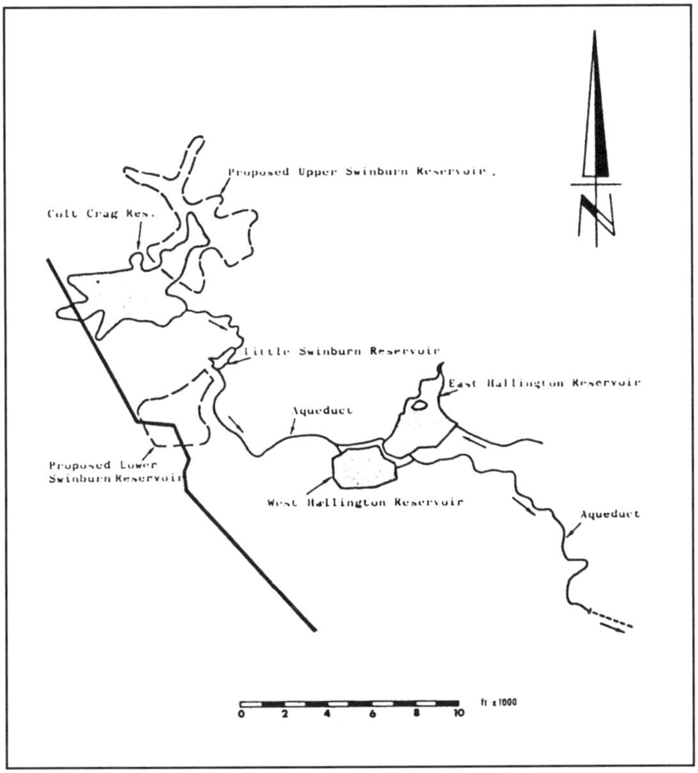

Fig. 8. Swinburn reservoirs: proposed and constructed, 1863–1889.

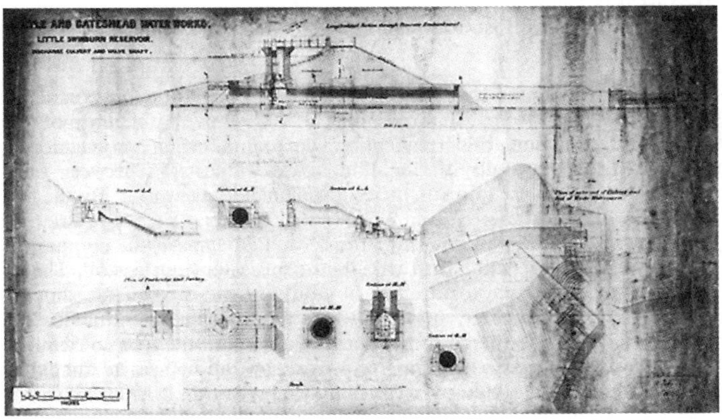

Fig. 9. Little Swinburn reservoir embankment: section, 1877.

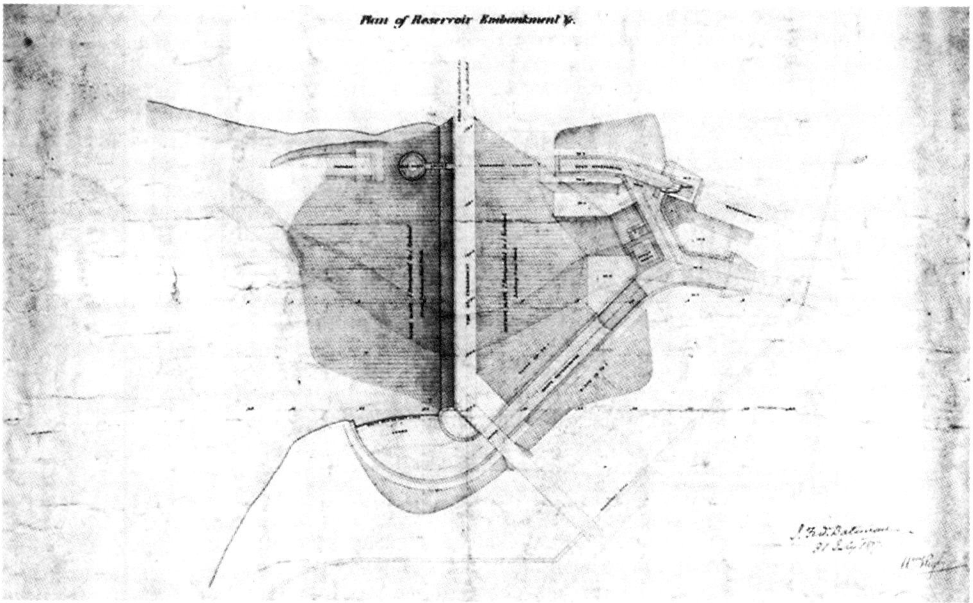

Fig. 10. Little Swinburn reservoir embankment: plan, 1877.

employed, at his own expense, a resident engineer, Brown. Bateman's fees were fixed at 6% of the value of work done, of which £750 had by that time been paid; the balance was to be made available at the rate of £200 per quarter subject to the 6% limit. Boreholes had been taken before designs had been finalised but, nevertheless, the puddle trench had to be taken down to depths greater than anticipated, a maximum of 40 ft. below original ground level being reached. The trench has a minimum width of nine feet and is filled with puddle clay found on site except where the outlet tunnel, which is supported by a concrete plug, crosses it. (Fig 11) Suitable building stone was also available on site for the construction of the overflow channel and walling, also for pitching. Ironwork was obtained from Joicey's foundry in Newcastle[41].

As work proceeded it was found that fissures were in evidence along the course of the Dry Burn, the stream to be impounded. The Geological Survey map of 1881 shows an outcrop of limestone running in a generally northerly direction along this stream and it was found during construction that many faults were present, attributed to the proximity of the Whin Dyke which runs between the Colt Crag and Hallington reservoirs; this outcrop was to cause many problems in later years. By the time the first year of the contract had passed the costs were given as follows: Colt Crag £4,534, Little Swinburn £1,023, aqueduct £2,057, extras £4,224;[42] by the time a further year had elapsed the comparative figures were £13,660, £4,094, £6,786 and £13,951 with a further £800 for miscellaneous costs[43]. The aqueduct had by now been completed and was available for use but, in addition, the first of the slips in the Colt Crag embankment had occurred. Unfortunately, no details are available but movements are reported until October 1880 and in March 1881 Bateman informed the directors that the reservoir could impound water to a depth of 40 feet. It would appear that slips took place on both faces of the embankment, as at Hallington, comprising an inner core adjacent to the puddle clay laid in 2 ft. layers with poorer material flanking the core and laid in 4 ft. layers swept upwards away from the centre; the embankments were 1 in 2 downstream and 1 in 3 upstream. To counteract the slips, berms, or offsets, were formed on both

THE IMPOUNDING RESERVOIRS OF THE NEWCASTLE AND GATESHEAD WATER COMPANY

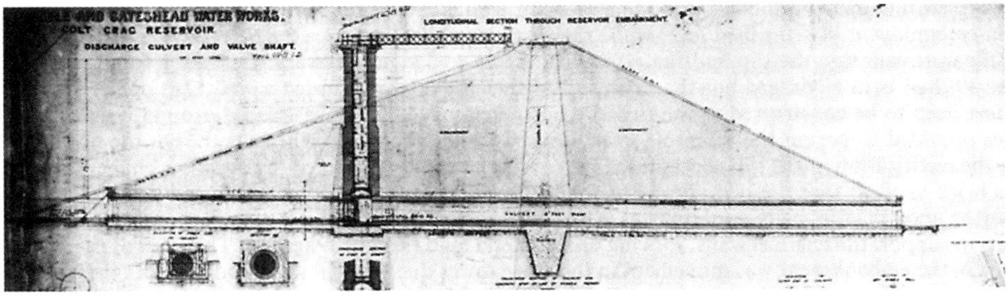

Fig. 11. Colt Crag reservoir embankment: section, 1877.

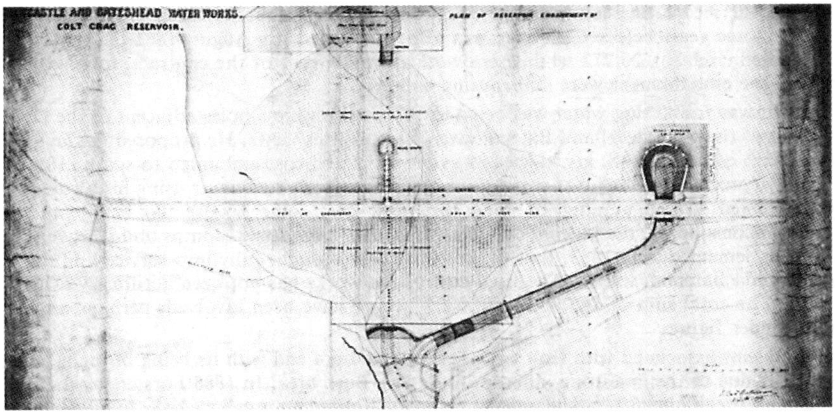

Fig. 12. Colt Crag reservoir embankment: plan, 1877.

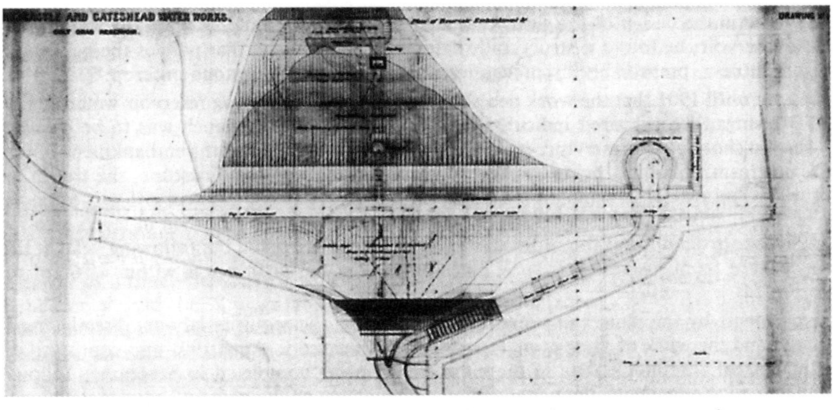

Fig. 13. Colt Crag reservoir embankment: plan as constructed.

faces consisting of dry rubble at the toe laid as an embankment in itself; fill placed between it and the main embankment was finished level while the slope immediately above the berm was reduced to 1 in 6. On the upstream side the slip had the effect of making re-design necessary. Initially a forebay (Fig. 12, Page 39) had been envisaged but the extension of the embankment caused a well 12 ft. × 8 ft. and some 20 feet deep to be constructed at the tunnel's upstream end. Dry fill was placed around it and drainage holes provided to permit the reservoir to be utilised to its original designed depth. On the downstream side the rectification of the slip necessitated the roofing of the lower part of the overflow channel by means of a brick vault so that the berm could be formed over it; the constriction which resulted was adversely reported upon in later years, especially as it had already been found necessary to provide arched masonry struts to support the channel walls, moving under lateral load (Fig. 13, Page 39). The third of the principal faults in the embankment was movement in the valve tower due to the unequal horizontal thrust imposed upon it as a result of its location within the embankment;[44] it was measured as being 14 ins. out of vertical but no steps were taken to rectify it. The principal problem which resulted was the adverse effect upon the operation of the sluices which were later complained of as necessitating four hours work to open and close.

Unlike Colt Crag, the Little Swinburn embankment caused no problems and, with the aqueduct, was brought into service during September 1879; the decision to retain it in the scheme was thus vindicated as it was to be some years before Colt Crag was to be completed. By August 1881 the capital expended on Colt Crag had risen to £20,272 while extra work, forming part of the contract, totalled £23,302 [45]; movements in the embankment were still causing concern.

In 1881 it was found that water was being lost through swallow-holes adjacent to the Dry Burn at the upper limit of the water level and Bateman was asked for his views. He proposed the laying of some four or five acres of concrete 12 ins. thick but as the estimated cost amounted to some £10,000 he was not permitted to proceed. Sealing was carried out by the Company's engineer using his own labour to do the work; executed in the autumn of 1881 the work cost some £600.[46] To enable the reservoir embankment to consolidate, the placing of fill was suspended for some months and it was only in1884, after further settlement during 1883, that the reservoir was brought fully into service and the accounts of both Rigby and Bateman settled. The final cost of the works has not been accurately assessed but it would seem that a total sum of £65,000—£70,000 [47] could have been involved, perhaps an increase of 80% on the tender figure.

The problems associated with Colt Crag reservoir did not end with its being brought into use and were principally due to the limestone outcrop in the Dry Burn area. In 1888 Forster, by then Engineer, reported that an area some 20 ft. square had fallen in, forming a swallow hole from which "a good portion of the water escaped".[48] The hole was subsequently filled with boulders and concrete and then finished by means of a layer of puddle clay. Problems continued until in 1894 Forster was instructed to report on the reservoir's condition. He responded within the month, commenting that the restrictions in the overflow channel, necessitated by the remedial earthworks, had rendered it inadequate and recommended that a new channel be built immediately.[49] As a result Charles Hawksley (1839–1917), by this time engaged in the design of the river Rede works, was asked to report upon the watertightness and safety of the reservoir; he found it structurally safe but recommended that part of the bank be raised to ensure watertightness, presumably by preventing leakage into the limestone outcrop.[50]

It was not until 1901 that the work necessary to render the Colt Crag reservoir watertight was commenced.[51] Drawings then prepared indicate that a puddle clay cut-off trench was to be provided along the south eastern shore of the reservoir, extending from the eastern end of the embankment for a distance of 1950 ft. upstream, its object being to prevent water passing into the limestone; the trench was to be six feet in width and extend to a maximum depth of 48 ft. A section of the cut-off was to be formed in the reservoir itself, this length being protected by an earth embankment formed in layers two feet thick. To ascertain in more detail what the problem was, boreholes were taken the following year, it then being found that leakage from the reservoir took place only when the surface was within 4′ 6″ of top water level. (See Fig 14)

Arrangements by this time had been made for housing the requisite labour, directly employed by the Company, and the work of excavating the trench commenced in April 1901 and continued until July 1904, the placing of the puddle clay in the main trench being completed in September following.[52] In May of the following year, with the water almost two feet above the point of previous leakage, Alfred L. Forster (1862–1924), who had in 1898 succeeded his father as Engineer, was able to report that in that year "with water at this level the flow from the Swin Well was 273,000 g.p.d. more than today",[53] considered by him to be a satisfactory measure of the efficiency of his remedial works. Concurrent with the formation of the cut-off trench were modifications to the overflow channel[54] and both works were completed in 1906, by which time £35,134—almost as much as the original cost—had been expended on the trench and some £7,000 on the overflow.

It was not until 1912 that work related to this reservoir's watertightness was completed. In 1908 it had been found that leakage was again in evidence at the end of the puddle clay trench and in 1911–12 this leak was made good by excavating a trench and filling with concrete in preference to puddle clay. Even this work did not completely cure the problem and the reservoir has sporadically given trouble since that time.

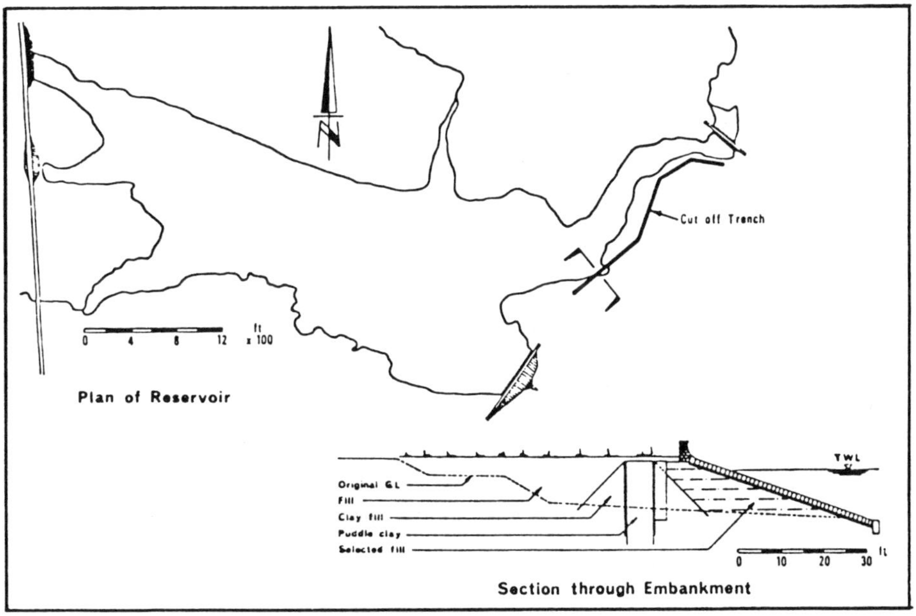

Fig. 14. Colt Crag reservoir: repairs, 1901-1904.

3.3 West Hallington. The fourth of the reservoirs in the Hallington group, West Hallington, was begun in 1884. Its construction had been authorised in the Company's 1876 Act but the delays at Colt Crag had the effect of postponing construction, the Company not considering itself able to build more than one major reservoir at a time. Dissatisfied with the efforts of Bateman and Hill, the directors resolved that its construction should be "under the sole supervision and direction"[55] of John Reay Forster (1821–1898) who had been employed by the Company since 1863; already receiving a salary of £500 p.a., he was awarded an additional £100 p.a. for the duration of the work.

The reservoir had first been suggested by Hill in his report of 1873, its purpose being solely to store water which would otherwise have run to waste from Colt Crag. It has no catchment of its own and is situated immediately to the west of the earlier reservoir, what was formerly the western embankment forming a causeway separating the two reservoirs; a 42" diameter pipe passing under this embankment connects them. The reservoir is formed by embankments on its eastern, southern and western sides, the outlet passing through the eastern embankment and the overflow channel running westwards from the eastern extremity of the southern embankment.

Envisaged by Hill as having a capacity of 655 mill. galls. the reservoir actually contains 743 mill. galls. at an elevation of 508 ft., some two feet higher than its neighbour. The maximum water depth is 35 ft. and the embankments are 750, 3,220 and 1260 ft. in length; at 124 acres its surface area is virtually the same as the earlier reservoir. In the details of its outlet and overflow arrangements the West Hallington reservoir is identical to the eastern one, the only difference being the fact that the overflow is of brickwork—the bricks were made on site—and is spanned by a wrought iron truss bridge. Forster could perhaps be critisised for using Bateman's detailing for the draw-off arrangement, potentially dangerous as they were, and are. In retrospect, however, it may be that he deliberately chose Bateman's earlier design on which to base his own; Bateman's Colt Crag design had by this time proved a problem—the valve shaft had tilted—and Forster simply adopted the earlier, simpler and cheaper design of East Hallington which had then given more than a decade of satisfactory service.

Partly due to the fact that the Company had carried out work in the Hallington area on its own account and perhaps because of some disenchantment with contract work, it was decided to construct the reservoir by means of directly employed labour under the supervision of a resident engineer, Charles George Henzell (1861–1934). As a result, records of construction details are available, added to which Henzell presented a paper on the reservoir's construction to the Institution of Civil Engineers.[56] From the records it has been possible to ascertain the problems encountered, and their solutions.

The site is underlain—as is Colt Crag—by the Bernician Series and by limestones. It is heavily faulted. Hard boulder clay overlies the rock formation and this is, in turn, covered "by boulder clay of a soapy nature, called the book-leaf marl, from its finely laminated structure"[57]. Following the taking of trial holes in the latter part of 1883 work began on the construction of the eastern embankment in April 1884 [58] and by July the outlet pipes had been laid and surrounded in concrete, at which time the south embankment was commenced. The outlet main is 24 ins. in diameter and at the crossing with the puddle trench was cast a 2" thick flange; to this flange is attached a wrought iron plate 12 ft. × 8 ft. × ½" thick. The pipe and flange are located in a concrete block. After the pipe had been laid the puddle trench was excavated, an average depth of 8 ft. being reached in the eastern embankment, although the maximum depth was 25 ft. due to book-leaf marl having been encountered. The embankment had a slope of 1 in 3 on the inside face and 1 in 2 on the outside, the puddle core being a minimum thickness of 8 ft. with an increasing batter of 1 in 12 on each face; the banks are 14 ft. wide at the top and the inner slopes protected with two feet of pitching. No problems were encountered until January 1885, after its completion, when the embankment settled some two feet over a length of 120 ft. above the site of the book-leaf marl.

The major embankment, the southern, was not so easily constructed, partly due to the fact that its line is crossed by the Small Burn; at this point the trench was excavated to a depth of 30 feet. The foundation for the whole length of this embankment was found to be poor with the book-leaf marl up to 10 feet in thickness; in addition wet sand was encountered, as was a length of black freestone, supposed to be a boulder although its extent was more than 150 ft. in length. Concrete was placed over the rock to form a seal before the placing of puddle clay was continued. In the early part of 1886 problems became apparent in the southern embankment when cracks were observed in the inside face; the damaged section was excavated, re-formed in thin layers and compacted, but still it moved and to prevent further slipping a concrete wall four feet wide was put in to a good base, a berm, or offset, being carried over it. No further movement was observed. To remedy further bulging of the embankment, timber piles 20 ft. in length were driven along the toe. The piles were left projecting two feet above the ground but within two days they had been overtopped; again the solution was to form a berm.

In August 1887, after other earth movements, the embankment alongside the overflow by-wash bulged so as to overtop the wall and to cause a depression some two feet deep alongside the puddle core; a further berm was formed to stabilise the earthworks. In the same year the embankment, at its western end, settled 6 ft. within two days, the toe heaving to a height of 4 ft. and again the same remedy was adopted. In all, seven berms were formed in the southern embankment, most with concrete walls or timber piling included. In every case of movement, whether apparent creep or slip, the book-leaf marl would seem to have been the principal contributing factor.

THE IMPOUNDING RESERVOIRS OF THE NEWCASTLE AND GATESHEAD WATER COMPANY

Like the previous embankment, the western one, too, had its troubles, the marl in places being 14 ft. thick. In addition, water proved troublesome and in August 1887 a spring yielding ¼ m.g.d. was encountered at a depth of 30 ft. below ground level,[59] necessitating the sinking of a wrought iron tank with an eight inch diameter pipe connected to it to bring the water, by pumping, to the surface. This section of the work also showed earth movements, cured as before. (See Fig. 15 for failure profiles).

The problems encountered are reflected in the cost of the work. Forster had originally estimated the reservoir construction as totalling £40,000 [60] and of the sum of £51,631 actually expended no less than £23,557 was classed as extras, work not envisaged when designs were first prepared. On the eastern embankment extras amounted only to an additional 3% over the original, on the southern 148% and on the western 104%. Material costs were small; stone was obtained from a quarry worked by the Company, situated some three miles to the north west, and connected by a narrow gauge railway; 750,000 bricks were burnt on site and incorporated in the works; and puddle clay was readily available. In all, the earthworks comprised 347,000 cu. yards, the puddle 58,000 cu. yards, stone pitching 43,000 cu. yards, cement 2170 tons and the bridge, from Joicey's, cost £190. Having begun to impound in May 1889, the reservoir was completed in January 1890 when Forster received an honorarium of £300 from the directors; Henzell received £50 for his supervision and organisation of the work.

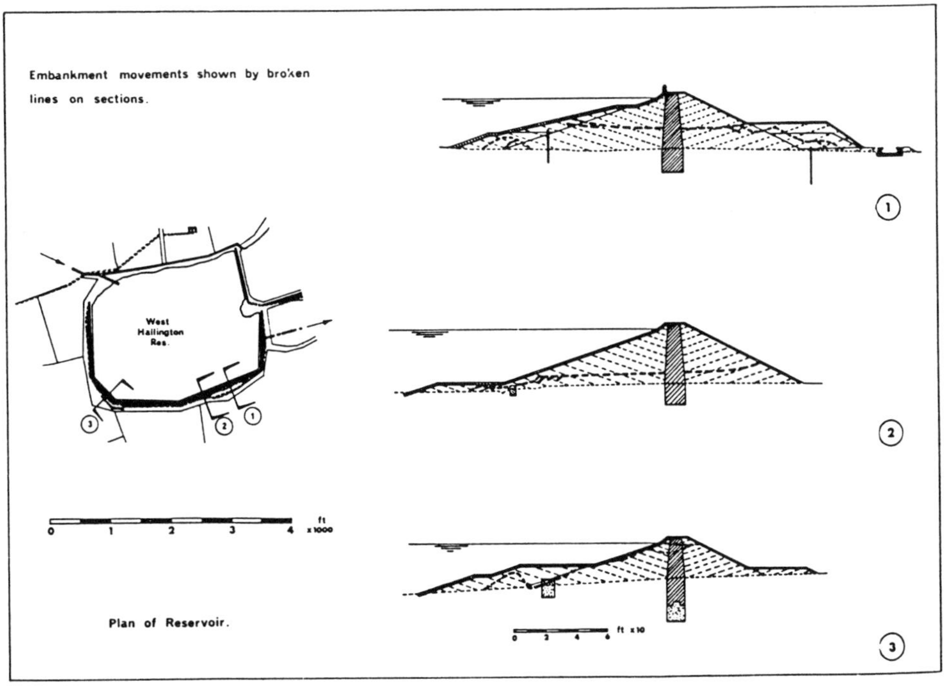

Fig. 15. West Hallington reservoir: slips during construction, 1884.

THE IMPOUNDING RESERVOIRS OF THE NEWCASTLE AND GATESHEAD WATER COMPANY

4. CATCLEUGH RESERVOIR

Although the complete scheme envisaged by Bateman and Hill—at least its first stage—had been completed, nevertheless the Company required yet more water to meet the needs of the area's rapidly growing population. In 1888 a report[61] was produced by Forster on the several sites in Northumberland available to the Company for the construction of a major impounding reservoir. At this time the consumption was almost 12 m.g.d., for a population of 320,000, and Forster propounded the view that a demand of 20 m.g.d. should be provided for. After considering eight potential sites Forster, like Hill, stated his preference for reservoirs on the Warksburn and Blackaburn of 1437 and 525 mill. galls. capacity respectively; the combined yield would be 10 m.g.d. He had instituted stream gaugings on the two catchments, the average flow over a 12 month period being 6.88 m.g.d. on the Warksburn and 6 m.g.d. on the Blackaburn; rainfall was not investigated.

Consequent upon Forster's report the directors resolved to ask Thomas Hawksley (1807–1893) for his views as to the Company's future development. Some doubts in the directors' minds had been evident in their decision to have water from both the river Rede and the Warksburn analysed but it was between the rivers Rede and Alwin that Hawksley's final choice lay; he eventually settled on a site on the Rede but considered that a yield of 6 m.g.d. would be sufficient[62]. The directors accepted his views, appointed him as Engineer—his fees were agreed at 5% of the cost of the reservoir and 3% of the pipeline, the Company to pay all supervisory staff—for the proposed works and resolved to seek Parliamentary powers for the construction of a small reservoir on the river Rede with a pipeline leading from it to Hallington. The subject of much discussion within Parliament—and without—the Bill finally became law in June 1889. Of the conditions which had been imposed upon the Company by the fishery interests, the Tyne Salmon Conservators, the main two were the construction of a fish pass and the prohibition of abstraction during February, March and April should the river's flow fall below 4.5 m.g.d.; for the remaining months the limit was to be 2 m.g.d.

So that the water of the Rede could be more quickly utilised, the laying of the 30 in. diameter main, 27 miles long, was the first section of the project to be put in hand, other than the sinking of exploratory boreholes at Catcleugh. Begun in 1892 it was brought into service in 1895, a small weir diverting water into it, upstream of the site chosen for the main dam, concerning which the Company had shown some prevarication as to its design. In 1890 following the preliminary site investigations, drawings had been produced which showed a masonry or mass concrete structure, its maximum depth of 20 ft. giving a capacity of some 40 mill. galls. (Fig. 16) This dam had a crest length of some 400 ft. with 264 ft. of it being available as an overflow weir three feet below crest level while it was founded, in effect, upon a concrete plug 15 ft. deep and 16 ft. wide. Two 12 ft. diameter apertures were to be formed in the dam, one to carry outlet and scour pipes and the other to form a means of rapidly drawing down the level. One of its features was a fish pass, an inclined channel 200 ft. long and ten wide, down which the water flowed past a series of interlocking part-width weirs; at its upper end the channel branched into three, passing into the reservoir at different levels. The fish pass was in 1891 the subject of model tests, carried out to ascertain its effectiveness[63] and it was the design of this part of the work which caused disagreement between Company and Conservators and which was eventually considered by the former as being so impracticable that £3,850 compensation was offered in its place[64]. This move did not find favour with the Conservators and the Company was forced into preparing a Bill for its abandonment, a measure rejected in Parliament. The matter was then discussed sporadically until it was resolved in 1893, following the submission by Hawksley of two reports to the directors.

Whereas the smaller reservoir was calculated as being able to produce a yield of 6 m.g.d.—presumably in conjunction with West Hallington—Hawksley now envisaged one yielding 7.5 m.g.d. in itself; of this figure some 2.5 m.g.d. would be "as compensation to the river and all parties interested therein"[65]. To provide this increased yield the reservoir capacity should be increased to 1400 mill. galls. and its surface level raised to 800 ft. The total volume of the earthworks involved in the embankment would be approximately 350,000 cu. yards and its height of 70 ft. would preclude the construction of a fish pass. The second report enlarged upon the rainfall available, estimated at 36 ins. p.a. with a three-dry-year figure of 30 ins., and foresaw a reservoir capacity of 2000 mill. galls. as being necessary to utilise fully the catchment of 10,000 acres; the yield would be 10 m.g.d. Hawksley did not recommend the construction of the larger reservoir but advised the directors only to provide storage of 1400 mill. galls.; he did, however, recommend compensation in money "with the total repeal of the Fish Pass clauses especially as each million gallons per diem will produce to the Company a gross revenue of £7,000 or

THE IMPOUNDING RESERVOIRS OF THE NEWCASTLE AND GATESHEAD WATER COMPANY

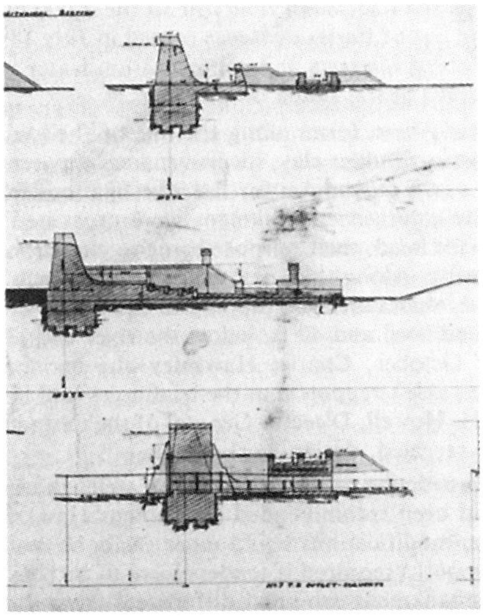

Fig. 16. Catcleugh reservoir, 1890.

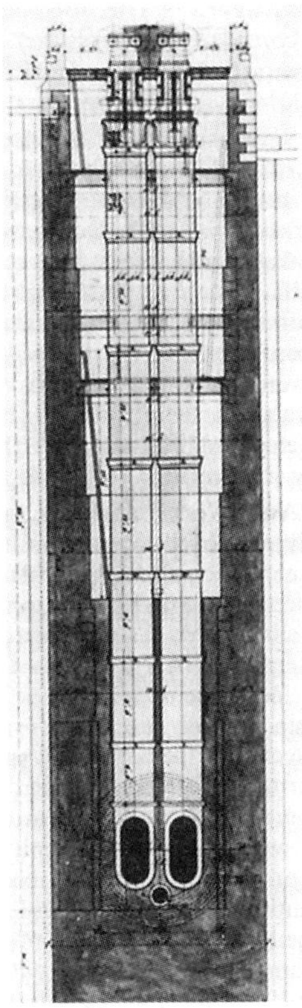

Fig. 18. (Right) Catcleugh reservoir valve tower, 1903. Each outlet is provided with duplicate valves.

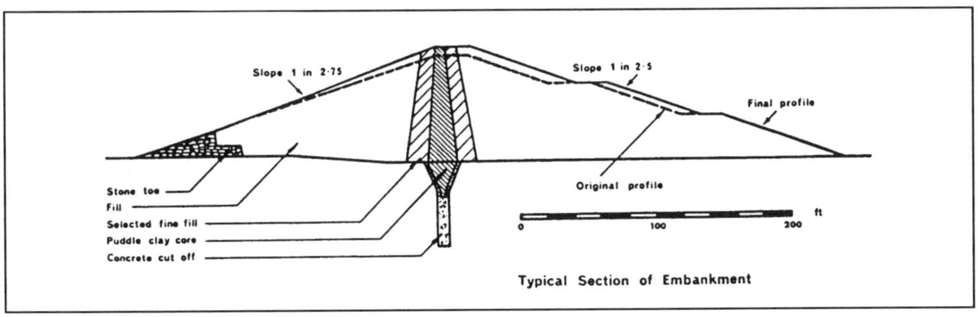

Fig. 17. Catcleugh reservoir embankment, 1896 and 1903.

THE IMPOUNDING RESERVOIRS OF THE NEWCASTLE AND GATESHEAD WATER COMPANY

£8,000 a year."[66] The directors agreed to increase the size of the Catcleugh reservoir to the 2,000 mill. galls. considered, but rejected, by Hawksley and the requisite Act of Parliament was passed in July 1894; compensation to the sum of £3,850 was to be paid to the fishing interests and compensation water was to total 4 m.g.d. or one quarter of the yield for the next seven years, based on gaugings.[67]

Seven boreholes to a depth of some 50 ft. had originally been taken along the line of the earlier dam in late 1889 and they had shown the drift to be underlain by boulder clay, its provenance apparently being the reason for the investigation. Following the passage of the second Act further investigations were undertaken in 1895. Six trenches parallel to the line of the enlarged embankment were excavated at intervals over the site of the reservoir stretching upstream to the head, their purpose being to cut through the drift, locate the watertight strata and prove their continuity. Along the line of the cut-off trench of the dam trial pits were dug down to the watertight limestone, shales and grit dipping to the north-west, the deepest pit—at that time—reaching 102 feet below ground level and 40 ft. below the river bed; 300 men were employed on this work.[68] By the beginning of October, Charles Hawksley—he had been appointed Engineer following his father's death in 1893—was able to report that the trial holes had been completed and in March 1896 they were inspected by Henry H. Howell, Director General of the Ordnance Survey in Scotland; following his approval as to the strata revealed, design work began.

As at West Hallington, the Company undertook the construction of the reservoir at Catcleugh using directly employed labour under Henzell. This procedure had been recommended by Thomas Hawksley on three principal counts: better workmanship would ensue; modifications could more easily be made; and the works could begin immediately, the estimated five months required if tenders were to be sought being eliminated. As designed by Charles Hawksley, the embankment exhibited differences from those of Bateman. Firstly, the cut-off trench was filled, in its lower part, with concrete to a thickness of eight to ten feet. The upper part of it, with sloping sides instead of the vertical of the lower, was lined with concrete on both faces, so forming a Y-shaped concrete cut-off which terminated at ground level. The puddle clay of the core-wall sprang from the invert of the junction. The second main difference was that the berms, or offsets, earlier resorted to in an effort to prevent embankments from slipping were now incorporated as a design feature, but on the downstream face only. The embankment was 1750 ft. in length with a maximum depth of water of 78 ft., a freeboard of seven feet—greater than Bateman's—being provided above the overflow level of 810 ft. On its upstream face the slope was 1 in 3 with a substantial toe of stone rubble incorporated; on the downstream face the slope was 1 in 2½ with two intermediate platforms 27′ 6″ wide. The embankment had a total width of some 510 ft. at its deepest point and the puddle core, with a minimum thickness of nine feet at the top, had batters on each face of 1 in 12; it was located with its centreline 11′ 6″ upstream of that of the embankment. As in Bateman's design a core of finer material was included on each side of the clay puddle; with batters of 1 in 6 this inner section had a minimum thickness of 23 ft. (Fig 17)

In principle, the outlet and overflow arrangements of the Catcleugh reservoir were similar to those at Colt Crag. Draw-off was by means of a 12′ 6″ diameter brick-lined tunnel 1443 ft. long which ran in undisturbed ground beneath the line of the embankment at its south western end. Control was by means of a substantial valve tower—the valves were set in a concrete block (Fig. 18) 36′ 0″ × 24′ 0″ and 20′ 0″ deep which formed a base for the brickwork, at the lower levels 4′ 6″ thick—controlling two pairs of sluice valves, located slightly upstream of the puddle core. One pair of valves discharged directly into the tunnel to enable the reservoir level to be lowered more rapidly; the other pair, in series, controlled the 30″ diameter outlet main which led to a gauge house, located downstream of the dam.[69] The reservoir's overflow was also at the south western end of the embankment, its crest of 310 ft. being adjacent to the valve-house. Passing under the embankment roadway, carried by means of a five span bridge, the channel reduced in width from 210 ft. to 105 ft. over its 2050 ft. length, terminating in the river Rede downstream of the works. All exposed facings were of masonry quarried on site. Hawksley estimated the cost of the embankment at £281,500; the directors approved his proposals and work began in April 1896.

Thomas Hawksley's first request on his appointment as Engineer had been that gaugings of both flows and rainfall should be arranged on the river Rede and its catchment and as a result two gauges had been brought into operation in 1889 and one in 1891. [70] The 1894 Act, however, had recognised that these records were probably inadequate, much work on this subject having by then been undertaken by George James Symons (1838–1900) who, since 1860, had recruited some 3000 observers, covering the country, measuring daily rainfall at 3404 stations and communicating their results to him. As a consequence it was enacted that the final assessment of rainfall on the catchment should be made more accurately and a period of seven years was allowed before compensation water needed to be determined. It was agreed that two arbitrators should be appointed to record rainfall and assess future compensation requirements; the Company made the suggestion—and it was agreed to—that they should be Hill and Symons.[71] The arbitrators were instrumental in having further gauges installed, 14 in number, and as a result of the recordings, Charles Hawksley suggested to the directors that it should be possible to increase further the capacity of the reservoir. Previously the catchment's rainfall had been assessed at 38 ins. p.a. but Hawksley now had no doubt that it could be taken as 42, if not 44 ins. p.a.[72], so resulting in more compensation water having to be provided and thus reducing the useful yield of the reservoir. The reservoir capacity should be increased to 2345 mill. galls. by raising the embankment, at that time formed to a height of 38 ft., by five feet. The extra cost would be £30,000, some 40,000 cu. yards of fill would be involved and the delay would be 10 weeks. His proposals were agreed to and to give effect to them a further Act was obtained in June 1902, in addition fixing compensation water at 3.01 m.g.d., to the Company a satisfactory outcome.

The resulting re-design of the embankment exhibited some elegance in that neither the puddle core nor the base of the embankment was affected. The stone toe, with its slope of 1 in 3 was allowed to remain but the upstream face of the dam was steepened to a slope of 1 in 2.75; on the downstream face the original slopes were maintained but the two horizontal platforms were reduced from their original 27′ 6″ to 20 ft., so providing the additional height of five feet required. The modifications resulted in an increase in length to its current 1830 ft. (Fig. 17, Page 45).

The construction of the reservoir is not well documented, other than by photographs and the progress drawings maintained as work continued. To enable the cut-off trench to be excavated the river was first carried over it by means of a timber trough, 50 ft. long with a cross section comprising two channels each 16′ 0″ × 8′ 6″ made up of 12″ × 6″ timbers; on completion of the tunnel in 1899 the river was diverted through it so that the embankment could be formed. Excavation for the cut-off trench began in June 1896 and continued into 1899 although by August 1898 it had been possible to begin the placing in it of the concrete membrane; the major part of this work was completed by 1901, the placing of the puddle clay below ground level having been commenced the year previous. Concurrent with the provision of the cut-off trench, 1890 ft. long and up to 160 ft. deep, was the construction of the lower parts of the valve tower while the overflow channel was begun, at its lower end, in 1899. The actual embankment began in 1900, together with the puddle core, and was substantially complete by the end of 1903; it was not until the following year that the valve house and overflow channel were completed. Due to the reservoir not being constructed by contract, no Specification or Bill of Quantities was prepared. However, notes for Henzell's guidance were prepared by the consultants and in them the methods to be used for carrying out the work are very strictly laid down, especially so far as the placing of clay puddle and fill were concerned. Puddle was laid in layers not exceeding six inches in thickness, placed by workmen standing, where practicable, side by side, so working the clay into the sections adjoining, watering it if so needed, and finally using puddling tools "to destroy every seam, joint or hollow in the puddling material whereby a leakage might occur—so as to ensure the utmost compactness and impermeability of the material"[73]. The inner layer of fill was also to be placed in layers not exceeding six inches in thickness while the outer section, depending on the method of placing, could be built up in layers up to three feet in thickness.

The construction of the reservoir employed a variable and seasonal number of men, recorded at times as being in excess of 600, mainly residing in a hutted village downstream of the dam. A narrow-gauge railway[74] ran from the site to Woodburn, there joining the North British Railway; a siding ran to the clay pit at Yatesfield some 10 miles south-east of the site. Stone was quarried on site, or in its immediate vicinity, bricks were obtained from William Foster of Croxdale, cement from Addison Potter & Co., timber from John Carr & Co., locomotives from Hudswell Clarke, Ruston steam navvies were used, ironwork was purchased from Abbot & Co. and valves from Glenfield & Co. at a cost of £5,400. As a result of eight years work some 62,000 cu. yards of material had been excavated from the cut-off trench and 386,000 from the overflow; a total of 65,000 cu. yards of concrete had been placed, as had a similar volume of puddle clay; and 560,000 cu. yards of fill had been placed in the embankment. The reservoir began impounding in June 1905 and was full within twelve months.

W. G. Armstrong

T. Sopwith

J. Simpson

R. Nicholson

J. F. Bateman

G. H. Hill

J. R. Forster

T. Hawksley.

C. Hawksley.

Fig. 19. The Company's Engineers, 1845-1905.

THE IMPOUNDING RESERVOIRS OF THE NEWCASTLE AND GATESHEAD WATER COMPANY

5. CONCLUSION

Of the impounding reservoirs designed for the Company during the period 1845 to 1905 none has remained unaltered; all have been modified—or modifications are in hand—in some form. In spite of this all of them have given satisfactory service and credit must be given to the designers, and to the contractors, for the generally high standard of workmanship; maintenance problems, taking a broad view, have been minimal. In assessing the performance of the reservoirs especial regard must be paid to two aspects of their design: firstly, the designers—and to a lesser extent the contractors—worked without the benefit of a knowledge of soil mechanics, a subject not formulated until some 50 years ago; secondly, although the science of hydraulics was well advanced, that of hydrology was not and it was only as a result of the work done by Symons and others, that its development took place. Taking these two factors into account errors have been few. (For Engineers' portraits, see Fig. 19)

It is not within the scope of this paper to detail the work carried out on the several reservoirs during the 20th century; the main works, however, may be listed as follows:

5.1 Whittle Dean. This group of reservoirs soon outgrew its original purpose of impounding water and came to be used virtually as a holding reservoir for untreated water from Catcleugh and the Swinburn group prior to its filtration. The outlet from the Great Northern reservoir was modified in 1930 to permit it to discharge directly into the Northern reservoir rather than into the by-wash while Nicholson's—or Armstrong's—outlet from the Great Southern reservoir was replaced in 1943 by a sluice valve with an inclined operating spindle running up the embankment; this, in turn, is soon to be replaced.

More serious was the failure of the scour pipe within the embankment of the Lower reservoir, Simpson's main point of supply. In 1936 a tunnel was driven into the downstream face of the embankment, the pipe renewed, a concrete plug inserted and the puddle clay core made good. Also of a potentially more serious nature was the inadequacy of the overflows. Consequent upon the *Reservoirs (Safety Provisions) Act, 1930,* all reservoirs came to be reported upon at intervals not exceeding ten years. Inspection at Whittle Dean revealed potential danger and a new overflow system with the final discharge channel by-passing the Company's principal filtration works, previously at risk, was completed in 1967.

5.2 The Swinburn Group.

5.2.1. East Hallington. The performance of the reservoir, the first designed for the Company by Bateman, has been more than satisfactory despite the fact that Bateman erred in placing the control valve on the main outlet pipe at the downstream extremity of the embankment. In spite of a century without problems it is now the Company's intention to install a valve on the upstream end of the pipe. As at Whittle Dean, overflow arrangements were found to be inadequate and were modified in 1947 by means of widening the channel.

5.2.2 Colt Crag. As during its construction, so in its performance has this reservoir been proved to exhibit the greatest problems. As already noted, additional work in making it watertight was executed very soon after its completion. Further work on the cut-off became necessary between 1908 and 1912 and since that time, at intervals, grouting has become necessary at the reservoir's eastern end; so persistent was the leakage that it is only within the past five years that it has been possible to fill the reservoir. Until then it rarely overflowed due to leakage taking place at high reservoir levels. Ironically, the decennial inspections showed that here, too, the overflow was inadequate; design work, with model tests, was subsequently undertaken and work was completed in 1951, a completely new alignment being adopted to eliminate the constricted channel.

As noted, the valve tower tilted during—or immediately following—its construction; since then it has given no further trouble other than the normal corrosion of ironwork. A house to afford protection when operation of the penstocks was taking place has been added to the shaft since its construction; aesthetically, Bateman's original design was to be preferred.

5.2.3. Little Swinburn. The principal problem at Little Swinburn has proved to be its overflow and this, too, was enlarged in 1948, the actual weir being both increased in length and amended in form to give a greater discharge. The overflow arrangements were complicated somewhat by the fact that any flood from Colt Crag passes into Little Swinburn, albeit with a lag effect caused by the intervening stream. Nevertheless it would seem that Bateman did not take this phenomenon fully into account, designing Colt Crag with an overflow of considerably greater capacity than that at Swinburn. Like Colt Crag, the valve tower has had a housing added to it, similar in design.

5.2.4. West Hallington. Due to its design being identical to that of the earlier Hallington reservoir similar proposals are in hand to install a valve on the upstream end of the outlet main. With no catchment of its own, it has not been found necessary to enlarge the reservoir's overflow.

5.3 Catcleugh. Problems associated with movements in the embankments had not been severe at the older reservoirs built by the Company, other than during construction. At Catcleugh, however, cracks became apparent between the overflow channel bridge and the valve tower, determined as being due to settlement in the embankment. Remedial works—not ascertained—were completed in 1910, work having continued over three summers as the replacement of considerable areas of pitching had been found necessary. In 1911 it was decided to by-pass the gauging basin at Catcleugh so as to enable the full reservoir head to be used to increase the discharge of the outlet main, in which problems in maintaining flow had been experienced. The measures were effective and the reservoir has subsequently supplied a daily flow varying between 11 and 13 m.g.d. depending upon the condition of the main, now scraped annually.

6. ACKNOWLEDGEMENTS

I am indebted to Mr. M. J. Lowther, B.Sc., F.I.C.E., F.I.W.E.S. and to Mr. H. D. M. Speed, B.Sc., M.Sc., F.I.C.E., F.I.W.E.S., Managing Director and Chief Engineer respectively of the Newcastle and Gateshead Water Company, for permitting me to use the Company's archives, from which virtually all the information used in this paper has been obtained. I would also like to thank Miss Joan Watt and Mr. P. Scorer for their assistance.

7. REFERENCES

1. For a full description of the Company's formation see R. W. Rennison, *Water to Tyneside,* (Newcastle upon Tyne, 1979) pp 22–46.
2. *Prospectus of.... the Whittle Dean Water Company* (Newcastle upon Tyne, 1845) p. 11.
3. Ibid, p. 8.
4. Ibid, p. 15.
5. T. Sopwith, *Diary of Thomas Sopwith, 1828–1879.*
6. Letter from Donkin, Stable and Armstrong to J. C. Blackden, 23 December 1844, *Duke of Northumberland Mss.,* Alnwick.
7. *Annual Report,* 24 July 1845.
8. *Prospectus....* op cit. (2), p. 10.
9. The present yield is taken as 2 m.g.d. In a report dated 18 February 1862, the same figure was arrived at by Bateman.
10. *Minute Book,* 11 February 1846.
11. *Contract for the execution of Reservoirs at Welton* (13 March 1846).
12. *Proceedings of the Managing Directors,* 2 May 1846. Four other tenders for ironwork were received, three of them identical.
13. *William Reid's Reports during Construction of Whittle Dean Reservoirs 1846 to 1847.*
14. *Annual Report,* 15 February 1849.
15. *Proceedings of the Managing Directors,* 12 July 1850. The reservoir was later said to be of great use for flow control. (See *Annual Report,* February 1852).
16. *Minute Book,* 14 March 1851. Four tenders, all from local contractors, were received varying from £3,025 to £5,298.
17. *Minute Book,* 31 July 1854. Six tenders were received from local contractors, varying from £17,497 to £32,150.

THE IMPOUNDING RESERVOIRS OF THE NEWCASTLE AND GATESHEAD WATER COMPANY

18. William Humber, *A Comprehensive Treatise on the Water Supply to Cities and Towns,* (1876), p. 131.
19. *Minute Book,* 22 May 1855. The appointment of Tone was recommended by Simpson.
20. R. W. Rennison, op.cit.(1), pp. 89–100 for details.
21. *Minute Book,* 4 October 1861.
22. Report from Bateman to Company, 8 October 1861.
23. Ibid. The rain gauges were installed immediately.
24. Report from Bateman to Company, 18 February 1862.
25. Ibid.
26. R. W. Rennison, op. cit. (1), pp. 102–3.
27. *Minute Book,* 6 July 1869. Four tenders from local contractors were received, varying from £19,900 to £33,442.
28. Note on *Contract Drawing No. 6,* signed by Bateman and McGuire.
29. *Minute Book,* 6 December 1870. Within 14 days the quantity impounded had risen to 200 mill. galls.
30. *Annual Report,* Feb. 1872.
31. Report from Hill to Company, September 1873.
32. The present-day figures are 26.52 and 29.62 ins. p.a.
33. The current figures for the three-dry-year rainfalls at Whittle Dean and Hallington are 21.72 and 24.99 ins. p.a. respectively.
34. R. W. Rennison, op. cit. (1), pp. 126–139 for details of the struggle between Company and Corporation.
35. Report from Hill to Company, 20 September 1875.
36. House of Commons, May 1876.
37. The reduction in elevation was no disadvantage; the initial intention was to discharge into the Dry Burn.
38. *Minute Book,* 17 April 1877.
39. This embankment was not envisaged initially.
40. Prices ranging from £36,845 to £88,714 were received from 14 contractors.
41. Four tenders were received, ranging from £607 to £778.
42. *Minute Book,* 13 May 1879.
43. Ibid, 13 April 1880.
44. *Engineer's Reports,* 9 October 1894.
45. *Minute Book,* 30 April 1881. No details are extant but it is probable that the major portion related to the Colt Crag embankment.
46. *Engineer's Reports,* 25 October 1881.
47. The cost in August 1881 was £57,181.
48. *Engineer's Reports,* October 1888.
49. Ibid, 23 October 1894.
50. *Minute Book,* 6 November 1894.
51. Ibid, 11 June 1901.
52. *Engineer's Reports,* 11 October 1904.
53. Ibid, 9 May 1905. A gauge had been installed on the Swin Burn to quantify leakage from the reservoir.
54. It is possible that the downstream section of the overflow channel was rebuilt as twin circular culverts.
55. *Minute Book,* 8 April 1884.
56. C. G. Henzell, "The West Hallington Reservoir", *Min. Proc. I.C.E.* vol. **cii**, Session 1889–90, Part iv.
57. Ibid, p. 3.
58. *Engineer's Reports,* 22 April 1884.
59. *Minute Book,* 23 August 1887.
60. *Engineer's Reports,* 11 September 1883.
61. Report from Forster to Directors, March 1888.
62. *Minute Book,* 4 September 1888. Hawksley considered the Company's safe yield at that time to be 12 m.g.d.
63. Ibid, 13 October 1891.
64. Ibid, 29 February 1892.
65. Report from Hawksley to Company, 31 July 1893.
66. Ibid, 8 September 1893.

THE IMPOUNDING RESERVOIRS OF THE NEWCASTLE AND GATESHEAD WATER COMPANY

67. R. W. Rennison, op. cit. (1), pp. 169–226 passim for a more detailed description. The generally accepted figure for compensation water was then one third of the yield.
68. *Proceedings of the Town Council of Newcastle upon Tyne,* 2 October 1895. The visit was made in September.
69. The Glenfield control valve was removed in 1911 to increase the flow in the outlet main by means of utilising the full reservoir head. A similar valve, purchased in 1892, is still in use at Whittle Dean filters.
70. Letter from Hawksley to Company, 5 October 1901.
71. Following the death of Symons in 1900 his place as arbitrator was taken by H. Sowerby Wallis.
72. Rainfall is now assessed at 44.14 ins. p.a. with the three-dry-year figure of 37.62 ins. p.a.
73. T. & C. Hawksley, *General Instructions.* Ms. undated.
74. H. D. Bowtell, "By Rail to Catcleugh", *Stephenson Locomotive Society Journal,* vol. **xxxix**, October 1963. The subject has since been further researched by him.

DISCUSSION

Mr. C. A. Noble opened the discussion by asking whether there was any further information about the ageing of the iron work of the Whittle Dean reservoirs. **Mr. Rennison** replied that the penstock illustrated had been installed in the period 1855–7. It had jammed due to lack of operation and was so firmly shut that the spindle bracket broke when opening was attempted; the penstocks themselves are still in mint condition.

Mrs. S. M. Rolt asked whether there had been any major objections from landowners to the construction of the reservoirs and had any of these objections resulted in the resiting of any of them? **Mr. Rennison** replied that in 1845, when the Whittle Dean reservoirs were being built, one of the landowners' conditions was that the work camps should be policed to reduce trouble both from poachers and from drunks on pay day. At Catcleugh, half a century later, it had been necessary to buy more than 10,000 acres of land from two landowners who were not satisfied with the Company's first proposals. The Company was fortunate, or farseeing, in appointing directors who were of the landowning class; landowners still hold fishing rights on some of the reservoirs.

Dr. R. A. Smith said that in the 1840s the main objections had come from millowners who had lost their water supplies and the compensation water was not necessarily sufficient to meet their requirements; he added that catchment areas were strictly patrolled for hygienic reasons. **Mr. Rennison** said that the Company had not been greatly involved with matters relating to compensation water although at Catcleugh there had been a long drawn out struggle with the millowners and the fishery interests; it ended with the Company providing three million gallons of compensation water a day with a supply to the consumers of 10 to 11 million gallons. Hawksley wrote to the Company at some length about this matter, suggesting that it would be much better to settle with the landowners and millowners financially rather than providing compensation water. Regarding pollution, it was probably not at first appreciated that it could occur but by the time that Catcleugh was built Hawksley had completed extensive filter beds at Whittle Dean; as a consequence there was no need to restrict access to the catchment area. The water supply must have been considerably contaminated as in Newcastle almost every household seemed to have possessed a charcoal filter and the press records references to the "mahogany colour" of the mains water. **Mr. I. McNeil** asked whether any reservoirs other than those listed were constructed by the Company and did Armstrong remain Chairman for the rest of his life? **Mr. Rennison** replied that all the impounding reservoirs had been covered by the paper. In answer to the second question, the Company was formed in 1845 and in 1863 it had changed its name to the Newcastle & Gateshead Water Company. Armstrong was Company Secretary for the two years from 1845 and then left, at the time of disapproval by the local press which accused him of using town's water, obtained too cheaply, to operate his hydraulic cranes. This accusation was false as the accounts show that he paid the same rate for water used as did everyone else. He returned to the Company as Chairman in 1853 and continued in that office until 1865.

Dr. R. J. M. Carr said that water had been obtained in the early days direct from the Tyne; was there any evidence to show that borehole pumping had been employed? **Mr. Rennison** replied that such pumping had been continued until about 1850, by then principally from old coal workings. In 1832—the year of the first cholera epidemic—the first pumping station on the river Tyne was built at Elswick. By

1853 the Company was trying to obtain further water from upland sources to augment Whittle Dean and when this proved unsuccessful the 1832 pumping station on the river was again brought into operation, so exacerbating the cholera epidemic; cessation of pumping virtually brought about the end of the epidemic.

Mr. J. W. Butler asked whether brickmaking continued on the reservoirs' construction sites. **Mr. Rennison** replied that there were many references to brickworks in the area and most big estates had their own brickworks or tileries. It would appear from the Company's records that bricks were purchased to build brick kilns at Hallington but the bricks used at Catcleugh were purchased from a Durham manufacturer. **Dr. M. D. Scott** asked whether the water was sold to the City or direct to the consumers? **Mr. Rennison** replied that the Company supplied the consumers direct, industrial consumers being charged for the quantity used and householders on the rateable value. Over the period covered by the paper water was also obtained direct from the river and in about 1865 the Company provided two supplies; most industrial users were supplied from the river and domestic consumers from the reservoirs. This state of affairs lasted until two years ago; if one went into a riverside factory one could still see taps labelled "not drinking water".

Mr. D. Adler reverted to Mr. Butler's question about brickworks and said that in London, Wormwood Scrubs Prison had been built by prisoners using bricks made from clay from the site. **Mr. D. H. Tew** added that such is the conservatism of the legal profession that even today on many housing estates each plot has a restrictive covenant against the digging out of clay for brickmaking; this is particularly so in South London.

Miss J. Elton asked why the Company had decided to dispense with contractors? **Mr. Rennison** replied that many works had been under-quoted and the contractors gave up the contract when they realised a loss would be incurred. The aqueduct from Hallington to Whittle Dean included a three-mile tunnel through limestone and it was completed only by the last of three contractors; the first two experienced financial problems and were released from their contracts.

Mr. M. T. Tucker asked what objections there were to taking water from the upper Tyne and could Mr. Rennison give any more information about the aqueducts? **Mr. Rennison** replied that the South Tyne was not considered suitable; in the 19th century it was reputed to be contaminated by lead from workings at the river's head. So far as the North Tyne was concerned it was thoroughly investigated in 1888 when every reservoir site was considered for development. The river North Tyne is now impounded by the Kielder Reservoir, brought into use in 1981. The earlier aqueducts were, in effect, contour canals six to eight feet across, of trapezoidal cross section and presumably—there are no record drawings—lined with puddled clay. The tunnel from Hallington was a major work, three miles long and six feet in diameter and through very hard limestone, excavated from 12 or 13 shafts. In 1905 it was augmented by Hawksley by means of an eight foot diameter tunnel which was mechanically excavated using compressed air breakers, the first so constructed for the Company; lighting was by means of electricity. From Whittle Dean the water was carried by a cast iron pipe which suffered from the fact that virtually the whole of the route was subject to mining subsidence, frequently causing fractures in the pipes. Automatic valves were installed to cut off the flow should a fracture occur; what was not foreseen was that the water hammer effect would cause problems and even further pipe fractures.

A visitor asked whether Portland cement had been used in any of these reservoirs? **Mr. Rennison** replied that in Simpson's works masonry was normally set in hydraulic lime. Roman cement was used, on account of its greater strength, where pipes passed through the masonry. Portland cement was used at Catcleugh but he was uncertain what was used between the two construction periods. One of the Company's directors was Wilkinson—he had patented a form of reinforced concrete construction in 1854—who had interests in a local cement works, Johnsons; another was Potter, who in addition to cement manufacture was in the gas and coke business. The Company's records show that the source of cement alternated between suppliers, depending on which directors were present at the board meetings.

The President asked how the puddled clay was kept moist and therefore impervious; **Mr. Rennison** replied that puddled clay was normally taken to about four feet from the top of an embankment so that it did not dry out. Puddling was a complicated procedure, generally performed by men wearing long heeled boots. As it was very labour intensive, the puddled core was made as thin as possible to keep down the cost. **Dr. R. A. Smith** said that puddling had first been demonstrated to a Committee of the House of Commons and was last used, he thought, on a large reservoir in 1966. **Mr. Rennison** said he had seen it used at about that time and **Mr. D. Adler** said he had witnessed its use in a lake constructed in 1967 at the University of Surrey. Mr. Rennison read Hawksley's work specification to the effect that puddled clay would be impervious only if it had been made completely homogeneous; if this requirement were not met water would pass through it.

The President called on Mr. J. H. Warden to propose the vote of thanks. **Mr. Warden** said that he had been impressed by the wide time scale of 140 years which had been covered by this paper which had been extremely well prepared and interestingly illustrated. It gave him much pleasure in thanking Mr. R. W. Rennison for this presentation and the vote of thanks was passed by acclamation.

8

The old Croton Aqueduct

George H. Rappole

The citizens of the City of New York, all 7,895,563 of them, are still putting their tax money into the construction of their water supply system, just as their forbears did at the beginning of the nineteenth century. In 1842 Mayor Robert H. Morris declaimed at the ceremonies opening the Croton Aqueduct that "pure and wholesome water will be our legacy to countless generations into the future."[1] The city's first public water supply system proved to be far more finite than his prediction, for the original Croton served the city for only fifty-one years from 1842 to 1893 when the New Croton Aqueduct was completed and brought into service. By 1968 the Old Croton was closed down and deeded by the City to the Taconic State Park Commission for use as a public walkway. In recent time, the efforts of interested people have resulted in the designation of the entire Old Croton Water System as a National Historic Civil Engineering Landmark.[2]

This aqueduct was regarded by its contemporaries, domestic and foreign, as one of the notable engineering achievements of its day. It remains of interest today not only for its structures but for the careers of the engineers who designed and built it; it was a practical academy for the founders of civil engineering in the United States.

The Physical Setting

Manhattan Island, with the Hudson River on its western side and the Harlem and East rivers on its eastern sides, had by the beginning of its growth into a city all but exhausted its natural sources of potable water. Its surrounding rivers could not be used because of the salinity of their waters. The Island itself is nearly solid rock with only a few surface ponds, streams and springs for use by its residents who numbered approximately 50,000 in 1800. These natural sources were inadequate and were suspected to be a source of disease. The only safe source was the water boats which brought barrels from springs in New Jersey and New York across the Hudson River. In addition, the supply of water for fighting fires was frequently insufficient.

One of the first organized attempts to solve the problem was the City's approval of a charter to The Manhattan Company, a group of business men led by Aaron Burr. In 1799, the Legislature of the State of New York approved the City's action by incorporating this company "to supply the City of New York with pure and wholesome water."[3] The Company did indeed build a small reservoir and lay some pipe, but this was a token action to preserve its charter; its main business became banking. Burr's real objective was to establish a bank to compete with the city's other two banks which were controlled by members of the Federalist Party. The Company survived its failure to supply water and is today known as the Chase Manhattan Bank.

The search for water went on and soon led to the rivers in Westchester County. Several members of the Common Council suggested tapping the Bronx River, a stream that flowed into the Harlem River at the northeast tip of the Island. In consideration of the limited flow of this stream, Francis E. Phelps of the Council proposed three different sources: Rye Pond (in Westchester), the Croton River (in northern Westchester), or the Passaic River above the Falls (at Paterson, New Jersey). Westchester County's terrain was well suited to the purpose as it has many natural streams and ponds, in hilly, well wooded land with good soil cover. The Croton River, a tributary of the Hudson some thirty-five miles north of the City, was known to have a heavy, reliable flow of water.[4]

[1] Charles King, *A Memoir of the Construction, Cost and Capacity of the Croton Aqueduct* (New York, 1843).

[2] Society for Industrial Archeology, *Newsletter*, Vol. V, No. 1 (Jan., 1976), p. 1.

[3] Alvah P. French, *A History of Westchester County* (New York, 1925), pp. 344-345.

[4] UTM Reference: Ossining 0041.6518. The author is indebted to Peter H. Stott for this and subsequent UTM Reference information related to The Old Croton Aqueduct. Personal communication, June 13, 1975.

George H. Rappole is a Program Specialist, Department of Human Resources, North Central Texas Council of Governments, Arlington, Texas.

Industrial Archeology

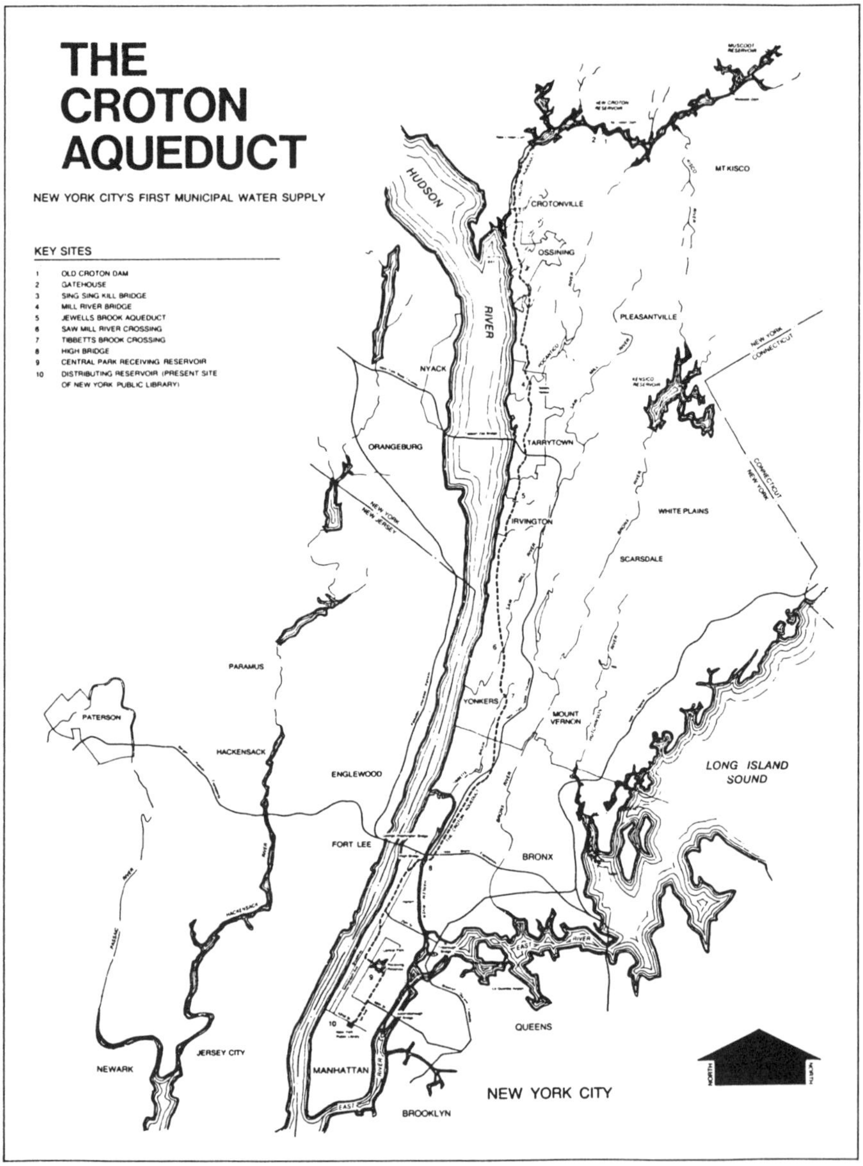

Figure 11. *Map of the Croton Aqueduct, prepared for this paper by Tom Shelton, Assistant Planner, Division of Management and Planning Coordination, North Central Texas Council of Governments, Arlington, Texas. The map is based on U.S. Geological Survey Quadrants and indicates the major structures and locations of the aqueduct.*

The first report on the Croton's possibilities as a source for the City was submitted to the Common Council in 1832 by De Witt Clinton, characterized in one account as a "competent engineer."[5] Colonel Clinton's report prompted the Council to petition the Legislature to establish a commission to make the decisions necessary to supply the City with water. One of the first actions of the Commissioners—who had been appointed by the Governor in early 1833—was to hire two engineers to make a complete study of the Bronx and the Croton Rivers. The men chosen, Canvass White and David B. Douglass, finished their survey and report later that year. On the basis of this information, the Legislature authorized a referendum by the voters of the City. The issue was approved in April, 1835, by a vote of 17,330 to 5,963. The cost would be $4,250,000.

Canvass White's career had begun with the surveys for the Erie Canal in 1816 when he was twenty-six.[6] Until his death in 1834, he was a principal engineer in the construction of the Delaware and Raritan Canal, the Chesapeake and Delaware Canal, and other similar engineering assignments, all of which brought him praise and a high reputation. His colleague on the Croton Survey, Major Douglass, was appointed Chief Engineer of the Croton Project by the Commissioners. Douglass (1790-1849) brought considerable experience to his new post.[7] A graduate of Yale in 1813, he had become Assistant Professor of Natural Philosophy at the West Point Military Academy in 1816 after serving in the War of 1812. In addition to his teaching duties, Major Douglass was a practicing engineer; he made surveys and estimates for several canals in Ohio and New Jersey. He resigned from the Academy to join the Morris Canal Company (New Jersey), for which he designed inclined planes rather than locks for lifting the boats, an innovation that drew much attention. In the year he and White studied the Croton River for the Commissioners, Douglass was a member of the faculty of the University of the City of New York. In 1841, he became President of Kenyon College in Ohio. Returning to New York in 1844, he continued his engineering activities. In his last years, he served as a Professor of Mathematics at Hobart (then Geneva) College. During his service to the Water Commissioners, he made a large part of the survey which located the line of the Aqueduct and drew up many of the preliminary structural plans.

[5]Frederic Shonnard and W.W. Spooner, *History of Westchester County, New York* (New York, 1900), pp. 548-553.

[6]Edward Wegmann, *The Water-Supply of the City of New York, 1658-1895* (New York, 1896), p. 235.

[7]*Ibid.*, pp. 235-236.

Plans for the Aqueduct

There must have been some urgency in the minds of Douglass and his co-workers as they worked on the surveys for the Aqueduct. New York City had had serious fires in 1828 and 1835. In 1832, a severe cholera epidemic caused 3500 deaths. For political reasons, Douglass did not remain long in his post as Chief Engineer. The Minutes of the Commission and its successor, the Croton Aqueduct Board, reveal only that the Major was thanked for his contributions and that a new Chief Engineer was appointed, John Bloomfield Jervis[8] (1795-1885).

Jervis wrote in later years that he had not sought the position, nor had he any particular knowledge of the project until the Commissioners made their offer to him.[9] Jervis gave full credit to Douglass for his preliminary work on the Aqueduct and its structures. Still, there was much work to be done on detailed cost specifications, several of the major structures and other necessary matters. Jervis wrote that he was the one person solely responsible for the success (or the failure) of the project; probably he did not consider himself proven in his decisions until Croton water flowed into Manhattan.[10]

Jervis' career in engineering had begun in the canal-building era. He was a native of New York, born in 1795 in Huntington, Long Island. His family moved to Rome, New York, where—following his elementary schooling—he became a surveyor's assistant, staking out a section of the Erie Canal. He moved up through higher positions to his appointment as a District Superintendent of the Canal. In 1825, he left this position to become to the Chief Engineer of the Delaware and Hudson Canal. His long career included not only canals and aqueducts but the development of the railroads: he was the inventor of the bogey-wheel locomotive pilot truck, which made feasible longer and heavier engines. He also wrote a treatise on railroad engineering.[11]

[8]City of New York, Water Commissioners, *Semi-Annual Reports, 1837-1842; Minutes of the Croton Aqueduct Board, July 18, 1849 to April 9, 1870*, New York, N.Y., 1903; *Quarterly Report of the President of the Croton Aqueduct Board*, to October 31, 1845.

[9]John Bloomfield Jervis, "Memoir presented October 18, 1876," *Transactions of the American Society of Civil Engineers*, Vol. 6, pp. 39-67.

[10]J.B. Jervis, *The Reminiscences of John B. Jervis, Engineer of the Old Croton*, edited with introduction by Neal FitzSimons (Syracuse, New York, 1971).

[11]J.B. Jervis, *Railway Property, A Treatise on the Construction and Management of Railways* (New York, 1861).

Industrial Archeology

Jervis began his duties on the Croton Aqueduct in 1837. His first problem was the design and construction of the dam on the Croton River, some six miles upstream from the Hudson River. The location of the dam was determined by the flow of water as gauged by Major Douglass in 1833. He found it to be a minimum of 44,120,924 gallons per day at any season. Horatio Allen, the principal assistant engineer under Jervis, rechecked this flow in the summer of 1838 in a period of a severe drought. Allen found it to be 27,562,280 gallons per day, an amount three times greater than that needed for the city's uses. Originally, a dam ninety feet long had been planned. A dam of this length built of the local granite would reach only part way across the river; an earth embankment would complete the structure. Work had progressed through the autumn of 1840, until winter weather brought it to a halt. In January, 1841, heavy rainfall on top of an eighteen inch cover of snow collapsed the embankment, flooding the valley of the Croton below the dam site. Several bridges were swept away, a number of buildings destroyed, and three lives were lost. In the subsequent law suits, the courts set the damages for which the city was liable at $75,000.[12] The loss of the first dam is vividly described by Bellamy Partridge in his novel, *Big Freeze*, a modest and pleasant fictionalization of the life and times of Jervis during the Croton era.[13]

Considering this brief setback, Jervis decided that the stone dam should be extended to a length of 270 feet with a short earth embankment. This dam is now under some forty feet of water but remains a reasonably sound-looking structure (Figure 1). This photograph and that of the original gatehouse were taken during a severe drought in the early 1960s. A concrete core, or armature of the dam, rests on the river bed which had been cleared of mud and boulders. This core was faced on both upstream and downstream sides with one-foot hemlock piers which, with white oak and stones in alternating layers, made a secure foundation for the body of the dam. This was made of hydraulic stone masonry onto which was laid the face work of cut granite with no joint allowed to exceed 3/16 inch. Jervis had observed that other dams were subject to undercutting from the erosive force of the flow of water. He added to the design a secondary dam three hundred feet downstream, to form a stilling basin which would break the body of water coming from the lip of the dam, and would keep the timber piers under water, preventing their decay. The face of the dam, the stilling basin and the second dam together made an elegant reversed or ogival curve. The top of the dam was fifty feet above the river bottom, allowing the river water to be raised forty feet and forming Croton Lake as the collecting reservoir. T. Schramke, a European engineer hired to work on the project, estimated that the lake made available 500,000,000 gallons which would allow the Aqueduct to discharge 35,000,000 gallons per day.

A tunnel, 180 feet long, was cut through solid rock beneath the bottom of the reservoir, leading to the back of the dam. The tunnel led directly to the gatehouse which was built on the solid natural rock. Water passed first through an arch in the bulkhead of the gatehouse, then through a screen-frame, a set of guard-gates, a set of regulating gates and finally into the main aqueduct. "The screen formed of oak-slabs, six inches by one, allowed a quantity of fish to pass through the one inch spaces into the Aqueduct; in order to prevent this, a fine brass-netting was put over the screen, through which only very small fish could pass, to prevent which other artificial preparations will be required."[14] In Figure 2, one may note the top of the control gates. The gates, with their frames, were made of cast iron. The two sets of seven each are called regulating gates and guard gates. The cast iron rods visible in the photograph are twenty feet long; at the bottom, these were attached to the actual gates. By turning the rod with a "key" or wheel, the rod would go up or down, opening or closing the gate.

From the gatehouse, the Aqueduct descends to Manhattan

[12] State of New York, Courts, "The Mayor, Aldermen, and People of the City of New York vs. John Bailey," 3 Hill 541, 1844.

[13] Bellamy Partridge, *Big Freeze* (New York, 1948).

[14] T. Schramke, *Description of the New-York Croton Aqueduct with Twenty Plates, in English, German and French* (New York and Berlin, 1846), p. 29.

Figure 1. *The Old Croton Dam at low water in the 1950s, exposing the main spillway. Photo courtesy of Charles S. Hay, Thurman, New York.*

for a distance of forty-one miles, at an average fall of fifteen inches per mile. In the process of bringing the water tunnel from the Croton Dam to the High Bridge in the Bronx, Jervis noted that approximately 400,000 cubic yards of rock were excavated. Sixteen tunnels were cut—varying in length from 160 to 1263 feet. One hundred fourteen culverts were built; thirty-three ventilating shafts at roughly one mile intervals; six waste-weirs; one major bridge crossing at Ossining (then, Sing Sing); and several other massive stone crossings at other points in Westchester.[15]

Figure 2. *The Gatehouse at the head of the Croton Aqueduct, during low-water in the 1950s. Photo courtesy of Robert M. Vogel, Washington, D.C.*

Where the Aqueduct was cut in level ground or side hills, a course of concrete was laid, three inches thick under the width of the masonry but if in soft or loose ground, twelve inches thick. The side-walls were carried up in brick to the spring-line of the inverted arch, and the arch made of brick to form the roof of the tunnel (Figure 3). "Hydraulic mortar," wrote Schramke, "had the proportions of one part cement to three sand, or in foundation walls, two and a half parts of sand." In the brick masonry, the proportion of cement to sand was one to two, and ". . . the mortar for vertical joints was put to the brick before laid, the brick forced into its bed in such a manner, that from horizontal and vertical joints the mortar is readily forced out like sausages; the superfluous mortar was then taken off and the joints smoothed immediately."[16]

Figure 3. *Portion of Croton Aqueduct exposed during construction by the New York Telephone Company of a new cable system, showing brickwork. Photo by author.*

When the section of the Aqueduct between Tarrytown and Hastings-on-Hudson was adapted in 1971 by the New York Telephone Company for cable lines, the author interviewed several of the engineers and, through their courtesy, inspected the work in progress. One observes (in Figure 4) evidence of the careful workmanship of the 1840s; a walking tour through many miles of the Aqueduct revealed that after one hundred and thirty years, the structure had remained sound. In only a few places, tree roots or the incursions of man had broken the smooth surface of the brick masonry work. At this time, the interior of the tunnel measured 7'5" x 8'5½".

Figure 4. *Interior of the Aqueduct. The plastic piping was installed to carry the cables of the New York Telephone Company. Photo by author.*

Along the Aqueduct south of the Croton Dam, the first structure of interest is the Sing Sing Kill Bridge (UTM; Ossining 9535.5726) in Ossining. To bring the Aqueduct across the gorge cut by a small stream, a simple arch bridge of masonry was constructed. It is a span of eighty-eight

[15] J.B. Jervis, *Description of the Croton Aqueduct from the Dam at Croton River to the Distributing Reservoir, New York* (New York, 1851).

[16] Schramke, *op. cit.*, p. 28.

Industrial Archeology

feet, twenty feet wide, with its top eighty feet above the small stream. The arch was constructed over a timber half oval thirty-three feet in height, four feet thick at the spring line and three feet at the keystone. Jervis noted that he employed the same method as used in the construction of Waterloo Bridge in London, centering the arch with timber construction. A survey by New York City's Water Department engineers in the 1940s found that this bridge had settled only three and a half inches.

The Mill River crossing is twelve and three quarters miles from the head of the Aqueduct (Figure 5). This massive construction is eighty-seven feet above the river (now known as the Pocantico), twenty-five feet wide, and 172 feet long. Washington Irving, whose home is near the line of the Aqueduct two miles south of Tarrytown, became acquainted with the project engineers and related one of the stories they told. In a letter to an unidentified friend dated March 17, 1840, he wrote of the workers newly arrived from Ireland:

> We have nothing new in these parts excepting that there has been the devil to pay of late in Sleepy Hollow; a circumstance by the bye, with which you of New York have some concern, as it is connected with your Croton Aqueduct. This work traverses a thick wood about the lower part of the hollow, not far from the old Dutch haunted church, and in the heart of the wood an immense culvert or stone arch is thrown across the wizard stream of the Pocantico, to support the Aqueduct. As the arch is unfinished a colony of Patlanders have been encamped about this place all winter, forming a kind of Patsylvania in the midst of a "wiltherness." Now whether it is that they have heard the old traditionary stories about the Hollow . . . ; or whether the goblins of the Hollow . . . have resented this intrusion into their solitudes by strangers of an unknown tongue, certain it is that the poor paddys have been most grievously harried for some time past, by all kind of apparitions. A waggon road cut through the woods and leading from their encampment past the haunted church and so on to certain whisky establishments, has been especially beset by foul fiends, and the worthy patlanders on their way home at night have beheld misshapen monsters whisking about their paths, sometimes resembling men, sometimes hogs, sometimes horses, but invariably *without heads;* which shews that they must be lineal descendants from the old goblin of the Hollow. . . . , the whole wood had become such a scene of *spuking* and *diablerie,* that the paddys will not any longer venture out of their shantys at night, and a whisky shop in a neighboring village, where they used to hold their evening gatherings, has been obliged to shut up for want of custom. . . . The Corporation of your City should look to it, for if this harrying continues I should not be surprised if the Paddies, tired of being cut off from their whisky should entirely abandon the goblin regions of Sleepy Hollow, and the completion of the Croton water works be seriously retarded.[17]

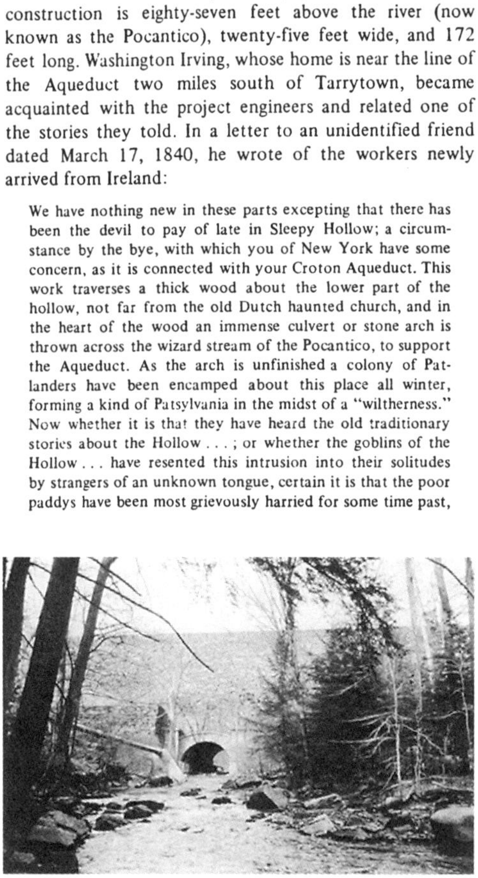

Figure 5. *The Mill River Crossing of the Aqueduct. This stream is now known as the Pocantico River. This site is in Sleepy Hollow Cemetery. Photo by author.*

Although Irving takes a whimsical view of the workers, the Aqueduct officials were not always comfortable in their relations with the 5,000 or so laborers who had been recruited from the ships arriving from Ireland and England. The Commissioners had forbidden the operation of whiskey shops within one mile of the line, but to little avail. There were several outbreaks among the workers who brought old grudges, one of these a donnybrook between the Corkites and the Fermanagh supporters. Philip Hone, who had served as Mayor of New York and was appointed to the Board of the Croton Water Commissioners, recorded in his diary in 1840 yet another reason for unrest: "There has been a flare-up amongst the Irish laborers on the Croton Aqueduct occasioned by the contractors reducing their wages from one dollar to seventy-five cents a day. Large numbers turned out, and marching down from Westchester to Harlem prevented others from working, and committed some acts of violence upon the workers."[18] This "flare-up" was serious enough to prompt the officials to call out the militia ". . . to suppress the rioters who had started out intent on destroying property and threatening the lives of those in charge of the work of construction."[19]

Below Tarrytown and above the crossing of the Harlem River, there are several structures of interest. At a ravine in

[17] Washington Irving, Letter, dated March 17, 1840, transmitted to the author in correspondence with John W. Harbour, Jr., Sleepy Hollow Restoration, February 22, 1973.

[18] Philip Hone, *The Diary of Philip Hone, 1828-1851,* editor, Allan Nevins (New York, 1936).

[19] French, *op. cit.,* p. 347.

Irvington, there is an arched masonry construction which is sixty-two feet high, fourteen feet wide, and one hundred forty-eight feet long (Figure 6). At four other brooks similar structures were built, all of them still in good repair. The crossing of the Saw Mill River in Yonkers–where the line of the Aqueduct runs directly east from the previous north-south line–necessitated the cutting of one of the longer tunnels six hundred eighty-four feet, partly through earth and, partly through rocks. (UTM, Yonkers, 9374.3209.) To bring the Aqueduct across the Saw Mill River, a bridge with two arches, each twenty-five feet wide was built: one arch for a roadway, the other for the river. (At this time, the fate of this structure is being deliberated by several highway departments whose planners consider this an obstruction to the free-flow of traffic in Yonkers.) Proceeding east from this point, the line of the Aqueduct comes to Tibbett's Brook valley. Here a tunnel 810 feet long, was driven through solid rock, to bring the Aqueduct onto high ground–along which it runs for two miles just under the surface to the bluff at the eastern bank of the Harlem River.

Figure 6. *Ravine crossing in Irvington-on-Hudson over a small stream and a roadway. Photo by author.*

The High Bridge

The Harlem River crossing presented Jervis with an opportunity to design a bridge of some grandeur. The State Legislature, looking to the possible use of the river for navigation, passed a law in May, 1839, which required that the Aqueduct be built either as a tunnel under the river or on a bridge with arches and piers, 100 feet above the highwater mark, with 80 spans. Jervis looked to Vitruvius and other Roman models and designed his bridge accordingly–and splendidly.

The ground above the river slopes, first at 20° to a section of table land twenty-five feet above high water, and then for 300 feet across the table land to water's edge. At that time, the river was 620 feet wide. On the Manhattan Island shore, the bank rises at a slope of thirty-five degrees and is here solid gneiss. To cross the river at this point, Jervis designed a bridge 1420' long, with 16 piers, 6 in the river and 10 on the land. These piers were 20' x 40' at the base and about 84' to the spring of the arch, tapering slightly as they rise in height. The arches had a span of about 80'. The land piers are proportionately smaller in size, their height varying with the ascending ground and their arches reduced to about 50'. From the base on river bottom to the top of the parapets, the total elevation is about 138'. The completed bridge is shown in an 1860 view (Figure 7). In 1923, the Navy Corps of Engineers removed the piers in the river to improve navigation, replacing the piers with a steel arch (Figure 8).

Figure 7. *High Bridge. View from the Bronx side looking north-west, c.1861. The addition of the new main pipe is in progress. Courtesy of The New York Historical Society, New York City.*

Figure 8. *High Bridge. The river piers were replaced by the steel arch at the Manhattan end of the bridge in 1923. The standpipe was built in 1869-1872. Photo by author.*

Industrial Archeology

The Aqueduct was completed in 1842 with water brought into Manhattan on a temporary crossing of the river. Construction on the bridge had begun but work was not completed until 1848 with one less arch than originally planned, and at a final cost of $963,428.

Schramke's account of the work is lengthy but parts of it are of interest as they indicate the nature of the construction:

> In order to reach the foundation of every pier, a coffer-dam was put down, the coffer of such a height as the depth of the water at its site required together with the thickness of the mud-deposit and three feet border above high-water; a course of side-beams framed (the pier), notched at the corners and shouldered together with the cornerposts and bolted to them; the corner-braces with their saddle-brace are for keeping firm the right angle;... the sides in general were kept in their places by the braces which were connected to them with iron straps.... those braces could easily be removed with the progress of the mason-work.... The box thus framed upon an inclined plane was launched, floated to its place and anchored, put horizontally by removing some sand from the bottom; the mud having been removed by the mud-machine. Round the box a sheet-piling was driven, the box then surrounded with earth, and by means of pumps worked by steam the water taken out of the interior of this coffer-dam. The sand of this interior was then dug out and afterwards, as the excavation proceeded, additional courses were laid down against the sheet-piling and braced off, thus continued until the marble-bottom was reached, which was then levelled off and the first course of masonry laid.... After a pier was carried up the wood-work was taken out as far as this could be done.[20]

One can understand the slow progress made on completing the High Bridge; it was tedious work, simplified with a few devices such as steam pumps.

To bring the Aqueduct south from the High Bridge crossing several tunnels had to be cut through the rock. The first of these, called the Jumel Tunnel because it traversed this notable French *emigre's* land, was cut through 234 feet of granite. Near 150th Street and Tenth Avenue (renamed Amsterdam), the longest tunnel of the Aqueduct was cut through 1215 feet of rock. The line was brought further south on Tenth Avenue through a "syphon" of thirty-six inch pipes at Asylum Hill, which is today the south side of the "valley" of 125th Street.

The next major valley was crossed with a series of masonry arches spanning 98th, 99th and 100th Streets. From here the Aqueduct was swung eastward through a ninety degree curve and through a final stretch of masonry tunnel into the Receiving Reservoir, which was located between 79th and 86th Streets, and what was then Sixth and Seventh Avenues. These avenues and the Reservoir were absorbed a generation later into the design of Central Park by Frederick Law Olmsted.

The Receiving Reservoir was built in two sections, divided with an earth embankment to allow for draining, inspecting and repairing. It covered thirty-seven and a half acres and, with an average depth of twenty-two and a half feet, was estimated to hold about 150 million gallons of water. Pipes were built through a set of vaults and gates at the southern end of the reservoir, and continued south along Fifth Avenue to the Distributing Reservoir on top of Murray Hill. This structure on the corner of Forty-Second Street and Fifth Avenue was designed by Jervis in the then-popular Egyptian style. The outer walls were four feet thick, the inner six feet, with an interior vault bricked and supported with spandrels. The length of the reservoir, measured on top, was 420 feet; its width, 386 feet. When filled to a depth of thirty-eight feet (four feet short of the top) it contained 21 million gallons. The structure was divided in half, so any needed inspection or repair could be accomplished without shutting down the whole system. A series of valves, gates and interconnected pipes provided for such activities and for drainage to carry away overflow or waste water from cleaning. Here as elsewhere in the system, Jervis planned carefully to protect the masonry from the effects of freezing weather. For those who went for a Sunday stroll, stone stairs were provided to climb to the top; to guard against any accidents, an iron railing was built around the edge of the water. Schramke notes that "... at the elevation of 119 feet above the level of the sea it commands a complete view of the west and east of Manhattan Island, as well as the south part of the city with New York bay, and at a greater distance in that direction, the Atlantic Ocean. This view is exceedingly beautiful and may be called one of the finest in the world"[21] (Figure 9). From this reservoir the water was piped to the lower Island where the city was then located, requiring 134 miles of pipes of all sizes between four and thirty-six inches. A hydraulic press, capable of 200 to 250 psi, was used to prove the pipes before they were laid.

Jervis completed the system within five percent of the estimated cost. Including the land purchased in Westchester County, the Aqueduct cost $8,575,000, to which was added $1,800,000 for the cost of the distributing pipes. This was funded at seven percent on the first two million dollars; five percent on the rest, loan costs additional, plus

[20] Schramke, *op. cit.*, p. 38-39.

[21] Schramke, *ibid., p. 1.*

Figure 9. *Croton Distributing Reservoir, Fifth Avenue and 42nd Street. The horse car is going west on 42nd Street. This site is now occupied by the New York Public Library. This photograph was taken in 1903-1904, shortly before the reservoir was demolished. Courtesy of the New-York Historical Society, New York City.*

interest paid during the construction. The total expense was $12,500,000. The water-tax on each house of "middle-size" was ten dollars; the commercial users were taxed according to their size and needs.

The Aqueduct Completed

The praise for the completed Aqueduct was nearly as voluminous as the flowing water. Schramke wrote that ". . . it is one of the greatest and most important works of the world whether we consider the boldness of its plan or the admirable manner of its execution."[21] Tower's fulsome prose: "The advantages, the comforts and blessings of this supply of pure water will be appreciated as the city extends the means for its use and the time is not distant when she will regard it as a treasure which was cheaply purchased and will proudly point to the noble work which she has achieved not only as an example of her munificence but as an illustration of what art and science can accomplish."[22] On the 8th and 9th of June, 1842, the Commissioners and the engineers made a final inspection of the Aqueduct, walking through the conduit from the Croton Lake Reservoir to the Harlem River, a distance of thirty-three miles. On June 22, water was admitted into the Aqueduct to a depth of eighteen inches. *The Croton Maid,* a boat large enough for four persons was placed in the current in Croton and arrived almost at the same time as the first flow of water at the Harlem River on June 23. During this odd voyage which was followed with much interest along the line, different persons took turns steering the boat. At this depth, it covered about one mile in forty minutes; when the water depth was increased to four feet, the rate of travel was two miles per hour. On June 27, water was admitted into the receiving reservoir and welcomed with a thirty-eight gun salute. The *Croton Maid* on its arrival from the short journey down Manhattan's section of the conduit was presented to the City's Fire Department by the President of the Water Commission.[23] Then, on July 4 water was let into the distributing reservoir.

Later in the year, a splendid celebration was mounted and noted by the two best known diarists of the time, George Templeton Strong and Philip Hone. Strong wrote of the events in October:

> Walked up to Union Square where the new fountain was playing. I saw the one in the Park at City Hall this morning. They're splendid, both, the latter looked magnificent with the sun streaming on it. I hope they'll play permanently. . . . (A few days later) October 14, the great day of the Croton Celebration. Not much done in Wall Street, of course; all the population of the city, and of the surrounding country too, I believe, seemed to have precipitated itself bodily into Broadway at about ten o'clock. Never saw the street so crowded. The procession was mighty fine. There was no end to the military men . . . no end to the firemen and their red shirts, nor the Temperance Societies.[24]

Hone, as a former Mayor, drew a place in the procession. He notes in his chronicle that

> Nothing is talked of or thought of in New York but Croton water. Political spouting has given place to waterspouts, and the free current of water has diverted the attention of the people from the vexed question of the national currency . . . how cheerfully our citizens acquiesce in the enormous expense which will burden them and their posterity with taxes to the latest generation.[25]

One item of note was a wagon on which was placed the identical printing press at which Dr. Franklin once worked. The press was used during the procession to print and distribute an ode written for the occasion by George P. Morris, a poet best known today for his stanzas which begin "Woodsman, spare that tree. . . ." The ode, set to music by M.W. Balfe, was sung in front of City Hall by a choir of two

[22] F.B. Tower, *Illustrations of the Croton Aqueduct* (New York, 1843), p. 123.

[23] Shonnard and Spooner, *op. cit.*, p. 558.

[24] George Templeton Strong, *The Diary of George Templeton Strong, 1835-1875,* edited by Allan Nevins and Milton Thomas (New York, 1952), pp. 188-189.

[25] Hone, *op. cit.*, pp. 624-626.

hundred. The ode begins this way and goes on for seven stanzas:[26]

Gushing from this living fountain,
Music pours a falling strain,
As the Goddess of the Mountain
Comes with all her sparkling train.
From her grotto-springs advancing,
Glittering in her feathery spray,
Woodland fays beside her dancing,
She pursues her winding way.

With the celebration over, New Yorkers soon settled down to the more mundane matters connected with Croton water. Its availability began to change domestic living arrangements: "In the 1840s advancing technology dramatically increased the comfort of New Yorkers living in fine row houses. In terms of such equipment as indoor plumbing and central heating, the New York row house of the 1840s more closely resembled a dwelling of the early twentieth century than one of the 1830s, just one decade earlier."[27] Bathing and toilet facilities soon became relatively common in most dwellings. Charles Lockwood observes that "In 1836 the luxurious Astor House hotel, on Broadway at Vesey Street, created a sensation because of the toilets and provisions for bathing on every floor."[28] Mass production of fixtures had not yet begun, but this did not stop the well-to-do from going into elaborate arrangements: one mansion had seven toilets, and eleven tubs and washbasins. Tower's prediction was soon proved to be accurate, for with this adequate supply of water the city grew at an astonishing rate. Croton water soon came to be a convenience which was taken for granted, but its free use presented many problems to the Water Commissioners and their workers. The minutes of their meetings have many comments on the waste of water and the costs added to operations by the practices of the water users.

In 1854, the Board appointed a special force of guards–naming them Water Police–to patrol the Aqueduct checking for leaks, breaks and other problems which would interfere with daily operations. They had to contend with such matters as livestock falling into the waste weirs, paying one claim of seventy five dollars for an ox which ended its existence in one of these drains. The Board was frequently called upon to take legal action against property owners who had built structures into the line of the Aqueduct. They also set a fine of two dollars if a person was caught using the Croton supply for street washing. By 1861 the force of Water Police had grown to thirty-eight.

The High Bridge drew not only pedestrians and admirers but quite another sort of spectator sport. In the minutes for September 14, 1852, there is this resolution:

> An exhibition has been advertised by public notices posted in the vicinity of High Bridge that a person is about to leap from the parapet of the Bridge into the Harlem River at three o'clock p.m. tomorrow, and whereas exhibitions of like character have heretofore drawn evil disposed and unruly parties to the damage of the work, be it resolved that such exhibitions at this and any future time be discountenanced and is forbidden by the Board.[29]

In 1856, the Board began consideration of additional pipes to carry more water into the receiving and distributing reservoirs. The pair of three foot pipes on the High Bridge was not sufficient for the needs of the rapidly growing city. Originally, Jervis had proposed four foot pipes but was overruled. Finally, the Board accepted the recommendations of the engineering staff and provided the funds for a ninety-inch diameter cast iron pipe to be laid on top of the existing pair. This construction work, underway in the 1860s, is shown in Figure 10. The engineer in charge of this addition was Alfred W. Craven, who had been appointed Chief Engineer in 1849 on the resignation of Jervis, and served in this position until May, 1868. Craven, in whose offices the American Society of Civil Engineers was founded on December 1, 1852, was yet another of the remarkable group of men who established civil engineering as a profession in the United States. Educated as an attorney, his interest in engineering led him first to surveys and construction of a railroad in Ohio. From there he moved to South Carolina to continue his new career in railroad engineering. Subsequently he worked on other projects in New York, Pennsylvania and New Jersey. With

[26] George P. Morris, *The Songs and Ballads of George P. Morris* (New York, 1843).

The New York Public Library has in its collections a copy of the handbill or broadside as printed and distributed in 1842. It is headed "The Croton Ode: written at the request of the Corporation of the City of New York by George P. Morris, Esq. and sung in front of the Park Fountain, by the members of the New York Sacred Music Society, on completion of the Croton Aqueduct." At the bottom of the sheet is this: "This ode was printed during the progress of the procession upon a press erected on a moveable stage. Nesbitt, Printer."

In the Music Library is the work of M.W. Balfe. This song is in a collection of "10 Songs by M.W. Balfe" published in Boston, 1857, by Oliver Ditson & Co.

[27] Charles Lockwood, *Bricks & Brownstone: The New York Row House, 1783-1929, An Architectural & Social History* (New York, 1972), p. 183.

[28] Lockwood, *ibid.*, p. 186.

[29] City of New York, *Minutes of the Croton Aqueduct Board* (New York, 1903).

Figure 10. *High Bridge. The addition of the 90-inch cast-iron pipe as it was seen from the west gatehouse looking towards the Bronx in 1861. Courtesy of The New-York Historical Society, New York City.*

the Croton assignment, he established his expertise in this field and was called on by other cities for advice on the development of their water supply systems. He traveled to Europe in 1878 to seek a cure for failing health but died in England in 1879. His nearly twenty years of service for the Croton was joined with other engineering duties for the City of New York. He set up the sewage system, as well as roads and streets.[30]

Jervis' career, following his service on the Croton, was directed almost entirely to the railroad industry, although he consulted occasionally on other water supply systems. He became President of the Chicago and Rock Island Railway in 1851, and ten years later took over the position of Superintendent of the Pittsburgh, Fort Wayne and Chicago Railway. In this post, he turned a nearly bankrupt operation into a profitable one, to the delight of the stockholders. He retired to Rome, New York, in 1866 but soon became the Secretary and President of the Rome Iron Mills. This occupation, together with his writings on capitalism and industrial economics, engaged him fully until his death in 1885.[31]

[30]Wegmann, *op. cit.*, p. 239.

[31]Jervis, *The Reminiscences.*

Bringing a never-ending flow of water into the metropolitan centers of the nation–New York, Philadelphia, Boston, Washington and Baltimore–made possible a geometrical change in the rate of growth of those cities. The technology which supported the beginnings of this urban expansion fed upon itself but, as in some imaginative folk tale, grew not weaker but stronger. The canal and the railroad builders contended with one another, enriching themselves and the country. They encouraged the flow of inventions from the drafting tables and the workshops of the new breed of engineer that is personified in the builders of the Croton. The processes developed, the materials employed and the scientific knowledge recorded in texts, treatises and journals came in a flow as plentiful as the natural resources of the land.

In writing about the Catskill Aqueduct which was added to the New York City water system in 1917, Professor Furman of the Stevens Institute of Technology speculated that the Stevens family "... in developing many detail inventions which made these larger enterprises successful, and in establishing manufacturing centers for carrying on their work, contributed to the more rapid growth of urban communities calling, in succession, for larger and larger sources of water supply and for greater and greater scientific knowledge in construction and conducting that supply."[32] The nation came to rely on this cycle; and in the yeasty atmosphere of the times no problem appeared which long lacked a solution.

As noted at the beginning of this paper, the residents of the City of New York are continuing to fund the expansion of their water supply with newly bored tunnels deep in the bedrock under Manhattan and Queens. The engineers of the original Croton would be amazed not so much by the character of the work, but by its scale and sophistication. The purpose remains the same today as it was in the 1840s: water and ever more water.

[32]Franklin DeR. Furman, M.E., "The Influence of the Work of the Stevens Family and of Graduates of Stevens Institute of Technology on Engineering Work Leading Up To The Construction of the Catskill Aqueduct," George Frederick Kunz, editor, *Catskill Aqueduct Celebration Publications* (New York, 1917).

9

Bringing water to the Crescent City: Benjamin Latrobe and the New Orleans waterworks system

Gary A. Donaldson

If anyone deserves the title of America's first great architect it is Benjamin Henry Latrobe. A professional among America's gentleman amateurs, Latrobe was responsible for establishing the professions of both architecture and engineering in the United States. Latrobe's designs of America's first great Greek Revival buildings, such as the White House and Capitol building, set the standards of early American architectural style, as well as created the great symbols of the United States government. Latrobe also had an illustrious career as an engineer, spending much of his time designing and building many of the country's first improvements. He was truly a builder of America.[1]

Latrobe's career as an engineer and architect ended in New Orleans, where he died in 1820 at the age of fifty-six. The rapidly growing, bustling, cosmopolitan Crescent City was Latrobe's last home, and the projects he designed and built there were his last. Latrobe came to New Orleans because he saw the city at the onset of a great growth period. At the mouth of the Mississippi River, New Orleans soon would become the major benefactor of the new steamboat technology being developed in the Northeast, and it was Latrobe's intention to be a part of the growth and wealth that was about to flow through the city. While in New Orleans, Latrobe took time from his

[1]Latrobe's principal biographer is Talbot Faulkner Hamlin, *Benjamin Henry Latrobe* (New York, 1955). On Labrobe's life and accomplishments, see also *Dictionary of American Biography*, 22 vols. (New York, 1928-1958), XI, 20-25; and Talbot Faulkner Hamlin, *Greek Revival Architecture in America: Being an Account of Important Trends in American Architecture and American Life Prior to the War Between the States* . . . (New York, 1944), pp. 23, 30. For an account of Latrobe's influence on the profession of architecture in America, see John Burchard and Albert Bush-Brown, *The Architecture of America: A Social and Cultural History* (London, 1967), pp. 3-64. See also, Fiske Kimball, *Domestic Architecture of the American Colonies and of the Early Republic* (New York, 1922), pp. 146. For an account of Latrobe's influence on engineering, see Lee William Formwalt, "Benjamin Henry Latrobe and the Development of Internal Improvements in the New Republic, 1796-1820" (Ph. D. dissertation, Catholic University of America, 1977), pp. 35-132, 190-241, 265-285, 302.

work to observe the city and its people, leaving an interesting and thorough record of early nineteenth-century New Orleans life and character.

Latrobe's greatest achievement in New Orleans was to design and build the New Orleans waterworks system, a project that took eleven frustrating years and ultimately cost Latrobe his life and the life of his son. The New Orleans system was based on the design of the Philadelphia waterworks system, probably Latrobe's most renowned engineering accomplishment. The first of its kind in the nation, the Philadelphia system was designed and built between 1799 and 1801. It was the success of that design that brought Latrobe to New Orleans.

The Philadelphia system was commissioned in response to a series of yellow-fever epidemics that plagued that city at the turn of the century. At that time, it was prevailing opinion that yellow fever was transmitted through impure water; fresh water pumped into the city, it was thought, would arrest the disease.[2] Latrobe's scheme in Philadelphia was to use steam pumps to move large volumes of water from the Schuylkill River through pipe made from hollowed-out oak tree trunks into a large settling basin. The water was then pumped through a tunnel to a second settling basin near the center of the city, and on to an elevated reservoir. From there, gravity allowed the water to be channeled to consumers.[3]

That project was something of an experiment, and the city of Philadelphia paid Latrobe a commission for his work. The system proved so successful in Philadelphia that in New Orleans Latrobe sought a franchise from the city. The plan was that he and a few investors would finance the project, reaping the profits from the completed system by selling water to its users. If the plan succeeded, Latrobe would make a great deal of money. He spelled out the possibilities of supplying water to New Orleans in a letter to his father-in-law in 1809:

[2]On New Orleans yellow-fever epidemics of the early nineteenth century, see Jo Ann Carrigan, "The Saffron Scourge: A History of Yellow Fever in Louisiana, 1796-1905" (Ph. D. dissertation, Louisiana State University, 1961), pp. 24-94. While in New Orleans, Latrobe commented in his journal of the perceived relationship between water quality and yellow fever. Benjamin Henry Latrobe, *The Journal of Latrobe* (New York, 1905), p. 237; hereafter cited as *Journal* (1905), to distinguish this source from a publication with a similar title published in 1980. See also, Nelson Manfred Blake, *Water for the Cities* (Syracuse, N. Y., 1956), pp. 6-11; and John Duffy, "Nineteenth Century Public Health in New York and New Orleans: A Comparison," *Louisiana History*, XV (1974), 331-333.

[3]The best description of the Philadelphia waterworks system is Darwin H. Stapleton, ed., *The Engineering Drawings of Benjamin Henry Latrobe* (New Haven, 1980), pp. 28-36, 186-192. Stapleton makes use of Latrobe's drawings to explain the Philadelphia system, which was similar to the system that Latrobe planned for New Orleans. It was Frederick Graff, a Latrobe student, who finally completed the Philadelphia system in the 1840s. See *Dictionary of American Biography*, VII, 467-468. For a less technical explanation of the Philadelphia system, see Hamlin, *Latrobe*, pp. 157-167, 171-173.

> . . . Nothing is there so much wanted, & at the same time nothing is so easily & cheaply accomplished. The first expense, before we begin to receive any compensation I estimate at 30.000$ [*sic*]—but to cover errors Say 50.000 [*sic*]. There are now 2.000 [*sic*] Houses which would take the Water . . . at 20$ a Year. Mr. Poydras the representative from N.O. an old French miser . . . says that we shall assuredly succeed, & be the richest men in America.[4]

It is difficult to determine when Latrobe considered undertaking the New Orleans project. He first mentions it in his writings in a letter to his brother-in-law, Samuel Hazelhurst, in December, 1809:

> I have . . . digested two schemes, which are . . . 'as yet a secret'. . . . One is to supply the city of New Orleans with water, on which object I have a communication from the Governor of the Territory, the other a canal around the falls of the Niagara, on which I am consulted by the Government. . . . The exclusive right of the supply of New Orleans to me & my associates is proposed if I effect it.[5]

However, in a letter to his associate, Robert Alexander, concerning the question of legal possession to the New Orleans franchise, Latrobe mentions that as early as 1807, Thomas Jefferson, then in his second term as president, suggested the idea after Latrobe had considered it even earlier:

> The idea of supplying the city of New Orleans [with water] was first suggested to me by a French gentleman [probably Poydras] whom I was exhibiting my works in Philadelphia. . . . But I dismissed the subject from my mind until it was renewed by a conversation with Mr. Jefferson while you [Alexander] were in N.O. building the custom house.[6]

Latrobe designed the New Orleans Customhouse and Alexander built it under Latrobe's direction in 1807.

The New Orleans waterworks system had many similarities to the

[4]Latrobe to Isaac Hazelhurst, December 17, 1809, Edward C. Carter, ed., *The Papers of Benjamin Henry Latrobe*, Microfiche Edition (Clifton, N. J., 1976).

[5]Latrobe to Samuel Hazelhurst, December 7, 1809, Carter, ed., *Latrobe Papers*.

[6]Latrobe to Robert Alexander, July 27, 1811, *ibid.*

Philadelphia system. A large steam engine would pump water from the Mississippi River through an airtight suction pipe into six elevated reservoirs made of wood. From there, gravity would send the water to the city's consumers through a system of wooden and iron pipes laid under New Orleans' streets from the riverfront to the Carondelet Canal at the back of the city. An 1811 drawing by Latrobe's son, Henry, shows a fountain on the riverfront square and designs for fire plugs and hand pumps at street intersections. Some pipes in the drawing are shown as made of wood, others are iron. About 5,000 feet of pipe was laid under New Orleans streets by the time the project was completed, but it is difficult to determine how many fixtures were installed at the intersections. The fountain was never built.[7]

The main suction pipe was one of Latrobe's biggest problems because it had to be airtight. Latrobe's original plan was to use cast-iron pipe, but he found it impossible to purchase in large enough diameters. He then turned to bored-out tree trunks like those used in Philadelphia, but he had difficulty in preserving the wood long enough to make that a practical solution. Finally, he settled on eight-inch, cast-iron pipe, about one half the size he had originally intended.[8]

Latrobe was not prepared to abandon his work in the Northeast to begin the New Orleans project, so in 1810 he sent his eighteen-year-old son, Henry, to represent him before the New Orleans City Council. A graduate with high honors in mathematics from St. Mary's College in Baltimore, Henry had worked as a surveyor for the National Turnpike Commission. He also spoke French, an important asset in dealing with the French-speaking New Orleans City Council.[9] Latrobe's confidence in his young son was obviously well founded, because on April 27, 1811, only a few months after Henry arrived in new Orleans, the city council passed an

> ordinance . . . to grant the Sieur Benjamin Henri Latrobe, and his associates the exclusive privilege to supply the City of New Orleans and its Faubourgs with water by means of one or several steam pumps. After various amendments and additions, the said

[7]For the best technical explanation of the New Orleans waterworks system, see Stapleton, ed., *Engineering Drawings*, pp. 36-41, 199-203. Latrobe's drawing of various parts of the New Orleans system is entitled, "Design of a pier to cover the Suction pipe of the Pump for supplying Water to the City of New Orleans." The original copy is in the Louisiana Division, New Orleans Public Library.

[8]Stapleton, ed., *Engineering Drawings*, pp. 199-203. See also, B. Latrobe to H. Latrobe, April 24, 1812, November 18, 1812, and July 17, 1814, in Carter, ed., *Latrobe Papers*.

[9]Hamlin, *Latrobe*, pp. 49, 601-602.

ordinance project has been put to a vote and adopted in its entirety. . . .[10]

This placed the project in motion, but Latrobe was far from prepared to begin work. He was notorious for accepting projects that he was immediately unable to undertake, and that was certainly the case of the New Orleans project. He left Henry in New Orleans to begin work on the project and act as his representative before the local government, but Latrobe himself would not break away from his other projects in Philadelphia and Washington to begin work in New Orleans until January, 1819, over eight years after he received the grant to build the system.

Besides Latrobe's tardiness, the project suffered a series of setbacks that continually delayed progress. One major problem was fickle investors. As war with England looked more and more likely, and later as the project floundered as a result of the war, Latrobe's investors began to drop away. Jacob Mark of New York, whose original investment of $40,000 got the project off to a sound start, remained steadfast, but Godfrey Haga of Philadelphia withdrew his $20,000 investment in 1811, as did Frederick C. Graff, a Baltimore investor of equal importance.[11] With Latrobe obviously struggling, the city of New Orleans agreed in July, 1812, to invest $6,000 in the project, but it was still not enough.[12] Latrobe needed more financial support to keep the program afloat. Finally, with a series of letters the next year, he was able to persuade Graff to reconsider his investment, and it looked as though the project would move forward.[13]

Not only did the waterworks project suffer periodic financial setbacks, but Latrobe had personal financial problems as well. Just as the war with England began, Latrobe completed work on the U. S. Capitol, bringing to an end a comfortably steady salary from the federal government. In an attempt to bring in money quickly to finance his various projects including

[10]New Orleans City Council, Proceedings, Louisiana Room, New Orleans Public Library, Session of April 27, 1811, p. 37.

[11]Latrobe to Godfrey Haga, August 20, 1811, and September 11, 1811; Latrobe to F. C. Graff, August 14, 20, 1811, September 11, 1811, November 5, 7, 1811, January 9, 1812, January 10, 11, 30, 1812, and February 5, 1812, in Carter, ed., *Latrobe Papers*. This is Frederick C. Graff, the Baltimore philanthropist, and should not be confused with Frederick Graff, Latrobe's student and assistant on the Philadelphia waterworks project.

[12]New Orleans City Council, Proceedings, Session of July 11, 1812, pp. 203-204. The council agreed to pay $3,000 immediately, and an additional $3,000 when the project was completed.

[13]Latrobe to Graff, February 10, 1812, February 12, 13, 17, 27, 29, 1812; March 17, 21, 1812, in Carter, ed., *Latrobe Papers*.

the New Orleans project, Latrobe became involved in a series of unsuccessful wartime schemes, each of which resulted in a significant loss of money. In one case, he attempted to sell German-made weapons and knapsacks to an American government bent on purchasing American-made products. In another, he endorsed notes for a cement manufacturer who joined the British Army, leaving Latrobe with the notes and lawsuits. He was apparently swindled out of several thousand dollars by a power-loom manufacturer from Massachusetts. He also became deeply involved with Robert Fulton, Robert Livingston, and Nicholas Roosevelt in a number of unsuccessful schemes to design and build steamboats.[14]

Another problem that delayed the New Orleans project was engine construction. When the engine builder in Philadelphia became too expensive and too difficult to deal with, Latrobe decided to build the engine he needed for the New Orleans project himself. When a lack of funds caused that project to fail, Latrobe packed up the castings for the incomplete engine and sent.them overland to Pittsburgh. There, he thought, he could have the engine and boiler built and shipped to New Orleans by river. But Latrobe's financial difficulties prevented the completion of that plan. Finally, following a series of small financial successes, Latrobe had the engine built in New York and shipped to New Orleans to 1819.[15]

During the war, while the New Orleans project sat idle in Pittsburgh, Latrobe took on other projects that were more likely to generate legitimate financial resources than the several schemes that had taken him to the brink of financial ruin. He designed several public buildings and private residences around the country and continued work on two large projects in Washington for the federal government. Each of these endeavors helped Latrobe earn the capital he needed to fund the New Orleans project, but they also delayed the project for several years.

The original grant from the New Orleans City Council expired after only two years, and in order to have the grant renewed it was necessary for Latrobe to approach the city fathers for an extension. He first wrote to the city council on April 23, 1813, citing money problems and various difficulties resulting from the war with England. The war, he said, made it

[14]Hamlin, *Latrobe*, pp. 364-366, 369-396. On Livingston, Roosevelt and Fulton and their schemes to monopolize the steamboat business on the lower Mississippi, see George Dangerfield, *Chancellor Robert R. Livingston of New York, 1746-1813* (New York, 1960), pp. 415-417. See also *Dictionary of American Biography*, VII, 68-72; XVI, 133-134, 320-325.

[15]The Philadelphia engine builder was James Smallman who built the engines for the Philadelphia waterworks project. See Stapleton, ed., *Engineering Drawings*, pp. 37-39. Robert McQueen was the New York builder who finally completed the engine for the New Orleans project. Latrobe to Robert McQueen, August 13, 1816, in Carter, ed., *Latrobe Papers;* Hamlin, *Latrobe*, pp. 357, 376-377, 415.

impossible to transport machinery from warehouses in Pittsburgh to New Orleans.[16] Five days later, the council considered a similar plea from Henry Latrobe and granted an extension of the agreement.[17]

In September, 1813, Latrobe borrowed $1,500, paid off his debts in Washington, and headed west to Pittsburgh. There, with Livingston and Fulton, he began building steamboats for the western rivers, and he hoped to begin work on the New Orleans steam engine. But the steamboat scheme, like so many other schemes in this period of Latrobe's life, ended in failure and again he lost large amounts of personal capital.[18]

Henry, however, kept the waterworks project alive in New Orleans during the war. He became the project engineer, breaking ground and completing much of the preliminary construction work such as building the engine house and accumulating and boring cypress pipe. His father often wrote that engine construction in Pittsburgh was underway, and that soon the New Orleans project would be in full swing.[19]

When the war with England came to New Orleans in 1815, Henry distinguished himself by risking his life to place a large cannon at a crucial spot during one of the strategic artillery exchanges prior to the January 8 battle. Henry also reportedly served as a "spotter," as the two armies jockeyed for position in the swamplands east of New Orleans.[20]

Henry Latrobe also took on several private projects during his stay in New Orleans. One project was to complete his father's plans to build a much-needed lighthouse at the mouth of the Mississippi River. An obvious necessity to navigation, a lighthouse in that strategic location had been a priority of the federal government since 1806 when Albert Gallatin, Jefferson's secretary of the treasury, took Benjamin Latrobe's advice and appointed a Latrobe pupil, Louis De Mun, to do a coastal survey of Louisiana for the purpose of finding a suitable island for the lighthouse. De Mun chose Royal Island at the mouth of the river as an appropriate site, and Latrobe drew up the plans of a masonry structure some 82 feet high with a diameter of 27 feet at the base and 12 feet at the top. The foundation was to be eight-and-one-half feet below ground, above pilings planted 20 feet below

[16]Latrobe to New Orleans City Council, April 23, 1813, Latrobe Papers, Tulane University Library.

[17]New Orleans City Council, Proceedings, Session of April 28, 1813, p. 70.

[18]Hamlin, *Latrobe*, pp. 395, 397-437.

[19]See particularly B. Latrobe to H. Latrobe, January 8, 1814, February 14, 1814, and July 17, 1814, in Carter, ed., *Latrobe Papers*.

[20]Arsène Lacarrière Latour, *Historical Memoir of the War in West Florida and Louisiana* (1816; reprint ed., Gainesville, Fla., 1964), pp. 120, 202.

that.[21] Henry determined later that Royal Island would not support the structure and the lighthouse was finally built on nearby Frank's Island in 1819.[22]

Henry Latrobe also designed and built New Orleans' Charity Hospital, then located on Canal Street between Baronne and University Place. The original hospital, built in 1782 on Rampart Street and called Hospital of St. Charles, was destroyed by fire in 1809. Henry completed the new structure six years later. The hospital was moved to its present location on Tulane Avenue in 1832.[23]

Another of Henry Latrobe's achievements while in New Orleans was to design and build the city's first French Opera House, completed in 1817. The opera house was touted as the first of its kind in North America, but its fame ended when it burned almost immediately after it was completed. Not until 1859 did New Orleans rebuild its famous French Opera House.[24]

The war with England reemphasized Benjamin Latrobe's place as the nation's foremost architect. The White House and the Capitol building, principally Latrobe projects, were burned by the British in August, 1814, and immediately, President Madison turned to Latrobe to repair the damage. While in Washington, Latrobe characteristically took on several jobs at once. He began designing (although it was never built) a national university Washington, St. John's Church, a large Washington hotel, and

[21]Formwalt, "Latrobe and Internal Improvements," 281. For De Mun's report on the coastal survey, see *American State Papers: Documents, Legislative and Executive of the Congress of the United States, from the First Session of the First Congress to the Second Session of the Thirty-Fifth Congress, Inclusive: Commencing March 4, 1789, and Ending March 3, 1859.* Vol. I: *Commerce and Navigation* (Washington, D. C., 1832), pp. 840-842.

[22]The Frank's Island Lighthouse was finally abandoned due to settling and eventually collapsed shortly after Latrobe's death in 1820. It was rebuilt three years later by Winslow Lewis. See Formwalt, "Latrobe and Internal Improvements," pp. 280-284; Charles H. Beardsley, "Franks Island Lighthouse: B. H. B. Latrobe, Architect" (unpublished research paper, Louisiana Room, Tulane University Library, 1960); Walter Prichard et al., eds., "Southern Louisiana and Southern Alabama in 1819: The Journal of James Leander Cathcart," *Louisiana Historical Quarterly*, XXVII (1945), 478. See also, *American State Papers, Commerce and Navigation*, II, 43-46, 460. A copy of Latrobe's drawing of the Frank's Island Lighthouse is in Edward C. Carter II et al., eds., *Latrobe's View of America, 1795-1820: Selections from the Watercolors and Sketches* (New Haven, 1985), p. 367.

[23]Benjamin Henry Boneval Latrobe, *Impressions Respecting New Orleans: Diary and Sketches, 1818-1820*, ed. by Samuel Wilson, Jr. (New York, 1951), pp. 41-42. See editor's note. When Latrobe arrived in New Orleans in 1819 he wrote in his journal only that he was pleased by his son's design of the hospital. See Latrobe, *Journal* (1905), p. 186. And Benjamin Henry Boneval Latrobe, *The Journals of Benjamin Henry Latrobe, 1799-1820; From Philadelphia to New Orleans*, ed. by Edward C. Carter II et al. (New Haven, 1980), pp. 192-193. Hereafter cited as *Journals* (1980), to distinguish this source form one with a similar title published in 1905. All quotes are taken from Latrobe's *Journals*, published in 1980. However, all three sources taken from Latrobe's original journals are cited here for each quote throughout the text.

[24]Sarah Searight, *New Orleans* (New York, 1973), p. 87; Nathaniel Cortland Curtis, *New Orleans: Its Old Houses, Shops and Public Buildings* (Philadelphia, 1933), p. 189.

the Hagerstown courthouse in Maryland, in addition to several homes for prominent Washingtonians. Again, New Orleans would have to be patient; Latrobe remained occupied in Washington for another two years.

By 1817, Henry had completed most of the preliminary work on the New Orleans waterworks project. All that remained was to install the engine and add the main suction line from the river. But the money to complete the project would have to be raised by Latrobe himself. The capital loaned by the original investors had been lost in the ill-fated attempts to build the steam engine in Pittsburgh, and his personal financial situation remained less than bright.

On July 22, 1817, he wrote his son that the engine was finally completed and on its way to New Orleans from the its new construction site in New York. He added that soon he would ask the New Orleans City Council for additional time to complete the project, since the last two-year extension had long passed.[25] Five weeks later, on September 3, the scheme suffered another blow when Henry died unexpectedly of yellow fever, the scourge he had come to New Orleans to help alleviate.

Latrobe arranged for one of Henry's associates to take charge of the project until he could make the trip to New Orleans and complete the job himself. Meanwhile, in October, he wrote to the New Orleans City Council requesting the extension, blaming his many problems on the war and his personal financial difficulties caused by the loss of the engine in Pittsburgh. ". . . It was a loss not only of money," he wrote the council, "but of the confidence of my friends, who would otherwise have assisted me."[26]

The council was not as quick to respond to the request as in 1813 when Latrobe first asked for an extension. The question was submitted to a committee, which recommended the following to the council in late November:

> . . . the reasons set forth by Mr. Latrobe in explanation of the delay in starting the work on his plant to supply the city with water are entirely satisfactory; consequently, the committee recommends to renew the contract entered into with Mr. Latrobe . . . extending the time of fulfillments . . . until January 1, 1819.

The council accepted the committee's recommendation.[27]

[25] B. Latrobe to H. Latrobe, July 22, 1817, in Carter, ed., *Latrobe Papers*.

[26] Latrobe to New Orleans City Council, October 13, 1817, Latrobe Papers, Tulane University Library.

[27] New Orleans City Council, Proceedings, Session of November 26, 1817, p. 87.

Even now, with the project in the uncertain hands of his son's subordinate, Latrobe was not ready to travel to New Orleans and supervise the program's completion. The problem, as always, was money. Finally, in December, 1817, Latrobe's financial situation became so intolerable that he was forced to declare bankruptcy and move his family from Washington to Baltimore. There he began work on the Baltimore Exchange, a project he had contracted several years before. That endeavor, Latrobe hoped, would generate enough money to complete the bigger, and potentially more profitable, New Orleans project.[28]

Latrobe completed the Exchange in 1818, leaving only the New Orleans waterworks project before him. The system was largely finished, but the usual lack of funds again halted construction. Latrobe was forced to sell some of his shares to an investor, Benjamin G. Orr, to bring the project back to life. In December, 1818, with the new financial resources, he boarded the ship *Clio* in Baltimore; the New Orleans waterworks project had finally attained priority.[29]

The *Clio* reached New Orleans January 10, less than one month's sail from Baltimore, a sail that was certainly a welcomed respite for Latrobe. He took a room in a hotel on the southwest corner of what is today Jackson Square, then called Place d'Armes; and from his window he could see the three most prominent buildings in the city at that time, St. Louis Cathedral, the Cabildo (know at that time as the Principal), and the Presbytery (presently, with the Cabildo, the Louisiana State Museum). He was not entirely impressed with the architectural style of these buildings, writing that "in detail [they] are as bad as they well can be. . . ." However, he added that their symmetry, proportions, strong relief and solid mass produced "an admirable effect when seen from the river or the Levee."[30]

Latrobe was a great observer of American life, society, and culture, and the cosmopolitan-frontier character of New Orleans offered an interesting comparative subject for his journal. He saw New Orleans as a rapidly growing city with three societies: French, American, and what he called "mixed." To him, the French society was dominant, but hundreds of Americans were pouring into New Orleans daily, and soon, he wrote, the American culture will overwhelm the French. To Latrobe, that was not the best course for the city's future. "In a few years," he wrote, "this will be an

[28]Hamlin, *Latrobe*, pp. 480, 486-497.

[29]Stapleton, ed., *Engineering Drawings*, p. 40.

[30]Latrobe's elevation drawing of the Principal and Cathedral is reproduced in Carter, ed., *Latrobe's View*, p. 371; and in *Journals* (1980), pp. 194-195. See also, *ibid.*, 192-193; and *Journal* (1905), p. 185.

American town. What is good and bad in the French manners and opinions must give way, and the American notions of right and wrong, of convenience and inconvenience will take their place."[31]

To Latrobe, the encroachment of Americans on New Orleans was a replacement of good taste by bad, a characteristic he considered particularly true of the city's architecture. The new American ways of construction, he wrote, were beginning to supplant the old ways of the Old World. Brick-fronted houses were gaining ground in New Orleans, he wrote, "and the suburb St. Mary, the American suburb, already exhibits the flat, dull, dingy character of Market Street in Philadelphia . . . instead of the motley and picturesque effect of the Stuccoed french [*sic*] buildings of the city." The Americans, he later added, include in construction of their houses "a sufficient number of holes for light and entrance." They "exhibited their red staring brick work, imbibing heat, thro' the whole unshaded substance of the Wall. . . . I have no doubt," he concluded, "that the American style will ultimately be that of the whole city. . . ."[32]

In other observations on New Orleans and its people, Latrobe spent considerable space in his journal commenting on the beauty of the city's women. He made his comments from the vantage point of a Carnival ball he attended just days after arriving in the city. A few of the city's women, he wrote, "are perfect, and a great majority, are far above the mere agreeable." One woman particularly fascinated him. Her "beautiful hair . . . was decorated with a single white rose, surmounting a figure exquisitely formed and moving with perfect grace, belonging to some young Lady apparently of 18, whom I am glad I do not know. . . ." He added that she was "perfect in all respects, as any thing I have ever seen in or out of Marble."[33]

An aspect of New Orleans that Latrobe detested was the treatment of slaves. He wrote disparagingly of his hostess, a Madame Tremoulet, who needlessly beat one of her slaves. "A few days ago, [the slave girl] failed (because it was impossible) to make the bed of a stranger, at the hour prescribed. In consequence of this fault Madame Tremoulet had her stripped quite naked, tied to a bedpost, and she herself . . . whipped her with a Cowskin till she bled." Another slave, Latrobe continued, was "made to whip her till she fainted. . . . I shall leave the house as soon as convenient

[31]Latrobe, *Journals* (1980), pp. 182, 185; Latrobe, *Impressions*, pp. 32, 35; Latrobe, *Journal* (1905).

[32]Latrobe, *Journals* (1980), pp. 194-195, 266-267; Latrobe, *Impressions*, pp. 42, 105-106; Latrobe, *Journal* (1905), pp. 187, 209-211.

[33]Latrobe, *Journals* (1980), p. 184; Latrobe, *Impressions*, p. 33; Latrobe, *Journal* (1905), p. 183.

to me."[34]

However, Latrobe had no great love for what he considered the uncivilized character of the African slave. While walking along St. Peter Street toward the Carondelet Canal he came upon Congo Square, a large open park where blacks traditionally congregated on Sundays for a day of religion, music and dancing, usually with an emphasis on their African roots. This one afternoon, Latrobe observed between five and six hundred blacks involved in what he described as "uncouth" acts. "I have never seen any thing more brutally savage," he wrote, "and at the same time dull and stupid than this whole exhibition." Latrobe goes on to describe in detail an African string musical instrument used to entertain the group at the square, in addition to African languages and native songs.[35]

It is interesting that Latrobe took care in his journal to describe and analyze the mosquito, the carrier of the virus that would eventually end his life. New Orleans, he wrote, would be "one of the most delightful abodes of affluence, and elegance in the world were it not for the Muskitoes. . . . A few are found, every warm day throughout the Year," he continued, "but from June to the Middle of October or beginning of November, their swarms are incredible." He observed four types of mosquitoes in his New Orleans room, unaware that of those four types one was the carrier of the disease he was attempting to arrest by supplying clear water to New Orleans. "As soon as the sun sets, the Muskitoes appear in Clouds, and fill every room in the house. . . ." Their noise, he wrote, "fills the air, and there is a character of occasional depression and elevation in it, like the concert of frogs in a Marsh."[36]

Those immigrants newly arrived from the North were the most susceptible to yellow fever, partly because of their inability to avoid mosquitoes. Northerners wearing the wrong attire, Latrobe pointed out, attracted swarms of the pests, as in the case of those foolish enough to wear black hats:

> At sunset, a black hat appears to have particular attraction for them. If a group of Men stand together on the Levee, those who

[34]Latrobe, *Journals* (1980), pp. 207-208; Latrobe, *Impressions*, pp. 53-54; Latrobe, *Journal* (1905), p. 183.

[35]Latrobe, *Journals* (1980), p. 204; Latrobe, *Impressions*, pp. 49-50; Latrobe, *Journal* (1905), p. 180. The best source on New Orleans' Congo Square is George W. Cable, "The Dance in Place Congo," *Century Illustrated Monthly Magazine*, XXXI (1886), 517-531. However, Cable's descriptions are from reminiscences of a later period of activity at the square than that observed by Latrobe. See also John W. Blassingame, *Black New Orleans, 1860-1880* (Chicago, 1973), pp. 3-5; Henry C. Castellanos, *New Orleans As It Was: Episodes of Louisiana Life* (New Orleans, 1905), pp. 90-100.

[36]Latrobe, *Journals* (1980), pp. 305-306; Latrobe, *Impressions*, pp. 141-142; Latrobe, *Journal* (1905), pp. 233-235.

> wear black hats are seen with a column of innumerable muskitoes ascending upward above their hats: those in white hats are not so distinguished.[37]

Latrobe's journal entries on New Orleans and its environment came to an end as the waterworks project occupied more and more of his time. One of Latrobe's first concerns when he came to New Orleans was to ask the city council to extend, for a third time, the 1811 grant which had expired January 1, 1819. In his letter to the council requesting the extension, Latrobe explained that many of the difficult problems that had plagued the project had been overcome, and that the system was on the verge of completion.[38] The council debated the question in almost every council session through February. Finally, on February 22, they reluctantly granted Latrobe another two-year extension.[39]

The steam engine finally arrived from New York in March, 1819, and was erected in the large octagonal building that Henry had built from his father's plans. Located on the levee near what is now the French Market, it had to be enlarged to house the boring machine that Latrobe had brought with him from Baltimore. The purpose of the new machine was to bore cypress trunks into pipes for the water distribution system that would be laid under New Orleans' streets. Henry had attempted some boring, but the finished pipe, allowed to dry out, had eventually split, rendering them useless. With the engine in place and the new pipe being completed, the project was finally beginning to fall into place.[40]

It was not in Latrobe's character to undertake only one project while he was in New Orleans. In May, 1819, he was commissioned to design and build the central tower of the St. Louis Cathedral in the Place d'Armes. In an unusual arrangement, the cost of the structure was divided between the city and the Catholic church, and Latrobe was forced to deposit shares in the waterworks project as security for the completion of the tower. The tower,

[37]Latrobe, *Journals* (1980), p. 306; Latrobe, *Impressions*, pp. 142-143; Latrobe, *Journal* (1905), pp. 235-236.

[38]Latrobe to New Orleans City Council, January 14, 1819, Latrobe Papers, Tulane University Library.

[39]New Orleans City Council, Proceedings, Sessions of February 9, 13, 16, 22, 29, 1819, pp. 23-30.

[40]The best explanation of the technical aspects of the New Orleans waterworks system is Stapleton, ed., *The Drawings of Latrobe*, pp. 40, 199-203. The engine was first used in July to run the boring machine. See *ibid.*, p. 40. Latrobe found pipe made from cypress tree trunks inadequate for the suction pipe from the river, and was forced to settle for small diameter cast-iron pipe. Some cypress pipe proved useful in other parts of the waterworks system. *Ibid.*, p. 203. See also, Hamlin, *Latrobe*, p. 517.

considered Latrobe's last architectural work, was barely completed at his death in 1820.[41]

Latrobe was also commissioned to design and build the Louisiana State Bank of New Orleans. He received the commission in the spring of 1820 and completed the drawings for the project later that summer. Although Latrobe died before work could begin, the building was constructed from his plans after his death. Located at the corner of Royal and Conti streets, the Louisiana State Bank building is similar to the Bank of Pennsylvania in Philadelphia which Latrobe designed and built between 1789 and 1800.[42]

By late summer, 1819, the waterworks project had again bogged down, and on September 17, Latrobe wrote to the city council to explain the delays. This time the problem was the yellow-fever epidemics that had slowed work on the project since the middle of the preceding summer. "The epidemics now prevailing, and the fear of it [*sic*], have dispersed almost all my workmen. I have been under the necessity of proceeding since the 4th of July last, with a very slow pace. . . ." Latrobe added that his own illness, a nervous disorder that plagued him occasionally, "has also contributed to retard my operations."

Latrobe also mentioned in his September 17 letter to the council that he was about to leave New Orleans to arrange for his family to join him from Baltimore.[43] He apparently felt that New Orleans held his destiny. New Orleans, he believed, would soon occupy an important place in America's future, and he planned to have a part in that future. As a leader in the emerging technology of steam-powered riverboats, Latrobe was one of only a few Americans who realized the potential of river transportation for America. He must have envisioned that soon the nation's rivers would be teeming with riverboats, moving both upstream and down, and that New Orleans, at the mouth of a great funnel, was about to become a wealthy, fast-growing, major American city. An architect such as Latrobe would be in great demand in such a rapidly growing environment. Had he lived, Latrobe certainly would have played an important part in the growth and development of the Crescent City.

Latrobe returned with his wife and children in April 1820, ready to complete the waterworks project quickly, to reap the rewards of his work,

[41]New Orleans City Council, Proceedings, Session of May 29, 1819, pp. 71-73. The cathedral was originally built in 1795, and the entire building, including Latrobe's tower, was remodeled in 1850 by the architect Louis Pilié. Curtis, *New Orleans*, pp. 77-78. See also, Allison Owen, "Latrobe's Last Commission," *Architectural Art* (July, 1907), 1-4.

[42]Curtis, *New Orleans*, pp. 8, 111.

[43]Latrobe to New Orleans City Council, September 17, 1819, Latrobe Papers, Tulane University Library.

and to begin new endeavors. By July, he had supervised the installation of some 5,000 feet of wooden pipe in the city, leaving only the placement of the main suction pipe into the river as the remaining accomplishment of the project. For that, he was given permission on August 26 by the city council to use the city's chain gang to dig the trench.[44] But after several days of orchestrating the work from the levee, Latrobe fell ill, and on September 3, exactly three years after his son had died, Latrobe succumbed to yellow fever.

Years later, Latrobe's wife wrote that her husband's health had been good through that summer in New Orleans. The work was nearly finished, and Latrobe was working long hours in the August heat to bring together the last components of the project. On August 28, she recalled, "he complained of headache from extreme fatigue and told me the works were now completely finished, having had a pipe laid from the Levee into the Channel of the River. . . ." Latrobe was attended by a physician, but to no avail. "I never for a moment considered that he might have the yellow fever," his wife wrote. "On the 3rd September he was taken from me."[45]

Following Latrobe's death, the New Orleans Water Works Company struggled to obtain financing and maintain solvency, but finally, as a result of a suit by the investor Benjamin Orr, the city took over the company, eventually buying the entire system for $10,000 in 1821. The surviving members of Latrobe's family received nothing from the project. The system remained in operation in New Orleans until 1840, when it was finally replaced by a system designed by Albert Stein.[46]

Latrobe was buried in New Orleans, but his grave site has long been lost to a system of burial ground neglect and reuse. On March 23, 1985, a plaque dedicated to Latrobe and his son was placed at the St. Louis Cemetery on Basin Street.[47]

Benjamin Latrobe had an important impact on the new American nation. As its first great architect, builder and engineer, he established a direction for early American architectural style and internal improvements that lasted for

[44]New Orleans City Council, Proceedings, Session of August 26, 1820, p. 105. Stapleton, ed., *Drawings of Latrobe*, p. 41.

[45]Latrobe, *Journals* (1980), p. 339. This memoir was written by Mary Elizabeth Latrobe many years after her husband's death. Latrobe returned home complaining of a headache on August 30, not August 28. See *ibid.*, p. 339n.

[46]New Orleans City Council, Proceedings, Session of June 23, 1821, p. 55. The actual date of purchase was June 26, 1821. See *ibid.*, Session of June 30, 1821, p. 58.

[47]Latrobe Collection, Louisiana Room, Tulane University Library.

many decades. He and his son also had an important impact on New Orleans. Although their impact was not based on the great architectural achievements that have brought Benjamin Latrobe a place in history, the New Orleans waterworks system was a much needed addition to the river town that was just beginning to emerge from its status as a remote colonial outpost to that of a major American city.

Latrobe believed that New Orleans soon would be a booming city, with a great deal of money to be made from architecture and internal improvements. Had he and his son survived yellow fever and remained in New Orleans, they certainly would have been at the forefront of the building spree that occurred in the city in the decades following their death. The result may have been a New Orleans with a different appearance.

10
Urban sanitation in preindustrial Japan

Susan B. Hanley

The sudden increase in Japan's urban population in the century and a half from the 1580s to the mid-1700s may well have had no parallel in world history prior to industrialization. Edo, renamed Tokyo in 1868, was a cluster of fishing villages around a castle in 1590, but, during the eighteenth century, it is readily acknowledged to have been one of the largest, if not the largest, city in the world. Even by 1700, Edo was certainly larger than any European city, including London at 575,000, and rivaled or exceeded in population the largest of the Chinese cities at the time, Peking. But urban growth was not limited to Edo; cities sprang up throughout the country from the late sixteenth century on and both Kyoto and Osaka had populations in the hundreds of thousands. By the late eighteenth century, Japan had about 3 percent of the world's population, but it is estimated to have had more than 8 percent of the people in the world who lived in cities of more than 10,000. By this standard, about 10 percent of the total population of Japan was urban in 1800.[1]

To support a metropolitan population, food and other daily necessities must be adequately supplied, and demands for housing, water, transportation, a monetary system, and numerous other requirements must be met. Even if all other requirements are

The author wishes to thank William B. Hauser, Joel A. Tarr, and Kozo Yamamura for helpful suggestions and Robert J. Smith for the invitation to the Social Science Research Council workshop on the Japanese city for which this research was begun.

1 Ōishi Shinzaburō, "Kyodai toshi Edo no jinkō wa dore kurai ka," *Rekishi to jimbutsu*, CX (1980), 76–81; E. Anthony Wrigley, "A Simple Model of London's Importance in Changing English Society and Economy 1650–1750," in Philip Abrams and Wrigley (eds.), *Towns in Societies* (Cambridge, 1978), 215; G. William Skinner (ed.), *The City in Late Imperial China* (Stanford, 1977), 29. Gilbert Rozman, *Urban Networks in Ch'ing China and Tokugawa Japan* (Princeton, 1973), 6, estimates Japan to have been 10% urban while Sekiyama Naotarō, *Kinsei Nihon no jinkō kōzō* (Tokyo, 1957), 239, considers 12% of the population to have been urban.

satisfied, however, if sanitation is inadequate, one would not expect the population of a city to reach a million, much less be maintained at this figure over centuries. Yet Edo, which lacked modern water supply and sewage systems, and which relied primarily on manpower or boats to carry away garbage, refuse, and other human wastes, is estimated to have had a population in the range of a million from about 1700 on. How did Japan manage to maintain such an enormous urban concentration of people for two centuries prior to industrialization?

Numerous studies have discussed urban sanitation and water systems in the West, as well as the negative rate of natural increase in the population of urban areas in premodern times. However, scholars in Japan have begun to research these subjects only in the last decade, and no one has attempted to evaluate as a whole the various aspects of sanitation in premodern Japanese cities and their significance in Japanese history. Here my focus will be on the problem of sanitation in the major metropolises of Japan from approximately 1590 to 1890 and will include an assessment of the sanitation standards in light of Western standards for the same centuries. This is not intended to be an exhaustive study that will provide full evidence, but one that provides a new perspective on Japan and, through comparison, on the West as well.

The lack of interest in these topics in Japan results primarily from the political-institutional focus that historians traditionally have had. Perhaps equally important is the fact that Japanese have not had to contend with the urban problems that Westerners faced in early modern times. Thus, the wealth of personal observations, proposals for improvement, and other materials so widely available in the West from at least the early nineteenth century are, by and large, lacking for Japan. But from Japanese laws, regulations, and political concerns, a good deal of information can be extracted about Edo's water supply and waste disposal systems and on Osaka's system for disposing of night soil.

This article examines how the premodern Japanese provided a water supply system for urban residents and developed systems to rid the cities of waste, ranging from garbage and rubbish to water and human excreta, or night soil. It also looks at Japanese customs that had an effect on levels of sanitation and disposal of night soil and compares urban conditions in Japan with those in the West. The article is divided into four main parts: 1) a brief summary of the growth and changes in the city population in the Tokugawa period (1600–1868); 2) a description of Edo's water supply system and brief comparison with how other Japanese cities were supplied; 3) an analysis of how the cities handled the crucial problems of waste disposal; and 4) a comparison of sanitation and hygiene in Japan from the seventeenth through the nineteenth centuries with sanitation levels in Western cities. The implications of the findings and their possible relationship to the course of industrialization in Japan are discussed in the conclusion.

THE URBAN POPULATION IN THE TOKUGAWA PERIOD The growth of Edo in the seventeenth century was the most remarkable event in Japan's urban history. In the sixteenth century, the site was occupied only by a number of small fishing villages and a castle, but it was strategically located on one of the largest bays in Japan and at the edge of its largest plain. Here, in 1590, the founder of the Tokugawa Shogunate began to build his new administrative capital. Edo grew so rapidly in the seventeenth century that the first survey of its population taken in 1678 listed a total commoner population of 570,361. In the period from 1734 to 1867, Edo's commoner population was always above 500,000, with a low of 501,166 in 1743 and a peak of 559,497 in 1844. However, there also existed a large unregistered population in Edo, estimated at 80,000 in the second half of the nineteenth century. Thus, a figure of 600,000 for the commoner population of Edo for much of the Tokugawa period is a reasonable estimate.[2]

Added to the commoner population must be that of the samurai class and their servants. Since the number of people connected with the warrior class was not included in the surveys of urban population, we have to "guesstimate" the number of people who occupied two thirds of the space of the city. The direct retainers of the Shogun numbered approximately 22,500 in the beginning of the eighteenth century and 26,000 at the start of the nineteenth. At a minimum, these retainers plus their households would have added 100,000 to Edo's population. But most of the space in Edo allocated to the samurai was taken up by the official residences of the 260-odd daimyo who each maintained a

2 See Takeuchi Makoto, "Edo no chiiki kōzō to jūmin ishiki," in Toyoda Takeshi et al. (eds.), *Nihon no hōken toshi* (Tokyo, 1983), II, 300; Ōishi, "Kyodai toshi Edo," 77–80.

residence and one or two additional establishments in the city. The daimyo had to spend half of their time in Edo to "pay attendance upon the Shogun," but in reality this requirement enabled the Shogun to keep an eye on them—a polite hostage system. Although some of the smallest daimyo probably had only a hundred or so people in residence in Edo, some of the largest had thousands. The largest daimyo, the Maeda of Kaga, maintained a regular staff of about 1,000 retainers, who with their families constituted at least 4,000 permanent residents. When the daimyo himself was in residence, it is estimated that as many as 8,000 people were in attendance. Thus even a conservative estimate of the daimyo population would require adding a few hundred thousand to the population of Edo, and this number does not include servants.[3]

To estimate Edo's population, from a base of 600,000 commoners, we would have to add a very conservative estimate of 100,000 for the Shogun's direct retainers, a minimum of 200,000 in the samurai class from the daimyo establishments, plus an unknown number of servants. Clearly a total of 1 million for Edo's population is not wild speculation. Edo was certainly much larger than London in 1700 and rivaled it in population in 1800.

Not only was Edo larger than any European city, but both Osaka and Kyoto were larger than the European capitals of Vienna, Moscow, and Berlin even at the beginning of the nineteenth century and were surpassed in size only by London and Paris. After the Battle of 1614/15, Osaka was built to become the commercial entrepôt of Japan. At its peak in 1763, it had a commoner population of 418,537. Even after Edo gradually took over many of the marketing and financial functions of Osaka, the city remained the second largest in the country with a population of 314,370 in 1858. Kyoto, the seat of the emperor, had a population recorded at 410,000 by 1534, and then, as other urban areas grew,

3 The samurai were originally warriors, but by the Tokugawa period this was the general term for the warrior estate or class. During the Tokugawa period, the daimyo were regional military lords who held domains assessed at 10,000 or more *koku* (a unit of rice, approximately 5 bushels). The direct retainers of the Shogun were his vassals: a minority of daimyo (*fudai*), the bannermen (*hatamoto*), and the housemen (*gokenin*). For information on the samurai, see Kozo Yamamura, *A Study of Samurai Income and Entrepreneurship* (Cambridge, Mass., 1974), 10; Naitō Akira, "Edo no toshi kōzō," *Edo jidai zushi* (Tokyo, 1975), IV, 169; Toshio G. Tsukahira, *Feudal Control in Tokugawa Japan: The Sankin Kōtai System* (Cambridge, Mass., 1966), 95.

its population fell into the mid-300,000 range from the late 1660s on.[4]

Japan's urban growth in the seventeenth century was not limited to the three largest metropolises; equally important in the long run was the proliferation of the castle towns in the same period. From the last quarter of the sixteenth century and into the seventeenth, numerous cities sprang up in the Japanese countryside. These originated as the headquarters of major daimyo and were known as castle towns, because at the center of each was the castle, which served as the domain's administrative center. Quartered here were the daimyo's forces, and to supply them there was a need for merchants and artisans. A domain's castle town usually maintained about 10 percent of the domain's total population. The commoner population of the major castle towns ranged from 10,000 up to more than 60,000 (Kanazawa) and, although their commercial activities began to be taken over by smaller towns in outlying areas in the late Tokugawa period, most of these towns remained vital centers. Half of the sixty most populous cities in Japan today originated as castle towns during the late sixteenth century.[5]

Japan also had a score of ports by the sixteenth century that ranged in size from 10,000 to 50,000, and an even larger number of cities or towns with populations in excess of 5,000. With the establishment of formal shipping routes in the Tokugawa period, the number and size of port towns increased. By the late eighteenth and nineteenth centuries, villages in the countryside had become towns, and towns had grown into cities. After the formation of the castle towns, Japan was no longer dominated by its capital, and two centuries later, with the growth of outlying centers, the castle towns were no longer the only urban areas in many domains. The Tokugawa period witnessed not only the rise of one of the largest cities in the world, but also the urbanization of Japan.[6]

4 Sekiyama, *Kinsei Nihon,* 220, 232; Osaka-shi Sanjikai, *Osaka shishi* (Tokyo, 1927), I, 482–483, 602, 880–882; (1928), II, 107, 180–181, 546, 758–759.

5 See John W. Hall, "The Castle Town and Japan's Modern Urbanization," in *idem* and Marius B. Jansen (eds.), *Studies in the Institutional History of Early Modern Japan* (Princeton, 1968), 169–188; Thomas C. Smith, "Pre-modern Economic Growth: Japan and the West," *Past & Present,* 60 (1973), 127–160; Jansen and Rozman (eds.), *Japan in Transition* (Princeton, 1986), 273–374.

6 For discussions of pre-Tokugawa urban development, see Toyoda, *Chūsei Nihon shō-*

6 | SUSAN B. HANLEY

EDO'S WATER SUPPLY SYSTEM When Tokugawa Ieyasu selected Edo for his capital, he recognized from the outset the problem of obtaining an adequate water supply and ordered a former retainer, Ōkubo Tōgorō Tadayuki, to construct a water supply system. In 1590, Ōkubo first went to Edo to assess the situation and make plans. The system he began was so large in scale and so successful that it has been compared to that of the Romans.[7]

Although Edo was strategically situated, much of it was on low, marshy ground near the sea. The potable water was originally obtained from ponds and underground springs, but wells had to be deep because of the depth of the aquifer. The first system constructed, the Kanda system, drew its water from the Inokashira spring east of the city. Water was carried to the city limits mainly in exposed aqueducts, and then in underground aqueducts or wooden pipes within the city. The Kanda system was over forty-one miles in length and had 3,663 subsidiary ducts which drew water from the main source and distributed it to various parts of the city. Whenever possible, natural waterflow was used, but the Japanese did increase pressure at selected points to lift water from one level to another. However, the Kanda system was designed for delivering water to the lower-lying sections of the city, so that natural waterflow was usually sufficient. A problem with this system was that the water from the Inokashira spring was limited in quantity and, if too much was pumped out, the quality suffered—the water turned muddy. By the mid-seventeenth century, this system was inadequate for the growing city's needs, and a new system was begun in 1652.[8]

The second system took its water from the Tama River. Well over fifty miles in total length, it was larger than the Kanda system and carried a far greater volume of water. The Tama River system brought water up to the Yotsuya gate of Edo castle, a distance of nearly twenty-seven miles, supplying first of all the Shogun and then the nearby areas, Kōjimachi, Yotsuya, Kyōbashi, and Aka-

---

gyō-shi no kenkyū (Tokyo, 1952), 364–386; Sasaki Ginya, *Shōgakkan Nihon rekishi* (Tokyo, 1975), XIII, 260–271; Yagi Akio, "Chūsei no shōkōgyōsha to toshi," in Nakamura Kichiji (ed.), *Taikei Nihon-shi sōsho* (Tokyo, 1965), VIII, 328–345.

7 Higuchi Kiyoyuki, *Edo* (Tokyo, 1968), IV, 233.

8 Descriptions of Edo's water supply system are to be found in Horikoshi Masao, *Ido to suidō no hanashi* (Tokyo, 1981); Sabata Toyoyuki, *Suidō no bunka: Seiō to Nihon* (Tokyo, 1983); Itō Kōichi, "Edo no suidō seido," in Nishiyama Matsunosuke (ed.), *Edo chōnin no kenkyū* (Tokyo, 1978), V, 283–308; Higuchi, *Edo*, IV, 231–247.

saka. When first constructed, there was sufficient water to serve the city and also to irrigate rice paddies in Musashino, a farming area west of Edo. However, as the city expanded, the use of the Tama River system's water for irrigation had to be curtailed.

The Tama River system proved to be inadequate, not so much because of the insufficiency of water, but because of the difficulties of raising the water to higher ground as the city expanded outward. A major stimulus for continued construction of water systems was also the problem of frequent and devastating fires in Edo, which was largely constructed of wood. In the great Meireki fire of 1657, approximately two thirds of the city was destroyed; deaths were estimated in the tens of thousands. Subsequently, a policy was implemented to decrease the density of the population at the center, and many samurai mansions and temples were relocated on the outskirts of the city, leaving open spaces within the city to serve as fire breaks. Four new water systems were added—the Honjo, Aoyama, Mita, and Sengawa—all of which relied on the Tama River for their water supply and were really subsidiary to the main system. These newer systems seem not to have been entirely satisfactory for, by the mid-nineteenth century, Edo relied primarily on the original Kanda and Tama River systems. Since the Kanda system now received some of its water supply from the Tama River, Edo's water supply came mainly from that one river.

During the century that it took to build Edo's water systems, the engineering techniques became increasingly sophisticated. In the early part of the seventeenth century, ditches were the most common conduit outside city limits. For its underground aqueducts, the Kanda system used square pipes of red pine, but other early systems had pipes made of other kinds of wood, stone, earth, and bamboo. By the time the Tama River system was constructed, the engineers designing it were using triangulation, which made accurate surveying possible and helped solve the problems created by differing heights of land. Siphons were used to draw water up into Edo castle from the ducts of the Tama River system. The newer the system, the more pipes it had rather than open channels, and thus much more of the later systems were underground.[9]

9 Higuchi, *Edo*, IV, 244.

The government was not only responsible for the building of the water systems supplying Edo, but it also tightly regulated their use. No individual was permitted to draw water directly from the main system unless he were a samurai of high status, usually a daimyo or a high-ranking retainer of the Shogun. The public was supplied with water from wells built into the aqueduct system; people were required to go to the nearest well and draw water there rather than tap the nearest duct themselves. This control ensured an adequate flow of water and proper maintenance of the system.

Clearly Edo was a special case. Its water supply system was designed at the same time the city was, and the system was expanded as the city grew. Edo's system was vastly superior to the water supply systems in all other Japanese cities, most of which relied on rivers and wells for their water supply. However, with the building of the castle towns from the late sixteenth century on, water supply systems were constructed in various regions, no longer primarily for irrigation but to supply drinking water to urban populations. The first major system built primarily to supply drinking water was the Kanda, followed by systems constructed for cities not only in central Japan but also in such far-flung places as the southern tip of Kyushu. Many were built in the early seventeenth century, followed by continued construction of new systems throughout the Tokugawa period. In addition to systems designed specifically for supplying cities with potable water, numerous irrigation projects supplied drinking water as well.[10]

WASTE DISPOSAL IN OSAKA AND EDO The methods developed for disposing of various kinds of wastes in both Edo and Osaka can be documented through regulations, contracts, and records of challenges or conflicts. The number and variety of the sources indicate that from the mid-seventeenth century on, waste disposal was a major concern for urban administrations. Osaka provides the best example of how different sewage disposal was in Japan, in contrast to Western nations, whereas Edo illustrates how complex waste disposal in a premodern Japanese metropolis was and how the problems were managed.

10 Horikoshi, *Ido to suidō*, 98–99.

JAPANESE SANITATION | 9

The most important difference between waste disposal in Japan and in the West was that human excreta were not regarded as something that one paid to have removed, but rather as a product with a positive economic value. The night soil of Japanese cities—and Chinese as well—was long used as fertilizer. With the growth of Japan's population, the limitation of the amount of arable land and the increasingly intensified use of land to feed the growing population, combined with the relative scarcity of animal wastes and other fertilizers, meant that human waste had a value as fertilizer that far exceeded its value in the West.

Long before Edo was even established, Osaka's night soil was used as fertilizer for the surrounding farm villages. Most of it was collected, loaded onto ships, and distributed to nearby farm areas. The huge volume brought to the wharves resulted in such an unpleasant odor that there were complaints. In the Tokugawa period, the magistrates deliberated upon these complaints but concluded that "it was unavoidable for the manure boats to come into the wharves used by the tea and other ships."[11]

In the early years of the Tokugawa regime, boats were sent into Osaka loaded with vegetables and other farm produce which were exchanged for the night soil of the city. But as the price of fish and other fertilizers rose, the value of night soil rose correspondingly, and vegetables were no longer sufficient to pay for it. By the early eighteenth century, with the increase in new paddies in the Osaka area, the price of fertilizer had jumped to the point that even night soil had to be purchased with silver.

The value of human wastes was so high that rights of ownership to its components were assigned to different parties. In Osaka, the rights to fecal matter from the occupants of a dwelling belonged to the owner of the building whereas the urine belonged to the tenants. Feces were considered more valuable and hence commanded a higher price. Generally speaking, the price of fecal matter from ten households per year amounted to between two and three *bu* of silver, or over one half a *ryō* of gold. This was a considerable sum since a *ryō* during much of the Tokugawa period was sufficient to buy all the grain staple one person would eat during a year. Rent was adjusted on the basis of how many tenants there were and was raised if the number of occupants dropped.[12]

11 Wakita Osamu and Kobayashi Shigeru, *Osaka no seisan to kōtsū* (Osaka, 1973), 127.

12 Although urine is usually higher in nitrogen and potash than solid excreta and is

With the rapid growth of Osaka in the seventeenth century, the city government found by mid-century that it had to step in and form guilds to insure that waste disposal was handled properly. As the price of fertilizer rose, by the end of the century farmers from neighboring areas were forming associations for the purpose of obtaining monopsony rights to purchase night soil from various areas of Osaka. Eventually fights broke out over collection rights and prices. In the summer of 1724, two groups of villages from the Yamazaki and Takatsuki areas fought over the rights to collect night soil from various parts of the city. Other disputes arose between the guilds in the city and farmers' associations, and examples exist for the neighboring provinces of Kawachi and Settsu as well, indicating that this type of conflict was neither a localized nor an isolated event.[13]

In the three major areas of Osaka, neighboring farm villages held the rights to collect night soil from households, but they either could not or did not want to collect all of the urine. The remainder was left to be collected by *shōben nakagainin,* literally, urine jobbers. The number of jobbers gradually increased, as did jurisdictional problems. Eventually they created their own association, and in 1772 they paid a fee to the Osaka authorities to establish a *kabunakama* (guild based on ownership of shares) with the authority to enforce jurisdiction and to set prices. However, the rights to collect the urine from containers left for passersby on the street corners in Osaka were given to an outcast village named Watanabe, but even though the price of urine was lower than for fecal matter, there were constant conflicts over these collection rights. Periodically, other people tried to get these privileges away from Watanabe, but the village managed to maintain its monopoly throughout the Tokugawa period, despite sabotage of its containers, challenges by others to its collection rights, and offers to buy the rights.[14]

especially useful as an activator in converting crop residues to humus, it is more difficult to transport and store than solid excreta, which probably accounts for its lower price. For the price of fecal matter, see Watanabe Minoru, *Mikaihō buraku-shi no kenkyū* (Tokyo, 1965), 297. The value of a *ryō* was calculated from Yamasaki Ryūzō, "Edo chūki no bukka dōkō to keizai hendō," in Harada Toshimaru and Miyamoto Matao (eds.), *Rekishi no naka no bukka* (Tokyo, 1985), 78.

13 Wakita and Kobayashi, *Osaka,* 128.

14 Watanabe, *Mikaihō,* 292–299. See also Osaka-shi, *Osaka shishi,* I, 866–868.

JAPANESE SANITATION | II

In Osaka, by the mid-eighteenth century, night soil was so clearly an economic good that ownership and monopoly/monopsony rights were assigned, the formation of officially recognized associations and guilds was permitted, and the price was determined by these organizations. The price was so high that the poorer farmers had difficulty in obtaining sufficient fertilizer, and incidents of theft began to appear in the records, despite the fact that going to prison if discovered was a real risk.

In Edo, judging from official records, disputes over night soil seem not to have been the problem that they were in Osaka, but the government did have to step in to handle problems of orderly waste disposal. Edo's waste, other than night soil, can be divided into four major types: 1) household waste, probably mostly kitchen garbage; 2) trash discarded along the roads and in the waste water drains; 3) junk floating in waterways—moats, rivers, and the harbor; and 4) waste from fires. In addition, Edo had the problem of disposing of its waste water since its ample water supply meant that enormous quantities of water had to be discarded somewhere.[15]

Regulations regarding waste began to appear in Edo during the mid-seventeenth century. The focus at this time was not waste collection per se, but the growing problem of keeping the streets, open areas, and drainage ditches free from rubbish. This was clearly viewed as a problem within the purview of city government. At the same time, problems relating to the disposal of night soil came to the attention of the authorities. In 1648, city regulations required small huts and toilets along the banks of rivers to be torn down. The repeated issuance of this and other regulations over the next half century indicates that Edo residents must have been slow to comply with the new, more sanitary arrangements for waste disposal.[16]

These regulations also indicate that demand for Edo's night soil was probably not high during the early Tokugawa period and much potential fertilizer must have been wasted. However, as truck gardening developed in the area around Edo and the city

15 Hayashi Reiko, "Kinsei ni okeru jinkai shori," *Ryūtsū keizai ronshū*, VIII (1974), 72–86.

16 For information on waste disposal in Edo, see Itō, "Edo ni okeru gomi, gesui, shinyō no shori," in Toyoda, *Nihon no hōken toshi*, 431–455; Kawazoe Noboru, *Uragawa kara mita toshi* (Tokyo, 1982), 152–190.

came to rely less on a supply of daily goods from afar, the demand for night soil rose. The market for the night soil of Edo was the farm villages surrounding the city within a radius of about ten miles. By the first half of the eighteenth century, the sudden withdrawal of a particular supply of night soil could be devastating to a farmer, as alternate supplies were not easily found. A village head in the neighboring district of Tama lost his supply of fertilizer in 1725 when the main residence of the daimyo of the Owari Tokugawa burned. Before the mansion could be rebuilt, the farmer suffered major crop losses.

Contracts were given to farm villages for the night soil from specific areas in Edo. Usually each daimyo contracted out these rights, with the price determined on the basis of the market demand. For example, the Hitotsubashi daimyo sold the rights to night soil from the residence in 1742 to one Hanbei from Tanashi village in the district of Tama for the price of 1,500 large *daikon* (large white radishes), or 2,000 middle-sized *daikon,* or two *ryō* in cash, whichever Hanbei preferred, to be paid at the end of the year. Each year this daimyo house decided the price by bidding out the rights. Each daimyo made different arrangements, sometimes with the payment made twice a year, and the price as high as six *ryō* or more.

As the price of night soil rose, entrepreneurs sought rights to place containers to collect urine on busy street corners in Edo, but these petitions were denied. Since Edo was the seat of government, officials were concerned with appearance more than they seem to have been in Osaka, but they also worried about the impeding of the narrow streets and the problem of odor. One particularly innovative petition in 1789 requested permission to set out soy sauce and *sake* barrels, which would be less of an eyesore than urinals. The petitioner argued that allowing the collection of urine from passersby would add to the quantity of fertilizer, thereby reducing the price of other fertilizers, enabling farmers to use more fertilizer and produce larger harvests, and ultimately would lower price levels in general.

At the same time that the first regulations about night soil appeared, the disposal of other forms of waste was also specified. By 1655, the people of Edo were ordered to dispose of their garbage and rubbish on the island of Eitai in Edo Bay, rather than just dump it in the rivers. During the decade from 1655 to 1665, basic disposal policies were gradually put into effect: collection points for refuse collection were established in each ward; transport was contracted to specific jobbers; and the wards were ordered to bear the costs. From the ward collection points, the rubbish was loaded onto boats and transported to Eitai Island. Although the original purpose of these measures was to keep the river channels open for commerce and traffic, they resulted in the establishment of an official dump for the city located outside city limits. These policies remained in effect for the next two centuries.

The designation of Eitai Island as a dump eventually resulted in the creation of new land from the swampy ground in eastern Edo. Several other landfills also resulted in the creation of fields, so that disposing of Edo's wastes became a very profitable business. By the 1820s at least eighty contractors were involved in collecting the rubbish from Edo, encouraged in part by the creation of official organizations, the *kabunakama*. Now the townspeople had only to deposit their refuse at a collection site within the ward and pay for its collection and transportation for final disposal by a contractor. Ward fees were assessed by length of frontage and location of the property, so that, in effect, people paid a property tax. Whether renters were assessed depended on the ward.

Finally, there was the matter of drainage and disposal of waste water in Edo, which had a number of rivers and canal-like moats flowing through the city. From the number of regulations issued, the major problem seems to have been keeping the moats and river channels free from rubbish, rather than providing for drainage itself. Detailed instructions were issued: people were ordered to construct drainage channels along the fronts of their houses, under the eaves. These ditches collected runoff from the streets and roofs, as well as people's waste water. The channels, about a foot wide and a couple of feet deep, partially covered by stones to prevent people from falling in, are still to be seen in many cities, including the outskirts of modern Tokyo. Archaeological excavations in Tokyo reveal clearly the fine network of drainage ditches even within the compounds of what would be considered crowded working-class housing at best and slums at worst.[17]

Edo and Osaka were not alone among Japanese cities in their emphasis on the proper disposal of waste material and the necessity of keeping streets and waterways clean and open. Even the

17 Koizumi Hiroshi, *Edo o horu* (Tokyo, 1983), 64.

main streets in most castle towns were relatively narrow, about twenty-four feet wide, but they were "extremely well maintained and immaculately clean." The regulations regarding the maintenance of public roads were detailed and infractions were reported. In Tottori, for example, streets had to be cleared and then sprayed with water (which probably reduced the incidence of respiratory disease). In Hirado orders were issued to the effect that all bridges, gutters, and waterways should be repaired, maintained, and cleared; to make certain that this was done, officers of the town were to inspect them constantly. "No corner shall be left uncleaned." Judged by regulations alone, cleanliness and proper sanitation were of high priority among Tokugawa urban administrators.[18]

WHY JAPANESE CITIES WERE MORE HYGIENIC THAN EUROPEAN CITIES

Metropolitan sanitation in Japan from the mid-seventeenth through the mid-nineteenth centuries was almost certainly better than in the West in terms of quality and quantity of the water supply and in terms of waste disposal, resulting in a healthier environment for urban populations as measured by the size of the population and mortality rates. But Japanese city life was also more sanitary than that in the West because of various customs concerning hygiene, food, and drink, and because of a lack of domestic animals. Finally, the government played a major role in setting and maintaining standards of sanitation in the cities. Evidence to support these hypotheses comes not only from descriptive material, but from comparisons made by observers from the West who visited Japan, and from the few quantitative measures we have on water quality, mortality, and life expectancy.

As already elaborated, an important difference between Japan and the West was that human excrement was an economic good in Japan, and was carefully collected for use as fertilizer, thus protecting the water supply in all phases, from source to urban pipes, and also preventing people from coming into contact with waste matter while walking on the street or in or near dumping grounds. In contrast, Westerners traditionally relied on pits in the ground, such as cesspools, for the disposal of human wastes, and

18 Murai Masuo, "Hōken-sei no seiritsu to toshi no sugata," in Morimatsu Yoshiaki et al. (eds.), *Taikei Nihon-shi sōsho* (Tokyo, 1965), XVI, 128.

the danger of polluting water supplies was ever present. Even in the 1880s, Cambridge, England, was described as "an undrained, river-polluted cesspool city." In the nineteenth century, "Leicester was typical of many towns in the way it tackled the problems of excrement removal. At mid-century it had almost 3,000 uncovered cesspits, covering 1¼ acres." Only by the end of the century had it managed to convert to a system of pails, which put an end to seepage into the sub-soil.[19]

Much has been made of the English invention of the water closet, but in the early years this system caused more problems than it solved. First, it required both a water supply and sewer system that could safely supply and remove the large quantities of water that the system used. When the water closet was first invented, Londoners flushed their wastes into the Thames, thinking that at last they were rid of a nasty problem in their houses. What they did not realize for decades was that the cause of the epidemics of infectious disease sweeping the city was the flushing of sewage into the upper Thames, since much of the city's water was taken from it downstream. Furthermore, faulty drains caused sewer gases to waft up into homes, and people with fixed basins in their bedrooms often had to cover them with towels at night, a rather primitive method of coping with this problem.[20]

Nor was the new world immune: "As late as 1849, physician John H. Griscom described the unhealthy sanitary state created on Manhattan Island by 'these thirty thousand cesspools studding it up and down, and filling the atmosphere with nauseous gases'." Stone comments that "even after the introduction of water supply systems, conditions in cities and towns remained unsanitary until properly engineered sewers replaced cesspools, beginning about 1850." Pinkney's assessment of Paris in the same year is that there was "one shockingly direct connection" between sewage disposal and the water supply. "The city drew part of its water supply from that main collector sewer, the Seine, and pumped it largely at points downstream from the mouths of sewers emptying into the river. Most of the remainder of the city's water supply came from sources little more inviting."[21]

19 Anthony S. Wohl, *Endangered Lives: Public Health in Victorian Britain* (Cambridge, Mass., 1983), 73–74, 95.

20 Thomas McKeown, *The Modern Rise of Population* (New York, 1976), 125.

21 John H. Griscom, *The Uses and Abuses of Air* (New York, 1854), 183, as quoted in

Japanese cities did not have the horrendous problems of sewage contamination of their main water supplies that London and Paris did, not only because they did not flush sewage into the rivers, but also because water in Japan tended to be purer at the source. Japan's four main islands are all dominated by mountain ranges, resulting in short, swift rivers. In addition, Japan's abundant rain, particularly during the typhoon season, helps flush out the rivers, preventing the pollution that occurs in slow, meandering water courses.

More than one scientific examination of Tokyo's water quality in the nineteenth century exists. In the mid-1870s Atkinson, an Englishman, tested Tokyo's water at various points in the system for the existence of solids, chlorine, ammonia, nitrogen, etc. At this time, it was not possible to test for bacteria content, or "germs" as they were called by Atkinson, but scientists could chemically test water for organic content which would indicate the approximate contact with raw sewage, and thus the likelihood that it would cause disease. Atkinson's survey of the Tokyo system revealed that the water was usually close to pure at the source, but the farther the samples were taken from the source, the more contaminated they proved to be. It is not surprising that his samples from surface water were contaminated; what is surprising is that early Tokyo had a water supply purer than did London. Remember that he was comparing the water supply system of a city yet to begin using modern technology with the largest city in Europe's first industrialized nation.[22]

Furthermore, Edo's system was so well designed that when it was modernized at the end of the nineteenth century, the only

May N. Stone, "The Plumbing Paradox," *Winterthur Portfolio*, XIV (1979), 292, 284. For good overviews, see Joel A. Tarr et al., "Water and Wastes: A Retrospective Assessment of Wastewater Technology in the United States, 1800–1932," *Technology and Culture*, XXV (1984), 226–263; Tarr, James McCurley, and Terry F. Yosie, "The Development and Impact of Urban Wastewater Technology: Changing Concepts of Water Quality Control, 1850–1930," in Marton V. Melosi (ed.), *Pollution and Reform in American Cities, 1870–1930* (Austin, 1980), 59–82. David H. Pinkney, *Napoleon III and the Rebuilding of Paris* (Princeton, 1958), 21.

22 R. W. Atkinson, "The Water Supply of Tokio," *Transactions of the Asiatic Society of Japan*, VI (1878), Pt. I, 96. In addition, O. Korschelt, in a paper read before the Asiatic Society of Japan on December 12, 1883, assessed the quality of wells in Tokyo and concluded that artesian wells were a suitable means of supplying water to areas of Tokyo not reached by piped in water. He was even more salutary in his assessment of Tokyo's water supply than was Atkinson.

major change was to replace the wooden pipes with impervious metal ones. Thus, the Japanese were able to use the main features of a system constructed in the seventeenth century when converting to a water supply system based on modern technology more than two centuries later. The engineering feat involved in building such a sound and large-scale system was remarkable, especially considering that dirt from the excavations for the channels and works was hauled away in baskets or straw mats.

Although London's "New River" was constructed in the same period as Edo's Kanda system, the success in bringing spring water to London in 1613 was scarcely the equivalent of the construction of the Kanda system in Edo in the same period. In London, most of the pipes were on the surface and watchmen had to be hired to deter tampering. As London expanded, the water supply became increasingly inadequate, but, unlike Edo, no ready solution was found. Pipes could be added but the supply was not increased, and so by the mid-eighteenth century Londoners could draw water only seven hours a day, three days a week. In contrast, Edo's pipes were not as strong, but they were buried and there was always a sufficient water supply twenty-four hours a day. Stoppage of water was so rare that Edo-ites made no backup arrangements for emergencies. Indeed, the Tama River system brought so much water into Edo that a waterfall in the center of the city was created from the excess, and when the Ebisu Beer Company was founded in the late nineteenth century, it was able to use the water from the Tama system.[23]

This is not to argue that the premodern Japanese water supply and waste disposal systems were without problems. When human excreta are used for fertilizer, there is always the danger of contamination—transmission of pathogens to the food supply and pollution of the water supply through runoff from fields or inadequate storage or transportation of the night soil. Although some human excreta were sold as fertilizer in the West, for the most part night soil was wasted, particularly in the largest urban areas after sewerage, or water carriage systems, were adopted. By the end of the nineteenth century, the West had some advocates of sewage farming, but public health officials, as well as the general public, had a strong bias against using human waste for

23 Sabata, *Suidō no bunka*, 32.

fertilizer. They "maintained that the raw sewage exposed farm employees to possible infection and that the vegetables grown on the farms could be the carriers of 'dangerous microbes or other parasites,' even though there was no clear evidence" of this. In fact, Asians did not merely dump raw night soil onto their fields, but they stored it at least a month in the knowledge that direct application of raw exreta was dangerous.[24]

The combination of heat and time necessary to kill various pathogens varies. Scientists also disagree or are uncertain as to the degree of contamination of water by an enteric virus that will infect a community. There are numerous documented cases of outbreaks of disease in both Asia and the West through the use of night soil or sewage for fertilizer. Thus it is difficult to determine today how safe treatments of human excreta or manure were 200 years ago. "The great majority of illnesses associated with sewage, however, appear to have been caused by application of raw or inadequately treated sewage wastewater, raw sludges, and night soil to crops which were consumed raw." And at least two studies indicate that "the predominant method of transmission of enteric viruses appears to be a direct fecal-to-oral route."[25]

It is difficult to assess the sanitation and hygiene standards of the populace in premodern times, but some information exists for elites, making possible a comparative evaluation for various nations. Although the sanitary conditions and the customs regarding personal hygiene of the elite cannot be considered representative of a society as a whole, they are indicative of what one would expect the highest standard to be. For example, in seventeenth- and eighteenth-century France, "the palaces of the Louvre, Vincennes or Fountainebleau in places became mere latrines." A house steward wrote: "In many places in the courtyard, in the upper passages, behind the doors and almost everywhere, one can see a thousand heaps of ordure, one can smell a thousand

24 Tarr, "From City to Farm: Urban Wastes and the American Farmer," *Agricultural History*, XLIX (1975), 610. F. H. King, *Farmers of Forty Centuries* (Emmaus, Pa., n.d.; orig. pub. 1911), 193–215; Reginald Reynolds, *Cleanliness and Godliness* (Garden City, 1946), 253.

25 Wylie D. Burge and Paul B. Marsh, "Infectious Hazards of Landspreading Sewage Waste," *Journal of Environmental Quality*, VII (1978), 7. Studies by G. Berg (1966) and J. W. Mosley (1972) cited in *ibid.*, 3.

unbearable stenches caused by the necessities of nature which everybody discharges there daily."[26]

In contrast, João Rodrigues, a Jesuit, who was in Japan from the late sixteenth into the early seventeenth century, noted that the Japanese

> provide their guests with very clean privies set apart in an unfrequented place far from the rooms. . . . The interior of the privies is kept extremely clean and a perfume-pan and new paper cut for use are placed there. The privy is always clean without any bad smell, for when the guests depart the man in charge cleans it out if necessary and strews clean sand so that place is left as if it had never been used. A ewer of clean water and other things needed for washing the hands are found nearby, for it is an invariable custom of both nobles and commoners to wash their hands every time after using the privy for their major and minor necessities.[27]

By the mid-nineteenth century, conditions in France were no longer so primitive, but the problems with sewage and contamination of the water supply indicated that major sanitation problems remained. In mid-nineteenth century Britain, even royalty was not immune from the effects of inadequate sewage disposal. Prince Albert, Queen Victoria's consort, is thought to have died of typhoid fever contracted from faulty drains. In contrast, in Japan in the 1870s in the privies in the "better class of private homes," an American visitor found "less annoyance and infinitely less danger . . . than are experienced in many houses of the wealthy in our great cities." His description of the privies is similar to Rodrigues' in the sixteenth century, except that attendants did not clean out the receptacle after every use, but rather it "was emptied every few days by men who have their regular routes." Morse was taken not only by the cleanliness of Japanese toilets, but also by the amount of artful carpentrywork that decorated them in homes that he visited. He was not describing the

26 Frantz Funck-Brentano, *The Old Regime in France* (London, 1929), 156. Nicholas de la Mare was house steward to the Comte de Vermandois.

27 João Rodrigues, as quoted in Michael Cooper (ed.), *They Came to Japan* (Berkeley, 1965), 221.

toilet facilities of the elite but those in the houses that he visited during his extensive stays in Japan.[28]

This is not to say that contamination from human wastes did not occur in Japan, but it was probably less frequent than in the West. And custom helped prevent Japanese from becoming ill even when their water supply was not free from impurities. The Japanese customarily drank their water boiled, usually in the form of tea, a custom remarked on by foreign visitors to Japan from the sixteenth century. With the exception of Japanese "pickles," preserved by fermentation and salt, the Japanese usually ate their food cooked, so that even if night soil was improperly applied as a fertilizer, it was less likely to make everyone sick. Within the household, each member had his own set of chopsticks, rice bowl, and teacup, which no one else used, so that it did not matter much that washing was perfunctory and in cold water. Food served outside the home was frequently finger food, and chopsticks used in restaurants were usually lacquered for easy cleaning and were not put in the mouth as were spoons in the West. By the mid-nineteenth century, disposable chopsticks had come into use.[29]

Moreover, the Japanese had strong notions about what was dirty and clean, many of which can be traced back to the Japanese native religion of Shinto and its concepts of pollution. Much of the pollution in Shinto is ritualistic, but what is considered polluting and what is purifying are related to contamination and cleanliness. Anything to do with blood, death, and illness—such as childbearing, menstruation, contact with a sick or dying person, and funerals—is considered unclean, and people used to be prohibited from participating in religious rituals, mixing with

28 The quotes are from Edward S. Morse, *Japanese Homes and Their Surroundings* (Rutland, 1972; orig. pub. 1887), 228, 231. See also, Wohl, *Endangered Lives*, 127; Morse, "Latrines of the East," *The American Architect and Building News*, XXXIX (18 Mar. 1893), 170–174.

29 Cooper, *They Came to Japan*, 198–199; *idem*, *This Island of Japon* (Tokyo, 1973), 263; Engelbert Kaempfer, *The History of Japan Together with a Description of the Kingdom of Siam, 1690–92* (Glasgow, 1906), III, 238–240. Although raw fish became popular during the Tokugawa period, particularly in large cities, it was most frequently partially preserved and fermented and not served as slices of fresh raw fish as is common today. In any case, fish came from the sea and would not have been contaminated by fertilizers. For Tokugawa dietary habits, see Watanabe, *Nihon shoku seikatsu shi* (Tokyo, 1964), 190–269; Hanley, "The Material Culture: Stability in Transition," in Jansen and Rozman (eds.), *Japan in Transition*, 454–461.

other people, and even returning home from a funeral without being purified. Salt, water, and fire were all considered purifying agents and were used both to perform religious rituals and to clean and purify. This emphasis on purification is not unique to Japan; it can be found throughout Asia, but it is particularly strong in Japan and has persisted into modern times. The strong avoidance of things dirty, while at one level almost superstitious, most certainly resulted in lack of bacterial or viral contamination for many, and impeded the spread of diseases and infections in Japan. At the everyday level, it resulted in the compulsory removal of footwear when entering a house or any other building raised from the ground, and the washing of hands after using the toilet.[30]

Japanese are known for their frequent bathing, a custom that became widespread in the Tokugawa period. Originally bathing was associated with temples, which often maintained public baths. Then, in the late sixteenth century, public baths began to appear in the largest cities, and these became so popular that by the early nineteenth century there were an estimated 600 public baths in Edo alone. The earliest baths worked on the principle of the sauna; people used relatively little hot water to steam their pores open and the dirt out. Gradually baths in which the bathers soaked in tubs of hot water began to predominate, and in time, the size of both the tubs and the bath houses increased as well. But how widespread the use of these baths was by the general public, especially women and children, is still open to question, as the bath houses had the same type of reputation as in Western cultures. Social aspects were often as important as the act of cleansing, and many of the women employed in the public baths were prostitutes. However, families that could afford them began to install baths in their homes, and the frequent mention of baths and bathing in popular literature and the depiction of them in drawings and paintings indicates how widespread the custom was. Regular laundering of clothing by the common people also began in the Tokugawa period, and the new emphasis on cleanliness must certainly have had a salutary effect on hygiene and sanitation.[31]

30 Ishige Naomichi, "Jūkyo to jū-seikatsu," in Umesao Tadao (ed.), *Nihonjin no seikatsu* (Tokyo, 1976), 29; H. Byron Earhart, *Japanese Religion* (Encino, 1974; 2nd ed.), 7; Emiko Ohnuki-Tierney, *Illness and Culture in Contemporary Japan* (Cambridge, 1984), 24.

31 Ochiai Shigeru, *Arau fūzoku-shi* (Tokyo, 1984), 71–85.

Although customs relating to hygiene within the Japanese family depended for effectiveness on individual conformity, at the public level maintenance of sanitation also depended on government. A major reason that clean streets and an adequate water supply of high quality could be maintained was the high degree of control that existed over the populace during the Tokugawa period.[32]

Government control was enhanced by two factors. First, the samurai as a class had a monopoly on government positions during the Tokugawa period. The raison d'être of this group was to govern Japan, and explicit in the neo-Confucian philosophy that the samurai adopted was the concept of rulers as benevolent, as responsible to the ruled, and as moral examples. Not only did the rulers expect such conduct from themselves, but there is evidence that the ruled expected it as well, and let the rulers know it. Second, by the mid-Tokugawa period there were more samurai than were needed to govern, and overstaffing resulted in numerous detailed regulations and sufficient officials to see that these regulations were carried out.[33]

The effect of this benevolent but thoroughgoing government can be found in the governance of Edo. The city was divided into *machi*, which were village-sized units responsible for government at the local level. This division enabled authorities to have tight control over the enormous population of the city. At large intersections in the city, the premodern equivalents of the police box were set up, not only to keep an eye out for criminal activities, but, among other things to ensure that no water pipes were leaking and that the streets were kept clear. In addition, the city authorities made use of the outcastes who lived within the city. These people not only served to keep the streets free from any dead animals, to carry away corpses, and to handle anything which ordinary residents would not touch, but also to report on anything suspicious that they found on their rounds. Thus, in the late nineteenth century, Morse could comment favorably on what

32 Hall, "Rule by Status in Tokugawa Japan," *Journal of Japanese Studies*, I (1974), 39–49.

33 Irwin Scheiner, "Benevolent Lords and Honorable Peasants: Rebellion and Peasant Consciousness in Tokugawa Japan," in Tetsuo Najita and Scheiner (eds.), *Japanese Thought in the Tokugawa Period* (Chicago, 1978), 46; Yamamura, *A Study of Samurai Income*, 47–48.

were the slums of the new city of Tokyo, less than a decade from when it had been called Edo. "In Tokio one may find streets, or narrow alleys, lined with a continuous row of the cheapest shelter; and here dwell the poorest people. Though squalid and dirty as such places appear to the Japanese, they are immaculate in comparison with the unutterable filth and misery of similar quarters in nearly all the great cities of Christendom."[34]

Finally, the Japanese did not have to cope with the density in cities that Westerners did. Tenements in Japan were one story high, not six as they commonly were in Europe. The families of the poorest daily laborers in Edo were often crammed into one-room apartments approximately nine feet square with a small entry for storing tools and footwear and for cooking. They shared toilets and access to water with other tenants in the block. But however densely these families were packed in, they were not on top of each other in multi-story buildings, nor did people live in basements. The ever-present danger from earthquakes and the light construction of Japanese housing precluded tall buildings, and the authorities forbade their construction by anyone who might be inclined to be so imprudent.[35]

Statistical evidence for the assertion that sanitation was substantially better in Japan than in Europe during the eighteenth and nineteenth centuries comes from estimates of life expectancy. Estimates for various village samples for Tokugawa Japan indicate that, by the nineteenth century, life expectancy was probably in the low forties, with five-year averages ranging from the low thirties to as high as seventy-five years. Even in the city of Takayama, for which records exist for the century from 1773 to 1871, birth and death rates were similar to the village rates: the average crude death rate for this century was 27.3 per thousand. As a metropolis, Edo probably had higher death rates than the farming villages in central Japan and certainly a lower birth rate, as the sex ratio was skewed due to the high proportion of samurai

34 Morse, *Japanese Homes*, 5–6. See Tamura Eitarō, *Chōnin no seikatsu* (Tokyo, 1966), 194–197; Harada Tomihiko, *Nihon hōken toshi kenkyū* (Tokyo, 1957), 438.

35 Robert Higgs and David Booth have found that "density effects on mortality were uniformly positive and statistically significant" in 17 American cities in 1890: "Mortality Differentials within Large American Cities in 1890," *Human Ecology*, VII (1979), 353. See also Tamura, *Chōnin no seikatsu*, 198–199; Nishikawa Kōji, *Nihon toshi-shi kenkyū* (Tokyo, 1972), 250. M. Dorothy George, *London Life in the Eighteenth Century* (New York, 1964), provides graphic descriptions of life in London.

without their families and males who migrated to the city to work. But had the death rate been significantly higher, the city would not only have had difficulty maintaining a population above a million, but contemporary Japanese would also have noted the high death rates.[36]

Tokugawa Japan and Europe in the same centuries had strikingly similar life expectancies. Life expectancy at birth in Europe in 1800 has been estimated as high as thirty-five to forty for some countries, a number higher than in the preceding centuries. Female life expectancy at birth was 42.18 in England and Wales in 1841 and 40.83 in France in 1817 to 1831. A composite figure for life expectancy for males in Western Europe in the nineteenth century, as calculated by the United Nations, is 30.6 in 1840, 41.1 in 1860, and 48.9 in 1900. Female life expectancy is estimated to have risen from 42.5 in 1840 to 52.1 by 1900. These estimates are similar to those we have for Tokugawa Japan.[37]

Although the demographic statistics available for Japan and Europe indicate a similarity in the figures, by the nineteenth century Western European nations had available modern technology and by 1850 were well into the process of industrialization with its concomitant rising standard of living. Japan, however, did not begin to use any modern technology until nearly the end of the century. Given the high proportion of Japanese who lived in cities, had sanitation been poor and the level of living low, this combination would have been reflected in high rates of morbidity and mortality.

In fact, Japan seems to have been surprisingly free from the devastating effects of epidemics. The plague never reached Japan, and cholera came only in the nineteenth century when it spread throughout the world. Intestinal worms and enteric infections—those that enter through the mouth and are spread through contamination of food and water—tended to be localized and to

36 Hanley and Yamamura, *Change in Preindustrial Japan*, 216–225; Dana Morris and Smith, "Fertility and Mortality in an Outcaste Village in Japan, 1750–1869," Hanley and Arthur P. Wolf (eds.), *Family and Population in East Asian History* (Stanford, 1985), 238–239; Yōichirō Sasaki, "Urban Migration and Fertility in Tokugawa Japan: The City of Takayama, 1773–1871," in *ibid.*, 137.

37 Joseph J. Spengler, "Demographic Factors and Early Modern Economic Development," *Daedalus* (Spring 1968), 440; United Nations, Department of Economic and Social Affairs, *The Aging of Populations and Its Economic and Social Implications* (New York, 1956), 54.

appear in endemic rather than epidemic form. This situation is what would be expected in a society which used human wastes for fertilizer. It was true for Tokugawa Japan, which would explain the relatively high death rates for children between the ages of one and four or five. Children after weaning were particularly susceptible to these diseases, but if they did not die early, they tended not to succumb to them. However, the fact that the cities with a single water source, such as Edo, Osaka, and Kyoto, did not experience rampant epidemics meant that the water supply must have been generally good.[38]

Based on the evidence available on water supply and sewage disposal systems in cities in Japan and in the West from the seventeenth through the nineteenth centuries, I argue that the level of sanitation was higher in Japanese cities than in Western ones during the same period. Not only was Japan able to provide better water and disposal systems, but Japanese customs led to better hygiene and sanitation than did Western modes of behavior. Thus Japan was able to maintain large urban populations from the sixteenth century because sanitation was better and because the Japanese had the control necessary to carry out large-scale engineering projects, to implement various systems connected with water supply and waste disposal, and to see that measures were enforced.

It might well be argued that none of the above systems, customs, or beliefs in themselves would necessarily have created more healthful conditions or been better than any sample found in the West. But the combination of them, combined with near universal application, resulted in more sanitary conditions in the city and more hygienic homes than were the norm in the West, either just prior to industrialization or in the first century of industrialization.

What has obscured the realization that the level of sanitation was higher in premodern Japanese cities than in the cities of the West in the same centuries is the fact that the situation has been the reverse in the twentieth century. In 1985, only 34 percent of Japanese communities had modern sewer systems and the resi-

38 Ann Bowman Jannetta, *Epidemics and Mortality in Early Modern Japan* (Princeton, 1987).

dents of Tamagawa Josui (Tama River Water Supply) were still without a sewer hook-up. Even after World War II, the Japanese continued to use night soil as a fertilizer and thus were seen as backward by Westerners. But it was the very success of the premodern methods for dealing with night soil that made the Japanese slow to modernize their toilet and sewage systems. In the first half of the twentieth century, the Japanese had neither the income—either private or government—nor the imminent need to spend the vast sums necessary to install flush toilets and construct water-carriage sewage systems to remove the waste water. Indeed, the very success of the premodern waste disposal system inhibited modernization in this area, for, despite the shortcomings of sanitation, the Japanese today have the longest life expectancy of any major nation in the world.[39]

The rapid urbanization of the Tokugawa period and the experience of the Japanese living in and administering urban centers may well have smoothed the transition of Japan from a preindustrial to an industrial society. The Tokugawa legacy left Japan with major cities that could be easily converted into modern governmental and industrial centers, towns of 10,000 to 40,000 scattered throughout the country at strategic locations for governmental and economic purposes, the experience of millions of Japanese who had migrated to work in these cities under contract, and the capacity to administer large populations. This legacy meant that Japan could be transformed within decades into a modern society and could adopt modern technology more readily than if the Tokugawa experience had not occurred. It is easier to set up a modern water system if the basic system is already in place and the major task is to put in metal pipes, just as it is easier to establish universal education if education already is widespread.

The Japanese have lived in a densely populated society for nearly a millennium and have learned to live with limited resources on the 15 percent of the land that is not mountainous. By the Tokugawa period, the Japanese had evolved a life style that enabled them to maintain large urban populations, and this life style may well be part of the key to the success that Japan enjoys today.

39 Susan Chira, "Most Japan Houses Still Lack Comforts of Those in the U.S.," *New York Times*, 30 Oct. 1985, 1, 41.

11
The development of Victorian infrastuctures: the example of Portsmouth

Robert A. Otter

Introduction

As a consequence of the industrial revolution, the major centres of Britain's urban population experienced an unprecedented and sustained growth throughout the nineteenth century. The average decennial growth rate for the principal towns was 24% with the greatest average expansion of 32% occurring between 1821 and 1831. Along with the consequent urban development came increased environmental health problems; disease was rampant with conditions ripening for the cholera epidemics to sweep the nation in the century's middle decades. When nationwide mortality statistics became available in the early 1840s annual death rates in excess of 25 per thousand head of population were not uncommon. It was to be a third of a century before any significant reduction became apparent. During this time nearly all towns and cities benefited from the development of an extensive water supply, effective main drainage, a relatively well maintained urban road network and a local authority empowered to control the process of urbanisation. Few localities in the 1840s could boast any of these features. Each urban centre can look back on a unique nineteenth century process of infrastructure development, yet many of the issues which faced all large communities were essentially identical. By reviewing the development of Portsmouth, the nation's premier naval base, sited on the South coast, many of the important trends can be identified.

Population Growth and Mortality Rate

Portsmouth's population growth between 1801 and 1881 is shown in Table 1 below. Because of its strong naval and military links Portsmouth tended to stagnate during times of peace but expand and prosper in times of war or a perceived threat of war. Hence, its growth after the Napoleonic conflicts was well below national averages but significantly above them in later decades when the threat of European war returned. The town's mortality rate was 25.37 in 1851 reducing to 21.9 in 1871 and then to 18.9 in 1891 which was the lowest for large towns.[1] This decline conforms to national trends.

The Establishment of Improvement Commissioners

At the start of the nineteenth century the vast majority of Portsmouth's population was confined to the Southwest tip of Portsea Island adjacent to the entrance to Portsmouth Harbour (see Map 1).

---

[1] W.E.C. Chamberlain, *A History of Portsmouth Drainage, 1865-1956* (Portsmouth, 1956), p.22.

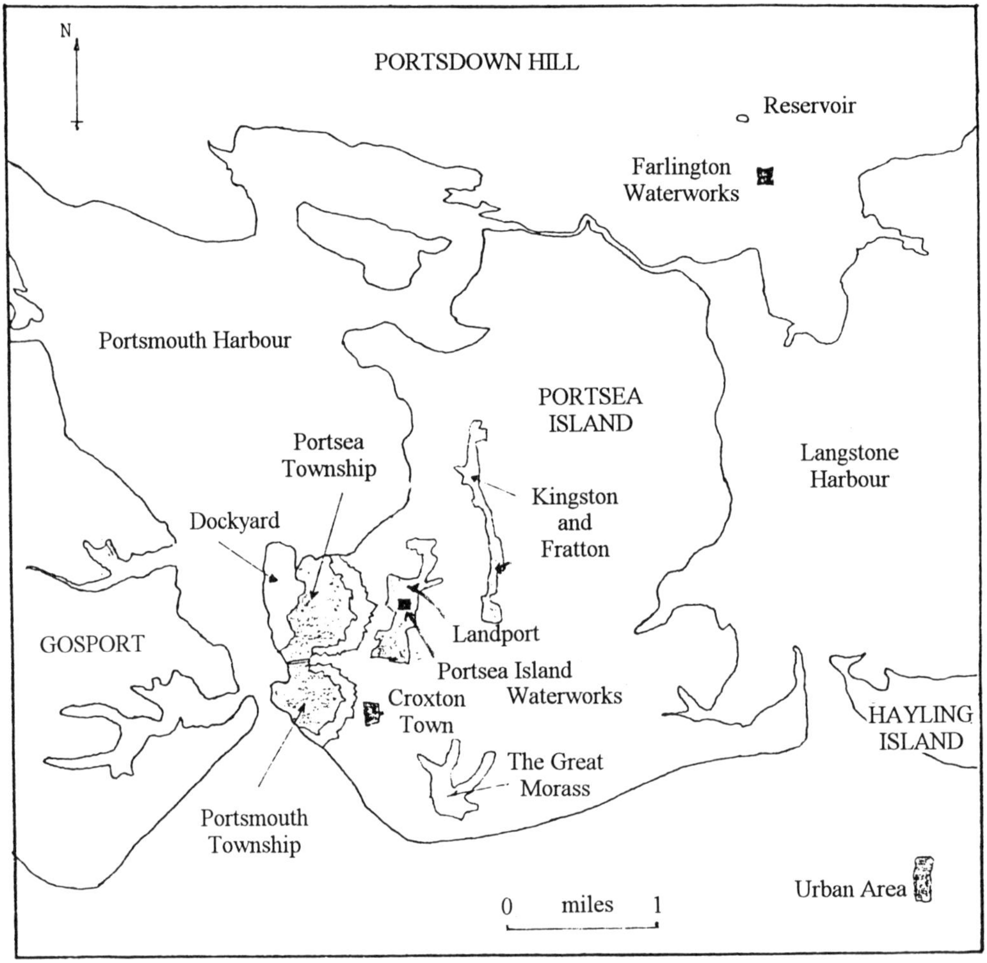

Map 1. Portsea Island. *c*. 1812. Showing Main Urban Areas and Waterworks

Table 1. Population Growth 1801-1881

		Decennial Percentage Rates of Change		
Year	Portsmouth's Population	Portsmouth	England & Wales	72 Principal Towns
1801	33,000			
		27.3	14.0	23
1811	42,000			
		11.9	18.1	30
1821	47,000			
		6.4	15.8	32
1831	50,000			
		6.0	14.3	26
1841	53,000			
		35.8	12.7	25
1851	72,000			
		31.9	11.9	18
1861	95,000			
		20.0	13.2	20
1871	114,000			
		12.3	14.4	21
1881	128,000			

Source: J. L. Field, 'Bourgeois Portsmouth, Social Relations in a Victorian Dockyard Town, 1815-1875'. (Doctoral Thesis submitted to Warwick University, 1979), p.10.

The inhabitants were divided into the distinct townships of Portsmouth and Portsea, each surrounded by a system of fortifications. Both had already experienced a significant development to their infrastructure. Local Acts of Parliament had established a body of Improvement Commissioners for Portsea in 1764 and for Portsmouth four years later. Their primary objective was to pave the townships; a task which was substantially completed over succeeding decades using both stone setts and unbound macadam.[2] The Commissioners were granted a variety of other powers including watching, scavenging, lighting, sewering, watering and control over buildings' waterspouts and openings, all of which centred upon their originally conceived responsibility of paving their areas of jurisdiction. Between 1748 and 1835 nearly every urban centre was brought under the jurisdiction of a similarly empowered body leading the Webbs (1922) to justifiably regard them as 'the starting point of the great modern development of town government'.[3] Their effectiveness was variable. Keith-Lucas

[2] A fuller account of their early activities is to be found in J. Noon, 'The Improvement Commissioners of Portsmouth and Portsea to *c.*1800' (Unpublished Moody Essay 1979), Available in P(ortsmouth) C(ity) R(ecords) O(ffice).

[3] S. and B. Webb, *Statutory Authorities for Special Purposes* (Reprinted 1963), p.235-6.

(1954) was able to suggest that the Manchester body acted in a manner far ahead of its time arguing that neglect of their achievements caused the public health agitation of the 1840s to be placed in a false perspective.[4] Hennock (1957) subsequently pleaded for the retention of the emphasis on the 1840s as the crucial era of sanitary reform.[5] Whatever view is ultimately taken of the Manchester body, whom the Webbs described as 'the best example of a really energetic and successful body of Improvement Commissioners',[6] the Boards for Portsmouth and Portsea are unlikely to attract similar academic debate. Portsmouth's meeting statistics alone suggest an unenthusiastic response to their problems. In 1805, for example, fourteen of their twenty-four arranged meetings were cancelled. The average attendance was between five and six and only four individuals attended more than 30% of the meetings.[7] Apart from their paving works, one of their other activities was the negotiation and supervision of scavenging contracts. To many Improvement Boards the disposal of refuse in the form of 'night soil', domestic waste or street sweepings did not present a major problem since contractors were prepared to pay for collecting and carting it away for agricultural or other industrial purposes. The early Portsmouth Act preserved the householder's right to ownership of their domestic refuse because of its economic value. This, however, did little to promote the health of the town for it encouraged the stockpiling of waste in the immediate vicinity of dwellings. As the volume of a town's waste grew and the carting distances increased the economic balance altered, changing the character of operations from one of 'recycling' to a rapidly increasing problem of 'waste disposal' which generally overwhelmed the capability of the Improvement Boards to respond.[8]

Early Water Supply Developments

A similar process can be identified in the development of water supply. Traditional sources, often based on shallow wells in close proximity to dwellings, proved utterly inadequate to cope with the increasing demands of the growing towns. One early, notable development in respect of water supply was the establishment, by local Parliamentary Act, of a great number of water companies in the first decade on the nineteenth century. Portsmouth found itself with two such companies who entered into several decades of intense competition to provide a somewhat limited piped water supply to a minority of the dwellings in the borough. One concern, The Portsea Island Waterworks Company, locally motivated and financed, proposed to construct an infiltration basin just outside the contemporary urban limits, pumping the water the relatively short distance to the towns. The other firm, confusingly named The Company of the Proprietors of the Portsea Island Waterworks, and promoted by a group of London

---

4 B. Keith-Lucas, 'Some Influences Affecting the Development of Sanitary Legislation in England', *Economic History Review,* 2nd Ser.,VI (1953-4), p. 290.

5 E.P. Hennock, 'Urban Sanitary Reform - Generations Before Chadwick?', *Economic History Review,* 2nd Ser.,X, (1957), p.113.

6 S. and B. Webb, op.cit., p. 256.

7 PCRO G/ICQ/1/2, Commissioners Minutes, pp. 109-25.

8 S. and B. Webb, op.cit., pp. 336-9.

financiers, proposed to utilise the springs at the base of Portsdown Hill, a chalk ridge some six miles to the North of the existing town. Using a 34 h.p. beam engine it was proposed to pump the water to a collecting reservoir on the hill's lower slopes and run the water through a 10 inch gravity main to the populated areas.[9] The completion of both schemes towards the end of 1811 was a significant step in the development of the town's infrastructure but since the supply only reached a minority of dwellings, and the pumps only operated for a short time on most days, these early schemes fell far short of the later ideal of a continuous high pressure supply to the whole population. As such they did little to relieve the overall shortage of water in Portsmouth; a problem common to most large towns. Twenty miles to the West, Southampton for example, had its supply administered by a board of Waterworks Commissioners established by Parliamentary Act in 1747. Whilst its inhabitants had long enjoyed the benefits of four public conduits supplied by a system based upon a Monastic scheme established in 1290, the Commissioners' early initiatives to increase the town's supply had met with little success. Attempts around 1805 to utilise surface run off from a large common area proved inadequate and a promising project started in 1836 resulted in a 1300 foot artesian well which absorbed many resources and generated much local friction but ultimately did little to contribute to the town's water supply.[10] Such unsuccessful municipal enterprises may be regarded as typifying local government's inadequate response to the developing infrastructure problems of the early nineteenth century.

Public Health Agitation and National Legislation

During the latter 1830s, those responsible for the administration of the New Poor Law of 1834 began to recognise the economic cost of illness and death resulting from the prevailing insanitary conditions. Surveys were undertaken on a national scale, the first of which was formally organised by the Poor Law Commissioners. After political opposition had attempted its suppression, it was published in 1842 as a 'private' view of Edwin Chadwick, thus clearly linking his name with the 'sanitary reform'. It graphically demonstrated the appalling magnitude of the problems[11] and was shortly followed by the establishment of a Royal Commission on the Health of Towns. Their first report published in 1844 considered fifty large towns, in only six was the water supply considered good, and drainage was everywhere bad. Enquiry was followed by legislative action, initially in the 'Clauses Acts' 1846-47 which provided model clauses related to aspects of the infrastructure which could be adopted by local authorities. The Commissioners Clauses Act would transform a self-perpetuating local body into an elected one; the Town Improvement Clauses Act contained many sound practical standards for the planning and development of towns, for example, a thirty foot minimum road width; and the Waterworks Clauses Act embodied the ideal of an extensive, high pressure, continuous supply. They were voluntary in nature but were shortly followed by the Public

---

[9] M. Hallett, *Portsmouth's Water Supply, 1800-1860,* Portsmouth Paper No.12 (Portsmouth, 1971) pp. 5-10.

[10] R. A. Otter, 'The Artesian Well on Southampton Common', *Industrial Archaeology,*17: 2-4(1983), p. 97.

[11] E.C. Chadwick, *Report on an Inquiry into the Sanitary Conditions of the Labouring Population of Great Britain* (1842).

Health Act of 1848 which, for the first time, introduced the concept of central government enforcing action on localities. It established a General Board of Health which had powers, seldom used in practice, to create a local Board of Health in localities where the death rate exceeded 23 per thousand. They could also act upon the basis of a petition from 10% of the ratepayers. Such legislation was not enacted without considerable opposition. Shaftesbury's diary lists the powerful vested interests arraigned against the sanitary reformers.

> The parliamentary agents are our sworn enemies because we have ... reduced their fees ... the civil engineers also because we have selected able men who have carried into effect new principles and at a less salary. The College of Physicians ... we... proved that many a Poor Law medical officer knew more than all the flash and fashionable doctors of London. All the Board of Guardians: for we exposed their selfishness ... The Treasury besides ... Then came the Water Companies whom we laid bare and devised a method of supply which altogether superseded them. The Commissioners of sewers - they hated us with a perfect hatred.[12]

Portsmouth in the 1840s

Portsmouth did not stand aloof from national trends. Both Boards of Commissioners obtained new Acts of Parliament substantially increasing their powers. The new Portsea Act of 1843 was influenced in its Parliamentary progress by Lord Shaftesbury's observations on its contents and as such may be regarded as one of the earliest pieces of legislation to be affected by the sanitary reformers. It was a fiercely contested local issue with a substantial group of Commissioners almost preventing its progress on the grounds that the town was already in debt to the sum of £1000 and that the proposed changes were not of sufficient magnitude for general benefit to result.[13]

The legislation gave the Board extensive powers but imposed financial limitations on its activities.[14] They could levy a paving rate of 8d in the pound and an improvement rate of 1s 4d for sweeping, cleansing, lighting, watering, draining, improving the streets and paying their officials. This proved sufficient for maintaining the paving and providing their services in broad harmony with the standards of the age, but it was these standards which were being challenged by the reformers. They were empowered to borrow a maximum of £3000, £1000 of which was an inherited debt leaving only £2000 for substantial improvement works. Over succeeding years some drainage work was completed but there remained little scope for capital improvement works.

The Portsmouth Board renewed its powers in 1847 incorporating some of the Clauses Acts of previous years. It was one of the first Acts to include this legislation but should not be regarded as a great triumph for sanitary reform since it was also severely constrained by

---

12 G.M. Binnie, *Early Victorian Water Engineers* (1981), p. 41.

13 PCRO G/ICP/1/7, Commissioners Minutes, pp. 131-2, 141-6, 151.

14 PCRO AP/31/A, Portsea Improvement Act Clauses, 55, 209, 210, 236, 237.

financial limitations. Its rating powers enabled 2s in the pound to be raised and its borrowing limit was only £2000 .[15] Unlike the Portsea Commissioners eight years earlier, the Portsmouth Board was united in its determination to renew its powers but it encountered ratepayer opposition which described the proposals as 'uncalled for, unnecessary'. One of the long serving Commissioners responded to this criticism with a candid appraisal of the Board's difficulties over previous decades.

> they were always in debt ... the streets were impassable ... badly lighted ... they were never able to carry out the old Act in the way it ought to have been.[16]

The opposition mounted against the passage of both of the borough's Improvement Acts of the 1840s illustrates one of the major problems associated with nineteenth century infrastructure development - the question 'who pays?' The full burden fell upon the ratepayers and their opposition to increased charges often proved a major stumbling block to the implementation of the ideals embodied in the various national and local Acts of Parliament. Whilst ratepayer opposition was overcome in respect of the passage of both Portsmouth's and Portsea's Acts, they did achieve a significant and far reaching local victory. It was proposed that the Portsmouth Act should be applied to the whole Borough rather than to the confines of Portsmouth township and a small area of Southsea known as Croxton Town. The Ratepayers from the latter claimed that they had paid rates in the sum of £580, had benefited from works conducted in the sum of £260 meaning that they had subsidised Portsmouth Township to the extent of £320 over preceding years, they also complained that their streets were seldom cleansed .[17] Their opposition to the extension of the Act prevailed which meant that the overwhelming majority of the borough's inhabitants, now living outside the defensive perimeters, were left with no Improvement Board to manage their infrastructure. There was a Highways Board established as a result of the Highways Act of 1835 which attempted to maintain about 20 miles of road. John Sheppard, its Chairman in 1850 candidly outline the problems faced by his Board.

> In 1835 the General Highway Act placed the control in a board of twenty persons, annually elected. From that time very great improvements have taken place; drains have been made; but there is no general system of sewers laid down; the roads have been repaired, partially with flints and granite, generally with sea-beach 'shingle'; but a vast deal remains to be done. As is frequent in Acts of Parliament, there are so many clogs, real and imaginary, in the Highway Act, that, from technical difficulties, no paving has been repaired, nor any regular cleansing of the streets. The facts of the whole board being subject to an annual election and the great amount of houses inhabited by persons in the humbler walks of life, can ill afford the outlay requisite, have prevented many improvements. Very many of the evils, of bad roads, dirt and filth, arise from the circumstances that parties have bought land, fields and gardens for building purposes, and without any plan of making

---

15 PCRO AP/A/5A, Portsmouth Improvement Act Clauses, 12, 33.

16 *H(ampshire) T(elegraph)* 21 February 1847.

17 *H.T.*, 23 January 1847.

> watercourses or drains, commenced building streets, rows of houses, and courts, throwing up the mould and clay dug out for foundations in front of such houses. This has been called making a road ...[18]

It is not surprising that public spirited residents from these 'outward' districts, as they were known, were able to mobilise 10% of borough ratepayers to petition the General Board of Health in London to enable Portsmouth to adopt the 1848 Public Health Act. On receipt of such a petition, the General Board sent an inspector to conduct an enquiry into environmental conditions in the locality.

Robert Rawlinson

Civil Engineer, Robert Rawlinson, came to the borough and took evidence at the George Hotel, Portsmouth. This was one of 243 such investigations conducted between 1848 and 1853 when political opposition destroyed the General Board. Many of its functions, however, were continued by various government departments and it was effectively re-established in 1871 as the Local Government Board, a forerunner of the Ministry of Health.

Rawlinson was one of a small team of civil engineering inspectors employed by the General Board, whose reports provide further detailed insights into prevailing environmental conditions. He proved to be the most eminent civil engineer closely identified with the Public Health movement and as such was not always in harmony with many of his professional colleagues. G. M. Binnie (1981) identifies William Cubitt, J. M. Rendel, Robert Stephenson and Joseph Bazalgette as opponents of aspects of sanitary reform, he describes Brunel's ideas on town drainage as 'very conservative and weak' and identifies Thomas Hawksley and James Simpson as pre-eminent figures in the campaign to destroy the General Board.[19] Rawlinson was born in 1810, the son of a builder, and commenced his professional career in 1831 as an assistant to Jesse Hartley engaged upon reconstruction works to Liverpool Docks. Five years later he joined Robert Stephenson engaged on the London-Birmingham railway. He returned to the North West in 1840 as Assistant Surveyor to Liverpool Corporation and then became Chief Engineer to the Bridgewater Trust during which time he proposed a major aqueduct as part of Liverpool's water system. He became an Associate of the Institution of Civil Engineers in 1848 and a full member in 1866. By this time he had become engaged in activities related to public service being selected by Chadwick as an inspector to the General Board. It was in this period that he presented his paper 'On the Drainage of Towns' to the Institution.[20] This provoked four nights of heated discussion regarding the viability of his proposals for an arterial system of relatively small size sewers with manhole access at changes of direction and gradient. He advocated soil pans or water closets directly fed by the mains and a system of back drainage, which obviated the necessity of carrying an individual pipe under the house

[18] R. Rawlinson *Report to the General Board of Health on the Sewage, Drainage and Water Supply of Portsmouth* (1850), pp. 36-8. Subsequently referred to as *Report* (1850).

[19] G. M. Binnie, op. cit., pp. 34, 36, 41, 42

[20] R. Rawlinson, 'On the Drainage of Towns', *Min. Proc. Civ. Engrs,* XII (1852), p. 25.

Fig. 1. The Drainage of Courts. *c.*1850

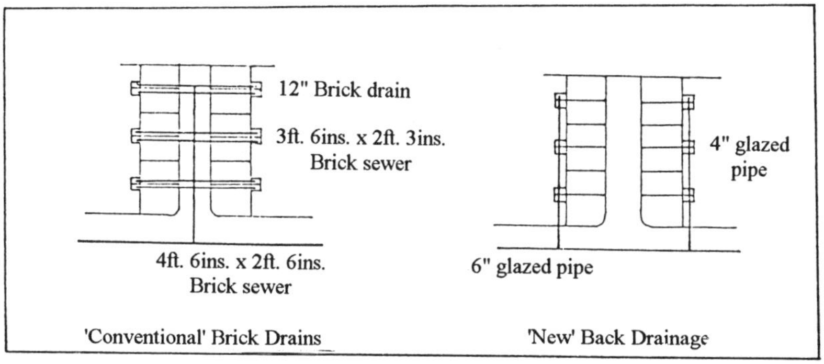

from the water closet at its rear to a branch sewer (see Fig. 1). The net effect of the proposals was to reduce the cost of drainage by about 60%. His system, now universally applied, was bitterly opposed by a group of engineers led by Joseph Bazalgette advocating the traditional brick built sewers. In the same year as Rawlinson's paper, the Board of Health published 'Minutes on House Drainage' embodying the same principles which also came under vitriolic assault from many engineers. However, as schemes constructed on the new principles and at substantially reduced cost began to operate successfully, public opinion was persuaded and professional opposition became increasingly muted.

After the demise of the Board of Health in 1854 Rawlinson served on several Government Commissions, visited the Crimea in 1855 and became the first Chief Engineering Inspector for the Local Government Board between 1871 and 1888. He received a knighthood in 1883, a KCB in 1888 and was elected President of the Institution of Civil Engineers in 1894 four years before his death.

Reports to the General Board

Rawlinson's first report on Portsmouth was published in 1850 and clearly described the deficiencies in the borough's infrastructure. One local doctor described the island of Portsea as one huge cesspool

> ... for there could not be less than 16,000 cesspools daily permitting 30,000 gallons of urine to penetrate into the soil. Just reflect on the character of the well-water of a district which became mixed every year with 365 times 30,000 gallons of urine, to say nothing of a host of other abominations too numerous to be mentioned on the present occasion ...[21]

In his further report of 1851 Rawlinson re-summarised his proposed measures of reform.

[21] R. Rawlinson, *Report* (1850), p. 94.

> One general system of underground drainage is required ... to remove ... all soil refuse or other waste fluids from the vicinity of dwelling houses ... to enable all owners of property to substitute water closets or soil pans ...
> All suburban and other roads should in the first instance be properly formed and duly maintained afterwards ... All courts, yards, passages, streets ... should be paved ... or provided with some form of artificial surface capable of being cleansed and resisting wear...The towns and suburbs should be properly lighted.[22]

The tone of these recommendations reflects the magnitude of the problems which Rawlinson advocated should be tackled by a single local authority, the Town Council acting as a local Board of Health within the framework of the Public Health Act. They would have the responsibility of preparing suitable improvement schemes and gaining the General Board's approval for borrowing the monies required. It was not unusual for the local Boards to engage the services of the Inspecting Engineers since they were employed by the General Board on a daily basis. Whilst this practice had advantages in that they had acquired a knowledge of the locality and its specific problems, it often led to accusations of corruption and jobbery. This problem was not relevant to Rawlinson at Portsmouth because the Public Health Act was not ultimately adopted but Rawlinson's colleague, William Ranger, who undertook a similar enquiry at Southampton, became embroiled in the politics of the town when he was engaged by the local Board of Health to advise on the town's water supply .[23] In this connection he was accused of plagiarising one of the specifications of Hawkesley who emerged as a vehement opponent of the consulting work of the inspectors.

> it further happens that the five Inspectors selected by the Board are permitted ... to obtain from Local Boards, as private engineers and for their own emoluments, the execution of as many as possible of the works proposed by local authorities ... the grossest jobbing prevails ... able professional men are wrongfully injured ... confidence in the integrity of public men is utterly destroyed.[24]

There was clearly some substance in these criticisms and in 1853 the Board's Inspectors were engaged full time and paid an annual salary.

Portsmouth's Rejection of the Public Health Act

Amongst other proposed works for Portsmouth, Rawlinson estimated that the principal elements of a suitable main drainage scheme could be financed at an annual cost of about £1,500, and would require a capital expenditure of about £14,000; a sum far in excess of the

---

22 R. Rawlinson, *Report to the General Board of Health on a Further Inquiry into the sewerage, drainage, and supply of water and the sanitary conditions of the inhabitants of Portsmouth* (1851), pp. 14, 15. Subsequently referred to as *Further Enquiry* (1851).

23 R. A. Otter, op. cit., pp. 104-9.

24 G.M. Binnie, op. cit., pp. 38-40.

combined borrowing powers of the existing Boards of Commissioners. Not all ratepayers were enthusiastic at the prospect of any, if relatively modest, increase to their rates.

> we resist, according to the prayer of the memorial presented to you, any endeavour to place the said district under the control of the Public Health Act, in so much as it would entail an enormous expense on the owners and occupiers of land, without producing any beneficial result.[25]

Thus there developed in nearly every town a 'clean party' and a 'dirty party' known in Portsmouth as the 'muckabites'. The Town Council's initial response was totally in favour of adopting the Act but gradually the 'muckabites' gained the ascendancy in the council chamber. Resistance to the concept of central government control and the inevitable increase in rates were both important issues but the dominant factor in Portsmouth appears to have been the fact that Government buildings were exempt from rates levied under the Public Health Act. There were approximately 5000 land based servicemen compared to a borough population of about 70,000 at this time, which when roughly translated into rate income meant that the issue was not oversignificant in financial terms, but its emotive effect was enormous. Rawlinson made a special plea for the government's exemption to be waived.

> With respect to the power for rating all classes of property, I have no doubt as to the justice of such an arrangement. The advantages will be mutual as betwixt the Government and town authorities, the benefit conferred by works of sewerage and drainage must be general throughout the district, and the power to rate equitably, ought to have as wide a range.[26]

No response came from the Government and the Council, in 1853, after three years of debate resolved not to adopt the Public Health Act. The 'status quo' of the borough was preserved leaving the water supply in the hands of a private company, the townships of Portsmouth and Portsea with the constrained financial powers of the Improvement Boards, a Town Council with virtually no environmental powers, the vast majority of the borough's population resident in the outwards with only the very limited powers of the Highways Board, and the infrastructure in the dreadful state described by Rawlinson.

The Borough of Portsmouth Waterworks Company

The first significant change to this depressing picture was in respect of the water supply. Economic and technical pressures had forced amalgamation of the two companies in 1840 with the united company basing its operations at the Farlington works. A breach of the sea wall surrounding these works combined with engine failures meant that the early years of the united company were a period of consolidation and recovery from debt. During the latter 1840s, expansion was an economic possibility but there was no movement by the company

---

25 R. Rawlinson, *Report* (1850), p. 69.

26 R. Rawlinson, *Further Inquiry* (1851), p. 7.

to transform its operations from the provision of an intermittently supplied 'luxury' commodity to the concept of a low cost necessity for the whole population. Company water was piped to about 40% of dwellings by 1850 but supplied only about 10% of the town's estimated demand. Over succeeding years the company did increase its output and reduce its prices but did little to rectify the overall water shortage.

Concurrently with their initial enthusiastic response to adopt the Public Health Act the council resolved to obtain powers to purchase the water company. Nothing was done, however, over the period when the 'muckabites' held sway in council, but the matter was again raised in 1855 when a committee's recommendation to proceed was postponed. In the same year a small group of individuals formed a provisional committee which engaged the services of civil engineer, Henry St John Diaper who recommended the purchase of the existing company and a substantial development of the water resources at the base of Portsdown Hill to include springs at Havant and Bedhampton, a little to the East of the Farlington works. These proposals obtained Council's approval in 1857 and the Borough of Portsmouth Waterworks Company was established by Parliamentary Act in July of the same year. It had power to raise £80,000, a little over half to purchase the existing company and the remainder for financing the necessary capital improvements. In concept, the new concern was to be a 'consumers company' and in harmony with this ideal its initial capital was to be raised in £10 shares. The new company was thus suitably empowered and financed to set about the task of providing a constant high pressure supply to the borough.[27]

The Landport and Southsea Improvement Commissioners

The second response to the state of administrative chaos was the establishment, in 1857, of an Improvement Board for the outwards to replace the Highways Board. The Landport and Southsea Improvement Commissioners, like others in the borough, were granted extensive powers and over the six years of their existence started a process of significant improvement in their districts. Amongst their officials they engaged the services of civil engineers as surveyors, firstly Mr Pineo who was sacked by the Board after a year or so on the grounds of alleged financial irregularities.[28] He was succeeded by Mr Longland selected from among nearly fifty applicants from afar afield as Sunderland, indicating that such municipal appointments held some attraction. The Board devoted about half of its annual financial resources to the improvement of the district's roads and put many of the principal thoroughfares into a respectable state of repair. The Board was only responsible for the carriageway and gutter, with adjacent owners being responsible for the pedestrian pavement and curbs. It was the common practice for the Commissioners to issue notices to owners to pave and curb in front of their property concurrently with a road improvement. Many owners allowed the Board to carry out frontage repairs but proved reluctant to pay the consequent charge. Two Commissioners carried such a debt for four years only settling their accounts

---

27 M. Hallett, op. cit., pp. 20-2, 24.

28 PCRO G/ICL/1/1, Commissioners Minutes, pp. 356, 357.

when legal proceedings were threatened, by which time the total sum owed was nearly £2500.[29]

The paving contractors were appointed for a specified term and then allocated individual jobs being paid according to the agreed rates for the measured work completed. This arrangement did not always run smoothly, there was at least one strike, there were problems related to progress and quality of work and inevitably arguments over measurement. Most carriageway construction was of the macadam type with the footways paved with 'square faced flatners' (small paving slabs). In 1861 there was an application to construct a footway in asphalt.[30] The Board agreed to the proposal and issued lists of streets where they would permit asphalt to be used for the guidance of householders. It is likely that they were seeking to avoid a 'patchwork' of construction types.

Most of the Board's remaining income was expended on the provision of street lighting, which was significantly extended and the management of scavenging contracts. Using their borrowing powers they constructed extensive drainage works based upon Rawlinson's arterial principle (see Map 2). Two networks were constructed, one leading to an outfall in Portsmouth Harbour, the other into the Solent. A further network was planned for the Eastern part of their area. The largest sewers were 3ft 6ins oval sections and the system also incorporated 15 and 12 ins glazed pipe drains. It was the intention of the Board that houses should be connected to the new sewers whilst construction was in progress to avoid unnecessary disruption to the road pavement. Only those benefitting from these works paid the special sewer rates which led to the creation of four drainage districts for this rating purpose.

The debt incurred as a result of roadworks, represented about 25% of the Board's £10,000 borrowing limit, which also financed the £1000 cost of obtaining their Act and the construction of new offices which expended about £2000.[31] This left under £5000 available for new drainage works leading their auditor to comment in the context of a financial irregularity that 'the Board had been expected to execute drainage works of magnitude with utterly inadequate resources'.[32]

One of the local newspapers, the Hampshire Telegraph, had identified this fundamental weakness in the Board's establishment, arguing that funds of the order of £40,000 would be required to effect necessary works; it also doubted the wisdom of entrusting such a sum to the Board[33]. This reflected Parliament's reluctance to grant extensive financial powers to local bodies unless there was central control to ensure its wise expenditure. Parliament's stance

---

29 Ibid., p. 64; PCRO G/ICL/1/2, Commissioners Minutes, p. 151.

30 Ibid., pp. 256, 257.

31 PCRO G/ICL/1/1, pp. 17, 96, 134, 135, 146, 149.

32 PCRO G/ICL/1/2, p. 489.

33 *H.T.*, 26 June 1858.

Map 2. Arterial Drainage Network of Landport and Southsea Improvement Board 1857-64

appears justified since the history of Improvement Commissioners, in general, does not present an impressive picture of sensible engineering judgement and sound financial control. None of the borough's three Improvement Boards appear to represent any exception to the general rule. Consequently the decade following the borough's rejection of the Public Health Act was a time when extensive powers remained invested in its local authorities yet severe financial restrictions prevented major progress towards an improvement of the infrastructure, except in respect of water supply where the new company was suitably empowered in 1857. This situation contrasted with towns which had adopted the Act enabling extensive works to commence. The Hampshire Telegraph, somewhat wistfully, published a description of improvement works in Worthing, 40 miles to the East, being carried out under Rawlinson's supervision within the framework of the Public Health Act.[34]

Further administrative change came as a result of growing public dissatisfaction with the behaviour of the Landport and Southsea Board. It was a politicised body, dominated by borough radicals for much of its existence. Its meetings were often conducted in a turbulent and bitter atmosphere with one committee session ending in a bout of fisticuffs which in turn resulted in a Commissioner being fined by the local Magistrates.[35] Disquiet at its activities was expressed by the ratepayers of Southsea, a well-to-do area of the borough, who resolved at a public meeting in 1863 to disassociate their area from the Board's jurisdiction. This initial objective became transformed into a more general move to bring the whole borough within the framework of current national public health legislation. It was only the Landport and Southsea Board which mounted any opposition to this proposal, expending £800 in an unsuccessful parliamentary campaign.[36] Therefore, on 29 July 1864 Portsmouth became subject to the Local Government Act 1858 and the history of Improvement Commissioners in the borough was terminated. All their powers were transferred to the Town Council acting as a local Board of Health and so for the first time, the borough had a powerful and adequately financed local authority to control and develop its infrastructure.

The Town Council as the Local Board of Health

Roadworks

Several Committees were soon appointed. The General Purposes Committee dealt with the licensing and regulation of Hackney Carriages and other miscellaneous matters. The Roads and Works Committee managed street paving, lighting, cleansing and watering and over succeeding decades gradually worked towards the ideals set out by Rawlinson in respect of the urban road network. Peacock (1975) cogently describes the initial organisation established to continue the steady improvement and maintenance of the borough's roads

---

[34] *H.T.*, 31 July 1858.

[35] *H.T.*, 1 Oct 185; 8, 29 Oct 1859.

[36] *H.T.*, 30 July 1864.

> that all repairs of pavements and crossings be performed by contracts at a schedule of prices for one year; that all other repairs be performed under a jobbing contract; that the Board should provide its own carts and contract for the supply of suitable horses, harness and drivers; that the making and repair of roads be performed under the supervision of a surveyor and road foreman, the estimated number of labourers to be employed as a permanent staff being about 25; the appointment of a person to be called a road foreman at wages of 36/- per week who should be nominated and selected by the Surveyor[37].

The borough also tended to favour a macadam type of construction among the many types of pavement in common use in Britain's towns during this period. In a paper presented to the Institution of Civil Engineers in 1879 G. F. Deacon attempted to compare the costs of various pavements, summarised in Table 2 below. Discussion ensued and disagreement inevitably emerged as to the best type of urban pavement construction. Such dissension, however, appears to have been in the context of sober professional debate rather than the vitriolic rantings which surrounded the drainage disputes in the 1850s. It is conceivable that Deacon's opening assessment of the status enjoyed by urban road builders had something to do with the

Table 2. Comparative Urban Road Costs

Pavement Description	Construction Cost per square yard	Annual Maintenance cost per square yard	Remarks
7¼ ins deep sets. Hand pitched foundations	12s 1d	1s 1½d	Older pavements
7¼ ins deep sets. Sundry concrete foundations	14s 11d	1s 1d	Suitable for heavy urban traffic
Bituminous Concrete	3s 9d	1s 10d	Suitable for light urban traffic
6 ins deep wooden blocks bituminous concrete foundations	15s 2d	2s 2d	
Macadam pavement, hand pitched foundation	6s 9d	2s 11d	Suitable for suburban roads

Source: G. F. Deacon, 'Street Carriageway Pavements', *Min. Proc. Civ. Engrs*, LVIII (1879), p. 24.

37 S. E. Peacock, *Borough Government in Portsmouth 1835-1974,* Portsmouth Paper No. 2 (Portsmouth ,1975), p.11.

lack of friction generated. 'As a branch of Civil Engineering street paving has never occupied a very high place'.[38] Its contribution to environmental health and the population's 'quality of life', whilst difficult to quantify, should not be under-estimated.

Drainage

The Drainage and Sewerage Committee's first and most pressing task was to provide an effective sewerage and drainage system. Lewis Angell, Borough Engineer 1863-1866, rapidly prepared and submitted a scheme for Council's consideration in 1864. It was approved by John Hawkshaw and since neither the War Office nor Admiralty raised objections, it was constructed over succeeding years. As a solution to the engineering problems of sewering the borough it was a bold, imaginative scheme and remains the basis of the city's drainage[39]. Angell proposed dividing the borough into a high level area, Landport and Upper Southsea and a low level area including the old townships and lower Southsea. Both main sewers (see Map 3) ran in a roughly West to East direction to Eastney and thence jointly to an outfall at the entrance to Langstone Harbour. The high level sewer was entirely gravitational but the low level system required a static lift of about 15 feet provided by two Clayton 25 h.p. beam engines at Eastney Pumping Station. The untreated effluent was discharged on the ebb tide through the 1500 yard, 5 foot oval brick outfall. The scheme was not designed to cope with surface water resulting in numerous storm water outfalls being constructed around the borough's shoreline. The major disadvantage of the scheme was the fact that the sewers acted as storage capacity for much of the tidal cycle which in turn resulted in large quantities of settled solids especially in the lower level sewer. It was estimated that about 1300 tons of solids were removed over a ten year period. This problem was tackled by 1882 by Consulting Engineer, Sir Frederick Bramwell, who shortly became President of the Institution of Civil Engineers. He recommended the construction of tanks providing 4½ million gallons of storage capacity adjacent to the outfall. This released the full capacity of the sewers to accommodate the town's increasing effluent discharge over succeeding decades.

Construction of the scheme commenced in 1865 and was substantially completed by 1870. The major technical problems proved to be the continued necessity to dewater the excavations by steam pump and the traversing of a bog area known as the Great Morass where a substantial bearing bed of concrete proved necessary for the sewers. On one occasion a squad of soldiers drove the navvies away from works adjacent to one of the forts when liaison broke down between the various authorities. The main sewers were brick ovals varying in height between four and five feet. They were laid to very flat gradients with a minimum slope of 6 ins in 1 mile. The branch sewers were similarly egg-shaped varying between 1 ft 9 ins and 3 feet in height, or glazed ware pipes to a minimum sewer size of 12 ins. Again gradients were small which meant that flushing and cleansing became somewhat of a maintenance problem.

---

[38] G. F. Deacon, 'Street Carriageway Pavements', *Min. Proc. Inst. Civ. Engrs,*LVIII (1879) p.1.

[39] The following account is substantially based upon W.E.C. Chamberlain's *A History of Portsmouth Drainage, 1865-1956* (Portsmouth, 1956).

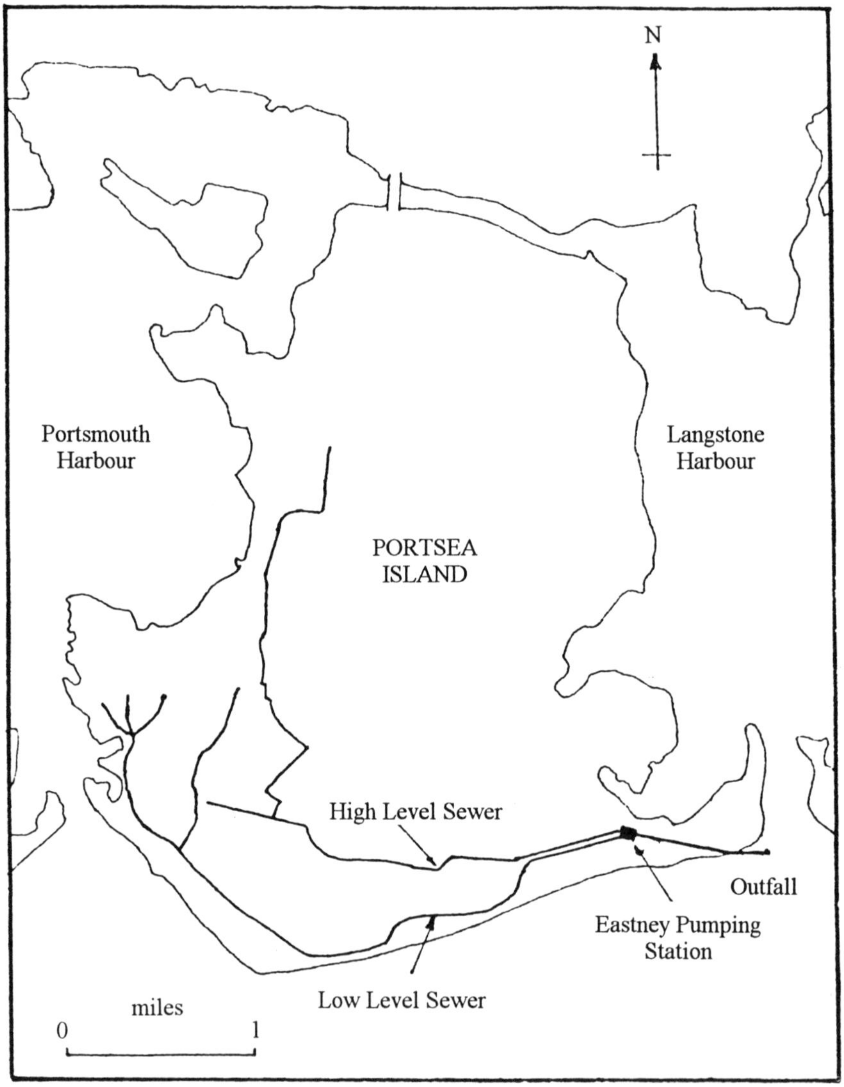

Map 3. Portsmouth Main Drainage 1870

The sewer system was completed by Mr J. E. Greatorex, Borough Engineer 1866-1879 and comprised:-

Works	Cost £
11 miles (approx) of Arterial Sewers	84,000
56 miles (approx) of Lateral Sewers	61,000
Eastney Pumping Station	7,650
Surface Water Sewers	7,000
Total	159,650

A Continuous High Pressure Water Supply

The borough's water supply was similarly improved over a slightly longer time span. The new company established in 1857 inherited works which were woefully inadequate to provide a constant high pressure supply. The Portsea Island infiltration basin was virtually ineffective, and the capacity of the Farlington Works was limited by the single 10 inch main conveying water to the town and the poor condition of the fifty year old machinery. The average daily supply was about 550,000 gallons and a given district received water for about two hours per day.

Investigations showed that the springs at Farlington did not have the potential to satisfy increased demand and under the civil engineering guidance of Joseph Quick (Diaper died in March 1860) the springs at Havant were developed. Collecting basins were constructed and two beam engines provided a pumping capacity of 4½ million gallons per day. These supplied the Drayton reservoir from which a new 20 inch gravity main supplemented the existing 10 inch pipe. These works, completed in 1860, substantially improved the supply, yet remained deficient in that the Drayton reservoir was too low to achieve the necessary combination of high pressure and high discharge. This problem was remedied by 1869 with the construction of two reservoirs higher up the hill slope with a combined capacity in excess of 5 million gallons. An additional beam engine was installed at Havant raising the maximum pump capacity to 6.75 million gallons per day, it supplied about 80% of the borough's dwellings under much improved pressure but the branch and service mains were operational for about 7 hours each day. The opening of the new sewer system at this time resulted in an increase in the borough's demand and also necessitated attention to be given to the problem of waste. This problem had always plagued water undertakings; despite technological improvements, poor joints, poor plumbing and water misuse resulted in significant wastage. When the supply was intermittent, the wastage was kept within tolerable limits but with the prospect of a continuous high pressure supply stringent regulations became necessary to mitigate the problem. In order to endorse these, the Company had to obtain additional parliamentary powers in 1873. These required the Council's consent but disagreement between the Council and Water Company resulted in protracted negotiations and a form of arbitration before regulations were introduced in 1876. Over succeeding years the Company introduced a constant supply on a district by district basis and by 1880 the town may be regarded as enjoying a constant piped water supply at a reasonable price to the vast majority of its dwellings and business establishments.[40]

Further Administrative Developments

Between 1875 and 1880, Portsmouth's three key aspects of urban infrastructure, roads, water and sewers, were substantially completed to a standard which significantly contributed to a reduction in the mortality rate. Additionally, the administrative and financial framework was established to enable Britain's towns and cities to continue their expansion without a repetition of the appalling conditions of the nineteenth century's middle decades. It was the great Public Health Act of 1875 which provided the foundation for the continued 'healthy growth' of urban centres. This was passed at the conclusion of a twenty year period which had witnessed volumes of enacted public health legislation resulting in administrative confusion in many locations. The second public health reform movement, as it is commonly regarded, was centred on a Royal Commission established in the latter 1860s. It was as a consequence of this enquiry that the consolidation acts of the early 1870s were passed. Portsmouth was not greatly affected by these developments although it was compelled to appoint a full time Medical Officer of Health in 1872 and in the same year its Board of Health became known as the Urban Sanitary Authority. The latter remained the centre of municipal control over the infrastructure until 1895 when the formal distinction between it and the Council was abandoned.

Portsmouth and National Trends

The general development of Portsmouth's water supply, drainage and urban road network shows a pattern common to most urban centres. The early establishment of Improvement Commissioners is entirely consistent with national trends but the presence of two bodies in close proximity is somewhat unusual in a provincial town. The very distinct separation of the Portsmouth and Portsea townships by the fortifications explains why they were regarded as individual administrative entities for nearly a hundred years. The establishment of a private water company in the century's first decade was a common occurrence but the intense competition between two such companies was less common outside of London. The renewal of the Commissioner's powers in the 1840s leaves at least one unanswered question, for it was a common tactic of local civil leaders to renew their local acts with enhanced powers in the hope of preventing the application of national legislation to their area. It is unlikely that this motive can be applied to the renewal of the Portsea Act in 1843 since, firstly, the contention about the Act was between those who wished to improve the township and those who viewed it as an additional unwelcome burden on the ratepayers. Secondly, it was passed when national reform legislation was at an embryonic stage. The renewal of Portsmouth's powers in 1847, however, could very well have been motivated by a wish to thwart the application of the 1848 Public Health Act to the township. Regrettably, conclusive evidence has not yet emerged to resolve the problem. The borough's reaction to the Rawlinson enquiries is consistent with other urban centres, like Birmingham, Newcastle and Hull, which successfully

[40] D. Halton Thomson, *A History of the Portsmouth and Gosport Water Supply* (Portsmouth ,1957), pp.35-42.

deferred sanitary reform. The principal issue of contention, the 'rating of Government properties', had no relevance to the vast majority of other localities. Brayshaw and Pointon (1983) do not include this issue in their consideration of Rawlinson's enquiry at Plymouth, which, like Portsmouth, included fortifications and barracks.[41] Portsmouth's rejection of the 1848 Public Health Act unquestionably resulted in the inhabitants suffering severe environmental hardships for a further decade or so but it did mean that when improvement came, the drainage and water supply were developed concurrently. In areas where development was sequential there was a probability of health risks being intensified for an interim period.[42]

The appointment of the Landport and Southsea Improvement Board in 1857 must be viewed as a 'stop-gap' administrative measure and meant that Portsmouth was very unusual in having three operational Improvement Boards in 1860. Their demise with the borough's adoption of the 1858 Public Health Act returned Portsmouth to harmony with national trends which remained true during the period of significant improvement to about 1880. The water company's re-establishment in 1857 was a little unusual in its concept of being a 'consumers concern' but its improvements were consistent with other private and municipal water undertakings of the period. It has remained a private company to the present, even surviving the creation, in 1973, of the Water Authorities. If current water privatisation proposals are brought to fruition it will again be restored to the norm rather than the exception!

Lessons of History

It is an intriguing fact of history that eminently sensible nineteenth century urban improvements were achieved against the background of intense opposition and obstruction. Portsmouth's history, as briefly outlined, illustrates many of the principal reasons for such opposition, which are listed below.

1. Inertia and conservatism of bureaucratic bodies
2. Sundry vested interests
3. Problems beyond the capability of current technology
4. Narrow parochial attitudes
5. Lack of co-operation between the organs of local and national government
6. Underfunding
7. Ratepayer opposition

If it is perceived that all of these have a modern day application and all to a greater or lesser extent stand in the way of current progress in the renovation and development of the nation's

41 M. Brayshaw and V.F.J. Pointon, 'Local Politics and Public Health in Mid-Nineteenth Century Plymouth', *Medical History*, 27:2 (1983), p. 162.

42 M. W. Flinn, Introduction to A. P. Stewart and E. Jenkins, *The Medical and Legal Aspects of Sanitary Reform* (Leicester University, 1969), pp. 8, 9.

infrastructure, it is worth reflecting that they were only overcome after long and sustained campaigning by public spirited men of vision and commitment.

Acknowledgement: The author wishes to acknowledge the assistance he has received in the preparation of this paper from the staffs of Portsmouth's Central Library and Records Office and also the staff of Portsmouth Polytechnic Library.

12

Edwin Chadwick and the engineers, 1842–1854: systems and antisystems in the pipe-and-brick sewers war

Christopher Hamlin

To the English sanitary reformer Edwin Chadwick, author of the famous *Report of an Inquiry into the Sanitary Condition of the Labouring Population of Great Britain* (1842), goes credit for recognizing the central importance of public works—waterworks, sewers, better-ventilated streets and houses—to public health. Chadwick's career as a public health official lasted only from 1848 to 1854, yet his influence was great. In a broad sense, the administrative structures, the sanitary sensibilities, and the technologies (e.g., indoor running water and water closets) he developed or endorsed were adopted, and on great scale: by 1905, local authority debt in England and Wales for waterworks and sewers was nearly one hundred million pounds.[1]

One might think engineers would have aligned themselves with Chadwick's programs—he brought them business. In fact, however, Chadwick's relations with engineers were wretched. For Chadwick, mid-century British civil engineers were part of the problems, not the solutions. He saw them as both loyal to a primitive laissez-faire and in cahoots with the most corrupt and irrational institutions of local government: the ancient municipal corporations, sewers commissions, and navigation trusts. He represented their works as hyperexpensive, uninformed by science, even dangerous. Worse, they clung to obsolete doctrines and rejected truths from outsiders. Historians, even those critical of Chadwick, have shared this view. They have seen

[1]See the "35th Annual Report of the Local Government Board," *Parliamentary Papers* (hereafter *PP*), 1905–6, vol. 35 [cd. 3105], p. cciii.

engineers as key actors in Chadwick's downfall, collaborating with tightfisted politicians to block needed improvement.[2]

At the center of Chadwick's troubles with engineers is an obscure technical controversy over sewer design, the "pipe-and-brick" sewers war of 1852–54. On one side were Chadwick and a handful of marginal engineers who advocated a novel system of small-bore pipe sewers; on the other, prominent members of the Institution of Civil Engineers, notably Thomas Hawksley, a leading water engineer, and Joseph Bazalgette, later builder of London's main drainage, who opposed this novelty.[3] Like many technical controversies, this one took place on many levels. Facts were in dispute: was there really mud in certain sewers? How fast did a sewer of given design discharge a given amount of water? There were also conflicts about the nature of expertise, the proper institutional and social framework for sewerage projects, and what constituted success or failure. The two sides even disagreed about what they were disagreeing about; only one side (the

[2]For example, M. W. Flinn writes of Chadwick as "sickened by the squandering of public money in purchasing the services of ill-qualified quacks . . . the main weight of his criticism fell upon the engineers." Likewise, Chadwick's pipe sewer is seen as "so sensible that its subsequent universal adoption has obscured its radical nature at the time." See M. W. Flinn's "Introduction" to Edwin Chadwick, *Report of an Inquiry into the Sanitary Condition of the Labouring Population of Great Britain,* ed. M. W. Flinn (Edinburgh, 1965), pp. 60–61. See also S. E. Finer, *The Life and Times of Sir Edwin Chadwick* (London, 1952), pp. 439–52; R. A. Lewis, *Edwin Chadwick and the Public Health Movement* (London, 1952), pp. 222–23, 295–300. Anthony Brundage's *England's "Prussian Minister": Edwin Chadwick and the Politics of Government Growth, 1832–1854* (University Park, Pa., 1988), is critical of Chadwick but does not significantly depart from earlier views of engineers (see p. 151). A recent engineering historian has also seen the dispute as about "prestige"; see G. N. Binnie, *Early Victorian Water Engineers* (London, 1981), pp. 31–42, esp. p. 36.

[3]It has not been recognized just how marginal Chadwick's partisans were. Of eighteen engineers and surveyors who advocated Chadwick's sewerage concepts, only seven appear to have held any class of membership in the Institution of Civil Engineers, while only nine appear in S. P. Bell's *A Biographical Index of British Engineers in the Nineteenth Century* (New York, 1975). The partisans can be divided into three groups: those who were only advisers (Morgan Cowie, Butler Williams, Edward Cresy, Sr., William Dyce Guthrie, James Vetch), those who were also surveyors on the Metropolitan Sewers Commission (John Roe, John Phillips, John Grant, Thomas Lovick, Joseph Medworth, J. L. Hale, Edward Gotto, George Donaldson), and those who were superintending inspectors for the General Board of Health (Henry Austin, Robert Rawlinson, William Ranger, William Lee, Thomas Rammell). Some achieved fame only later, while others achieved it in areas other than sewerage. Several rejected aspects of Chadwick's doctrine during his years in power. On concepts of marginality, see Ian Inkster, "Aspects of the History of Science and Science Culture in Britain, 1780–1850 and Beyond," in *Metropolis and Province: Science in British Culture, 1780–1850,* ed. Ian Inkster and Jack Morrell (Philadelphia, 1983), pp. 39–43.

Chadwickians) saw itself in a doctrinal "war." To Chadwick's biographers the quarrel was between progress and stagnation (literally and figuratively), right and wrong, even good and evil. Yet it ended not, as they presumed, with the triumph of Chadwick's system, but with an affirmation of the flexible, client-driven practice that characterized British engineering.

Using the pipe-and-brick sewers war as a focus, this article reassesses Chadwick's relations with orthodox engineers. I hope to make clear that the engineers' opposition was far more than a matter of professional jealousy or personal pique, for when one focuses on what S. E. Finer called the "insufferably tedious" technical literature of the controversy, many of their criticisms appear well founded.[4] But at the heart of the disagreement were conflicting views of good engineering: Chadwick's allies thought in terms of "systems" while the orthodox engineers took an explicitly decentralized and anti-systems approach.[5]

* * *

While historians of technology have given little attention to sanitary technologies, urban historians and historians of public health have had much to say about the conditions that led to the need for new water supplies and sewers. Growing industrial cities found traditional sources of water and means of removing wastes inadequate. Unplanned growth meant there was often no coordinated drainage (or even no drains at all). The new water closets and macadam pavements strained what sewers there were, increasing the input of both foul water and sediment. Contemporary medical theory gave prominence to environmental causes of disease, while fear of revolution height-

[4]Finer, pp. 448, 451.

[5]I use "system" much as the concept has been developed by Hughes and several commentators. See Thomas P. Hughes, *Networks of Power: Electrification in Western Society, 1880–1930* (Baltimore, 1983), pp. 2–17, and "The Evolution of Large Technological Systems," in *The Social Construction of Technological Systems: New Directions in the Sociology and History of Technology*, ed. Wiebe E. Bijker, Thomas P. Hughes, and Trevor J. Pinch (Cambridge, Mass., 1987), pp. 51–82; Bernward Joerges, "Large Technical Systems: Concepts and Issues," in *The Development of Large Technical Systems*, ed. Renate Mayntz and Thomas P. Hughes (Frankfurt, 1988), pp. 9–36; B. Barnes, "Review of Hughes, *Networks of Power*," *Social Studies in Science* 14 (1984): 309–14. The engineers' opposition to technological systems is especially interesting as it is in an area of technology that might be expected to exemplify such systems. The term "network," explicitly applied to the technologies considered here, carries many of the same connotations as "system"; see Joel Tarr and Gabriel Dupuy, eds., *Technology and the Rise of the Networked City in Europe and America* (Philadelphia, 1988).

ened the need to do something for (or to) the industrial working classes, and cholera epidemics lent immediacy to all these matters.[6]

Such was the situation confronting Edwin Chadwick in the late 1830s. A disciple of the utilitarian Jeremy Bentham, Chadwick was already well known as a poor-law reformer and as secretary to the Poor Law Commission. The poor law provided the initial context for his concern with sanitary engineering, which was predicated on the argument that preventing pauperism by preventing disease was cheaper than supporting paupers.[7] After making his *Sanitary Report* of 1842, Chadwick went on to become in 1848 chief member of the General Board of Health (GBH), established to administer the Public Health Act his work had inspired. He was deposed in 1854, a victim of old vested interests and opponents of centralized government.

Although Chadwick is often seen as a dogmatist, his views on sanitary engineering changed significantly over the years. The *Sanitary Report* focused on the old association of disease with dampness and was more concerned with the provision of drains than with their size or construction (there was some concern with flushing them of obstructions).[8] In the 1843–45 investigations of the Royal Commission on the Health of Large Towns and Populous Places (Chadwick was not a member but helped organize the inquiry and write the reports), that focus had shifted from class to urban conditions. Many English towns had public sewers and cesspools, which removed surface water but did not carry it to an outfall. As a result, sewage stagnated underground, generating unpleasant and presumably pestilential gases.[9] This led Chadwick to think in terms of sewage flow rather than sewer capacity and to the evolution of his "arterial-venous" conception of a city in which water constantly moved in, through, and out, removing all wastes to the country for recycling.[10]

[6]A good summary is Flinn's "Introduction" to Chadwick's *Sanitary Report,* 1965 ed. (n. 2 above), pp. 3–21; or see Anthony Wohl, *Endangered Lives: Public Health in Victorian Britain* (Cambridge, Mass., 1983).

[7]Flinn, ed., "Introduction" (n. 2 above), pp. 42–44. Compare F. Barker and J. Cheyne, *An Account of the Rise, Progress, and Decline of the Fever Lately Epidemical in Ireland,* 2 vols. (London, 1821), 1:24.

[8]Chadwick, *Sanitary Report,* 1965 ed., pp. 80–81, 88–89, 99–104, 109–10, 380, 424. He had already commissioned John Roe, surveyor to the Holborn and Finsbury Sewers Commission, to investigate use of cheap ceramic pipes for sewers.

[9]Royal Commission on the Health of Towns, *First Report on the State of Large Towns and Populous Districts* (hereafter HOT, *First Report*), *PP,* 1844, vol. 17 [572.], p. x. Chadwick's contribution has been reevaluated by Brundage, pp. 93–96.

[10]Lewis, pp. 52–54. Chadwick's associate James Kay-Shuttleworth claimed credit for the concept. See Frank Smith, *The Life and Work of Sir James Kay-Shuttleworth* (London,

The claims made here for "system" were new to Chadwick. What systematicity there had been in the *Sanitary Report* had been mainly administrative—how much easier it would be if all streets and all understreet infrastructure were under one authority: pipe-and-main laying and repairing and street breaking and building could be done in a coordinated manner; a scientific municipal administration by well-trained practitioners would become possible.[11] By contrast, the approach that took shape between 1843 and 1845 was driven by economic, technical, and physical considerations. Underlying it was the hope that the agricultural use of urban sewage could finance much urban improvement. If one took this view, end-use considerations dictated upstream components of the system. For example, Chadwick, believing that the best sewage was fresh and dilute, would increasingly insist on rapid removal of wastes. Other design features followed: the kinds of sewers to be used (velocity-augmenting pipes), the layout of the network and locations of outfalls (arranged for easy access by farmers), the material to be used as road pavement (that which would minimize grit), the control of what was allowed to enter sewers (useless road grit and storm water had to be kept out), and the rate at which it entered (effective flushing had to be ensured).[12]

This was very much then a technological system as Thomas Hughes has conceived such systems; it was centrally controlled, and its development impelled by a momentum, both social and physical, that was subject to impediments—in Chadwick's case, failure to secure control of water supplies would be the principal one—that resulted in "reverse salients."[13] So mutually necessitating were the system's com-

1923), p. 33. The analogy to the city as a body was sometimes explicit. See C. Fowler in discussion of J. Thornehill Harrison, "On the Drainage of the District, South of the Thames," *Minutes of Proceedings, Institution of Civil Engineers* 13 (1853–54): 107; "Review of General Board of Health, Minutes on Sewer Manure," *Builder* 10 (1852): 77–78. See also Graeme Davison, "The City as a Natural System: Theories of Urban Society in Early Nineteenth-Century Britain," in *The Pursuit of Urban History*, ed. D. Fraser and A. Sutcliffe (London, 1983), pp. 349–70.

[11]Chadwick, *Sanitary Report*, 1965 ed., pp. 380–96.

[12]For early expressions of the system, see James Smith of Deanston in discussion of James Green, "The Sewerage of Bristol," *Minutes of Proceedings, Institution of Civil Engineers* 7 (1848): 87–88.

[13]Like the inventor-entrepreneurs of electrification, Chadwick had to consider minute details of the system: gratings that removed road grit, household fixtures suited to high-pressure water service, even kilns and molding machines to supply cheap and uniform pipes. See General Board of Health, "Minutes of Information Collected with Reference to Works for the Removal of Soil Water or Drainage of Dwelling Houses and for the Sewerage and Cleansing of the Sites of Towns" (hereafter "Minutes on House Drainage"), *PP*, 1852, vol. 19 [1535.], pp. 98–124, 185–90.

ponents that one could start with any one and deduce the others: one could start with the imperative of cleansing cities with water and deduce the details of drainage and outfall works, or one could start with the imperative of removing decaying matter and deduce the requisite stream of water and the system of drains needed to remove that water. The sewage farms would still be necessitated as the only adequate means of purification. Equating technological with economic rationalization, the Chadwickians also claimed that their system was economically optimal.[14] Appeals were made to all these arguments—as they all led to the same conclusion, there was no tension among them.

The prospect of developing a coordinated hydraulic system led Chadwick to seek engineering expertise. Unlike the *Sanitary Report,* the Health of Towns reports were full of engineers' opinions, from Robert Thom, William Hosking, Henry Austin, William Dyce Guthrie, Edward Cresy, Thomas Wicksteed, Thomas Hawksley, and J. Butler Williams. Chadwick's star engineer was John Roe, who had gone from canal building to become surveyor to the Holborn and Finsbury Sewers Commission. Roe was full of ideas. Egg-shaped sewers (pointed end downward) would increase hydraulic mean depth for a given flow, thereby increasing velocity and consequently carrying capacity. Tangential junctions of branch sewers into a main sewer would diminish turbulence and prevent deposition. Most important, a system of flushing dams could be installed in existing sewers that would allow one to accumulate the head of pressure needed to flush out sediment in the sewers below.[15]

In contrast with Roe's enlightened practice stood the corrupt and inefficient technical and administrative practices of other sewers administrations, particularly the seven Greater London sewers commissions. As Chadwick would blame engineers for much of what he found wrong there, it is well to consider what the problems were. Each of the commissions (for the City, Westminster, Holborn and Finsbury, Tower Hamlets, Surrey and Kent, Ravensbourne, and Poplar) consisted of a large number of appointed members (e.g., 220 in Westminster in the early 1840s). Some took no part in the business (some were dead, it was noted), while others—architects, builders, and surveyors—were all too active, using their membership to advance their own careers. Lavish dinners and cumbersome procedures made

[14]Chadwick, *Sanitary Report,* 1965 ed. (n. 2 above), p. 130.

[15]For engineering testimony, see HOT, *First Report* (n. 9 above), question nos. 108–72, 369–95, 807, 1075–1113, 2094–2130, 4480–4528, 5345, 5423–46, 5827–46. On Roe, see his obituary in *Minutes of Proceedings, Institution of Civil Engineers* 39 (1874–75): 297–98.

the commissions targets for criticism, and, beginning early in the century, calls for reform had led to amendments to their charters and to construction of a significant number of new sewers.[16] Yet Chadwick's surveyors still found sewers that appeared

> not to have been conducted on any system, nor are the capacities of these lines proportioned to the requirements of the drainage. The sections are of all shapes and sizes, from squares nearly to circles, and to various modifications and combinations of these forms, differing greatly in very short distances, that of the outfall being frequently one of the smallest parts,—they are generally without artificial bottoms,—the sides, in parts, are built upon or supported by piles, which project to a great extent. Their falls are frequently from, instead of to, the outfall; pits or cesspools are thus formed, in which there must always be accumulations of deposit.[17]

"Sewers of vicious construction" was Chadwick's label for such works. Aside from generating deadly gases of decomposition (and sometimes undermining nearby buildings), such sewers required cleaning out by hand, a task Chadwick regarded as inhumane, unhealthy, and unduly expensive.[18]

Corrupt and incompetent the commissions may have been, but Chadwick was also judging them on new criteria of his own. Founded as quasi-judicial bodies to apportion costs among householders and settle drainage disputes, the commissions were ill suited to construct or administer sewers. Their borrowing powers were minimal; when funds were gone, building stopped until a new rate was in. They did not build, nor did they control all sewers in their districts; many sewers were private. Conceiving of sewers as *systems* was foreign to them; sewers were built bit by bit, here and there. And they were

[16]Select Committee on Sewers in the Metropolis, *Report, PP,* 1823, vol. 5 (542.), pp. 6–17, 33–35; Select Committee on Metropolis Sewers, *Report, PP,* 1834, vol. 15 (584.); Select Committee on the Health of Towns, *Report, PP,* 1840, vol. 11 (384.), evidence of J. W. Unwin, B. Drew, J. Peeke, W. Baker; HOT, *First Report,* evidence question nos. 2303, 2318; Royal Commission on the Health of Towns, *Second Report of the Commissioners of Inquiry into the State of Large Towns and Populous Districts* (hereafter HOT, *Second Report*), *PP,* 1845, vol. 18 [602.], pp. 7–10; Metropolitan Sanitary Commission, *First Report of the Metropolitan Sanitary Commission, PP,* 1847–48, vol. 32 [888.], evidence pp. 24, 33–39, 74–75.

[17]T. Lovick and J. L. Hale, "Report on the State of Sewers in the Surrey and Kent District, near Borough," in *Times* (London) (August 7, 1848), p. 8f.

[18]"Minutes on House Drainage" (n. 13 above), pp. 27, 29–30.

built, not to move waste to the countryside, but to drain the surface.[19] Nor was there a close link between engineers and the commissions. The surveyors, their chief technical officers, tended to come to their work from building or architecture. Professional engineers, like John Rennie, who had advised the Westminster Commission in 1807, shared Chadwick's concern for finding a good gradient to avoid deposition.[20] Because most engineers recommended larger sewers than he did, however, Chadwick equated their views with the uninformed practice of surveyors and architects.

* * *

When the Health of Towns inquiries ended in early 1845 Chadwick was still on good terms with the engineering profession. Roe's innovations were well received by engineers if not always by sewers commission surveyors.[21] In Hamburg, the English engineer William Lindley was working along similar lines; he sought Chadwick's patronage while remaining independent. Indeed, in the coming controversies, engineers would claim both Roe and Lindley as their own, as proof that they were masters of sewer building.[22] Chadwick was satisfied with the profession's competence. Requiring government inspectors to sanction engineers' plans of sanitary works would only "clog procedure," he wrote; the guarantor of quality was to remain "the ordeal of local scrutiny."[23]

The first threat to this harmony was Chadwick's break with the water engineer Thomas Hawksley, who, with J. F. Bateman, would

[19]Sidney Webb and Beatrice Webb, *English Local Government*, vol. 7: *Statutory Authorities for Special Purposes* (London, 1922). Borrowing powers were increasing; see Metropolitan Sanitary Commission, *First Report*, evidence pp. 91–92.

[20]John Rennie, reports 1807–8, WCS 873 (WCS is a designation for reports to the Westminster Sewers Commission), Greater London Record Office (hereafter GLRO).

[21]See Braithwaite, Wicksteed, and Walker in discussion of Green (n. 12 above), pp. 102–7. The chief anti-Chadwick polemic, *Engineers and Officials: An Historical Sketch of the "Health of Towns Works" (between 1838 and 1856) in London and the Provinces* (London, 1856), pp. 44–47, also saw a consensus until 1847.

[22]See the obituary for Lindley, "William Lindley," *Minutes of Proceedings, Institution of Civil Engineers* 142 (1899–1900): 363–70; Lindley to Chadwick, October 1842–February 1844, Chadwick MSS, no. 1235, University College, London (hereafter Chadwick MSS). On the reception of Lindley's work, see J. Walker in discussion of Green, p. 104; Fowler, Bidder, Manby, and Simpson in discussion of Harrison (n. 10 above), pp. 78–81; "Health of Towns: The Government and the Profession," *Civil Engineer and Architect's Journal* 11 (1848): 17–18.

[23]Chadwick to Duke of Buccleuch, December 17, 1844, Chadwick MSS, no. 2181/6. Compare HOT, *Second Report*, pp. 6, 20–21.

dominate that field for the next forty years. Impressed with Hawksley's development of constant-supply water service in Nottingham, Chadwick had recruited him as engineer to the Towns Improvement Company, a for-profit company Chadwick was organizing to sell cities integrated gas, water, sewerage, and sewage recycling systems. Their split had many causes: incompatible understandings of whether Hawksley was co-promoter, regular consultant, or key employee; contrasting estimates of how feasible were the technologies Chadwick advocated; and incompatible conceptions of the company's mission. For Chadwick, sanitary improvement was a moral obligation backed by the security of utilitarian proof (and, if needed, by natural theology). He would guarantee the validity of the system, was unconcerned with details, and reluctant to have the company sell partial services (e.g., water supplies) to particular cities. Hawksley saw the company (and himself) as selling services in a free market. Far from trusting Chadwick's system (he doubted the profitability of sewage recycling), he saw his reputation as being at stake in every estimate he approved. For Chadwick, the client was an abstract notion of public good (embodied in himself, no less); for Hawksley, the client was a real person, group, or public authority, whose problem had to be solved within unique constraints.[24] These differences would loom large in the pipe-and-brick sewers war a decade later.

By January 1846, the Towns Improvement Company was fast fading and Chadwick and Hawksley were no longer speaking. For Chadwick, the episode would do much to confirm a distrust of engineers; he saw Hawksley as the archetype of the hypocritical professional who would endorse any position for a fee.[25] Still, there was no irreparable break with the engineers. What changed was not the engineers' views but Chadwick's power base. After two difficult years (1845–46) at the Poor Law Board and the Towns Improvement Company, Chadwick went on, within about a year, to hold three significant positions in the engineering world, as key member of the Metropolitan Sanitary Commission, the General Board of Health, and the new Metropolitan Commission of Sewers that replaced the local commissions. It became clear that he meant not just to suggest new technologies but to orchestrate their introduction and dominate the profession that would put them into place.

Established in early 1847 to investigate sanitary conditions and to reform London's sewer administration, the Metropolitan Sanitary

[24]See Brundage (n. 2 above), pp. 101–12; Lewis (n. 2 above), pp. 120–21; Binnie (n. 2 above), pp. 14–30.

[25]Lewis, p. 133.

Commission was dominated by Chadwick. The attack on engineers began in the commission's first report in November 1847.[26] True sewerage principles were now known, it insisted; they were "demonstrated with a degree of clearness which admits of no misapprehension by well-informed minds earnestly directed toward the attainment of the object."[27] Yet engineers of high repute ignored them.[28] Their opposition could be easily explained, Chadwick observed; because their fees were a percentage of project costs, economy and efficiency were not in their interest. Conventionally trained engineers also were incompetent to build sewers, he added: sewer building was a recondite science that "could not be reasonably expected to be dealt with incidentally, or collaterally to ordinary occupation, or even to connected professional pursuits, but require[d] a degree of special study which not only place[d] . . . [it] beyond the sphere of the discussions of popular administrative bodies, but beyond that of ordinary professional engineering and architectural practice."[29]

This stronger tone reflects Chadwick's growing conviction that the gap between what was and what might be was even greater than he had thought. That conviction had been imparted by a new star engineer, John Phillips, who was displacing Roe as his chief sewerage theorist. A bricklayer, Phillips had taught himself hydraulics and worked his way up to become a clerk of works and, in 1846, surveyor to the Westminster Commission.[30] His stay in Chadwick's camp would be brief—he was purged in mid-1849—but he was the main source of the pipe-sewers dogma Chadwick would defend so bitterly.

Phillips's effect on Chadwick was as much personal as technical. He was bright and ambitious but tactless and quick to take offense. He encouraged Chadwick to equate technical error with moral failing, to believe that there could be no legitimate opposition to their views on

[26]Chaired by Lord Robert Grosvenor, the commission's members were Chadwick, Dr. Thomas Southwood Smith, the anatomist Richard Owen, and Richard Lambert Jones, a City of London politician. Owen and Smith were orthodox Chadwickians, Grosvenor sympathetic and pliable; only Jones was independent.

[27]Metropolitan Sanitary Commission, *First Report* (n. 16 above), p. 24; see also pp. 2, 49.

[28]The target here was James Walker, longtime president of the Institution of Civil Engineers, who as surveyor to the Poplar Sewers Commission had endorsed flat-bottomed sewers. Walker protested that it was years ago (1834) when no one was interested in self-flushing sewers. See Metropolitan Sanitary Commission, *First Report,* evidence pp. 36, 41; Walker in discussion of Green (n. 12 above), p. 105. By contrast, Roe credited Walker with having recognized the desirability of curved-sided sewers; see the Clowes edition of E. Chadwick, *Report of an Inquiry into the Sanitary Condition of the Labouring Population of Great Britain* (London, 1842), app. 1, p. 373.

[29]Metropolitan Sanitary Commission, *First Report,* pp. 2, 24, 42–43, 51.

[30]On Phillips, see ibid., evidence p. 42; and his obituary in *Minutes of Proceedings, Institution of Civil Engineers* 64 (1897–98): 203.

sewerage and that ostensibly technical criticisms hid ulterior motives. For example, to Phillips, the Westminster Commission's rejection of his innovations could only be explained in terms of its corruption.[31] In 1842, Chadwick had suggested that poor administration went hand-in-hand with poor technology, but he had been concerned more with inefficiency and incompetence than with corruption and conspiracy. Yet by 1854, he had come to share Phillips's outlook: opposition to pipe sewers could only reflect the persistence of old vested interests, expelled from power but not annihilated.

A key task of the Metropolitan Sanitary Commission was to design a sewerage system in accord with the principles of 1845—that is, to determine how small sewers could be and how much water they could carry. Phillips (and Chadwick) found Roe's approach deficient in two respects. First, they objected to Roe's emphasis on flushing sewers on the grounds that even temporary deposits of sewage would endanger the public. Phillips proposed a self-flushing system in which there would be no deposition, no emanations, hence, no disease.[32] There was also to be a change in sewer construction: the egg-shaped brick sewers of Roe were to be superseded by glazed earthenware pipes, not only cheaper but better overall (by concentrating flow and thereby augmenting velocity, they would keep sediment in suspension even at low gradients). Chadwick called this scheme a "second revolution" in sewerage (the brick, egg-shaped sewers of Roe had been the first).[33]

The key to the effectiveness of self-flushing sewers was ensuring that sewage always flowed at the same rate, Phillips asserted. Constant velocity could be achieved by controlling the input of sewage and by arranging converging (and diverging) lines so that sewage from upper districts would flush lower districts. To control flow in this way would require controlling the input of water, closely matching sewer size to discharge (and minimizing size to maximize velocity), and keeping extraneous water (whether rainfall or soil moisture) out of the sewers—in other words, significantly redefining what a sewer was for.[34] The vision was thus one of a finely tuned system, virtually a

[31]Metropolitan Sanitary Commission, *First Report,* evidence pp. 42–48, 73. Roe, by contrast, was more tactful if no less aware of the inadequacies of the sewers commissions.

[32]Ibid., pp. 23–24, 44, evidence p. 130; "Minutes on House Drainage" (n. 13 above), p. 59.

[33]General Board of Health, *Report on the Supply of Water to the Metropolis,* app. 2, engineering reports and evidence, *PP,* 1850, vol. 22 [1282], pp. 113–15.

[34]Metropolitan Sanitary Commission, *First Report,* pp. 49–51; cf. L. C. Hertslet in discussion of G. Donaldson, "An Account of the Drainage of the Town of Richmond, Surrey, under the Authority of the Metropolitan Commissioners of Sewers," *Minutes of*

perfectly engineered city, and it entranced Chadwick. In practice he was forced to compromise on almost all of its particulars, but he maintained the ideal of theoretical perfection. The engineers who would oppose him would use the opposite approach: one started, not by stating an ideal and backing down from it, but by gradually optimizing actual conditions.[35]

The chief concession the Chadwickians had to make was in control of input. In London they did not secure control of the water supply (though they did in some towns sewered by the General Board of Health), and there were usually pressing reasons to let sewers receive street drainage, soil moisture, storm runoff, and industrial effluents, all of which had to go somewhere. In practice, therefore, the Chadwickians focused on increasing velocity, the hope being that the full carrying capacity of rapidly flowing sewage would never be called upon. As Roe had shown, in most sewers there was great room for significantly increasing velocity by decreasing diameter. As long as sewers were still less than half full this approach was uncontroversial. But Chadwick tried to apply it to sewers flowing full; it is hard to reconstruct his thinking (various rationales were given), but the argument was that in a full sewer a further decrease of diameter would increase velocity almost infinitely, with discharge remaining constant and no significant back pressure.[36]

To observations that there seemed a limit to how fast water flowed and that sewers did back up, the Chadwickians replied that much more water could be put through if there were additional branches entering. Some saw the influx of each successive branch into a descending sewer as delivering an increment of velocity (see fig. 1); others believed sewage would accelerate so that a pipe initially full would become partly empty during flow, leaving room for more sewage or even drawing it in by vacuum.[37] As the Reverend Morgan Cowie,

Proceedings, Institution of Civil Engineers 11 (1851–52): 414; R. Rawlinson, "The Drainage of Towns," *Minutes of Proceedings, Institution of Civil Engineers* 12 (1852–53): 31.

[35][Thomas Wicksteed], "A Copy of a Report by Thomas Wicksteed, CE, on the State of the Works of Drainage and Sewerage in the Town of Croydon," *PP,* 1854, vol. 61 (450.), p. 5.

[36]Metropolitan Sanitary Commission, *First Report,* evidence pp. 133–34. He was already thinking of this in 1842; see Chadwick, *Sanitary Report,* Clowes edition (n. 28 above), app. 1, p. 379n. Compare Gibbs in discussion of Rawlinson (n. 34 above), p. 95, and Hale in "First Report of Mr J. L. Hale on Some Experiments already made to Ascertain the Practicability and Advantages of substituting small Pipes in the Places of existing large Sewers and Drains," December 1848, p. 8, GLRO MCS 476.

[37]See Medworth's explanation in "Report of the Trial Works Committee," December 1849, GLRO MCS 193, pp. 20–21; Cowie in Metropolitan Sanitary Commission, *First*

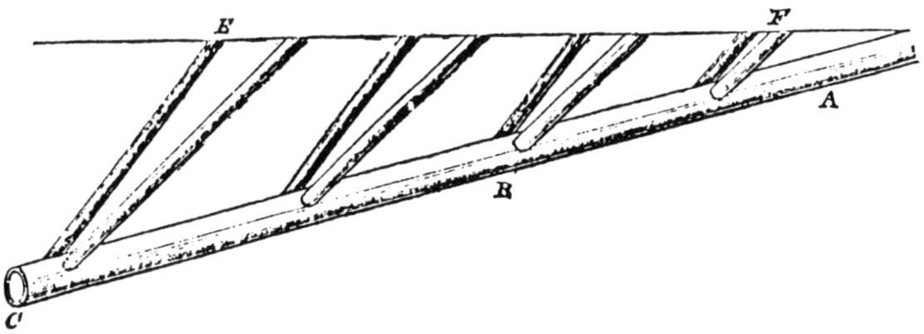

FIG. 1.—Rev. Morgan Cowie's diagram illustrating how additional inputs in a descending sewer would accelerate flow in the main sewer. The greater steepness of the branches would cause their contents to deliver an increment of velocity to the sewage flowing in the main sewer, Cowie argued. (Metropolitan Sanitary Commission, *First Report of the Metropolitan Sanitary Commission, Parliamentary Papers,* 1847–48, vol. 32 [888.], evidence p. 158.)

senior wrangler in 1839 and principal of the Putney College of Civil Engineering, put it, in theory one need not increase the aperture of a main sewer pipe no matter how much sewage from how many branches it received. The only limit was structural: the outlet had to be strong enough to withstand the jet of sewage that would issue forth.[38]

To justify this ambitious proposal, the Chadwickians tried to claim the high ground of hydraulic theory. The first report of the Metropolitan Sanitary Commission carried pages of theorems from treatises of the post-Galilean hydrologists. These were impressive but largely irrelevant. Hydraulic science was in dreadful shape, the Chadwickians argued; equations describing flow through pipes were inconsistent with one another and too conservative. Worse still were the tables derived from them by British engineers: "blindly put forward" and "as blindly followed."[39] Hawksley's equation relating velocity to head of pressure and diameter of pipe was their particular target. It was consistent with Continental authority and widely used by British engineers, but surely incorrect, Chadwick insisted. Were Hawksley right, sewers would be too expensive and "extensive voluntary adoption of works of sanitary improvement" would be impossible. He

Report (n. 16 above), p. 159; Rawlinson in Public Records Office, London (hereafter PRO), MH 13 59, no. 887/54.

[38]Metropolitan Sanitary Commission, *First Report,* pp. 158–59. On Cowie (who became chaplain to Queen Victoria), see J. Venn, *Alumni Cantabrigienses,* pt. 2, *1752–1900* (Cambridge, 1944), 2:157.

[39]H. Austin in Metropolitan Sanitary Commission, *First Report,* evidence pp. 127–28; see also in *First Report,* evidence Phillips, pp. 53–56; Cresy, pp. 141–49; Roe, pp. 81–82; Cowie, pp. 156–59; "Minutes on House Drainage" (n. 13 above), p. 65.

began to decry "hypothetical dogmas founded upon a display of algebraic signs and quantities" as no substitute for "common and empirical observation," and he called for a vast program of experiments to demonstrate what he knew to be true.[40]

Such experiments were undertaken during the summer of 1849 by a second Chadwick-dominated agency, the Metropolitan Sewers Commission, established (on recommendation of the Metropolitan Sanitary Commission) to build and administer a coordinated system of sewers for Greater London. Under the direction of its Trial Works Committee, discharge rates were determined on various sizes of pipe, laid at various inclinations with various heads of pressure and arrangements of branches. It was found, claimed Chadwick, that a single head of pressure produced discharges about a third greater than Hawksley's equation predicted, while with multiple branches discharge could be increased from three to eight times further.[41]

The experiments were put forward as proof that Chadwick's sewerage doctrine had an empirical basis, unlike that of the engineers. Yet neither the raw data nor any explanation of experimental procedure was ever published, and for good reason. Roe was in charge of the research, but most of the time he was ill and absent. Throughout the summer, unsupervised and unskilled technicians neatly plotted the inconsistent, even absurd, results they were getting. In one case, doubling steepness of a sewer from 1:480 to 1:240 led to a 30 percent decrease in velocity. When the commission was forced to resign in October 1849, the experimenters, Joseph Medworth, J. L. Hale, and Thomas Lovick, produced progress reports. Medworth, in charge of the most troublesome experiments, had extracted data that gave the sort of curve one would expect. But the Trial Works Committee, chaired by Cowie, felt itself "unable to . . . place confidence in the comparative results . . . the experiments . . . do not seem to have been

[40]"Minutes on House Drainage," pp. 62, 67. For an assessment of mid-19th-century hydrology, see A. K. Biswas, *The History of Hydrology* (Amsterdam, 1970), pp. 264–69. Among Chadwick's stratagems, his attempt to outflank the engineers on the physics of fluid flow caused the most anger. See the discussion of James Leslie, "Observations on the Flow of Water through Pipes," *Minutes of Proceedings, Institution of Civil Engineers* 14 (1854–55): 273–317. Even Rawlinson ridiculed Chadwick's theorizing: according to theory, house drains "might be reduced . . . to the size of a quill, as more water will pass through such an aperture, in the course of twenty-four hours, than is used in a cottage." See Rawlinson, p. 35.

[41]"Minutes on House Drainage," pp. 38, 67, 69. Compare "Trial Works Committee Papers," report of Medworth, May 5, 1849, GLRO MCS 193, with Medworth's explanation in "Report of the Trial Works Committee," December 1849, p. 22, GLRO MCS 193.

conducted with sufficient care . . . and are thus contradictory when compared with each other."[42]

The appalling quality of the raw data did not become public at the time. Had this happened, Chadwick would have been unable to mount the criticisms he did, but what ensued was a controversy over authority in hydraulic science. The engineers cited Continental authority: Prony, duBuat, Eytelwein, whose approaches were far more consistent than Chadwick's advisors would admit. They argued that something had gone wrong with the sewers commission experiments, even if they could not say what.[43] Chadwick, on the one hand chastising Medworth for incompetence, was, on the other, flaunting the experiments as proof that Hawksley was overthrown.[44] He sent the raw data to Roe, who was to use them to derive tables for sewer design (see fig. 2) and submit these to the General Board of Health, the agency charged with facilitating the building of sanitary works and Chadwick's last stronghold. Eschewing both formulae and the useless data, Roe produced tables based on his twenty years' experience. These indicated how many acres could be drained by a given size of sewer at a given inclination and were, after bitter correspondence between Roe and Chadwick (like Hawksley, Roe expected to be paid), acquired by the board.[45] They indicated the necessity of sewers substantially larger than some of the board's staff advocated and not significantly smaller than those indicated by Hawksley's tables.[46]

[42]See "Report of the Trial Works Committee," December 1849, and raw data, in "Trial Works Committee Papers," undated graphs (probably September 1849).

[43]Even strong opponents like Hawksley, who tried to cast suspicion on the experiments, had little specific information; see discussion of Leslie, pp. 291–93, 315; J. Bazalgette, "The Drainage of London," *Minutes of Proceedings, Institution of Civil Engineers* 24 (1864–65): 355–56. But see *Engineers and Officials* (n. 21 above), pp. 49–52. See also Bateman in discussion of Bazalgette, p. 340; Murray in discussion of Rawlinson (n. 34 above), pp. 54–57; Hawksley and Bidder in discussion of Harrison (n. 10 above), pp. 116–17; discussion of Leslie (n. 40 above), pp. 273–317.

[44]General Board of Health, *Water Supply Report* (n. 33 above), pp. 185–93. See also W. Lee in "Communication from the GBH and Reports from the Superintending Inspectors to the Board, made to the Secretary of State in relation to the Reports of the Metropolitan Sewers Commission in Respect to the Operation of Pipe Sewers" (hereafter "Reports from the Superintending Inspectors"), *PP,* 1854–55, vol. 45 [1891.], pp. 18–21.

[45]"Minutes on House Drainage" (n. 13 above), pp. 67–69; Roe to Cowie, February 19, 1849, GLRO MCS 192; Roe to Austin, February 9, 1852, June 19, 1852, Chadwick MSS (n. 22 above), no. 206; Chadwick to Roe, December 19, 1851, Roe to Chadwick, March 2, 1852, March 9, 1852, Chadwick to Roe, March 30, 1852, Chadwick MSS, no. 1704.

[46]Roe to Chadwick, December 1, 1852, December 8, 1852, Chadwick MSS, no. 1704; Hawksley in discussion of Rawlinson, p. 57. Medworth felt some of his results confirmed Hawksley; see "Report of the Trial Works Committee" (n. 37 above),

TABLE 2.—SHOWING THE QUANTITY OF PAVED OR COVERED SURFACE FROM WHICH CIRCULAR SEWERS (WITH JUNCTIONS PROPERLY CONNECTED) WILL CONVEY AWAY THE WATER COMING FROM A FALL OF RAIN OF ONE INCH IN ONE HOUR, WITH HOUSE DRAINAGE, AS ASCERTAINED IN THE HOLBORN AND FINSBURY DIVISIONS.

	Diameter of Pipes and Sewers in inches.											
	24	30	63	48	60	72	84	96	108	120	132	144
	acres	acres	acres	acres	acres	acres	acres	acres	acres	acres	acres	acres
Level	38¾	67¼	120	277	570	1,020	1,725	2,850	4,125	5,825	7,800	10,100
¼" in 10' or 1 in 480	43	75	135	308	630	1,117	1,925	3,025	4,425	6,250	8,300	10,750
½" in 10' or 1 in 240	50	83	—	355	735	1,318	2,225	3,500	5,100	7,175	9,550	12,400
¾" in 10' or 1 in 160	63	113	203	460	950	1,692	2,875	4,500	6,575	9,250	12,300	15,950
1" in 10' or 1 in 120	78	143	257	590	1,200	2,180	3,700	5,825	7,850	11,050	14,700	19,085
1½" in 10' or 1 in 80	80	165	295	670	1,385	2,486	4,225	6,625				
2" in 10' or 1 in 60	115	182	318	730	1,500	2,675	4,550	7,125				

FIG. 2.—John Roe's table for determining sewer sizes, based on his twenty years' experience, was all that could be salvaged from the Metropolitan Sewers Commission's 1849 experiments. They indicated sewers significantly larger than Chadwick preferred. (Reprinted in E. C. S. Moore, *Sanitary Engineering: A Practical Treatise on the Collection, Removal, and Final Disposal of Sewage* [London, 1898], p. 35.)

While the sewer-flow experiments were going on, the "arterial-venous" approach was itself being tested. During the nearly two years his faction dominated the Metropolitan Commission of Sewers (December 1847–October 1849), Chadwick was forced to recognize the great difference between designing an ideal system and implementing it. His biographers attributed his failures to the persistence of the old order that retained enough power to clog the procedure of the new commission.[47] But the Chadwickians held a working majority until their deposition; their failure to achieve anything significant was, as the *Times* recognized, due more to bewilderment about how to begin than to opposition.

So well integrated were the components of Chadwick's "arterial-venous" city that there was no clear place to begin a design. Cowie

pp. 15–16. Roe backed off earlier claims that sewers could be made significantly smaller; see Metropolitan Sanitary Commission, *First Report,* evidence p. 81.

[47]Finer (n. 2 above), pp. 356–78; Lewis (n. 2 above), pp. 216–37.

argued that one had first to think about end use, about sewage recycling. It made no sense to design upstream features of the system until one knew whether sewage was to be recycled in liquid form or as solid precipitate. Chadwick wanted to start with house drains and street sewers, but the *Times,* and Phillips, protested that until the location of the outfall(s) had been decided on it was not clear where the street sewers ought to go. John Leslie, a progressive on the old Westminster Commission but Chadwick's bane on the new commission, invoked Chadwick's own dictum that the success of pipe sewerage depended on control of the water and urged this as a priority. Detailed planning had to await completion of a topographic survey, the Chadwickians insisted, but here the question arose of whether the surveyors should concentrate on the sort of fine-grained survey needed to sewer individual houses and streets or a general survey for determining the line of the main drainage. Temporary drainage works were out of the question: they would represent wasteful, perhaps deadly, expenditure.[48] Thus, a large force was employed for nearly two years on work that brought little immediate benefit to Londoners: a survey, the hydraulic experiments (and experiments on sewage utilization, sewer ventilation, and pipe manufacture), a "subterranean survey" of existing sewers, and a few small, local sewerage projects.

The years from 1850 to 1854 were the main period of application of Chadwick's sewerage program. The Board of Health's superintending engineering inspectors, all loyal Chadwickians, sanctioned towns' plans for sanitary works (and, as they were only inspectors part-time, often undertook to build them as well). But Chadwick's "system" remained a concept rather than a set of rules. The main "how-to" manual, the board's "Minutes on House Drainage" (1852), was aspecific and frequently lapsed into platitudes or into polemics against orthodox engineers.[49] Just how open-ended the doctrine was is clearest with regard to the key issue of sewer size. How much surface runoff should one allow for, for example? Sewers big enough to handle a heavy rain would in drier times be too big to maintain the rate of flow needed to prevent deposition. Unwilling (except in the case of Phillips) to endorse the expense of separate sewerage, the

[48]*Times* (London) (July 21, 1849), p. 6b, (July 24, 1849), p. 8d, (July 27, 1849), p. 6f, (July 30, 1849), p. 3b, (July 31, 1849), p. 7c, (August 4, 1849), p. 4e, (September 24, 1849), p. 3f, (September 28, 1849), p. 5b, (October 4, 1849), p. 6c, (October 20, 1849), p. 4e. Compare Metropolitan Sanitary Commission, *First Report* (n. 16 above), pp. 25, 39–40, 49, evidence p. 61 (Phillips).

[49]"Minutes on House Drainage" (n. 13 above). See also "Report of the Surveyors on House Drainage," GLRO MCS 198.

Chadwickians downplayed the likelihood of heavy rainfalls or suggested alternative outlets for excess rain.[50] Having determined how much discharge to allow for, one faced the problem of calculating sizes. Should one believe the great claims made on authority of the unpublished sewers commission experiments, or Roe's less radical tables, or find a good rule of thumb? When Robert Rawlinson drained 550 acres at Hitchin through a 20-inch sewer, Roe protested. His own tables, based on actual gaugings, indicated a 60-inch sewer. To get the requisite discharge with a 20-inch pipe would require sewage to flow 4 miles per minute, he claimed.[51]

Even when the principles seemed clear, Chadwick's engineers frequently departed from them. Chadwick held that sewers became blocked because they were too large and that, the flatter the sewer, the more important it was to use a small-bore pipe. He also insisted that in properly working pipe systems no ventilation was necessary; rapid flow would create a downstream current of air and remove sewage before it could emit dangerous gases.[52] On all these issues his own engineers broke ranks. Whatever the hydraulic merits of 2-, 3-, and 4-inch pipes, most Chadwickian engineers preferred 6- and 9-inch pipes. Even John Grant, who endangered his career by supporting Chadwick, favored using pipes only for short, steep runs of sewer. And ventilation was admitted as a practical necessity, even by Austin, the most doctrinaire of the Chadwickians.[53]

[50]"Minutes on House Drainage," pp. 14, 62; "Report of the Trial Works Committee," pp. 3–4.

[51]Rawlinson (n. 34 above), p. 41. Roe to Chadwick, December 1, 1852, December 8, 1852, Chadwick MSS, no. 1704; Roe in discussion of Rawlinson, pp. 96–98. Compare the criticisms of Heywood in discussion of Rawlinson, p. 52. Rawlinson and Austin developed their own rules of thumb that did not depend on hydraulic formulae; see Rawlinson in discussion of Bazalgette (n. 43 above), p. 317; Austin in "Trial Works Committee Minutes," pp. 47–52, GLRO MCS 192.

[52]"Metropolitan Sanitary Commission," *First Report,* evidence p. 69 (Phillips).

[53]On sewer sizes, see General Board of Health, *Water Supply Report* (n. 33 above), pp. 108–11 (Lovick), 136 (Grant), 156 (Gotto); Rawlinson and Heywood in discussion of Rawlinson (n. 34 above), pp. 35, 51; Page, in "Reports by Neil Arnott, MD, and Thomas Page CE on an Inquiry ordered by the Secretary of State Relative to the Prevalence of Disease at Croydon and to the Plan of Sewerage," *PP,* 1852–53, vol. 96 [1648.], pp. 39–40. Even Austin admitted that the pipes in Croydon were too small; see "Statement of the Preliminary Inquiry by T. Southwood Smith, Esq., MD and John Sutherland, Esq., MD on the Epidemic at Croydon; together with Reports by R. D. Grainger Esq. and Henry Austin Esq. to the General Board of Health on the Circumstances connected with the Epidemic Fever at Croydon" (hereafter "First Croydon Report"), *PP,* 1852–53, vol. 96 [1683.], pp. 39–40. On Grant, see General Board of Health, *Water Supply Report,* pp. 135–36. On ventilation, see "First Croydon Report," p. 44; and Henry Austin, "Further Report from the Consulting Engineer to the General Board of Health on the Croydon Drainage" (hereafter "Second Croydon

* * *

Prior to 1852 there was little reaction by orthodox civil engineers to Chadwick's campaigning. Land surveyors had fought his proposal to employ military engineers on sanitary surveys, but, despite Chadwick's regular attacks on their competence, the elite at the Institution of Civil Engineers had been remarkably quiet.[54] As the Board of Health's inspectors began roaming the country, however, condemning the plans of leading engineers, undermining their relations with clients, displacing them, and even, it was alleged, stealing their plans, it became impossible to maintain a dignified silence.[55] The inspectors' harassment was no accident. The board's "Instructions . . . to the Superintending Inspectors" (1848) warned inspectors to be on their guard as they visited towns. They would be presented with sewerage schemes not in accord with correct principle, and they were not even to consider these. It was hard to ignore the 10,000 copies of the board's "Minutes on House Drainage," which informed towns that they could have excellent sewers for 40 percent of what orthodox engineers estimated. The board also warned local boards that they might be prosecuted under the Nuisances Removal Act if they built conventional sewerage systems.[56]

Report"), *PP,* 1852–53, vol. 96 [1009.], pp. 9–10; Rawlinson, p. 39; "First Report of Hale" (n. 36 above), p. 6.

[54]But see J. Simpson in discussion of Green (n. 12 above), p. 95. On surveyors' responses, see F. M. L. Thompson, *Chartered Surveyors: The Growth of a Profession* (London, 1968), pp. 115, 119–20; "The Civil Surveyors and the Military Engineers," *Civil Engineer and Architect's Journal* 11 (1848): 198; comments on J. Newlands, "Report to the Health Committee on the Sewerage [of Liverpool]," ibid., pp. 278–79; *Times* (London) (May 26, 1848), p. 8f, (June 10, 1848), p. 8f.

[55]Hawksley and Bidder in discussion of Leslie (n. 40 above), pp. 295, 312; [Wicksteed] (n. 35 above), p. 4. For defense of this practice, see General Board of Health, "Report by the GBH on the Measures Adopted for the Execution of the Nuisances Removal and Diseases Prevention Act, and the Public Health Act up to July 1849" (hereafter "First Report on the Public Health Act"), *PP,* 1849, vol. 24 [1115.], pp. 44–45. See also "Return of all cases in which Superintending inspectors of the GBH have been employed as Engineers in carrying out the works which have been executed under the Authority of the Public Health Act, 1848," *PP,* 1852–53, vol. 96 (512.), p. 1.

[56]"First Report on the Public Health Act," pp. 64–65, 131–34; "Minutes on House Drainage" (n. 13 above), pp. 143–44; General Board of Health, "Report from the GBH on the Administration of the Public Health Act and the Nuisances Removal and Diseases Prevention Acts" (hereafter "Second Report on the Public Health Act"), *PP,* 1854, vol. 35 [1768.], p. 40. Pipe sewerage cost one-quarter as much as brick according to Austin; see "Second Croydon Report," p. 7. See also Thomas Hawksley, "Letter to the Most Hon Marquis of Chandros M.P. in Relation to the Exercise of Some of the

Not until the end of 1852 were there enough completed works to allow assessment of Chadwick's approach. The following months saw investigations of the new works in Croydon, first town to be sewered under GBH doctrine, as well as in London and elsewhere. The Croydon investigations revealed both technical problems in pipe sewerage and procedural problems in the board's oversight of sanitary engineering. In 1849, William Ranger, GBH superintending inspector, approved a plan for Croydon sewerage submitted by the erstwhile Chadwickian George Donaldson and Thomas Cox, Croydon's surveyor. Ranger soon took over from Donaldson and altered his plans, using even smaller sewers—most 6-inch sewers became 4-inch, most 9-inch became 6-inch.[57] The sewers were finished in mid-1852, and within weeks there were reports of breakages and blockages. Fever broke out. Chadwick ordered an investigation. His medical men found that the sewers had not caused the fever, while Austin, his chief engineer, attributed the defects in sewerage to poor installation: they were Cox's problem.[58]

Chadwick's mistake was to seek the exoneration of an independent Home Office inquiry. This was assigned to Dr. Neil Arnott, his aging Benthamite crony, and Thomas Page, a Board of Trade engineer. Arnott was brief, but Page's engineering report was detailed and damning. The sewers were simply too small and too thin. Page saw no grounds for Ranger's changes nor any rationale for the design as a whole. There were numerous changes of gradient, and no attempt at securing the steady flow Chadwick called for. But what alarmed him most was the GBH's attempt to avoid responsibility. The GBH held that Ranger acted privately in designing the sewers and that its inspection (by Ranger) could not take in "every minute portion of the plans." But Cox and the contractors could not be held responsible, Page maintained; as the GBH insisted on a particular approach, it was responsible for the satisfactory working of the sewers.[59] In fact,

---

Most Extraordinary Powers Assumed by the General Board of Health and the Superintending Inspectors," in Chadwick MSS (n. 22 above), no. 960.

[57]Page, "Reports" (n. 53 above), pp. 26–27. Ranger denied the systematic downsizing he was accused of, but this denial was not made part of the GBH's official response to the investigation. According to Ranger, he significantly revised the Donaldson-Cox plan, and what had been private drains became small public sewers. See Ranger to GBH, September 19, 1853, PRO MH 13 59. Compare Lewis (n. 2 above), pp. 314–17; Finer (n. 2 above), pp. 447–48.

[58]Austin, "First Croydon Report," pp. 38, 42.

[59]Page, "Reports," pp. 33, 46–47, 51–52. Austin, "First Croydon Report," pp. 38–40, "Second Croydon Report," pp. 8, 10–11.

stipulations in the Public Health Act for rigorous inspection did not appear to have been met.[60]

In late 1852, in the midst of the Croydon controversy, the Metropolitan Sewers Commission, no longer part of Chadwick's empire, sent Bazalgette, its engineer, to investigate pipe sewers in other GBH towns and in London, the latter installed chiefly during the period of Chadwickian dominance in 1848–49. Bazalgette too found blockages and breaks. In February 1853, he made an unannounced inspection of 122 London pipe sewers. Of these, 66 had greater than 2 1/2 inches of deposit, while 47 had deposits less than 2 1/2 inches deep and 23 were cracked or broken. (Or so Bazalgette claimed: it was rumored that the sewers had been opened for inspection by a contractor hostile to pipe sewerage—at best Bazalgette had been tricked.) By 1855, the Chadwickians had regained influence on the commission and the district (assistant) surveyors were asked their views of pipe sewerage. Two of them, John Grant and Thomas Lovick, denied Bazalgette's facts (see fig. 3). Grant, who would later become Bazalgette's chief assistant at the Metropolitan Board of Works, accused his superior of blocking access to records that would show the truth. The episode ended with Bazalgette threatening Grant with legal action.[61]

It is in the Chadwickians' responses to the charges made in these reports that one finds most clearly the image of the righteous few hounded by a conspiracy of reactionaries. Austin accused Page, for example, of biased observation or willful misrepresentation. Chadwick, who was preparing an exposé of corruption among engineers, wanted to take the offensive. Rawlinson and Austin counseled moderation, but the 1854 report of the GBH depicted engineers as part of a cadre of opponents to sanitation.[62]

Yet it is not clear that the critics saw themselves as a united opposition. They approached pipe sewers, not as a doctrine, but as an

[60]"First Report on the Public Health Act," p. 63.

[61]"Copy of the Reports of Mr Bazalgette to the Metropolitan Commission of Sewers, relating to the Application, State, and Examination of Tubular Pipe Drains and Sewers," *PP*, 1852–53, vol. 96 (668.), pp. 12–13; Lee, in "Reports from the Superintending Inspectors" (n. 44 above), p. 17; "Copies of Reports to the Metropolitan Commission of Sewers on the Working of Pipe Sewers, of the District Engineers, Messrs Lovick, Grant, Cooper, Donaldson, and Roe," *PP*, 1854–55, vol. 53 (281.), pp. 3–6, 29–33, 36–40, 42; *Engineers and Officials* (n. 21 above), pp. 76–77.

[62]Austin, "Second Croydon Report" (n. 53 above); "Second Report on the Public Health Act," pp. 45–53; Lee, in "Reports from the Superintending Inspectors," pp. 46–52. For the corruption, see Lee to Chadwick, November 9, 1852, Chadwick MSS, no. 1201; for Rawlinson's caution, see Rawlinson to Chadwick, January 26, 1854, Chadwick MSS, no. 1645.

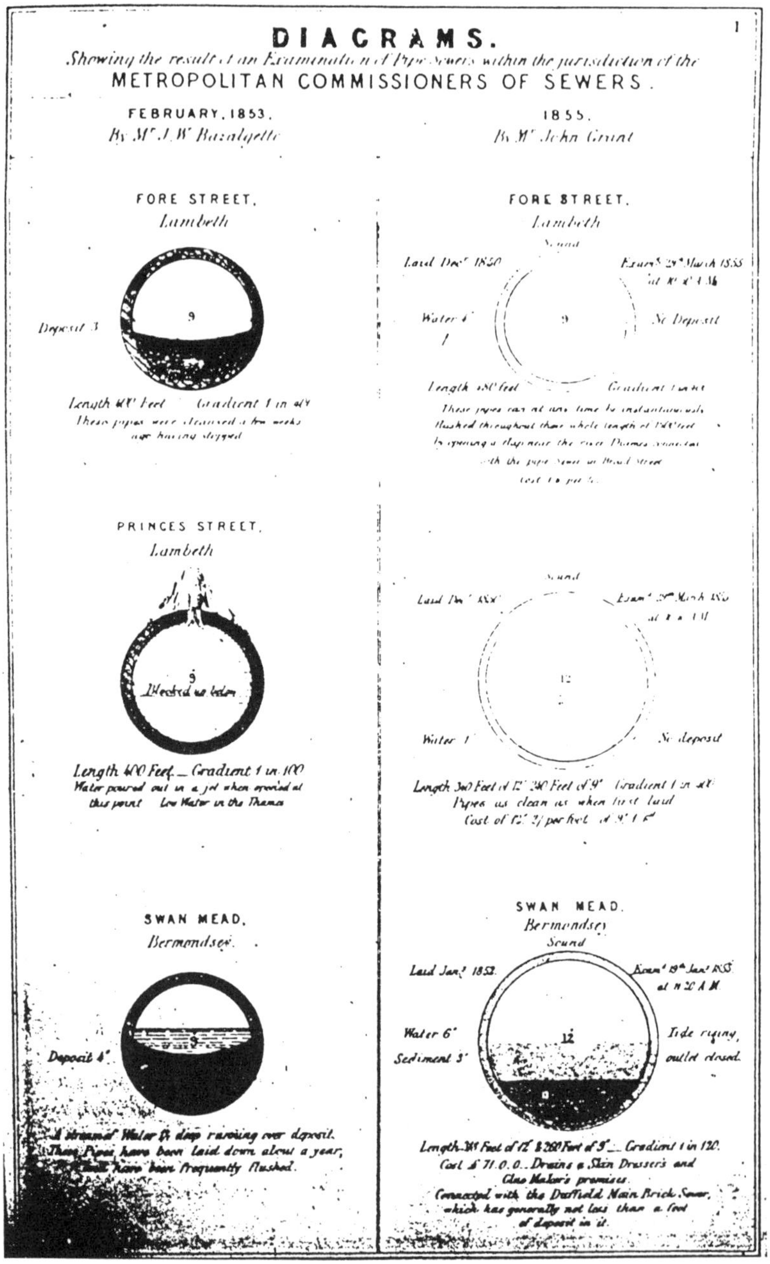

FIG. 3.—Included in John Grant's defense of pipe sewerage were cross-sectional drawings of the sewers in question. The left column reproduced drawings accompanying Bazalgette's 1853 report; the drawings on the right showed the same sewers as Grant found them in 1855. ("Copies of Reports to the Metropolitan Commission of Sewers on the Working of Pipe Sewers, of the District Engineers, Messrs Lovick, Grant, Cooper, Donaldson, and Roe," *Parliamentary Papers,* 1854–55, vol. 53 [281.], p. 45.)

attractive new technique whose applicability had to be worked out.[63] Their perspective is evident in four papers on sewerage read at the Institution of Civil Engineers between March 1852 and February 1855, each generating lengthy discussion of Chadwick's doctrine. The first author, George Donaldson, a Chadwick-appointed assistant surveyor to the Metropolitan Sewers Commission, described the commission's sewerage of the suburb of Richmond. Donaldson had been told to use pipe sewers as far as possible and ended up building about 10 miles of pipe sewers and about 2.5 miles of brick sewers. He opposed pipes except for short, steep runs.[64] Eight months later the general question of "The Drainage of Towns" was taken up by Rawlinson, the most conciliatory and successful of Chadwick's engineers. Yet Rawlinson took much the same line: small pipe and large brick sewers each had advantages, disadvantages, and appropriate uses.[65] In 1855, John Thornehill Harrison's paper considered the main drainage of low areas south of the Thames, and James Leslie discussed flow-through pipes.

The theme of these papers was the engineers' opposition to a single, systemic solution. The GBH had done "great evil . . . endeavouring to suggest general systems which were to prove panaceas, when the very nature of the work . . . required variations in levels, sizes of sewers, and materials, not only varying with different localities, but also in each locality," complained Thomas Wicksteed.[66] Page contrasted two ways of designing sewers, one according to "the maxims of the General Board of Health, which point to the sizes of pipes, to their use as conducts [*sic*] for rubbish, to the ventilation in the direction of the outfall, to the outfall itself, and to the economy in the first cost of the work," and the other (preferable) "depending upon the state and requirements of the population, and so arranged that its operations shall leave nothing to be desired that can practically be effected for their convenience and health, a result which does not depend on the adoption of pipe sewers or brick sewers, but upon the use of such sewers as shall completely, and economically, and enduringly effect the object."[67] Almost all en-

[63]Bazalgette's criticisms were mild; see "Bazalgette to the Metropolitan Commission of Sewers" (n. 61 above), pp. 1–2, 11–12. They led to a Metropolitan Sewers Commission decision to use pipe sewers no smaller than 9 inches in diameter, at a gradient of at least 1:200; see Page, "Reports" (n. 53 above), p. 43.

[64]Donaldson (n. 34 above), pp. 408–9, 421. Donaldson was even more antipipe by the end of the year; see discussion of Rawlinson (n. 34 above), pp. 42–43.

[65]Donaldson, pp. 407–21; Rawlinson, pp. 25–109.

[66][Wicksteed] (n. 35 above), p. 5.

[67]Page, "Reports," pp. 46, 48.

gineers used pipe sewers, leaders of the profession insisted, but not indiscriminately.[68]

One can read such statements as routine assertions of professional autonomy. Yet they reflected the context in which British engineers worked. In the reports they wrote as consultants on sanitary matters, engineers considered problems as they were defined by clients.[69] As a bureaucrat with power, Chadwick could ignore clients; his inspectors could impose a vision of a properly sanitized town on all the towns that came under their consideration.

One implication of the engineers' perspective was that there was no single technical solution. Wicksteed, one of Chadwick's nemeses, observed that larger towns able to afford "a permanent work, [would probably] prefer a large-sized [brick] sewer"; smaller places "might postpone the larger work until they could better afford it" and prefer the cheaper pipes. The choice was "to a great extent, a ratepayers' question, not an engineering one."[70] One was to think not of right or wrong but of advantages and disadvantages of particular designs in particular circumstances. The engineers argued further that many factors other than hydraulics should enter into sewerage design.[71]

[68]Simpson, in discussion of Rawlinson, p. 94. See also Bazalgette and Stephenson, in discussion of Rawlinson, pp. 67, 84–85; Page, "Report," pp. 38–39. Rawlinson placed his own gentle advocacy of pipe sewers in this context; see Rawlinson, pp. 36, 100.

[69]On the structure of the British engineering profession, see Gareth Watson, *The Civils: The Story of the Institution of Civil Engineers* (London, 1988); Charles M. Norrie, *Bridging the Years: A Short History of British Civil Engineering* (London, 1956); F. R. Conder, *The Men Who Built the Railways: A Reprint of F. R. Conder's Personal Recollections of English Engineers*, ed. Jack Simmons (London, 1983); *The Education and Status of Civil Engineers, in the U.K. and in Foreign Countries*, compiled from documents supplied to the Council of the Institution of Civil Engineers (London, 1870); R. A. Buchanan, "Engineers and Government in Nineteenth Century Britain," in *Government and Expertise: Specialists, Administrators, and Professionals, 1860–1919*, ed. Roy M. MacLeod (Cambridge, 1988), pp. 41–58. Compare Edwin T. Layton, *The Revolt of the Engineers: Social Responsibility and the American Engineering Profession* (Baltimore, 1986), esp. chaps. 1, 3, 5.

[70][Wicksteed], pp. 1–5; PRO MH 13 59, nos. 1950/53, 2061/53. This difference in context might account for what to Chadwick seemed hypocrisy on the part of engineers. He could not understand, except in metaphors of heresy, how Hawksley, who favored constant water supply in one case, could advocate intermittent supply in another or how Roe could speak against pipe sewers. But both justified themselves on the grounds of assessing each case independently. See Chadwick to Roe, December 19, 1851, Chadwick MSS (n. 22 above), no. 1704. By contrast, Chadwickian engineers like Rawlinson insisted that there was a single right answer; see Rawlinson, pp. 27–28.

[71]Robert Stephenson suggested the analogy of a steam engine in developing a sewer design: just as one judged an engine not only by its power output per unit fuel but by the "facility for its erection, for its being kept in order, and for its rapid and effective repair," so must one judge a sewer system. See discussion of Rawlinson, p. 86.

Especially important were the questions of sewer capacity, strength of works, and blockages. Chadwick held that any clog in a pipe sewer would generate the pressure needed to remove it and also that a clog indicated too low a velocity due to too large a pipe or too little water. But hydraulic arguments aside, most engineers were unwilling to use pipes smaller than 6 inches. Small pipes clogged too easily, they felt, and it was difficult to unclog them. They continued to insist on flushing provisions despite the GBH claim that these were unnecessary in properly designed sewerage and continued to require steep gradients where pipe sewers were used.[72]

Chadwick's strategy for keeping sewers clear centered on controlling their use by inspecting household connections and instructing the public in the proper use of the new sanitary appliances. But in Croydon and London, failure to address these matters led to blockages. Cloth, paper, hair, and more exotic fare such as a bullock's heart blocked sewers. The Chadwickians believed that, with sufficient effort, the strategy of inspection and instruction could work.[73] But engineers were skeptical. Given the vast number of connections to a public sewer, even if accidents were rare they would cause great damage. To locate a blockage, a long length of pipe might have to be exposed; pipes were difficult to re-lay and often became reblocked at the same spot.[74] And even if it were possible to keep out illegitimate detritus, legitimate materials like hair, grease, and road grit could block sewers. These might resist even high-pressure flushing.[75] Grit-minimizing pavements were not enough; Roe, Rawlinson, and others

[72]"Minutes on House Drainage" (n. 13 above), pp. 33–34; Austin, "Second Croydon Report" (n. 53 above), pp. 14, 22; "Second Report on the Public Health Act" (n. 56 above), p. 23; "Reports from the Superintending Inspectors" (n. 44 above), pp. 25–27, 75, 78, 97. Some early GBH towns (Exeter, Tottenham) did have very flat pipe-sewer systems (1:1,000); "Reports from the Superintending Inspectors," pp. 22–23. Many engineers, including some Chadwickians, insisted that tubular drains required a gradient on the order of 1:60, which significantly restricted their use. See also John Fulton, "Description of the Drainage of the Borough of Dundee," *Minutes of Proceedings, Institution of Civil Engineers* 22 (1862–63): 264; "Reports of the District Engineers" (n. 61 above), pp. 20, 27.

[73]Metropolitan Sanitary Commission, *First Report* (n. 16 above), pp. 41, 122; Page, "Reports" (n. 53 above), p. 49; Austin, "Second Croydon Report," p. 23.

[74]Discussion of Rawlinson (n. 34 above), pp. 46–49; in discussion of Harrison (n. 10 above), p. 91. See also Donaldson (n. 34 above), p. 421; Fulton, p. 268. The GBH experience confirmed these views; see Rammell, in "Reports from the Superintending Inspectors," pp. 57, 66–67.

[75]Donaldson, pp. 410, 413; Hawksley, in discussion of Donaldson, p. 417; Fulton, pp. 262–72, esp. p. 270; Alfred Williams, "Description of the Sewerage and Drainage Works of Newport, Monmouthshire," *Minutes of Proceedings, Institution of Civil Engineers* 22 (1862–63): 273–304, esp. 277.

taxed their ingenuities in designing grit-removing traps as well as traps for house drains. For this reason many engineers continued to favor brick sewers large enough to be cleaned or repaired manually, though not *designed* to require such attention.[76] Since things might go wrong, there was need for access to sewers. The Croydon sewers had only five manholes in 17 miles of sewer, but by 1852 even Rawlinson was calling for a manhole every 100 yards.[77]

Blockages raised the question of system boundaries. Were such problems intrinsic and justly held against the system or extrinsic (somebody else's problem) and therefore irrelevant to its assessment? The two sides might agree on the facts but assess them differently. Were "difficulties, failures, and disappointments . . . to be looked upon . . . as so serious . . . and so necessarily inherent in the system, and that system itself so worthless, as to deter from the attempt to overcome them?" asked Rammell, one of the GBH inspectors. No, insisted Austin: "failures have not arisen from causes inherent in the system, but from causes palpable and preventable. . . . [In] pipe drainage, fairly and properly executed, [these] may be . . . entirely avoided." The fact that pipe sewers in St. Giles had failed in part because the area was inhabited by "the lowest and most filthy of the Irish" was a valid explanation.[78] It had always been made clear that pipe sewerage required a large and constant input of water and control over what went down the drain. That problems arose when this was not provided reflected no error in doctrine and was not the responsibility of its proponents.

Bazalgette, by contrast, held that one had to accept the full weight of circumstance in judging pipe sewers: if they "required more perfect workmanship and greater care than is ordinarily obtainable to make them effective, and that the absence of this degree of perfection subjected them to frequent temporary failures, these facts might become sound reasons for a more limited application of pipe sew-

[76]The Chadwickians misrepresented orthodox views on this issue. They assumed that sewers large enough to permit manual cleansing were intended to be cleansed manually. Rawlinson and Chadwick argued that requiring workmen to clean sewers manually was like forcing little boys to climb chimneys, a practice inhumane as well as uneconomical and (so they claimed) unhealthy. For Chadwick, this was part of the meaning of "sewers of vicious construction." See Rawlinson, p. 28; "Minutes on House Drainage," pp. 26–30.

[77]Page, "Reports," p. 49; Rawlinson, p. 38.

[78]Rammell, in "Reports from the Superintending Inspectors" (n. 44 above), p. 74; Austin, "Second Croydon Report," pp. 5, 11–13; "First Croydon Report" (n. 53 above), p. 40. On St. Giles, see Lee in "Reports from the Superintending Inspectors," pp. 9, 17. Grant protested use of "failure" to refer to a "local or temporary accident"; see "Reports of the District Engineers," pp. 29, 37.

ers."[79] There was no point calling problems "avoidable" if they were not avoided; it might be possible to obtain constant flow, teach people what not to throw down the drains, and supervise every installation, but the designer of sewers had to allow for accident and error. In essence, the question was whether technology was master or servant. For the engineers, a satisfactory system was designed for society as it was; for Chadwick, it was necessary to alter society to suit the system. Page put it simply: "as the population can not be hastily fitted for the sewerage, the sewerage must be fitted for the population."[80]

As to capacity, to determine this one had to decide what sewers were for. Chadwick, as we saw, sought to change their purpose from removing surface and soil moisture to spiriting away wastes. But both jobs needed doing. Reluctant to acknowledge a need for separate sanitary and storm sewers, the Chadwickians held that a single network could accommodate both house waste and modest rainfall, though separate tile drains might be needed to carry off groundwater. In practice, GBH engineers often relied on old sewers to remove storm and groundwater, and critics pointed out that if these supplementary systems really were required their cost should be acknowledged in GBH claims of economy. Brick sewers did all three jobs, they noted. Leaving roof or side bricks unmortared allowed sewers to drain groundwater.[81]

If sewers had to receive some rain, how much should be planned for? Should one plan for averages or extremes? Hawksley accused Chadwick of thinking in terms of a quarter- or a half-inch of rain per day even though there were cases of 2 inches falling in an hour. The Chadwickians saw such observations as excuses for wanton overbuilding, but if sewers really were to have a flood-prevention role, there was a warrant for concern.[82] In essence, Hawksley was criticizing the premium Chadwick put on the fine-tuning of components to a single use; he believed sewer systems should serve as many functions as

[79]"Reports of the District Engineers," p. 34. Finer (n. 2 above), accuses Bazalgette of having "little conception of scientific method" in thinking that a few failures "discredited the system" (p. 449).

[80]Page, "Reports" (n. 53 above), p. 46.

[81]Walker, in discussion of Green (n. 12 above), p. 107; Rawlinson, pp. 40–41; Bazalgette, in discussion of Rawlinson, p. 68; "First Report on the Public Health Act" (n. 55 above), pp. 130–31. Phillips saw need for a triple system, separating sanitary sewage from storm and street drainage. Austin and Lee boasted, however, that pipe sewers could act as subsoil drains, with groundwater entering through cracks in the junctions. Their claim did not sit well with the usual GBH claim that pipe sewers were impermeable. See "Second Croydon Report" (n. 53 above), p. 20; "Reports from the Superintending Inspectors" (n. 44 above), pp. 29, 41.

[82]Hawksley, in discussion of Donaldson (n. 34 above), p. 418; cf. Phillips and Austin, in Metropolitan Sanitary Commission, *First Report* (n. 16 above), pp. 53, 118; Rawlinson

possible. Rawlinson, by contrast, thought sewers should be "adapted, exclusively, for removing all the liquid and soil refuse from houses."[83]

Furthest from Chadwick's experience were problems of structural integrity. He worried about pipe strength but could not decide how strong pipes had to be and had no way to ensure uniform quality, either in manufacture or installation. The early pipes did break often. As rigid structures they were less able to respond to distorting force than were brick sewers; heavy overhead traffic could break an unevenly supported pipe. Prudent engineers, including GBH inspectors, found it expedient to strengthen pipes by jacketing them in concrete, laying them in puddled clay or hard-packed earth, or protecting them with an overlying brick arch. Bazalgette refused to use pipes in deep excavations because he worried both about the damage to structures and the expense incurred by the need to dig down and repair them.[84] The Chadwickians claimed that even if pipes had to be replaced yearly they would still be cheaper than a brick manual-cleansing system. But engineers took a broader notion of public good. Wicksteed opposed use of pipes on busy commercial streets because of the disruption of commerce caused by the need to dig them up.[85] Most engineers, including Rawlinson, would not use pipes greater than 15–20 inches in diameter; manufacturers simply could not produce good big pipes.[86] They also worried about joining pipe segments (either mortar squeezed into the interior, blocking flow, or the joints leaked) and about internal wear: road grit, carried along at the high velocities, would erode the relatively soft earthenware, suggested James Simpson.[87]

(n. 34 above), p. 31; and Heywood, Hawksley, and Roe, in discussion of Rawlinson, pp. 46, 58, 99; Rawlinson, in discussion of Fulton (n. 72 above), p. 298. This was one of the few issues to be quickly resolved, Bazalgette admitting in 1865 that Rawlinson, who held that even with lots of gullies most rain ran off on the surface or was absorbed by the soil, was right. See Bazalgette (n. 43 above), p. 292.

[83]Rawlinson, p. 29. Compare Lavers, in discussion of Donaldson, p. 420; Hawkshaw, in discussion of Bazalgette (n. 43 above), p. 329.

[84]Rawlinson, p. 41; Donaldson, p. 410; discussion of Donaldson, p. 416; Donaldson and Parker, in discussion of Rawlinson, pp. 43, 66; Rammell and Ranger, in "Reports from the Superintending Inspectors" (n. 44 above), pp. 75, 77.

[85]P. H. Holland, in discussion of Rawlinson, p. 76; [Wicksteed] (n. 35 above), p. 5.

[86]Rawlinson, p. 37. In Manchester, pipes up to 30 inches in diameter had been used successfully, but these could be made with very thick walls (almost 4 inches) owing to use of a local clay. See Rawlinson, pp. 42, 45; Cliff, in discussion of Rawlinson, pp. 78–79. Doulton, principal manufacturer of pipes, admitted problems in quality control; see Rawlinson, p. 60.

[87]Donaldson, Doulton, Parker, J. Cliff, and Simpson, in discussion of Rawlinson, pp. 43, 60, 66, 78–79, 92–93; Simpson, in discussion of Harrison (n. 10 above), p. 94; [Wicksteed], p. 5.

* * *

Advocacy of the pipe-sewerage system died out after Chadwick's expulsion from the General Board of Health in 1854. Chadwick expected his engineers to follow him into exile, but most did not.[88] Subsequent sewerage projects generally used both pipe and brick sewers. It was usually felt that for diameters larger than 20 inches (some even said 12 inches) brick was cheaper. Diameters were reduced, though less than Chadwick advocated; manual cleaning died out, though provisions for ventilation, inspection, and repair became commonplace.[89] Chadwick remained unrepentant, but in 1857 Austin apologized: "to confess that some amongst us, over-zealous in the pursuit of new doctrines, should have urged their tenets beyond legitimate limits, and that some partial errors in the earliest practice should have been the consequence, is to admit only that we have not differed from all previous promoters of improved views."[90]

Rather than reflecting the Chadwickian rationalization of engineering, post-Chadwick sanitary administration reflected the engineers' approach. In large part, local authorities decided what projects to undertake, engineers worked out plans with them, and central government inspection focused more on acceptability than optimality.[91] Pipe sewers remained tools to be used in particular client-defined

[88]Rawlinson to Chadwick, August 28, 1854, Chadwick MSS, no. 1645.

[89]Rawlinson, p. 45; Williams (n. 75 above), p. 277; Donaldson, pp. 410–11; Roe, in Metropolitan Sanitary Commission, *First Report,* evidence p. 84. For subsequent sewerage practice, see Baldwin Latham, *Sanitary Engineering: A Guide to the Construction of Works of Sewerage and House Drainage,* 2d ed. (London, 1878). Three decades later the controversy would be repeated in America with respect to Waring's sewerage theories. See Joel A. Tarr, "The Separate vs. Combined Sewer Problem: A Case Study in Urban Technology Design Choice," *Journal of Urban History* 5 (1979): 308–39.

[90]H. Austin, "On a Few Points in Relation to the Drainage of Towns," *Transactions of the National Association for the Promotion of Social Science* 1 (1857): 422–29. Compare Chadwick, in discussion of Bazalgette, pp. 333–40. He admitted pipes had been "often made inexactly, and laid ignorantly, and badly jointed, so as to let out the fluid, and detain the matter which it was the object to discharge; but on the whole they were a success" (p. 334). The principal anti-Chadwick polemic took much the same view as Austin; see *Engineers and Officials* (n. 21 above), pp. 12–16. As the key argument of the work (that inspectors are themselves a vested interest) is uncommon, appearing only once during the public controversy (in similar terms), there is reason to think the pamphlet's author was G. P. Bidder. See discussion of Leslie (n. 40 above), p. 312.

[91]Consider Christine Bellamy's characterization of the Local Government Board, eventual successor to the Board of Health in 1871. It adopted a "political-diplomatic" approach that relied on "general guidelines which are not so much task objectives as parameters for long term exploration of influences and for the conduct of negotiations. . . . Agents seek movement rather than achievement, an acceptable outcome rather than the technically correct solution." The approach is "characterized by

situations, not a system imposed from on high. In the long run, water and sewerage matters were rationalized in Britain, but to a significant degree rationalization remained the achievement of local government and Parliament, not of central administration.

In opposing Chadwick's system and calling for situation-specific technology, these engineers espoused principles of decentralization reminiscent of modern appeals for "appropriate" technology. Yet they did not see themselves as endorsing a philosophy of technology but only as advocating good engineering, which included designing sewers with local needs in mind and making allowances for human error, heavy storms, or town growth. Such an outlook was a professional necessity, for few towns wanted integrated systems; they wanted lengths of sewer here or there or partial waterworks. They expected engineers to respect their concerns and often asked them to change their designs in all sorts of ways and for all sorts of reasons. Chadwick had invited engineers to discard carefully constructed relationships and reputations for the security of a hierarchical bureaucracy, but, with a few exceptions, engineers rejected the risk.

To see the pipe-and-brick sewers war as a legitimate technical controversy is not to deny that it was political as well. The engineers' championship of on-site judgment was incompatible with Chadwick's centralization. Their perspective coincided with that of Chadwick's enemy Joshua Toulmin Smith, champion of local autonomy, but there is no reason to think they were Smith's "creatures" in any sense.[92] Nevertheless, to discover good technical reasons for opposing Chadwick does affect our understanding of his rise and fall: it means we need to look less for sinister motives among his opponents and more for the reasons his adherents followed him. It also raises new questions about urban improvement. If we recognize that technical controversies were not concocted to block progress but reflected legitimate uncertainty, then we uncover a new framework for asking why people advocated the solutions they did and how decisions came to be made.

---

flexibility, discretion, and influence." See Christine Bellamy, *Administering Central-Local Relations, 1871–1919: The LGB in Its Fiscal and Cultural Context* (Manchester, 1988), p. 117. See also C. Hamlin, "Muddling in Bumbledom: On the Enormity of Large Sanitary Improvements in Four British Towns," *Victorian Studies* 32 (1988): 55–83.

[92]In discussion of Rawlinson (n. 34 above), pp. 70–75. Clearly, Toulmin Smith tried to recruit them, arguing that Chadwick's administration was "reducing the engineers . . . to . . . mere clerks of works" (p. 75).

13

The separate vs. combined sewer problem: a case study in urban technology and design choice*

Joel A. Tarr

INTRODUCTION

Public capital-spending projects involving technologies such as water supply, street paving, and sewers were characteristic of American cities during the late nineteenth and early twentieth centuries. All involved questions of technology adoption and design considerations. The decision by a municipality to construct a sewerage or water carriage system of waste removal involved a critical design decision between two forms of sewer technology—the combined sewer and the separate one. In normal use the combined system carries both household wastes and storm water in one large pipe. The separate system theoretically provided two sets of pipe—a small-diameter pipe for household wastes, called a sanitary sewer, and a larger-diameter pipe for

Author's Note: *This article is based on research supported by the National Science Foundation under Grant No. ERP-75-08870. Any opinions, findings, and conclusions or recommendations expressed in this publication are those of the author and do not necessarily reflect the views of the National Science Foundation. The author would like to thank James McCurley, Frances Clay McMichael, Clay McShane, Terry Yosie, and Edward W. Constant II for their comments and criticisms of earlier drafts of this paper.*

storm water from streets, roofs, and yards. In many cases, however, cities with separate systems only constructed sanitary sewers and made no provision for underground removal of storm water.

This paper focuses upon the elements involved in decisions about sewerage system design choice in the later nineteenth and early twentieth centuries. In order to clarify the dimensions of these choices, I will first consider the factors leading to the decision to replace the existing privy vault-cesspool system of waste removal with the sewerage or water carriage system.

THE DECISION TO ABANDON THE PRIVY VAULT-CESSPOOL SYSTEM AND ADOPT WATER CARRIAGE REMOVAL

The method of human waste collection replaced by sewerage technology in American cities in the late nineteenth and early twentieth centuries was the privy vault-cesspool system. The privy vault and cesspool were essentially holes in the ground, often lined with stone, located close by residences or even in cellars. Although a good part of the contents of a well-constructed cesspool or privy was often absorbed by surrounding soil, the receptacles still needed periodic emptying. In some cities the contents of the vaults or cesspools were removed by scavengers or farmers, often under contract to the municipality. In many locales, however, householders merely covered the full vaults with dirt and dug new receptacles. The privy vault-cesspool system, therefore, had the following characteristics: it was locally based, removal was inefficient and labor intensive, and the system was largely privately maintained. Whatever sewers existed in cities were for storm water removal. Most municipalities utilized surface gutters for this purpose, while some larger cities had underground channels. Regulations, however, usually prohibited householders from placing excremental matter in them.[1]

As cities developed and grew in the nineteenth century, a combination of demographic and technological factors caused the privy vault-cesspool system to become increasingly inadequate to deal with waste disposal problems. Urban population and density growth enlarged the pressure on existing facilities, increased the frequency of cleaning, and necessitated the digging of new privies in urban alleys, backyards, and cellars. Located close together and serving larger populations, these receptacles often overflowed, causing nuisances and aesthetic problems. The night soil carts of the scavengers, utilized to contain and remove the wastes from the privies, created similar difficulties. Soil saturated with fecal wastes contaminated groundwater supplies and wells and ponds, often badly polluting sources of water supply.[2]

But while demographic factors were an important element in increasing the nuisances and problems caused by the privy vault-cesspool system, the adoption by American cities of the new technology of piped-in water was equally critical. Cities built water systems to obtain larger supplies of potable water for drinking and household purposes, to fight fires, and to use for street flushing at times of concern over epidemics. Philadelphia constructed the first water works in 1802, and by 1860 the 16 largest cities in the nation had followed its example. In 1860 there were a total of 136 systems in the country, a number that had increased to 598 by 1880.[3] The availability of a constant supply of water in the household caused a rapid expansion in per capita use, as demand interacted with supply. Chicago water usage, for instance, went from 33 gallons per capita in 1856 to 144 in 1882; Cleveland increased from 8 gallons per capita in 1857 to 55 in 1872; and Detroit from 55 gallons per capita in 1856 to 149 in 1882.[4] These figures include industrial and other non-household uses, but they are still indicative of greatly increased household water consumption over a relatively short span of time.

But while hundreds of cities and towns installed waterworks in the first three-quarters of the nineteenth century, no city simultaneously constructed a sewer system to remove the water (some short, private sewers were built). In most cities with waterworks, wastewater was initially diverted into cesspools or existing storm water sewers or street gutters, often causing serious problems of flooding and disposal.[5] The adoption by urban households of the water closet and the discharge of large quantities of fecally polluted water into cesspools and privy vaults designed for much smaller volumes greatly exacerbated the problems.[6] Soil became saturated, cellars were "flooded with stagnant and offensive fluids," and vaults needed much more frequent emtying. Urban public health officials, especially those imbued with a belief in the anticontagionist "filth theory" of disease, were especially concerned. They warned that overflowing cesspools and privies with water closet connections posed an especially dangerous threat to a healthy environment.[7]

The adoption of the two new technologies of piped-in water and the water closet, therefore, combined with higher urban densities to cause the breakdown of the privy vault-cesspool system of waste removal and to increase its productivity of both nuisance and of real and perceived health hazards. City councils, sanitary committees, and health and engineering groups throughout the nation engaged in discussions concerning the various alternatives to the privy vault-cesspool system.[8] Most groups ultimately agreed, although not without dissent, that the so-called water carriage or sewerage system of waste disposal provided the most benefits and the lowest costs compared to other options.

In contrast to the privy vault-cesspool system, sewerage systems were capital rather than labor intensive and required the construction of large public works. They operated in automatic fashion, almost eliminating the need for human decisions and actions to remove the wastes and offered greatly improved conditions of convenience, cleanliness, and the elimination of nuisance. And, sewerage systems solved both the collection and transportation problems, moving the wastes in a wastewater stream from the immediate to a distant locality.[9] Proponents of the system maintained that municipalities should adopt them for primarily three reasons: (1) the capital and maintenance costs of sewerage systems would represent a saving over the annual cost of collection and cleaning with the privy vault-cesspool system; (2) sewerage systems would create greatly improved sanitary conditions and result in lowered morbidity and mortality from infectious disease; and (3) because of improved sanitary conditions, cities that constructed sewerage systems would attract population and industry and grow at a faster rate than those that did not.[10]

Starting in the 1850s with Brooklyn and Chicago, and increasingly in the late nineteenth and early twentieth centuries, city after city made the decision to abandon the privy vault-cesspool system of waste collection and to construct water carriage systems. An integral and inseparable part of the discussions and decisions over system change was the design choice over whether to build combined or separate sewers.

The decision between types of sewers would appear to be one of simple engineering design, based upon an evaluation of the needs of a city from the perspective of some rational model of engineering choice. Such a rational model, based mainly on rudimentary cost-benefit calculations, was primarily utilized in making the choice for sewerage. No such model, however, existed for choice of design. The absence of such a model is partially explained by the relative newness of sewerage systems and the absence of a community of practitioners who agreed on basic criteria for implementation. As Leonard Metcalf and Harrison P. Eddy argued in the 1914 edition of their classic work on sewerage, "American Sewerage Practice is noteworthy among the branches of engineering for the preponderating influence of experience, rather than experiment, upon the development of many of its features."[11]

The debate over design criteria, however, concerned more than the newness of the field. Also involved were basic differences over theories of disease etiology, the removal of storm water, and the disposal and treatment of sewage. Complicating the efforts to arrive at a solution to these questions were the important roles played by a charismatic individual, Colonel George E. Waring, Jr., and a crisis event, the Yellow Fever epidemic that ravaged the Mississippi Valley in 1878 and 1879. Waring's success in publicizing his own ideas about sewerage retarded the development of a "rational" engineering design model for implementation of the alternative sewerage systems. (By "rational" model, I mean one based upon empirical considerations, such as costs and rainfall and traffic conditions necessitating storm water removal, rather than considerations of disease etiology not empirically verifiable.) The final development of this "rational" model, however, and the adoption of one system or another according to its formula, were to have unanticipated consequences of a magnitude requiring modification. In addition, these unexpected impacts necssitated the evaluation of a public policy regarding water quality issues.

EARLY EXPERIENCE WITH THE SEPARATE AND COMBINED SYSTEMS

The design of sewerage systems in the United States during the first period of their implementation in the latter half of the nineteenth century was largely based upon previous European and British experience. While technology transfers were modified by local conditions, there was no development in American sewerage that was not previously anticipated in Great Britain. The first combined sewerage systems originated in both Europe and Great Britain when households, newly supplied with running water, tapped into existing drainage sewers to dispose of household and water closet wastes. In some cases, this required alteration of statutes forbidding the placing of excremental wastes in the sewers. The American record in regard to the original combined sewers replicated the European experience.[12]

Small pipe separate sewers seem to have been advocated first by the British sanitarian Sir Edwin Chadwick in 1842. Chadwick's motivation was two-fold: to remove excremental matters from the household speedily and with a minimum effort, and to remove them in a form that permitted their agricultural use as fertilizer. The first motivation stemmed from a belief that human wastes, if left to putrefy, produced disease-propagating miasmas; the second from the conviction that the valuable ingredients in human wastes should be preserved and that their utilization would make a system of arterial sewers self-financing.[13]

When the city of London finally constructed a sewerage system in 1858, however, it was based upon the plans of engineer Joseph

Bazalgette providing for a combined system constructed of brick that would accomodate both household wastes and storm water. The concept of the separate system was rejected because of the need to remove storm water from the surface of crowded London streets and because of the argument that, primarily because of animal excrement, "rain falling upon the streets is as much polluted as sewage, and ought to be treated as such."[14]

The debate in London between advocates of the separate and combined systems had a profound impact upon sewerage building in the United States. The two major sewerage systems constructed in the United States before the Civil War—that in Chicago, constructed by E. S. Chesbrough, and that in Brooklyn, built by Julius W. Adams, were of the combined type. In his 1857 report to the Brooklyn Commissioners of Drainage, Adams referred to the London debate and rejected the separate system on the grounds that only the intended agricultural use of sewage justified its construction. He argued that if the combined system were built, the sewage could be disposed of in the ocean and storm water would also be eliminated from the streets.[15] After examining European sewerage systems in 1859, Chesbrough also decided against the separate system. He told the Chicago Board of Sewerage Commissioners that the separate system entailed great expense and that "it would not result in freeing the sewers intended for surface water from the introduction of substances that render them offensive."[16]

In the 1860s and the early 1870s, a number of cities in the United States constructed sewerage systems but always of the combined type. The choice of this technology was dictated by several factors: (1) there was no European or American precedent of a successful separate system, and engineers were reluctant to experiment with large capital works; (2) from a cost-benefit perspective, the combined system was cheaper than a separate system that provided separate pipes for house wastes and rainwater and more practical for densely built-up cities that required street drainage; and (3) engineers rejected the concept of the agricultural use of sewerage which required a separate system because they maintained that the costs of the system outweighed the benefits to be derived from resource recycling. They believed that sewage could be safely deposited in nearby waterways because of the theory that running water purifies itself. The methods of chemical analysis of water quality used in the 1860s and 1870s gave this theory scientific validity.[17]

COLONEL GEORGE E. WARING, JR., THE "SEWER GAS" THEORY, AND THE MEMPHIS SEPARATE SEWER SYSTEM

George E. Waring, Jr., was the most noted sanitarian of the late nineteenth century. Trained primarily as a scientific agriculturist with some work in engineering, he had lectured to farmers on agricultural methods, managed several large farms, and published books on husbandry before the Civil War. In 1857, Waring was appointed drainage engineer of New York's Central Park, and during the Civil War he served as a calvary officer. In the period after the Civil War, Waring became actively involved in the sanitary movement, constructing sewerage systems and writing and speaking extensively about his theories.

Waring combined a belief in the agricultural utility of human wastes with a devotion to the theory that "sewer gas" produced by putrefying fecal wastes was the cause of "zymotic" or infectious disease. In the latter half of the nineteenth century, medical theory was in a state of confusion revolving around contagionist and noncontagionist theories of disease etiology. From the 1860s to the 1870s, the dominent belief was the so-called "filth" or "pythogenic" theory, which held that disease evolved *de novo* from putrefying organic matter. This theory was critical in stimulating the sanitary movement and in arousing the public to the dangers of poor sanitary facilities.[18] Waring wrote prolifically in popular and professional journals about the importance of sanitary reforms, reflecting the contemporary state-of-the-art in terms of disease etiology.

Waring's early writings on sanitation focused on the importance of drainage and the agricultural use of wastes. In 1867, he published *Draining for Profit and Draining for Health*, and in

1868, he began promoting the use of the earth closet as an alternative to the "sewer gas"-producing water closet. The earth closet, however, secured little public acceptance due to the greater efficiency and self-working nature of the competing water closet, and in the 1870s, Waring became a promoter of the separate system of sewers in conjunction with improved water closets. Waring's rationale for the adoption of separate as opposed to combined sewers was that they provided for swifter removal of wastes from the household and the city. He argued that unless human feces were transported out of the household in a "fresh condition," they would "undergo putrefaction and give off objectionable gases." These sewer gases would be lethal to the extent that the fecal matter had decomposed. Waring maintained that the separate system was more sanitary and efficient than the combined system, and cheaper to build and to maintain.[19]

During the mid-1870s, Waring published a number of influential articles on sanitation in such popular journals as the *Atlantic Monthly, Harper's Magazine*, and *Scribners*, as well as in more specialized journals, such as *The American Architect and Building News*. In these articles he described the dangers of sewer gas and espoused the adoption of separate sewer systems with flush tanks as the surest means of guaranteeing a healthy sanitary environment. In addition, he argued for agricultural use of sewage as a means of economy. In 1875-1876, he constructed a separate system in the small Massachusetts resort town of Lenox. The Lenox system was not only the nation's first separate system, but also had the first disposal plant, a subsurface irrigation system.[20] Waring's greatest fame, however, was to come from his work in Memphis, Tennessee.

The city of Memphis is located on the east bank of the Mississippi River, halfway between St. Louis and New Orleans. In 1880, it had about 40,000 inhabitants and covered approximately four square miles. A private water company supplied the city with piped-in water, but most people obtained their water from cisterns or wells. Prior to 1880, there were only four miles of private sewers in the city; all surface water was delivered to the river or nearby bayou by street gutters. Human and household wastes were deposited in about 7,000 privy vaults and cesspools, many located in the basements of houses and filled to overflowing. Rotten wooden streets in the densely settled areas and frequent flooding in the low-lying sections worsened the city's sanitary condition.[21]

In both 1878 and 1879, yellow fever ravaged Memphis, as well as other cities in the Mississippi Valley, causing over 5,000 deaths in Memphis alone. At the request of the city officials, the National Board of Health, a short-lived body founded in 1879, appointed a special commission, including George E. Waring, Jr., to make recommendations for sanitary improvements. The commission considered a number of plans for the sewerage of Memphis, including one by Waring for a separate system at a cost of approximately $200,000, and another for a combined system at a cost of $1,000,000, drawn up by E. S. Chesbrough, the engineer of the Chicago sewerage system, and George Hermany, a Louisville civil engineer.[22]

Over the opposition of a number of civil engineers, the commission accepted Waring's plan, and it was constructed within a matter of months. The Waring system had the following characteristics: (1) it provided for house sewage only, excluded all rain water, and made no provisions for storm water; (2) the lateral sewers were six-inch vitrified pipe and the house branches were four inches in diameter; (3) Field's automatic flush tanks of 112 gallon capacity were placed at the head of each branch; (4) manholes were omitted and inspection provided through occasional hand-holes; (5) sewers were ventilated through an open soil pipe at each house and fresh air inlets; (6) subsoil drainage was provided by agricultural tiles laid in the same ditch with the sewer; and (7) the sewage was discharged through an outlet into the Wolf River. The Memphis system differed from other separate systems in that it excluded all rainwater, including that from roof gutters, used automatic flushing tanks for cleaning the sewer, and had no manholes for inspection or to remove obstructions.

While a great deal of controversy developed over these design features of the Waring separate system, the construction of sewerage as well as other sanitary improvements caused a marked improvement in the health of the town. By transforming a "fever-plagued spot into a healthy city," and by doing it at a cost affordable to a nearly bankrupt city, Waring greatly publicized the benefits of sewerage technology to the nation at large, and centered attention on himself.[23]

THE HERING REPORT AND THE ATTEMPT TO PROMULGATE A RATIONAL ENGINEERING MODEL

The construction of the Memphis separate sewer system and the considerable improvements it created in sanitary conditions within the city stimulated a great debate in the engineering press and in engineering circles about the virtues of the separate and combined systems and of the special features incorporated by Waring. This debate was complicated by questions of personality, professional ethics, and medical theory as well as matters of engineering design, although few questioned the benefits of water carriage removal. The important part played by nonengineering factors in the decision to implement sewerage technology and the confusion over design are characteristic of the early stages of new technology implementation.

In 1881, the National Board of Health, undoubtedly concerned with its role in the separate vs. combined system controversy, sent sanitary engineer Rudolf Hering to Europe to report on sewerage experience there. Unlike Waring, Hering was a trained engineer and had graduated from the German Royal Polytechnical School in 1867. He had worked from 1876 to 1880 as assistant city engineer in Philadelphia in charge of bridges and sewers and had written a number of well-regarded articles on sewerage problems.[24]

In his exhaustive report, Hering carefully explored the variations in European sewerage, and suggested a model for the choice between combined and separate systems. Neither system, he said, had a greater sanitary value. Implementation, therefore, should depend upon local conditions and financial considerations, not on a supposed sanitary advantage. Hering maintained that the combined system was best suited and most economical for large densely built-up cities that had to be concerned with stormwater runoff as well as household wastewater. In these cities, the combined system would permit underground removal of stormwater, thereby eliminating a traffic hazard. In smaller cities, on the other hand, where traffic was not a problem and where household wastes were the chief concern, the separate system without underground removal of storm water could be constructed. "The factor," he concluded, "which will mainly govern a preference is less the sanitary value . . . than the cost of construction and maintenance."[25]

Hering had attempted to put the question of sewer design and implementation on a rational engineering basis within a cost-benefit framework. The personalities involved, questions of professional ethics, and confusion over theories of disease etiology, however, prevented acceptance of the model until nearly the end of the century. In this next section, the issues which were exogenous to the rational engineering design model but which intruded upon its formulation, will be considered. In the subsequent section the paper will examine factors that were endogenous to the engineering model, such as design features, provision for storm water, and disposal.

EXOGENOUS FACTORS AFFECTING THE RATIONAL MODEL: PERSONALITY, PROFESSIONAL ETHICS, AND DISEASE ETIOLOGY

PERSONALITY

George Waring, Jr. was a great publicist as well as a great sanitarian. His achievement in Memphis received attention throughout this country and in Europe, and Waring did not hesitate to capitalize on it. He spoke before many medical, sanitary, and other professional groups, and he published numerous popular and professional articles describing his system and defending it against its detractors. He developed a loyal following, including engineers and public health authorities, who defended his principles and pushed adoption of his system. Even though many engineers resented his fame and believed that he made unwarranted claims for his system, his ability to convince decision-makers of the correctness of his position hampered the adoption of the rational model.[26]

PROFESSIONAL ETHICS

Equally as controversial as Waring's design was his patenting of the features of his separate system and his formation of a company, the Drainage Construction Company, to sell it to the cities of the nation, even though the general features of the system had been known for some years. At a time when the engineering profession was attempting to establish its professional status, many engineers considered Waring's patents as reprehensible behavior. In 1881, one civil engineer wrote to the *Engineering News* that Waring's actions were "of serious importance to the profession and the public generally, because if this instance is made a precedent and the idea should become fixed, that no city or town can be properly sewered without having recourse to Mr. Waring or his patent rights, the profession is unfairly and heavily handicapped, and the ignorant public imposed upon."[27] In addition, in his report to the National Board of Health, Hering observed that there were "direct personal interests concerned" in the espousal of the separate system and that those who advocated it exclusively, were "generally not concerned in directing, managing, nor are always familiar with the engineering question in a large and populous city."[28] As in the case of Waring's personality, his patents and financial involvement in his system amplified difficulties of engineers and policy-makers in reaching a rational engineering approach to the problems of sewerage implementation.

DISEASE ETIOLOGY

The third exogenous factor involved was that of disease etiology. Historians and sociologists of history and technology have argued about whether or not complex technological change is based upon previous scientific "discoveries."[29] The public health and sanitation movement of the late nineteenth century rested to a large extent on a belief in the filth theory of disease and stimulated the construction of sewerage systems of all types in order to achieve local health benefits. The case of the Waring separate system, however, furnishes a clear example of a direct rather than a general relationship between a theory of disease etiology—the sewer gas theory—and a specific form of technology design—the Waring separate system.

The link between sewer gas theory and the Waring separate system had a direct impact on public policy in regard to the construction of sewerage systems in the 1870s and the 1880s. In 1882, for instance, the New York State Board of Health recommended that towns adopt the separate system of sewerage with flush tanks (the Waring system) because it avoided the "evils of sewer gas." In Stamford, Connecticut, a town which suffered from a high death rate from infectious disease in the 1870s and 1880s, industrialist and engineer Henry R. Towne, a local influential, became a devout follower of Waring. In 1885, he persuaded the town council to hire Waring to construct a separate system for the city because of its superior health benefits and elimination of sewer gas. In other cities where Waring constructed systems, such as Norfolk, Virginia, and Newport, Rhode Island, the policy issue was phrased in the same terms.[30]

For most engineers, however, the varying theories of disease etiology debated by physicians and scientists were irrelevant to the question of good engineering design. They regarded matters of cost, need for storm water drainage, and for sewage disposal as the important considerations in their choice between systems. They agreed, as Hering observed in his report, that both systems were equally sanitary if they were well designed and adequately maintained. Sir Robert Rawlinson, a pioneer British sanitary engineer, cogently expressed this lack of concern with disease theory in an 1882 letter to the *Sanitary Engineer*. After noting that he owed "little of my knowledge to tables or to books," Rawlinson observed:

> As to the so-called causes of disease, by the growth of low forms of organic life, "bacteria," etc., in sewers and drains, I am contented to leave their study to medical men, chemists, and microscopists. My aim is to construct sewers and drains of absolute truth in line, grade, and sectional form, having smooth and vitreous surfaces, and to so proportion them to the flow of sewage and flushing that in work they shall remain absolutely clean. I then leave rats, bacteria, germs, and other organisms to take care of themselves, as I know by experience that such cannot find any abiding place in sewers and drains so constructed and so managed.[31]

A well-known American civil engineer who expressed similar sentiments in 1883 went on to note that "nine of ten of the best engineers of the United states" believed that neither the separate nor combined system was superior from a sanitary perspective.[32]

But although the great majority of engineers believed that the two systems had equal sanitary capabilities, elements of the public health profession and the public continued to be concerned about the health dangers of combined sewers and to advocate the building of separate sewers to avoid sewer gas hazards. By 1892, 22 towns and small cities had constructed Waring separate systems, while public health authorities had advocated these for large cities, such as Pittsburgh and Kansas City. While it is not clear that those municipalities who built separate rather than combined sewers did so because of concern over sewer gas, it was obviously an issue in some cases. The concern over the health dangers of combined sewers therefore, impacted public policy on both a local and a state level in regard to system implementation and retarded full acceptance of the rational model.[33]

ENDOGENOUS FACTORS AFFECTING THE RATIONAL MODEL: WARING SYSTEM DESIGN FEATURES, COSTS, STORM WATER DISPOSAL, AND SEWAGE TREATMENT

This section will examine those factors that might be labeled endogeneous to those seeking to make decisions concerning the implementation of one form or another of sewerage technology. The elements that will be considered are the special design features of the Waring system, comparative costs, storm water disposal, and sewage treatment.

WARING SYSTEM DESIGN FEATURES

Much of the criticism of the Waring separate system revolved around the special design features that distinguished it from the state-of-the-art characteristics of the sewerage systems that had been formerly constructed. Initial comment focused upon the small size of the sewer laterals and mains, the lack of manholes to remove obstructions, the inadequacy of the flush tanks to scour the sewers, and the absence of provision for storm water. Several engineers predicted that the mains would become clogged and that without manholes, the streets would have to be dug up to remove the obstructions. The extent to which Waring had deviated from the state-of-the-art of engineering practice was reflected by a letter written in 1881 to the *Sanitary Engineer* by noted Boston civil engineer, Edward S. Philbrick: "Colonel Waring launched out upon new and unbeaten paths, pursuing methods which had been discarded as bad ones by nearly all engineers who had had much experience in the construction and maintenance of city sewers during the past generation, both in this country and in England."[34]

In 1887, seven years after installation of the Memphis system, sanitary engineer Rudolph Hering subjected it to a thorough evaluation. Hering found that all of the original criticisms made by engineers of the special design features of the system had proven valid and that none had worked as Waring had predicted. The small pipe sewers did not flow smoothly and were subject to frequent stoppages. Since there were no manholes, the streets had to be torn up to clear blockages, and the city installed 43 manholes in 1883. Hering found that the flush tanks, justified by Waring as necessary to speed the sewage to its outlet before putrefaction, were inefficient and unnecessary because of the slope of the sewers. And, while Waring had intended to keep all rainfall from his sewers, this had proven impossible in practice because of surreptitious connections of drains by householders. The city, therefore, was forced to construct overflows and an intercepting sewer to accommodate excess water volume.

Hering concluded his study by making a strong statement defending the state-of-the-art of sewer design and implicitely justifying the rational model he had presented in his 1881 report to the National Board of Health: "Every one of the deviations made by Colonol Waring from the principles previously adopted," said Hering, "have . . . proved to be either of minor importance or wholly objectionable." Whatever success the system had, was due to "those principles of sewer construction which had been known and practiced elsewhere for many years," as well as to a competent maintenance force.[35] The implications of Hering's criticisms were clear—the "novelties" popularized by Waring had retarded progress toward the development of a rational model to help engineers and municipal policy-makers determine which particular sewerage system design suited their city's needs.

COSTS

In comparison with the alternative system of combined sewers, the separate system with only sanitary sewers offered the advantage of eliminating privy vaults and cesspools for a comparatively low cost. This was one of the principal selling points of the Waring system, rivaling its boasted health benefits. There is evidence, for instance, that the Memphis decision-makers who chose the Waring plan found its cheapness as attractive as its sanitary qualities.[36] A separate system that only provided small diameter sanitary sewers and made no provision for underground removal of storm water would obviously be cheaper than a combined system that required pipes large enough for both sanitary wastes and storm water. At the time of the Memphis crisis, however, because of lack of precedent, no other engineer besides Waring had the daring to suggest such a system.

Waring made the most of his system's low costs, and glossed over the lack of provision for storm water. He argued, for instance, that there was "no instance recorded of the greater cost of the sewerage of a city by the separate system than by the combined system, and it is doubtful whether one-half of the cost has ever been reached."[37] The comparative examples he cited—$7,000 per mile in Memphis and in Keene, New Hampshire, for separate sanitary sewers and $25,600 per mile in Brooklyn and $34,550 per mile in Providence for combined sewers—neglected to note that the towns with separate systems had no storm water sewers. Sanitary Engineer Benezette Williams, who had himself installed a separate system at Pullman, Illinois, but with storm sewers for underground removal of storm water, pointed out the fallacy of Waring's position in an 1885 article:

> Neither Col. Waring nor any other advocate of the separate system as cheaper at all times and in all places, has ever advanced an argument tending to alter the conclusion, that a given quantity of sewerage and a given quantity of storm water can be carried at less expense in one conduit than in two, and that all arguments used for the all-time and all-place cheapness of this system start with the fallacy of . . . ignoring storm water drainage entirely.[38]

The confusion over relative costs and benefits, however, persisted into the 1890s and complicated the decision-making process for municipalities trying to make policy choices regarding type of system to install. In 1886, for instance, two engineering followers of Waring published a text on the *Separate System of Sewerage* that went into a second edition in 1891. In the text, the authors argued that taxpayers should find the separate system attractive because of its low cost compared with the combined system. They maintained that the only reason for cities to continue to build combined systems was "that engineering precedent carries great weight with it among engineers, and a venerable error, even, is hard to put down."[39] The appearance of such arguments in reputable engineering texts made more difficult the acceptance of a rational cost model by both practitioners and policy-makers.

STORM WATER DISPOSAL

The problem of storm water had essentially two dimensions. The first concerned the actual physical nuisances created during heavy rains, while the second involved the chemical composition of storm water. Among the physical difficulties created by storm water were street and basement flooding and the disruption of traffic. These problems were amplified when cities paved their streets, a developmet usually viewed as an improvement. Paved streets, as compared with unpaved streets, resulted in less ground absorption, rapid runoff, and the flooding of low-lying neighborhoods.[40] Municipal policy-makers facing the storm water problem usually considered building combined sewers or leaving rain to runoff in surface channels, only constructing underground pipes in heavy traffic areas.

Waring maintained that storm water should be kept on the surface as long as possible in order to restrict the production of sewer gas, although he admitted that in a few cities heavy traffic necessitated underground removal. Even in these cases, however, he argued that "very shallow sewers would be as effective as deep ones."[41] Hering, on the other hand, maintained in his report that in densely built-up districts and rapidly expanding cities, storm water had to be removed and that the combined system did this most expeditiously. Most engineers agreed with Hering's position. In small cities, however, and especially those with unpaved streets where mud and silt would be washed into the sewers, engineers maintained that storm water should be removed on the surface.[42]

The question of the chemical composition of storm water posed a different set of problems. For some time sanitary engineers had observed that storm water flowing through the streets accumulated all kinds of waste and debris. These wastes included organic materials, such as animal excrement and food particles, as well as mud, leaves, and grit. The first sewers constructed were intended for the drainage of surface waters, and they naturally became the receptacles for the wastes collected by street runoff. Occasionally, households surreptitiously connected their water closets to these storm sewers, increasing their organic content. Because of their large size, poor design, and slow flow, the accumulated wastes decayed, producing, according to the medical theory of the day, disease-laden "sewer gas." "A sewer was synonymous," noted Hering, "with all that is repulsive, filthy, and disgusting,"[43] and they were often referred to as "elongated cesspools."

While the construction of smaller-size and better designed sewers, as well as an increase in water flow, reduced the accumulation of organic wastes, the polluted character of the street runoff still posed problems. Chemical analysis of street runoff in London showed that the first flush from the streets during a storm was approximately as polluted, due mainly to animal excrement, as ordinary sewage. The use of catch basins, intended to prevent debris from entering and clogging the sewer, increased street runoff pollution. The catch basins retained organic materials which became septic and which heavy rains washed into the sewers. The close comparison in chemical nature between household sewage and street runoff was often used as an argument for building a combined sewerage system.[44] Waring argued, however, that the solution to the polluted character of street runoff was to clean the streets effectively rather than leaving that function to storm water. Storm water could then drain on the surface into adjacent water courses without any adverse effect.[45]

The overflow problem was a further argument for surface removal of storm water. Rainfall figures were imprecise, and to build combined sewers large enough to cope with any possible amount of storm runoff would have been prohibitively expensive. Combined sewers were, therefore, provided with storm overflows which acted automatically to remove sewage from the pipe above a certain volume. These overflows were usually located on sewer mains, on the intercepting sewer, or before the treatment plant, if one existed, and the overflow led directly into nearby water courses. Since this overflow consisted of storm water mixed with household sewage, it created both water pollution and nuisances in parts of rivers that were normally free of a high content of wastes. Since overflow drains were often close to water intake pipes or heavily populated areas, a serious hazard could be created.[46]

To a large extent, however, the rational model disregarded the potential dangers of the overflow condition because of the belief that the sewage would be very diluted. While Hering described the construction of overflows in his 1881 report, he did not regard them as a pollution danger and observed that it was "generally permissable to let a slight amount of sewage flow into the river together with a large amount of storm water."[47] Waring took a more skeptical view, and in 1886, he warned of the dangers of the "relief overflow." Most engineers, however, viewed overflows as a means of dealing with the probability of very heavy rains and to restrict the size and costs of combined sewers. From the purely technical perspective, they were a mechanism to prevent the type of backup flooding from storms that had marked the early combined sewers.

SEWAGE TREATMENT

Waring and other proponents of the separate system often maintained that it simplified and lowered the costs of sewage pumping and treatment. Waring believed that his system facilitated sewage purification by land treatment, and his Lenox plan of 1876 had this feature. Other advocates of the separate system maintained that it made possible sewage farms, from which revenue could be produced to pay for maintenance. In 1881, for instance, engineer C. S. Latrobe, in a report to the Baltimore city council, took this position. By 1892, 27 cities and towns treated their sewage, 21 by some method of land treatment and five by chemical precipitation; of these 27, 26 had separate sewers.[48]

Up until approximately 1890, however, engineers and urban policy-makers viewed sewage treatment primarily as a means to avoid nuisance. They recognized that locating a sewer outlet near a water intake often posed a health danger, but they also believed that separating the two by a reasonable distance could resolve this problem. What distance was safe, however, was unclear. The justification for disposing of raw sewage into streams was the theory that running water purified itself. This theory had been challenged in both Great Britain and the United States in the 1870s, but it was still accepted by most American engineers. In 1877, for instance, in their report on a sewerage system for Quincy, Illinois, John Nichol and Charles Macritchie noted that:

> with its rapid current [the Mississippi River], will quickly carry away the discharge from the sewers, and from the quantity and nature of the river water in constant and rapid motion, will quickly dilute and deodorize the sewage, so that all traces of it will disappear in a short distance below the port of discharge.

Even Massachusetts, a pioneer in water pollution control, permitted waste disposal in streams 20 miles above any public use and exempted large rivers from the regulation.[49]

Thus, even though the separate system did facilitate sewage treatment, the rational engineering model held that treatment was only necessary in a few special cases of inland cities not located on large streams. There was no universal need for treatment because of the theory of the self-purification of streams, and city policy-makers could feel free in adopting combined sewers if their outlets were into comparatively large bodies of water.

THE RATIONAL MODEL AND THE CHALLENGE OF WATER POLLUTION

By 1890, ten years after the construction of the Memphis system, most engineers, with the support of the major civil engineering publications, had accepted the rational model of design choice articulated by Hering in his 1881 report. They believed that since neither the separate nor combined system had unique sanitary qualities, implementation decisions should be based upon the needs of local areas within the parameters set by costs. The various exogenous factors that had been created by the Memphis experience and Colonel Waring's philosophy and style had ceased to play a role in decisions involving sewerage system choice. The so-called "novelties" utilized by Waring at Memphis and in the other systems he constructed had been largely discredited. As for Waring himself, while some engineers regarded him as the exemplar of bad engineering design, others gave him credit for removing "old prejudices" against small pipe sewers and emphasizing their "manifest advantages."[50] To the public at large he retained the image of the great sanitarian of his time.

The concept of the rational model of design choice, however, rested upon the assumption that large cities that constructed combined sewers could safely dispose of their sewage in adjacent waterways. At approximately the same time that the rational model was winning wide engineering acceptance, new evidence was emerging that challenged its validity. During the late 1880s, bacteriologic research supplemented chemical analysis in the determination of water quality and gradually resulted in the substitution of the germ theory of disease for the filth and sewer gas theories. In the early 1890s, the work of William T. Sedgwick at the Massachusetts Board of Health laboratories, as well as other bacterial researchers, clarified the etiology of typhoid fever and its relationship to sewage polluted waterways. This clear identification of the water-born nature of communicable disease raised a serious challenge to the rational model and to the safety of discharging sewage into streams utilized for water supply by downstream cities. It also caused increased interest in methods of sewage "purification."[51]

Outputs from combined sewers were the cause of the pollution of most streams and, by 1893, leading engineering opinion was calling for reexamination of the separate system because of its utility in facilitating purification.[52] As sanitary engineers Moses N. Baker and George Rafter noted in their 1893 path-breaking work, *Sewage Disposal in America*, purification was necessitated by "the recent extensions of knowledge of the causation of typhoid fever, and the other water-born communicable diseases."[53] Engineering supporters of the rational model had previously rejected the concept that one system provided more sanitary benefits than another, but, in ways unforeseen, this assumption was being challenged by the concern over treatment. The new bacterial findings posed a direct challenge to the further utility of the model as a means by which engineers and municipal policy-makers could decide between sewerage alternatives.

Although the vulnerability of the rational model was apparent in the 1890s, engineers and municipal policy-makers continued to make sewerage decisions by its precepts. (By the 1890s, mainly due to the work of Rudolph Hering, the hypothesis of the self-purification of streams had been replaced by the concept of dilution. Dilution rested upon the empirical evidence that under certain conditions, streams would actually be self-purifying through natural processes. However, in the late nineteenth century, estimates as to the exact conditions under which dilution took place altered. In addition, increases in miles of sewerage systems, in population connected to sewers, and in municipalities discharging into particular waterways overwhelmed the dilution capacity of many streams.[54]) From 1890 to 1905, the total mileage of sewers of all types increased from 6,005 (cities over 25,000) to 19,460 (cities over 30,000). Of the 1905 total, there were 14,856 miles of combined, 3,756 miles of sanitary, and 847 miles of storm water sewers (see Table 1). Thus, under the precepts of the rational model, cities continued to build many more miles of combined than separate sewers.[55]

The rational model had always been sensitive to cost considerations in its choice between sewerage systems, but it had assumed the safety of stream disposal. If, however, stream disposal was limited for health or other reasons, and if treatment became a necessity, the installation by large cities of higher cost separate systems became a possible public policy alternative. (Waring had proposed the installation of separate systems for New York, Philadelphia, and Washington in the 1880s.) But since large cities already had combined systems and the costs of replacement

TABLE 1

Sewar Mileage by Type and Population Class, 1905, 1907, 1909

1905

Population Group	Sanitary Sewers	Storm Sewers	Combined Sewers	Total Mileage by Pop. Group
1. >300, 000	335.2	157. 0	8, 229. 9	9, 422. 6
2. 100, 000–300, 000	809. 0	120. 5	2, 961. 0	4, 101. 5
3. 50, 000–100, 000	965. 8	242. 8	2, 491. 1	3, 709. 7
4. 30, 000–50, 000	1, 313. 4	326. 6	1, 507. 4	3, 147. 7
Total by Sewer Type	3, 756. 2	846. 9	14, 856. 5	20, 381. 5

1907

Population Group	Sanitary Sewers	Storm Sewers	Combined Sewers	Total Mileage by Pop. Group
1. >300, 000	554.8	352. 0	9, 242. 3	10, 149. 1
2. 100, 000–300, 000	1, 300. 0	262. 9	3, 690. 5	5, 253. 4
3. 50, 000–100, 000	1, 097. 1	181. 6	2, 627. 1	3, 905. 8
4. 30, 000–50, 000	1, 611. 3	383. 9	1, 562. 9	3, 558. 1
Total by Sewer Type	4, 563. 2	1, 180. 4	17, 122. 8	22, 866. 4

1909

Population Group	Sanitary Sewers	Storm Sewers	Combined Sewers	Total Mileage by Pop. Group
1. >300, 000	789. 5	349. 9	9, 834. 3	10, 973. 7
2. 100, 000–300, 000	1, 404. 4	284. 2	4, 405. 8	6, 094. 4
3. 50, 000–100, 000	1, 831. 5	384. 2	2, 615. 5	4, 831. 2
4. 30, 000–50, 000	1, 232. 9	333. 8	1, 505. 9	3, 072. 6
Total by Sewer Type	5, 258. 3	1, 352. 1	18, 361. 5	24, 971. 6

SOURCE: Bureau of the Census, U.S. Dept. of Commerce and Labor, *Statistics of Cities Having a Population of over 30,000: 1905,* Moses N. Baker, "Appendix A: Sewerage & Sewage Disposal," (Washington, DC, 1907), 342-247; Bureau of the Census, U.S. Dept. of Commerce and Labor, *Statistics of Cities Having a Population over 30,000: 1907, "Special Reports" (Washington, DC, 1910), 458-463; Bureau of the Census, U.S. Dept. of Commerce, General Statistics of Cities with Populations Greater Than 30,000:* 1909, "Special Reports" (Washington, DC, 1913), 88-93.

were extremely high, cities continued well into the twentieth century to build combined sewers and to discharge their untreated wastes into adjacent waterways.[56] The incentive to construct sewage treatment facilities, except in cases of obvious nuisance, was actually diminished by the development of methods of water filtration and chlorination, which reduced the hazard from water-born infectious disease. As sanitary engineer Allen Hazen commented in his 1907 book, *Clean Water and How to Get It*, "the discharge of crude sewage from the great majority of cities is not locally objectionable in any way to justify the cost of sewage purification." Rather, said Hazen, downstream cities should filter their water to protect the public health, and sewage purification should only be utilized in order to prevent nuisance.[57]

THE RATIONAL MODEL AND SEPARATE SYSTEM PROBLEMS

Cost considerations also constrained the model in regard to smaller cities and towns. In these municipalities, engineers created a priority of goals that caused them to ignore underground removal of storm water and to construct sanitary sewers only. These limited systems were especially characteristic of inland towns that had to treat sewage because of lack of access to large waterways. But these partial solutions came to haunt the policy-makers who had chosen the least costly option. As sanitary engineer Harrison P. Eddy maintained in 1922, "neglect or postponement of provision for prompt removal of storm and surface water while offering a happy solution of immediate financial problems, may have been a policy of doubtful merit."[58]

One of the chief problems resulting from the construction of separate sewers without provision for storm water, Eddy noted, derived from the unintended discharge of storm and ground water into sanitary sewers. This discharge occurred when ground water infiltrated through inferior sewer joints or when roof and cellar drains and street inlets were connected with the sanitary sewer. In some cities, storm and ground water so surcharged sanitary sewers that the sewage flowed back into house cellars or through manhole covers into the streets. In others, the volume of water overwhelmed the capacity of treatment plants, resulting in the diversion of raw sewage into streams. "Great care," concluded Eddy, "should be exercised in the selection of the type of sewer system to be installed. It should not be assumed that separate sewers are more advantageous, but conclusions should be reached only after careful study of all the conditions."[59]

CONCLUSIONS

The case of combined vs. separate sewers provides useful insights into the complicated interplay between changing scientific theories, engineering practice and choice of technological design, and impacts upon cities. During the last quarter of the nineteenth century, sanitary engineers struggled to define a rational engineering model for sewerage system design choice within the parameters of municipal budgetary constraints and conflicting scientific theories about disease etiology. George E. Waring, Jr.'s advocacy of the separate system with special design features, based upon the sewer gas theory of disease etiology, complicated attempts to define questions of design choice. Also important in the debate were probability problems, such as the expected amount of storm runoff, predicted levels of traffic, and anticipated volumes of water usage.

By the last decade of the century, however, the rational engineering model, based upon empirical considerations of urban need, had triumphed, and a community of engineering practitioners had evolved who followed its precepts in design implementation. Essentially, this meant that large cities having a need for subsurface storm water removal as well as the removal of household wastes constructed combined sewers, while smaller cities built separate systems with sanitary sewers only and surface removal of storm water. By 1909, cities over 30,000 population had 18,361 miles of combined sewers but only 5,258 miles of sanitary sewers and 1,352 miles of storm; in cities over 100,000, miles of combined sewers outnumbered those of sanitary sewers by about a seven-to-one margin (see Table 1).

But while the rational model provided a means to choose the sewerage design best suited for the local needs of a particular city, it created problems for downstream communities. Again the question of disease etiology was crucial. Engineers had originally maintained that sewage disposal into waterways was safe because of the theory that running water purified itself. Bacterial research during the 1880s and 1890s, however, refuted this theory and generated concern over the role of sewage-polluted water in generating typhoid epidemics. In addition, this research stimulated a reappraisal of the separate system because of its facilitation of sewage treatment. Large cities, however, having built thousands of miles of combined sewers, resisted the pressure of public health authorities to convert to separate sewers because of the costs involved. Engineers and municipal authorities maintained that the dilution capacity of streams should be utilized to its fullest, and that downstream cities should adopt retrofits, such as filtration and chlorination, to protect water supplies against infectious disease.

While the "pendulum" of engineering opinion (pressed by directives from state boards of health) swung somewhat in the direction of the construction of separate sewers in the second and third decades of the twentieth century, thousands of miles of combined sewers remained in place.[60] They constituted a heritage to future policy-makers seeking to deal with problems of water pollution and as testimony to the earlier triumph of the rational model of sewerage construction.

NOTES

* A later version of this essay appears in Joel A. Tarr, *The Search for the Ultimate Sink: Urban Pollution in Historical Perspective* (Akron University Press, 1996).

1. B. A. Segur, "Privy-Vaults and Cesspools," *Papers and Reports of the American Public Health Association* (hereafter cited as *APHA*) 3 (1876), 185-187; Henry I. Bowditch, *Public Hygiene in America* (Boston, 1877), 103-109; Mansfield Merriman, *Elements of Sanitary Engineering* (New York, 1906), 139-142; Leonard Metcalf and Harrison P. Eddy, *American Sewerage Practice*, 2 vols. (New York, 1928), I: 15-19.

2. All of the reports cited above make comments about fecal pollution. See also, George E. Waring, Jr., "The Sanitary Drainage of Houses and Towns, II," *Atlantic Monthly* 36 (October 1876), 434.

3. Nelson M. Blake, *Water for the Cities* (Syracuse, 1956); "ownership of American Water Works," *Engineering News* (hereafter cited as *EN*) 27 (January 23, 1892), 83-86.

4. J. T. Fanning, *A Practical Treatise on Hydraulic and Water-Supply Engineering* (New York, 1886), 625.

5. See, for example, E. S. Chesbrough, "The Drainage and Sewerage of Chicago," *APHA* 4 (1878), 18-19; Town of Pawtucket, Committee on Sewers, *Report, 1885* (Paw-

tucket, 1885), 15; Town Improvement Society of East Orange, *The Sewerage of East Orange* (East Orange, NJ, 1884).

6. In Boston, for instance, in 1863 (population ca. 178,000) there were over 14,000 water closets out of a total of approximately 87,000 water fixtures; in Buffalo, in 1874 (population ca. 118,000), there were 5,191 dwellings supplied with water and 3,310 with water closets. By 1880, although the data are imprecise, it can be estimated that approximately one-quarter of urban households had water closets while the remainder depended on privy vaults and cesspools. See, Boston, Cochituate Water Board, *Annual Report, 1863* (Boston, 1864), 43; City of Buffalo, *Sixth Annual Report of the City Water Works, 1874* (Buffalo, 1875), 47. The estimate for 1880 is based upon information in U. S. Department of the Interior, Census Office, *Tenth Census of the United States, 1880, Report of the Social Statistics of Cities*, comp. by George E. Waring, Jr., 2 vols. (Washington DC, 1887).

7. Charles V. Chapin, "The End of the Filth Theory of Disease," *Popular Science Monthly* 60 (January 1902), 234-239; George E. Waring, Jr., "The Sanitary Drainage of Houses and Towns, III" *Atlantic Monthly* 36 (November 1875), 535-551; Benjamin Lee, "The Cart Before the Horse," *APHA 20 (1895), 34-36.*

8. Azel Ames, "The Removal and Utilization of Domestic Excreta," *APHA* 4 (1877), 65-80.

9. These characteristics of water carriage are a summary of information from many sources. See, for example, [Julius W. Adams], *Report of the Engineers to the [Boston] Commissioners of Drainage*, September 10, 1857 (Brooklyn, 1857); J. Herbert Shedd, *Report on Sewerage in the City of Providence* (Providence, 1874); Ames, "The Removal and Utilization of Domestic Excreta," *APHA* 4 (1877), 69-70; Rudolph Hering, *Report on a System of Sewerage for the City of Wilmington, Delaware* (Wilmington, 1884).

10. These benefits and costs are described in more detail in Joel A. Tarr and Francis Clay McMichael, "The Evolution of Wastewater Technology and the Development of State Regulation: A Retrospective Analysis," in Joel A. Tarr, ed., *Retrospective Technology Assessment—1976* (San Francisco, 1977),. 174-178.

11. Leonard Metcalf and Harrison P. Eddy, *American Sewerage Practice: Volume I—Design of Sewers*, 3 vols. (New York, 1914), I: 1.

12. Rudolph Hering, "Report of the Results of an Examination Made in 1880 of Several Sewerage Works in Europe," *Annual Report of the National Board of Health 1881* (Washington, 1882), 101-104 (hereafter referred to as *Hering Report, 1881*).

13. Edwin Chadwick, *Report on the Sanitary Condition of the Labouring Population of Great Britain*, ed. by M. W. Flinn (Edinburgh, 1965), 58-73.

14. Francis Sheppard, *London 1808-1870: The Infernal Wen* (Berkeley, 1971), 253-258.

15. Adams, *Report of the Engineers to the Commissioners of Drainage*, 4-7.

16. E. S. Chesbrough, *Chicago Sewerage: Report of the Results of Examinations Made in Relation to Sewerage in Several European Cities 1856-7* (Chicago, 1858), 66.

17. "Errors in Engineering," *The Plumber and Sanitary Engineer* 2 (June 1, 1879), 181; "Alfred S. Jones to Editor," *The Plumber and Sanitary Engineer* 2 (August 1, 1879), 27; "Separate Drainage and Sewerage," ibid. 2 (October 1, 1879), 350. For self-purification of streams, see William T. Sedgwick, *Principles of Sanitary Science and the Public Health* (New York, 1918), 128-129, 233-246.

18. For an appreciation of Waring as an environmentalist, see Martin V. Melosi, "Pragmatic Environmentalist: Sanitary Engineer, Col. George E. Waring, Jr.," *Essays in*

Public Works History 4 (Washington, April 1977). For a more critical appraisal of Waring, see James H. Cassedy, "The Flamboyant Colonel Waring: An Anti-Contagionist Holds the American Stage in the Age of Pasteur and Koch," *Bulletin of the History of Medicine* 36 (March, April 1962), 163-176.

19. See, Waring, "The Sewering and Draining of Cities," *APHA* 6 (1879), 34-40.

20. "Sewage Purification in America," *EN* 28 (July 14, 1892), 33.

21. For descriptions of Memphis see, "The Sanitary Condition of Memphis," *Bulletin of the National Board of Health* 1 (Washington, DC, December 13, 1879), 187-189; "Memphis," *Tenth Census of the United States, 1880, Report of the Social Statistics of Cities*, comp. by George E. Waring, Jr., 2 vols. (Washington, 1887), 2: 144-147; Waring, *The Sewerage of Memphis, U.S.A.*, (London, 1881).

22. Ibid., "Memphis," *Social Statistics of Cities* 2: 144-147.

23. William Paul Gerhard, "A Half-Century of Sanitation—II," *American Architect and Building News* 43 (March 4, 1899), 67.

24. Frank A. Taylor, "Rudolph Hering," *Dictionary of American Biography*, ed. by Dumas Malone (New York, 1933), 8: 576-577; "Rudolph Hering," Committee on History and Heritage of American Civil Engineering, American Society of Civil Engineers, *A Biographical Dictionary of American Civil Engineers* (New York, 1972), 58-59.

25. *Hering Report, 1881*, 115. See, also, Rudolph Hering, "Sewerage Systems," *Transactions of the American Society of Civil Engineers* 10 (November 1881), 361-386; and "Report on European Sewerage Systems, with Special Reference to the Needs of the City of Philadelphia," *Journal of the Franklin Institute* 114 (July,, December 1882), 186-201. Another statement of the model is Howard A. Carson, "Separate vs. Combined Sewers," *Sanitary Engineer* 4 (February 15, 1881), 142-143.

26. See the discussion, for instance, following Frederick S. Odell, "The Sewerage of Memphis," *Transactions of the American Society of Civil Engineers* (1881), 32-52.

27. "Col. Waring's Sewerage Patents," *EN* 8 (October 22, 1881), 425-427.

28. *Hering Report, 1881*, 112. Hering made reference to European systems, such as the Shore or Liernur Compressed Air Systems.

29. See, for instance, "The Interaction of Science and Technology in the Industrial Age," *Technology and Culture* 17 (October 1976).

30. James T. Gardiner, *Report to the State Board of Health on the Methods of Sewerage for Cities and Large Villages, in the State of New York, No. 42* (1882); "New York State Board of Health on Sewerage," *Sanitary Engineer* 5 (March 23, 1882), 337-338. For Stamford, see Estelle F. Feinstein, *Stamford in the Gilded Age* (Stamford, 1973), 166-185.

31. Robert Rawlinson, "Sewerage and Drainage," *Sanitary Engineer* 5 (April 27, 1882), 450.

32. Charles E. Chandler, *Plan of Sewerage for Greenville, Conn.* (Norwich, 1883), 5. Chandler also noted, "The germ theory of disease, the generation of bacteria, and the comparative safety of a large sewer with an intermittent flow, and a small sewer with a nearly uniform flow, are subjects which I freely confess my inability to discuss to advantage. It is my opinion, however, that either of the three systems properly carried out, can be relied on to safely and satisfactorily perform the offices for which they are designed, provided the plumbing and house drainage is properly designed and honestly constructed."

33. George E. Waring, Jr., "The Memphis System of Sewerage at Memphis and Elsewhere," *APHA* 18 (1893), 153-168.

34. See discussion, following Odell, "The Sewerage of Memphis," 32-52. See also,

letters to the editor, *Sanitary Engineer*, 4 (April 1, 15, May 15, and June 15, 1881), 202-203, 231, 277, 326, and 5 (April 27, 1882), 450-451.

35. Rudolf Hering, "Six Years' Experience with the Memphis Sewers," *Engineering & Building Record and the Sanitary Engineer* 16 (November 26, 1887), 738-739.

36. Odell, "The Sewerage of Memphis," 26. See also, "Memphis," *Social Statistics of Cities* 2: 145-146.

37. See the previous citations on Memphis by Waring as well as, "The Memphis System of Sewerage at Memphis and Elsewhere," *APHA* 18 (1893), 153-168.

38. Quoted in Benezette Williams, "The Separate Vs. Combined System of Sewerage," reprint from *Journal of the Association of Engineering Societies* 4 (1885), 17-18; see also, "Separate Versus Combined Systems of Sewerage," *Sanitary Engineer* 11 (April 16, 1885), 413-414.

39. Cody Staley and G. S. Pierson, *The Separate System of Sewerage* (New York, 1891), 50.

40. Henry N. Ogden, *Sewer Design* (New York, 1899), 70-71.

41. Odel, "The Sewerage of Memphis," 45-46.

42. *Hering Report, 1881*, 108.

43. *Hering Report, 1881*, 103; George E. Waring, Jr., "The Sanitary Drainage of Houses and Towns, II," *Atlantic Monthly* 36 (October 1875), 437-442

44. *Hering Report, 1881*, 105; George W. Rafter and M. N. Baker, *Sewage Disposal in America*, 150-151.

45. Waring, "Storm Water in Town Sewerage," 326-327; Staley and Pierson, *The Separate System of Sewerage*, 39-42.

46. George E. Waring, Jr., "Mechanical Appliances in Town Sewerage," *Journal of the Franklin Institute* 121 (1886), 276-277.

47. *Hering Report, 1881*, 107, 124-126. *The Sanitary Engineer* claimed that overflow from combined sewers was not a serious problem since "the amount of sewage actually coming in fresh to be mixed with the rain fall, is so small in comparison with the latter, and the dilution so enormous, that no appreciable pollution of the open channels actually occurs." See, "Errors in Engineering," *Plumber and Sanitary Engineer*, 181.

48. George Waring, Jr., "Out of Sight, Out of Mind," *Century* 27 (1893-1894), 943-946; George W. Rafter and M. N. Baker, *Sewage Disposal in the U.S.* (New York, 1893), 361-559. For Baltimore, see, "The Sewerage of American Cities and Towns," EN 8 (December 3, 10, 1881), 486-487, 496-497. See also, John H. Gregory, "Separate and Combined Sewers in Relation to Sewage Disposal," ibid., 70 (October 1913), 875-876, for a later statement of this position.

49. Charles Macritchie and John Nichol, *Report on the Sewerage of the City of Quincy, Illinois* (Milwaukee, 1877), 4. Jay Slater, "The Self-Purification of Rivers and Streams: A Study of the Transformation of Water Safety Standard in the Late Nineteenth Century," *Synthesis* 2 (Autumn 1974), 41-45 Barbara Rosenkrantz, *Public Health and the State* (Cambridge, 1972), 81.

50. See, for instance Gerhard, "A Half Century of Sanitation," 67.

51. Rosenkrantz, *Public Health and the State*, 103-106.

52. Sewage Purification and Storm and Ground Water," *EN* 28 (August 25, 1892), 180-181; Gregory, "Separate and Combined Sewers in Relation to Sewage Disposal," 875-876. Hering held that combined sewerage systems were as suitable for sewage treatment as were separate systems. He advocated treating the dry weather flow plus the first flush from the streets in case of a storm, allowing the remainder of the storm water to be

diverted to nearby rivers by an overflow. He warned, however, that local conditions would ultimately determine the best system. See *Hering Report, 1881*, 107.

53. Rafter and Baker, *Sewage Disposal in America*, 150.

54. Hering, "Notes on the Pollution of Streams," *APHA* 13 (1888), 272-279. For a useful brief description of dilution and problems with changing standards, see, Milton J. Rosenau, *Preventive Medicine and Hygiene* (New York, 1927), 1009, 1099-1103.

55. For data on cities with separate or combined systems, see Table 41—"Length and Classes of Sewers," Department of Commerce and Labor, Bureau of the Census, *Statistics of Cities Having a Population of Over 30,000: 1905* (Washington, DC, 1907), 342-347; Gregory, "Separate and Combined Sewers in Relation to Sewage Disposal," 875-876.

56. Moses N. Baker, "Sewerage and Sewage Disposal," Department of Commerce and Labor, Bureau of the Census, *Statistics of Cities Having a Population of Over 30,000: 1905* (Washington, DC, 1907), 103-106; "Sewers and Sewer Service," Department of Commerce and Labor, Bureau of the Census, *General Statistics of Cities: 1909* (Washington, 1913), 31. In 1914 engineers from the Dept. of Public Works in Philadelphia prepared a report on collection and disposal of sewage for that city. In this report they argued that even though the separate system had advantages in terms of treatment and protection of water courses from pollution, a number of factors militated against its installation. These factors were (1) the city already had 1,300 miles of combined sewers; (2) sewage in the combined system was not as concentrated as in the separate; (3) the combined system occupied less space in the streets; (4) citizens would have to rearrange their plumbing if the separate system was installed; and (5) the streets would have to be torn up and repaved at a great cost if the separate system was installed. See, Department of Public Works, Bureau of Surveys, City of Philadelphia, *To Protect The Public Health* (Philadelphia, 1914), 32-35.

57. Allen Hazen, *Clean Water and How to Get It* (New York, 1907), 34-37. In 1903, the *Engineering Record* editorialized that, "it is often more equitable to all concerned for an upper riparian city to discharge its sewage into a stream and a lower riparian city to filter the water of the same stream for a domestic supply, than for the former city to be forced to put in sewage treatment works." See "Sewage Pollution of Water Supplies," *Engineering Record* 28 (August 1, 1903), 117.

58. Harrison P. Eddy, "Use and Abuse of Systems of Separate Sewers and Storm Drains: Can Their Failure Be Prevented?" *Proceedings of the American Society for Municipal Improvements* 29 (October 1922), 128-129.,

59. Ibid., 137; W. H. Dittoe, "Prevention of Misuse of Sewers," *Proceedings of the American Society of Civil Engineers* 47 (December 1921), 642-644; "Poor Sewers Worse Than None," *EN* 177 (March 22, 1917), 193.

60. In 1923 Harrison P. Eddy wrote, "It is an interesting fact that human thought moves to extremes, like a pendulum. At first sentiment was strongly in favor of combined sewers; later it swung to the other extreme, favoring separate sewers and tending to discourage the construction of combined sewers. But improper use of separate sewers and a restudy of the economics of the two systems indicate that the time may be approaching when the pendulum will swing in the other direction." See, Eddy, "Present Status of Sanitary Engineering." U.S. Public Health Service, Reprint No. 831, from the *Public Health Reports* (April 20, 1923), 826-837.

14

Sir Joseph William Bazalgette (1819–1891): engineer to the Metropolitan Board of Works

Denis Smith

> Londoners who can remember the state of London and of the Thames about thirty-five years ago, before those vast undertakings of the Metropolitan Board of Works, the system of main drainage and the magnificent Thames Embankment, which have contributed so much to sanitary improvement and to the convenience and stateliness of this immense city, will regret the death of the able official chief engineer, Sir Joseph Bazalgette.
>
> *Illustrated London News*, March 1891

Although the family name is well-known, Bazalgette remains a somewhat shadowy figure in Victorian engineering. He was an eminent engineer, respected by his contemporaries and responsible for public works on an unprecedented scale in the largest city the world had ever known, yet there is no published biography. This may be explained by the absence of personal letters or a diary which would illuminate his family life or reveal relationships with his circle of friends and professional colleagues. This paper draws on contact with the family, on his published reports, on printed Minutes and on the technical press to assess both the man and his work in developing the virtually new field of municipal engineering.

The Bazalgette family is of French origin and Sir Joseph's Grandfather, Jean Louis Bazalgette (1750–1830), was born at Ispagnac in southern France. Jean Louis arrived in England in 1784, via the West Indies, a wealthy man. He acquired British citizenship in October 1792 when living in Grosvenor Street, London.[1] Jean Louis married three times and from his first marriage Sir Joseph's father, also Joseph William, was born. (see Fig. 1). Jean Louis died on 6 February 1830 at Eastwick Park, Bookham, Surrey, which he had purchased in 1809. Joseph William Bazalgette (1783–1849), Sir Joseph's father, entered the Royal Navy in October 1796, was engaged in the Napoleonic Wars, and was severely wounded in action in 1809 whilst boarding a French vessel. After a distinguished career he retired in July 1814 with the rank of Commander[2] and became the first Secretary of the Army and Navy Bible Society. Joseph married Theresa Philo Pilton by whom he had nine children; she died in June 1850.[3] In 1819 he was living in Enfield, Middlesex, and in 1831 moved to Hamilton Terrace, Regent's Park,[4] where he died in 1849. He is buried in Brading Churchyard, Isle of Wight.

Joseph William Bazalgette (1819–1891), the fourth child and only son of Joseph and Theresa, was born in Enfield on 28 March 1819. Little is known of Joseph's childhood although we are told that he was educated privately.[5] By the age of nineteen he was 'pursuing a course of study to qualify him for the profession of a civil engineer'.[6] On 20 February 1845 Joseph married Maria Keogh at St. Margaret's Westminster. Maria was also born in 1819, at New Cross, County Wexford. Her father, Edward, was a magistrate of Kilkenny but, at the time of the marriage, was described as a 'Gentleman of Ham, Surrey'. In October 1845 Joseph and Maria moved to a newly-built house at 12 Westbourne Park Road (now No. 28). By 1851 they had moved to Morden in Surrey where they and their young family occupied a house in Morden Lane (now Central Road). The 1851 Census lists Joseph as 'civil engineer, born Enfield, aged 32', together with Maria his wife, their sons Joseph William (aged 5), Charles Norman (aged 4) and Edward (aged 2) all born in London, and a daughter Theresa Philo (aged 7 months) born in Morden.[7] Joseph and Maria were eventually to have ten children. The eldest, Joseph William, went into the Army, Edward became a civil engineer and other descendants followed in the profession. Charles Norman became a lawyer and embraced his father's interests to the extent that he wrote on the legal aspects of municipal engineering.[8] In 1873 Joseph built a house, called 'St. Mary's', in Arthur Road, Wimbledon, opposite the parish church. He had been Rector's Churchwarden at St. Lawrence, Morden and he was also Churchwarden at St. Mary's, Wimbledon. Sir Joseph died in the Wimbledon house on Sunday 15 March 1891 after a brief retirement and is buried in the family vault in St. Mary's churchyard.

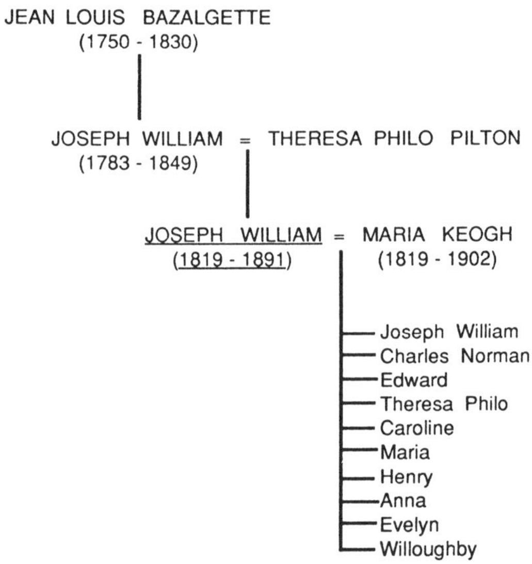

Fig. 1. Part of the Family Tree.

BAZALGETTE'S ENGINEERING CAREER

Before considering Bazalgette's works with the Metropolitan Board, it is important to consider his early career and set it against the background of sanitary reform which characterised the middle years of the nineteenth century. The resulting legislation provided the appropriate administrative structure and sufficient funds for massive civil engineering schemes in public health works.

Bazalgette began his engineering career in 1836 when he became an Articled Pupil of John McNeill, the eminent Irish civil engineer. Bazalgette was Resident Engineer for him on land drainage works at Loughs Foyle and Swilley in Londonderry. This work was the subject of the young Bazalgette's first paper to the Institution of Civil Engineers,[9] of which he was elected a Graduate on 13 March 1838.[10] In 1842, aged only 23, he set up in private practice as a consulting engineer with an office in Great George Street, Westminster. He was elected a Member of the Institution on 17 February 1846 when he had:[11]

> served a regular period of pupilage under Sir J. McNeill, was for 2 years Resident Engineer on works in Ireland, 1 year laying out lines for the Railway Commissioners & has been upwards of 2 years in business for himself as a civil engineer.

During 1847-8 the pressure of work took its toll and he had a severe breakdown in health and left London to recuperate. Later he said 'I began my work when the great railway mania broke out, and nearly killed myself before I joined the old Commission of Sewers'.[12]

SIR JOSEPH WILLIAM BAZALGETTE (1819–1891)

THE COMMISSIONERS OF SEWERS

The consequences of inadequate drainage became serious as London expanded during the early Victorian period and small, local attempts to deal with the problems 'produced results as multifarious as they were discordant'[13] and central government intervention became inescapable by the 1840s. In 1847 a Royal Commission was established to 'Inquire whether any, and what special means might be requisite for the improvement of the health of the Metropolis, with regard more especially . . . to better house, street, and land drainage'.[14] They proposed amalgamating the seven existing drainage districts and brought into being the first Metropolitan Commission of Sewers under an Act of 1848.[15] The recognition of the need for a body with a broader view of London's sanitary problems marked the beginning of a new era. It made possible the full-time employment of salaried professional civil engineers on urban drainage issues with the consequent accumulation of expertise, design data and site experience.

Under the 1848 Act the Commissioners were responsible for all matters relating to drainage and flood prevention within an area including:[16]

> all such places or parts in the counties of Middlesex, Surrey, Essex, and Kent, or any of them, not more than twelve miles distant in a straight line from St. Paul's Cathedral in the City of London, but not being within the City of London or the liberties thereof.

The exclusion of the City reflects the disproportionate influence of the "square mile" in metropolitan matters, and an 1851 guide book to London wondered:[17]

> why the effete corporation of the city should be allowed to retain its privilege of exemption from the effect of every general measure for the public benefit; it is as though it were desired to form an Egypt of darkness in the midst of the enlightened progress of the rest of London.

The duration of each twelve-man Commission was limited to two years and there were six such Commissions between 1848 and 1855.

Having recovered from his illness, Bazalgette returned to London in 1849 and on 16 August was appointed 'Assistant Surveyor' to the second Commission at a salary of £250 a year. Thus began his engineering career in local government. The technical staff of the Metropolitan Commissions were based at No. 1 Greek Street, Soho, and the main drainage scheme, as implemented, was planned there. The City Corporation had their own Commissioners of Sewers and the two bodies obviously had to be in communication. The Surveyor (later Engineer) to the City was Colonel William Haywood (1821–1894),[18] who was appointed in 1846 and served until he retired in March 1894. Bazalgette and Haywood were to have a long professional relationship and a few disagreements. Apart from their salaried staff, both the City and the Metropolitan Commissioners occasionally employed consulting engineers including Robert Stephenson, I. K. Brunel, Sir William Cubitt, Thomas Hawksley and George Parker Bidder. By 1851, the Metropolitan Commission, under their Engineer, Frank Forster (1800–1852),[19] had clearly established the broad outline of London's proposed main drainage scheme, with its essential components of separate systems north and south of the Thames, and west-east intercepting sewers with remote outfalls to the east of London. However, Forster died, the victim of 'harassing fatigues and anxieties of official duties',[20] in 1852 when Bazalgette was appointed Engineer and remained so to the last two Commissions. The engineering team of the last Metropolitan Commission in 1855 comprised:

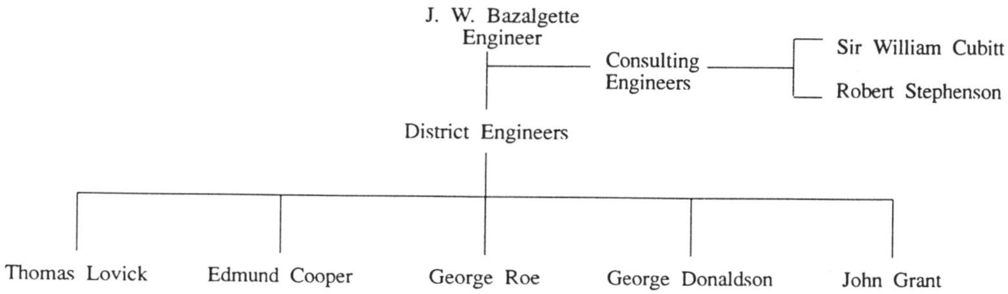

SIR JOSEPH WILLIAM BAZALGETTE (1819-1891)

One of the issues which drew Bazalgette and Haywood into close collaboration, and in fact, united them against a common enemy, was the proposed *Great London Drainage Company* to bring private capital into the funding of the drainage works. The promoters hoped to exploit a profitable agricultural fertilizer market. During 1853 these proposals occupied much of Bazalgette's and Haywood's time. The Company had Thomas Wicksteed (1806–1872) as their engineer and Haywood's diary for 29 March 1853 records; 'With Bazalgette to Leicester',[21] where they met Wicksteed and stayed three days. The working relationship between Bazalgette and Haywood, or of the Metropolis and the City, immediately before the formation of the Metropolitan Board of Works is revealed in letters written during 1853. On 18 March Bazalgette writes:[22]

Principal Office: 1 Greek Street
Soho Square

Dear Haywood,
I am to meet Mr. Stephenson and Sir Wm. Cubitt on Tuesday at 12 o'clock. Could you manage to meet me here say at 10 that morning & go down there with me. It will be an advantage our all meeting together, and though you would not attend officially but at my invitation it can do no harm and it will afford you an introduction which will help your transfer at the Instn.

That Bazalgette intended to keep Haywood abreast of Metropolitan developments is clear from his letter of 10 October:[23]

My Dear Haywood,
I have been driven on in the preparation of plans for works upon which we are to borrow money under our Consulting Engineers, step by step till the Southern drainage is designed, and the Hackney Brook on the North Side, without the opportunity of consulting you in the matter. I regret this has been so, but I have had no choice in the matter, for I have been acting directly under the instructions of Sir William Cubitt who has taken up the subject himself and has devoted a great deal of his time to it; and I have been so fully engaged that I have scarcely had breathing time till now.

By November 1853 the effect of the proposed underground railway on the drainage plans had to be considered, and on 3 November Bazalgette writes to Haywood:[24]

My Dear Sir,
I have just written to Mr. Fowler to request him to lay his views upon the diversion of the sewerage by the North Metropolitan Railway before me as soon as possible. This we must have I think before we can progress with the main drainage.

The foregoing serves to illustrate Bazalgette's considerable involvement with the London Main Drainage from 1849 up to the formation of the Metropolitan Board of Works in 1855.

THE METROPOLITAN BOARD OF WORKS

The Metropolitan Board of Works was created by the Metropolis Management Act of 1855.[25] The Board acquired powers on 1 January 1856 and were responsible for the built environment within the capital's area of 117 square miles. The technical journal *The Engineer* was launched in the same week as the Board and said 'we confess that we take a great interest in the proceedings of this Board'[26] and its Editorials provide an interesting critical commentary on the progress of the works. In March 1856 it described the Board's function:[27]

> the public works of the metropolis are in its charge, metropolitan improvements are its special function; under the earth and above the earth its authority is now established . . . metaphorically speaking, the Board is appointed physician to the metropolitan organism, and has accepted the duty of restoring it to health and promoting its future growth, of giving strength to its muscular, and vitality to its arterial system, roundness to its limbs, and beauty to its face.

The Board, comprising some 45 members, had no offices in which to meet and for the first three meetings hired rooms in Burlington House. Neither did the Board have any technical staff and, in January 1856, decided that:[28]

> In order to prevent the injurious consequences to the public which might result from any interruption in the supervision of sewers, Mr. Bazalgette be requested to take charge of all the main sewers and other works . . . with power to obtain such temporary assistance as may be necessary to discharge such duty until the Board otherwise direct.

So Bazalgette continued to act in this temporary capacity from his old office in Greek Street.

SIR JOSEPH WILLIAM BAZALGETTE (1819–1891)

The press took a keen interest in the appointment of an Engineer to the Board emphasising the rivalry between the City and the new Metropolitan body. *The Observer* thought the City aimed to gain maximum power 'so that it might be spared any innovation' and that 'Their great object is to get the engineer to the late Commission of Sewers re-appointed. Hence their anxiety to have him nominated for a month. In this they have succeeded, and have certainly stolen a march upon the metropolitan party'.[29] The extent of public interest in the selection of an Engineer can be gauged from their concluding remarks:[30]

> Upon the appointment of this officer hundreds of thousands of the ratepayers' money depend; and they, therefore, look to their deputies to weigh carefully all the pros and cons of the case, and to select only the very best possible man that can be procured, and pay him most liberally and ungrudgingly. The proceedings of the next fortnight are pregnant with the fate of the great measure now in operation.

A week later the Board formally decided to appoint an Engineer, that 'advertisements as to such election be inserted in all the daily papers,'[31] and that the election should take place on Friday 25 January at 12 o'clock. On the Sunday before the election *The Observer* remarked:[32]

> The number of candidates will doubtless be considerable but the contest, most probably, will virtually lie between Mr. Bazalgette and Mr. Robert Rawlinson . . . Of Mr. Bazalgette it is unnecessary to say anything, as his saying and doings have but too frequently been discussed in the public press . . . Between the two candidates the Metropolitan Board of Works must now judge. It is to be hoped that they will select the right one.

In the event nine candidates applied for the post.[33] On the day, the Board proceeded, by a show of hands, to reduce the list to four. The letters of application and testimonials were then read. At this point the Secretary announced that he had that morning received a letter claiming that Bazalgette's statements in his application relating to his work on the Tame Valley Canal, Portsmouth Dockyard, and other works were 'at variance with the facts'.[34] Bazalgette was called in to address the Board, and must have been convincing since at the next vote he was elected Engineer outright. He was appointed at a salary (including travelling expenses) of £1000 a year, and he quickly had three Assistant Engineers appointed. They were John Grant, Thomas Lovick, and Edmund Cooper, who had each worked with Bazalgette at the Metropolitan Commission of Sewers. The salary of each Assistant Engineer was £350 a year plus £50 for travelling expenses. Grant was put in charge of the area south of the Thames and Lovick and Cooper shared the northern area.

THE ENGINEERING WORKS OF THE BOARD

In addition to main drainage the Board was also responsible for a range of other works concerning the built environment of London. These included street improvements, street lighting, new roads, river bridges, ferries and tunnels, flood prevention, and the Metropolitan Fire Brigade. There was also much Parliamentary business in connection with the Board's own Bills and with those of others. The Parliamentary deadline for submission of proposals for consideration in the ensuing session was 30 November and Bazalgette said 'We were working generally night work in November to get the plans and book of reference ready for deposit on the 30th.'[35] When living in Morden, Bazalgette said 'I often used to drive down there from my office in a cab at twelve or one o'clock in the morning'.[36] Even then the pressure of Parliamentary work did not ease as Bazalgette added:[37]

> It is my duty every year after November, as soon as possible after the bills are deposited, to report as to the effect upon the public interests of all private enterprises contemplated in the coming session; then to appear before Parliament and give evidence if necessary on behalf of the public.

He was thus involved with Railway and Tramway Acts, the Gas and Petroleum Acts, Hydraulic Power, Waterworks and Docks Acts, among many others.[38]

The Board met every Friday and Bazalgette made an engineering progress report on the first Friday of each month. He had to collect data, produce design proposals and submit them to the Board. The proposals having been agreed, the Engineer would produce the contract drawings and specification and advertise for public tenders. The documents were printed and issued for a fee of five pounds, 'to secure that none but *bona fide* tenderers shall send in their tenders'.[39] Bazalgette added:[40]

> I have always adopted the plan of having a lithographed circular sent out to every person who is supplied with plans in order that every contractor may have the same information before him before he tenders.

SIR JOSEPH WILLIAM BAZALGETTE (1819-1891)

When tenders were received a meticulous procedure was always followed:[41]

> The tenders are deposited in a box on the Thursday afternoon before a Board meeting. They are opened at the Board meeting, or they are opened by the clerk before the Board meeting and handed to the Chairman, who announces them at the public board.

The Board did not always accept the lowest tender and, when questioned, Bazalgette replied:[42]

> it stands to reason that if there is a contractor whom they well know, and whose tender is not materially above the lower one, they will naturally prefer it. It is a very great mistake to employ a contractor who has not the means of carrying out his contract thoroughly. It always leads to constant wrangling, difficulty, stoppage, and very often eventual failure.

Having let the contract, the Engineer's team supervised and measured the work for certified interim monthly payments. A list of the principal constructional contracts is given in Appendix A. In estimating, Bazalgette allowed 10% for contingencies and on completion the contractor sent in an account. When there were disagreements, Bazalgette said:[43]

> I then have the assistant engineer and the clerks of works with the contract drawings and specifications before me, and the contractor and his agents, and I hear what each has to say, and look to the specifications and drawings and determine those points. After that is done the account is sent to the accountant, certified as to quantities, and the accountant compares the prices with the schedule of extra works . . . and eventually signs it as correct. Then I sign the certificate for the Board.

Being in the service of local government and responsible for spending exceptionally large sums of public money, Bazalgette spent a great deal of his time explaining his schemes and procedures to lay members of Boards and Commissions. Nevertheless, a prodigious amount of construction was successfully completed during the thirty-three years of the Board's existence. It is not possible here to describe in detail the extensive engineering works which were to shape the London we now know and the following account of the principal works uses, as far as possible, Bazalgette's own words.

THE MAIN DRAINAGE

Concern for public health was the primary motive in setting up the Board and, in January 1856, Bazalgette dramatically described the nature of the problem which it had inherited:[44]

> the whole of the sewage passed down sewers from the high ground at right angles to the Thames into the low grounds adjoining the Thames, where at high water it was pent up in the sewers, forming great elongated cesspools of stagnant sewage, and then when the tide went down and opened the outlets that sewage was poured into the river at low water at a time when there was very little water in the river.

Furthermore, this sewage 'kept oscillating up and down the river, while more filth was continually adding to it, until the Thames became absolutely pestilential'.[45] On 8 February 1856 the Board, 'impressed with the necessity of at once proceeding with the works necessary for the complete interception of the sewerage'[46] asked Bazalgette to report as soon as possible. His report, of 3 April, was for an interception system which was fundamentally that of the last Metropolitan Commission. Then began an enormously protracted public discussion of this and other schemes involving central government, referees, and the press. Eventually, in November 1857 the Board asked for a final report by Bazalgette in conjunction with G. P. Bidder and T. Hawksley as consultants and this was published in April 1858.[47] Bidder and Hawksley were paid £2754 16s. 4d. for 'professional services and disbursements'.[48]

In August 1858 the Board obtained its enabling Act[49] and site work began on the northern mid-level sewer on 31 January 1859.[50] The difficulties of constructing sewers in dense urban areas made cost-control a problem and as early as May 1859 Bazalgette was taken to task by a member of the Board about cost variations and he replied:[51]

> I have learned, after estimating and constructing some hundreds of miles of sewer in London, that it is not possible to frame standard estimates which shall be generally applicable to other cases, and any attempt to do this must prove delusive and lead to great and serious errors.

Bazalgette received a vote of confidence from the Board and in July received his first salary increase.[52]

The brick-built intercepting sewers were laid to a fall of 2 feet per mile from west to east and it was, therefore, necessary to lift the sewage into the outfall sewers with pumping stations sited in east London. Progress was the more rapid south of the river and in February 1859 it became necessary to advertise for tenders for the engines, boilers, and other plant for Deptford pumping station. Bazalgette had only limited experience of this type of work and it was decided to invite a committee of engineers to examine and comment on the designs sent in. This eminent ad hoc committee comprised Bazalgette, Robert Stephenson, Joshua Field, John Penn, Thomas Hawksley and George Parker Bidder,[53] and in July the Board thanked those who 'with a degree of liberality and public spirit . . . gratuitously undertook this important and laborious duty'.[54]

The work progressed towards completion, at least on the south side, during the summer of 1864 and Bazalgette's eighth annual report of 4 July captures something of the scale and character of the task:[55]

> It is with satisfaction that I am enabled to state that fair progress has been made in the construction of the Main Drainage Lines of Sewers, and with numerous and extensive works in course of execution therewith at the several Pumping Stations, Reservoirs, Outlets, &c. during the past year, the weather having, as in the preceding year, been favourable to their progression, without any material interruption from long continued frosts or heavy rains. It is also satisfactory to me to be able to state that, whilst a large amount of tunneling has been completed on the South Side of the River, under canals, railways, houses, and through treacherous soils, filled with water, from the precautions taken by the several Contractors for the safety of their respective works, and from the skill and judgement evinced in their execution, no one section of these works has failed, whilst the damage to property, allowing for the nature and magnitude of the work, and the extent of the Metropolis traversed by it, has been unimportant, and the casualties to the workmen not numerous, and with but few of a fatal character.

In 1865, at the peak of the drainage works, the Engineering Department comprised:

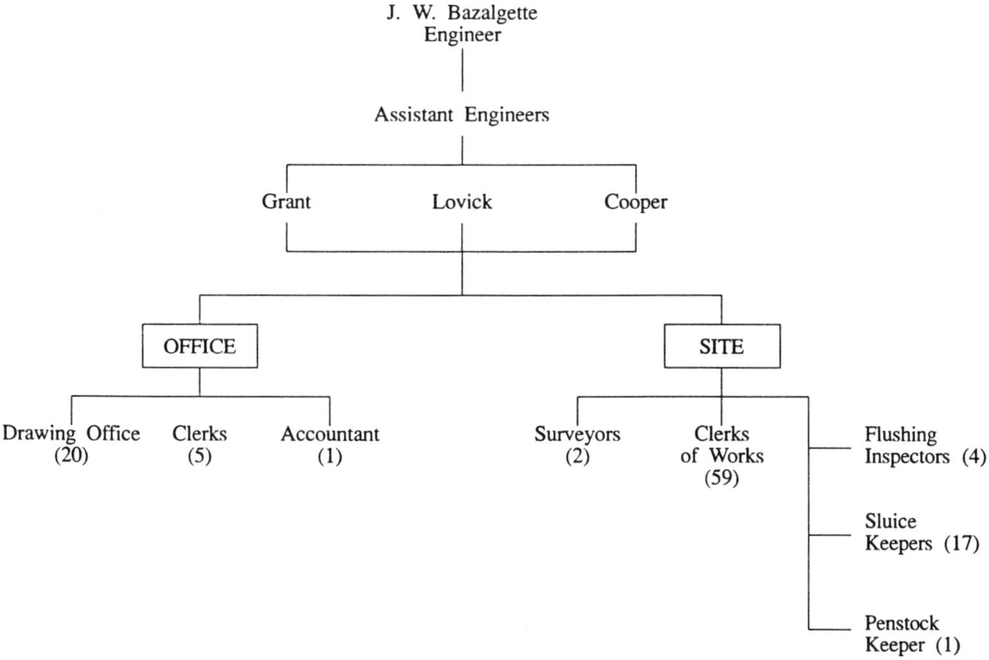

To cater for the interest of those concerned, annual visits to the drainage works with groups comprising Board members, the City, Parliament, and District Vestries were arranged. Such a visit was made on Monday 25 July 1864 when 'The party numbered between 500 and 600 strong, and were conveyed by two steamboats, which started from Hungerford early in the morning'.[56] These visits foreshadowed the ceremonial opening of the drainage works. This took place on Tuesday 4 April 1865 when amongst others, the Prince of Wales, Members of Parliament, the Archbishops of Canterbury and York, and the Lord Mayors of London and Dublin 'proceeded by steamboat down the river',[57] where:[58]

> the Royal party landed at the Northern Outfall, at Barking, and after a brief inspection of the works at that place resumed their passage to the Southern Outfall at Crossness, where the general company had already assembled . . . they were conducted over the works; after which the Engineer explained the general principles and engineering details . . . the four pumping engines were then succesively set in motion by His Royal Highness the Prince of Wales, which completed the opening of the works . . . the company then partook of luncheon.

After lunch, the Prince proposed 'Success to the great national undertaking' and congratulated 'Mr. Thwaites, the Chairman of the Metropolitan Board of Works, and the eminent and skillful Engineer, Mr. Bazalgette, upon the great success which they have achieved'.[59] The Crossness works were designed to impound sewage in a 6½ acre covered reservoir until it was discharged into the Thames on the ebb tide.

In July 1866, perhaps in recognition that the bulk of the drainage works were complete, the Main Drainage Committee of the Board was re-designated the 'Works and General Purposes Committee'.[60] Work north of the Thames was delayed by the complication of accommodating the low-level sewer in the Victoria Embankment. But Abbey Mills Pumping Station, the Northern Outfall Sewer, and the Beckton Outfall works were completed by 1868. The fourth, and last of the original sewage lift works was the Western Pumping Station in Pimlico, opened in 1875. Technical details of the pumping stations are given in Appendix B.

Once the system was commissioned, Bazalgette's department had a continuing responsibility for staffing, maintenance and replacement, and for renewable contracts for coal and other supplies. The 80 miles of intercepting sewers, and the two outfall sewers, were maintained by the Board's engineers and although the other sewers were managed by the local District Boards and Vestries, Bazalgette explained that:[61]

> in order to make it one complete system, they are bound to send in plans for any sewers they propose to construct to the Metropolitan Board of Works and get its approval of those plans. By this means one complete system is always maintained although the works are constructed by a number of different district boards.

The system worked well and was adequate for about twenty years until the growth of London made the Board explore new means of treating and disposing of the sewage at Crossness and Barking. In 1888 Bazalgette remarked:[62]

> the subject of the treatment of sewage by various chemical agents has occupied the attention of the Board, and their chemical advisers, for a great number of years.

Precipitation works were planned for the outfalls and Bazalgette had experimented with pressing the residual sludge to reduce its bulk, but having failed to find a market for it as fertilizer said, 'We are having hopper vessels constructed for taking the sludge right out to sea'.[63] But this was all too late; it was in the last days of the Metropolitan Board of Works and, with Bazalgette's retirement imminent, the full realisation of these plans fell to the London County Council.

THE THAMES EMBANKMENTS

A long-cherished plan was achieved when the Board built the Victoria, Albert and Chelsea embankments. They succeeded, where others failed, because sufficient funds and powers were made available to them.[64] The issue was discussed at length in the early 1860s when it was suggested that the embankment works would be better managed by special Commissioners rather than by the Board. However, the need to integrate the low-level sewer into the works ensured control by the Board. Bazalgette's department undertook design and supervision of this major project although he later confessed, 'I get most credit for the Thames Embankment, but it wasn't anything like such a job as the drainage'.[65]

The Victoria Embankment is a mile and a quarter in length running between Westminster and Blackfriars bridges. It follows the line laid down by James Walker in 1840 and was the most complex

SIR JOSEPH WILLIAM BAZALGETTE (1819-1891)

of the three built. Protracted negotiations were necessary with private property owners, coal wharf operators, the City Gas Company, and the railway authorities before work could begin in 1864.[66] Bazalgette had to reconcile matters he considered to be 'of small importance to a private company, but of great consequence to the public interests'[67] and in April 1864 a barrister at the Temple confided to his diary, 'Never more after this year will any one look down from our walks straight into the tide. Oh Public Good, what private wrongs are committed in thy name.'[68] There were three contractors[69] and difficulties arose with the Thames Conservancers. The contractors were required to take a certain amount of material dredged from the Thames as fill behind the new wall and the Conservancers found them reluctant. Bazalgette was involved in numerous meetings and massive correspondence to keep the works running smoothly. The supply of granite for facing the wall also proved difficult. Bazalgette had specified stone from the Dalbeattie quarries but explained, 'We could not get it in fast enough, and eventually we had to get it from a number of different quarries'.[70] The embankment was opened by the Prince of Wales on 13 July 1870. The massive scale of this work is given by the quantity of material used:

Granite	650,000 cubic feet
Brickwork	80,000 cubic yards
Concrete	140,000 cubic yards
Timber (for cofferdam)	500,000 cubic feet
Iron caissons (for cofferdam)	2000 tons
Earth filling	1,000,000 cubic yards
Excavation	144,000 cubic yards
York paving	125,000 square feet
Broken granite	50,000 superficial yards

The Albert Embankment between Westminster and Vauxhall bridges was constructed under an Act of 1863.[71] Work began in September 1865 and it was opened to the public in May 1868. The contract was let to William Webster and the Resident Engineer was John Grant. The reclaimed land provided a site for the building of St Thomas's Hospital. The Chelsea Embankment, about three-quarters of a mile long, was also constructed by Webster under the Board's Act of 1868.[72] Work began in July 1871 and was completed in May 1874. The engineering of both these works is generally similar to that of the Victoria Embankment. Bazalgette summarised the contribution of the three embankments by saying they 'extended for a distance of 3½ miles along the river, and they had reclaimed about 52 acres of land.'[73]

FLOOD PREVENTION AND THE RIVER

The Thames River, Prevention of Floods Act of 1879 imposed on the Board the duty of implementing its requirements and Bazalgette considered this 'one of the most difficult and intricate things the Board have had to do'.[74] He described the scale of the project as:[75]

> 40 miles of river frontage, and every wharf has had to be dealt with and plans prepared carefully for each particular wharf after conference with the persons in occupation to see the nature of their business and how it was conducted and to devise those works in such a way that the business could be carried on with as little interruption as possible.

His system comprised 'moveable tide boards which have to be inspected and kept in order'.[76]

Another aspect of the régime of the river which involved the Board's Engineer was the claim made by the Thames Conservators that mudbanks in the Thames at Crossness were the result of deposits from the sewer outfalls. The Conservators were responsible for dredging, but demanded that the Board should remove the obstructing mudbanks. The case went to arbitration in the spring of 1880 and both parties were represented by Counsel. The arbitrators were Captain Douglas Galton, appointed by the Conservators, and F. J. Bramwell, appointed by the Board; Sir Charles Hartley was appointed Umpire by the Board of Trade. The inquiry lasted 25 days and in May 1880 the Arbitrators reported in favour of the Board:[77]

> the three banks cannot be said . . . to have arisen from the flow of sewage at the outfalls; and we are, therefore, of opinion that the Metropolitan Board of Works should not be called on to remove, or contribute any portion of the expense of removing, the three banks, or any of them.

The two parties each had to bear their own costs and the press thought it possible that 'Sir Joseph Bazalgette bears a heightened reputation as an engineer in consequence of the decision'.[78]

RIVER CROSSINGS

In 1854 a Select Committee had reported on the bridges over the Thames[79] and the Board inherited a growing traffic problem where 'the greatest obstructions in the way of distributing the traffic are the tolls and the insufficient approaches'.[80] But it was not until 1877 that the Metropolis Toll Bridges Act[81] enabled the Board to purchase the river bridges from their private owners and free them from tolls. Bazalgette observed, 'Ever so small a toll tends very much to restrict the use of a bridge',[82] and:[83]

> The loads going over a bridge when under the control of a toll keeper with gates and so forth are very different to what they are when you throw it open free, and it is crowded as much as possible.

Twelve bridges were to be freed; Hammersmith, Putney, Wandsworth, Battersea, Albert, Vauxhall, Lambeth, Waterloo, Deptford Creek and Chelsea, together with the Charing Cross and Cannon Street railway footbridges. They were all owned by bridge companies with the exception of Chelsea, built by the Commissioners of H.M. Works, and the footbridges owned by the South Eastern Railway Company. Bazalgette described his task:[84]

> Before the bridges were purchased it was my duty to see what the condition of the bridges was, and what in my judgement would have to be done to them, and present that to the arbitrator to show him if the Board purchased the bridge in its then condition what expenditure would have to be laid out on it in all probability to make it fit for public use.

Bazalgette published his very thorough report in May 1888.[85] Having agreed the value, the Board acquired the bridges for just under £1,500,00.

Work was done on nearly all the bridges, beginning with the underpinning of Rennie's Waterloo Bridge. The acquisition led to three major bridge works to Bazalgette's designs, namely the replacement of Putney, Hammersmith and Battersea bridges.[86] The bridge at Putney was of timber and was built in 1729; Bazalgette replaced it with the present five-arched granite structure, opened on 29 May 1886. In February 1882 the Fulham District Board complained that William Tierney Clark's suspension bridge at Hammersmith was 'dangerous and unfit for general traffic'.[87] Bazalgette examined the old structure and decided to retain the abutments but to rebuild the masonry towers in metal and replace the wrought iron chains with steel links of a new design; together with a new deck, the whole was estimated to cost £80,000. The new bridge was opened on 18 June 1887. At the same time, Battersea Bridge was 'an old wooden structure, unsightly in appearance, inconvenient to passengers over it, and still more to steamboats under it'.[88] To replace it, Bazalgette designed the present five-span arch structure with iron ribs. It was still under construction when the Board was replaced by the London County Council and Bazalgette lived just long enough to see it opened in 1890.

After the up-river bridges had been freed from tolls using Metropolitan funds, demands arose from east-Londoners for river crossings and Bazalgette pointed out that whilst three-fifths of Londoners had access to twelve bridges above London Bridge, the two-fifths below that bridge had none. Various proposals were discussed, resulting in Bazalgette designs for a bridge at the Tower, a ferry at Woolwich and a tunnel at Blackwall—his last major project before retirement.

Bazalgette produced three designs for Tower Bridge in 1878[89] but his preferred scheme was for a 'braced arch bridge, crossing the river by one span of 850 feet'.[90] Moreover, he proposed using steel and justified this by saying:[91]

> Steel is a new material to use in a bridge of such a span, and in point of fact is the only material which would render the construction of such a bridge practicable, taking the whole thing in one span.

However, the City Corporation effectively excluded Bazalgette and the Board by proposing to pay the whole cost of the bridge from their wealthy Bridge House Estates Trust.

The Board obtained an Act for constructing Blackwall Tunnel in 1887[92] and Bazalgette said of the possible cost: 'the parliamentary estimate was one million and a quarter . . . I can only give the works. I cannot state what the property will come to'.[93] His design comprised 'three subways, two 25 feet in diameter, and one 15 feet' and he feared:[94]

> there is very considerable difficulty to be encountered in constructing that tunnel. It goes through treacherous ground and water. We shall do it by pneumatic pressure.

SIR JOSEPH WILLIAM BAZALGETTE (1819-1891)

Although the Board's demise was imminent, it pressed on and was about to advertise for tenders when wound up in the Spring of 1889. The tunnel was eventually constructed, to a different design, by the London County Council. However, the Metropolitan Board did initiate the other east London river crossing and in July 1888 Bazalgette could say 'our steam ferry at Woolwich we are constructing will cost £113,000'.[95]

STREET IMPROVEMENTS

When the Board took control in 1856, connections between the main-line London railway termini entailed journeys in horse-drawn road vehicles across the Capital. New traffic routes were urgently required. Bazalgette took up the challenge and explained his role: 'Each street improvement being determined upon, it is my duty to lay out the line of street, and consider the width of street, and report upon it and prepare it'.[96] He worked in conjunction with the Architect's and Solicitor's departments to acquire the necessary property. This all had to be done before work could begin: 'the contractor must have room to put his men into the work; we must have the ground cleared to lay out the street in fact'.[97] Garrick Street was the first new street constructed by the Board and Bazalgette seized the opportunity of making provision for underground services:[98]

> we have introduced the system of subways in the main thoroughfares in order that the gas and water pipes may be laid in the subways under those thoroughfares and the streets may not be pulled up from time to time afterwards. It will only do in important main streets because it is costly, but it is a great improvement.

Many new streets were built from the 1860's until the end of the Board's period in 1889. The principal examples and opening dates include; Southwark Street (January 1864), Queen Victoria Street (November 1871), Northumberland Avenue (March 1876), Shaftesbury Avenue (June 1886) and Charing Cross Road (February 1887). The Board did not take on subsequent responsibility for new roads and 'All new streets are laid out by the Board, paved and completed, and then handed over to the district boards'.[99] Bazalgette explained that 'Only the Thames Embankment and the approaches'[100] were maintained by the Board. Street lighting by gas and, later, electricity also came within the power of the Board. Bazalgette was involved with the first significant step taken by a public authority in connection with electric lighting. In October 1878, he got his Committee to accept an offer from a firm in Paris to instal an experimental system on the Victoria Embankment. As a result:[101]

> The Victoria Embankment was illuminated by electricity for the first time on 13 December 1878. The lights, which were along the river wall . . . consisted of 20 Jablochkoff lamps spaced about 45 yards apart. . . . just west of Charing Cross Bridge, was a wooden shed containing a semi-portable steam engine . . . capable of developing about 60 I.H.P. at 160 R.P.M. . . . It was belted to a countershaft from which were driven a direct current Gramme dynamo at 650 R.P.M. and its separate exciting dynamo at 700 R.P.M.

The arrangement was terminated in 1884 and the Embankment reverted to gas lighting.

Bazalgette's department had reported on the plans for about 3000 new streets and the total cost of street improvement works was about £14,000,000 and, as he correctly observed, 'Had we not kept pace with the increased want of communication we should have been in a state of block in our streets.'[102]

BAZALGETTE AS CONSULTING ENGINEER

An interesting aspect of Victorian civil engineering in local government was the freedom enjoyed by the salaried professional engineer to engage in private, fee-earning, consulting work and even to take private Articled Pupils at a premium. In August 1859 Bazalgette reported to the Board that he had been asked to:[103]

> receive a pupil who might gain experience upon the Main Drainage Works now being executed, and whose services would shortly become available upon those works, and would thus effect a saving to the Board.

He was granted permission to take the pupil and receive the premium. Through one such pupil, who joined Bazalgette on 1 January 1865 and stayed four years, we learn something of his Articles and of life and work in the Engineer's office at Spring Gardens.[104] He was H. P. Boulnois, who went on to a successful career in municipal engineering.[105] His mother paid Bazalgette a premium of £420, and his Articles describe Boulnois as 'an infant of the age of 19 years', and as Bazalgette's 'pupil and apprentice

SIR JOSEPH WILLIAM BAZALGETTE (1819–1891)

in the profession, business, or employment of a civil engineer'[106] At that time there were five other pupils in Bazalgette's office all working on the Main Drainage and the Thames Embankments. Portland cement was extensively used in these works and Boulnois reminds us of research undertaken by the Board which was to make the material more widely adopted by engineers:[107]

> I remember that at one period of my pupilage I assisted in carrying out some experiments, at Messrs. Eastwood's works, into the strength of concrete beams of various descriptions, by subjecting them to certain strains and pressures, carefully noting the results, which were handed to Mr. Grant.

The pupil also recalled that 'The head of the drawing office was a Mr. Gunyon, an extraordinarily clever man, and at the same time of a most kindly and considerate disposition'.[108] He also remembered that the surveyors' chainmen:[109]

> were not on the permanent staff, but were merely "odd" men who sat in a corridor until wanted, and were then paid a shilling an hour whilst the job lasted. Some of them would thus be out of work for days, and as they often looked half starved they were the recipients of many small tips from some of us.

The term of his pupilage with Bazalgette expired in 1869 but:[110]

> he kept me on for a year longer as an assistant on various private jobs that he had in hand. Amongst others I surveyed and levelled the Cray Valley for a joint sewerage scheme, and also Oxford and Weston-super-Mare, but whether these schemes were carried out or not I do not remember.

Throughout his career as Engineer to the Board Bazalgette was consulted and reported on the drainage of towns at home and abroad as well as on many other matters. The drainage schemes were mostly prepared by the local authority's engineer and Bazalgette was asked for his comments. Occasionally, the Local Authority would hold a competition for a design for a drainage system and Bazalgette was asked to judge the entries. The Minutes of the Board are punctuated with Bazalgette's requests for permission to engage in such activities. Appendix C gives an idea of the scope of his work. One or two examples will serve to illustrate the range of his clients. In July 1862 he was asked to serve as Juror for the prizes awarded at the International Exhibition in London.[111] In September 1864, in connection with the Metropolitan District Railway, he reported that as:[112]

> the superintending so much of the Railway as shall be constructed through the Thames Embankment . . . was cast upon the Engineer to the Board . . . Mr. Fowler, the Engineer to the company, had agreed, subject to the approval of the Board, that the Engineering Fee to be paid by the Company for this portion of the work should be divided between them, in equal moieties.

The Board agreed. In July 1867 he arbitrated a disputed account between Brighton Corporation and a contractor.[113] In November 1868 he was consulted by the Crown Agents for the Colonies and asked to report on the drainage of Port Louis, Mauritius[114] and, early in 1869, Bazalgette sent his son Edward out there to make surveys.[115] On another occasion, in January 1869 the Board resolved:[116]

> That the sum of £210 forwarded by the London and South Western Railway Company to the Board, in respect of the Engineer's professional fee for examining, revising, and reporting upon their plans of proposed sewers and bridges in connexion with their Kensington and Richmond line, be received by the Board, and handed to the Engineer, in pursuance of the resolution of the Board of the 19th. February last, allowing him to charge a fee for the works in question.

MAIDSTONE BRIDGE

The replacement of the old bridge over the Medway provides further evidence of Bazalgette's external consultancy work. He was introduced to the proposal in September 1875 when 'requested by the Charity Commissioners to advise them respecting a proposed expenditure of their surplus funds in forming an improved communication across the river Medway'.[117] The Board again agreed to Bazalgette's involvement. The Town Council promoted the work and, in January 1877, resolved to apply for an Act. Bazalgette was retained as Engineer and he joined in the public debate as to whether the bridge should be of iron or masonry. However, in July Bazalgette sent in his Report saying:[118]

> the bridge should be constructed of stone, the abutments, piers, and arch stones, cornice and coping being of granite; and the spandrils, wing walls and parapets of Kentish rag . . . Its interior and foundations would be of concrete. In form it would consist of three flat segmental arches . . . The span of the centre arch is 54 feet, and that of each of the side arches 47 feet 6 inches.

Bazalgette estimated the cost at £32,000 of which the Wardens of Rochester Bridge would contribute

SIR JOSEPH WILLIAM BAZALGETTE (1819-1891)

half. The Maidstone Bridge Act[119] was passed in August 1877 and the Local Board advertised for Tenders for the:[120]

> ERECTION and COMPLETION of the proposed STONE BRIDGE . . . and the maintenance thereof for twelve month after the completion of the works. The Plans, Sections, Specification, Bill of Quantities, and Form of Tender . . . can be had on payment of £3 3s., at the office of HENRY LAW, Esq C.E., 18 Adam Street, Adelphi, and from whom, or from Sir Joseph Bazalgette, Spring Gardens, London further particulars may be obtained.

Even at this late stage the iron versus stone argument continued and the Mayor published a letter explaining the bridge committee's dilemma and pointing out that if they:[121]

> were now to throw over SIR JOSEPH BAZALGETTE; why, they would have to pay him his commission for the work, in full probably, as if it had all been executed, and so sacrifice £2000 or £3000 . . . You would lose the services of an eminent Engineer, and instead begin afresh with an engineer of a lower order.

However, four tenders were received on 3 October 1877 and that of William Webster for £22,300 was accepted, 'as advised by Sir J. Bazalgette'.[122] The first pile was driven on 24 October 1877 and the bridge was opened with due ceremony on Wednesday 6 August 1879.[123] Bazalgette and William Webster received the balance of money due to them in June 1881 when the Bridge-building Account was closed.[124]

THE ODESSA CONTRACT

Bazalgette's relationship with a contractor currently working for the Board and with another engineer was called into question in 1863. Sir John Rennie had negotiated a concession with the Russian government for the drainage and paving of the city of Odessa. In February 1862 Rennie consulted Bazalgette about a suitable contractor for the work, suggesting George Furness, and offering Bazalgette a share in the profits of the concession for his advice. Bazalgette introduced Rennie to Furness, who was subsequently awarded the Odessa contract. A tripartite agreement was made between Rennie, Bazalgette, and Furness. Furness was to pay Rennie 5% of the gross contract price and Rennie agreed to share 3½% equally with Bazalgette and the remaining 1½% with others. The Vestries were angry and the Board unhappy about this and the Board set up a Committee of Inquiry comprising the Chairman and 24 Board members. They reported, unanimously, in November 1863 in the following terms:[125]

> the Committee disapprove of their Engineer having been concerned in any transaction with any contractor without the consent of the Board, and in contravention of the terms of his appointment, although in respect of works to be performed in a foreign country.
>
> That they find that Mr. Bazalgette, from the first, expressed his readiness to relinquish all his interests in the Odessa Contract, which course he has adopted.
>
> That, after a most anxious inquiry into the whole facts of the case, they are satisfied that there has been nothing reflecting in any way on Mr. Bazalgette's professional or personal honor, and that he has throughout discharged his duty, both to the Board and their Contractors, with ability, impartiality, and integrity.

At the beginning of this affair, Furness was constructing the Northern Outfall Sewer and, during the Inquiry, was awarded contracts for the Northern Outfall reservoirs and the Victoria Embankment. Bazalgette's reputation did not suffer as a result of this episode. He was made a Companion of the Bath in 1871, was Knighted by Queen Victoria at Windsor in 1874 and elected President of the Institution of Civil Engineers in 1884.[126]

THE LAST DAYS

As the end of the Metropolitan Board drew nearer, Sir Joseph decided to retire and, on 8 February 1889, he wrote to the Board:[127]

> My Lord and Gentlemen,—After forty years of arduous and responsible work in the public service, I feel that the time has arrived when I may request to be allowed to retire from the office I hold under your Board, and, in doing so, I would ask you to award me such pension as my services appear to you to deserve.

A week later the Board resolved to pay him a 'retiring allowance' of £1333 6s. 8d. per annum (two-thirds of his salary), 'on condition that he should resign his said office on the 25th. March next'.[128] His retirement was regarded as 'something more than an ordinary event . . . in the management of the

Metropolis'.[129] Thomas Lovick and several other principal assistants in the Engineer's Department also resigned at this time. In July 1889 the newly-formed London County Council appointed Joseph Gordon, previously Engineer and Surveyor to the Borough of Leicester, Chief Engineer to the Council at a salary of £1500 a year.[130]

Sir Joseph continued to live in his house in Wimbledon and retained some consultancy and Parliamentary business. In August 1890 he was interviewed at home and the resulting publication gives a rare and fascinating glimpse of Sir Joseph's home life in retirement. His house was 'a large square-built, roomy-looking dwelling, with lawn and shrubberies in front of it, and twenty acres, or so, of land behind it'.[131] He was described as doing 'a little bit of farming—keeps one or two cows, and makes a little hay in summer'.[132] Sir Joseph said, 'I ride a good deal—usually two or three hours a day. I find it splendid exercise for counteracting the effects of a sedentary life'.[133] Of Sir Joseph, the reporter said:[134]

> Among his colleagues at Spring Gardens, the veteren engineer was, I believe, generally regarded as a quiet man of a somewhat cold and phlegmatic temperament, but there is something very pleasant and genial in the smile that he turns upon his visitor . . . and though he was born in 1819, he gives you the impression of a man with very many years of life and activity yet before him.

However, Sir Joseph died, aged 72 years, at home, on Sunday 15 March 1891, just two years after the Board ceased to exist. At the Institution of Civil Engineers on the following Tuesday evening the President, Sir John Coode, spoke of 'our old and esteemed colleague' whose works 'lie, as it were, at our very doors; and perhaps there is no member of our profession who has had his great works brought so much under the eye of the whole world as Sir Joseph Bazalgette'.[135] The official resolution of sympathy and condolence with Lady Bazalgette and the family, said:[136]

> his life had been given, to considerations affecting public health and welfare in all the large cities of the world, and his works, as the engineer for many years of the Metropolitan Board of Works, will ever remain as monuments of his skill and professional ability.

Sir Joseph's death was widely reported in both the national and technical press. *The Engineer* said, 'the late chief engineer had a very quiet way of discharging his functions' and 'Sir Joseph Bazalgette never appeared to be an ambitious man, and yet he became identified with undertakings which the most ambitious might have been proud to accomplish'.[137] *The Times* commented on another sphere of his life's work:[138]

> In the Parliamentary committee rooms there were few witnesses better known or more influential than Sir Joseph Bazalgette . . . there are few witnesses, it may be added, who will be so much missed as the late engineer of the Metropolitan Board of Works.

An engineering career in Local Government, with responsibility for huge sums of public money, provides daily opportunities for alienating politicians, committee members and ratepayers. Sir Joseph's career with the Metropolitan Board, spanning thirty-three years, shows him to have been an administrator of great ability and tact. He also had the gift of selecting able engineering colleagues without prejudice.[139] Bazalgette was also very supportive to members of his staff, however lowly, and the Board Minutes contain many examples of his persuasive pleading on their behalf. Although the Board ended its days somewhat under a cloud,[140] Bazalgette and his department retained their unimpeached integrity.

Sir Joseph was described as 'very slight and spare, and considerably under the average height; but his face, with its prominent acquiline nose, its keen grey eyes, and its grey whiskers and black eyebrows, gives you the impression of a man of exceptional power'.[141] This impression is captured in his bronze memorial, unveiled in 1901,[142] on the Victoria Embankment. Bazalgette certainly left his mark on London and, like Sir Christopher Wren, it could surely be said of Sir Joseph: 'If you seek his memorial, look around you'.

ACKNOWLEDGEMENTS

The author gratefully acknowledges his debt to Colin Glover, the staff of the Library of the Institution of Civil Engineers and, in particular, to Tom Bazalgette and Rear Admiral Derek Bazalgette, for their kindness and invaluable help.

SIR JOSEPH WILLIAM BAZALGETTE (1819–1891)

APPENDIX A

PRINCIPAL M.B.W. CONTRACTS

JOB	CONTRACTOR	DATE	AMOUNT
Southern Outfall Sewer	Webster	1860	310,000
Northern Outfall Sewer	Furness	1860	669,000
Northern Mid-level Sewer	Brassey	1861	358,000
Crossness Works	Webster	1862	395,000
Southern Low-level Sewer	Webster	1863	237,000
Northern Outfall Reservoir	Furness	1863	172,000
Thames Embankment North (1)	Furness	1863	500,000
Thames Embankment North (2)	Ritson	1864	247,000
Northern Low-level Sewer	Webster	1864	253,000
Thames Embankment South	Webster	1865	293,000
Isle of Dogs Branch Sewer	Webster	1866	120,000
Thames Embankment North (3)	Webster	1867	127,000
Chelsea Embankment	Webster	1871	132,000
Western Pumping Station	Webster	1872	188,000*
Putney Bridge	Waddell	1882	222,000
Sewers	Waddell	1882	140,000
Battersea Bridge	Williams, Son & Willington	1886	143,000
Northern Outfall: Precipitation Works	Mowlem	1887	406,000

*Includes Engines, boilers, etc.

SIR JOSEPH WILLIAM BAZALGETTE (1819-1891)

APPENDIX B

MAIN DRAINAGE PUMPING STATIONS (M.B.W.)

PUMPING STATION	DATE OPEN	CONTRACTORS		ORIGINAL PLANT	TOTAL STAFF
		BUILDING	ENGINE		
Deptford	1865	Aird & Son	Slaughter, Gruning & Co.	4 Beam Engings Total HP = 500 10 Cornish boilers	23
Crossness	1865	Webster	James Watt & Co.	4 Beam Engines Total HP = 500 12 Cornish boilers	51
Abbey Mills	1868	Webster	Rothwell & Co.	8 Beam Engines Total HP = 1136 16 Lancashire boilers	26
Western	1875	Webster	James Watt & Co.	4 Beam Engines Total HP = 360 8 boilers	19
Effra Creek	1879			1 Steam engine Vertical spindle pump (68 in. dia.)	
Falcon Creek	1879			1 Steam Engine Vertical spindle pump (72 in. dia.)	
Isle of Dogs	1888	Perry & Co.	James Watt & Co.	2 Horizontal engines.	

APPENDIX C

BAZALGETTE'S DRAINAGE REPORTS 1858-1875

	SUBJECT
5 Aug 1858	EPSOM: Drainage of the Royal Medical Benevolent College.
1858	LUTON: Drainage.
13 Jan 1859	NETLEY: Royal Victoria Hospital—best means of disposing of its sewage.
10 Oct 1862	FELTHAM: Drainage of Middlesex Industrial School.
Jun 1862	SHREWSBURY: Sewage, Pipe sizes, Estimate.
8 Jun 1863	BRISTOL: Drainage.
14 Oct 1863	CHELTENHAM: Sewage Deodourization.
20 Jun 1865	WESTON SUPER MARE: Drainage.
Nov 1865	OXFORD: Drainage.
10 Jan 1866	HASTINGS & ST. LEONARD'S: Drainage.
Mar 1866	FOLKESTONE: Drainage.
1 May 1866	OXFORD: Drainage.

SIR JOSEPH WILLIAM BAZALGETTE (1819-1891)

Aug 1866	CAMBRIDGE: Improvements.
1866	DON VALLEY: Drainage (Joint Report with John Fowler).
1867	TOTTENHAM: Evidence on Lea Valley Drainage.
Mar 1868	WINDSOR CASTLE: Drainage, Estimate.
1868	GLASGOW: Sewerage.
20 Jul 1868	GLASGOW: Sewerage (Joint Report with Bateman).
6 May 1869	HERNE BAY: Drainage.
1 Jun 1869	PEST (Hungary): Drainage.
1871	NORTHAMPTON: Drainage.
1871	SKIPTON: Sewage Irrigation Works.
4 Mar 1872	BIRMINGHAM: Drainage.
15 Mar 1872	SCARBOROUGH: Sewerage.
5 Dec 1872	DORKING: Drainage.
Nov 1873	CROYDON: Drainage.
Nov 1873	BECKENHAM: Disposal of Sewage.
15 Dec 1873	NORWICH: Drainage.
1874	FARNHAM: Pollution of River through Moore Park.
11 Feb 1874	NORWICH: Drainage.
2 Apr 1874	HAMPSTEAD: Burial Board, Drainage.
1874	MARGATE: Drainage.
27 Feb 1875	RUABON: Drainage.

REFERENCES AND NOTES

Abbreviations

Min. Proc. I.C.E. Minutes of the Proceedings of the Institution of Civil Engineers.

M.B.W. Minutes Minutes of the Metropolitan Board of Works, Greater London Record Office.

1. *Journal of the Huguenot Society*, Quarto Series, Vol. **27**, p. 199, "Letters of Denization and Acts of Naturalization for Aliens in England & Ireland, 1701-1800" (Guildhall Library).
2. William R. O'Byrne, *Naval Biographical Dictionary*, (John Murray, London, 1849), p. 59.
3. *Gentleman's Magazine*, August 1850, p. 227, Obituary "June 20 Mrs. Theresa Philo Bazalgette, of Tunbridge Wells, widow of Capt. Joseph William Bazalgette, R.N." She is buried in St. Lawrence Churchyard, Morden (See also Ref. 7).
4. The house is now numbered 17 and a GLC blue plaque was unveiled by Peter Black, Chairman of the Thames Water Authority, in October 1974.
5. Obituary of Sir Joseph William Bazalgette, C.B., *Min. Proc. I.C.E.*, Vol. **CV** (1890-91, Part III), pp. 302-8.
6. J. W. Bazalgette's Graduate Membership certificate, Muniment Room, I.C.E. His address is given as "48, Hamilton Terrace, Regent's Park".
7. See W. J. Rudd, "J. W. Bazalgette in Modern", *Bulletin of Wandle Group*, No. 9 (1981) pp. 2-3, for an interesting account of the family in Morden.
8. See C. N. Bazalgette, "The Sewage Question", *Min. Proc. I.C.E.*, Vol. **XLVII** (1876-77, Part II), p. 105, and Charles Norman Bazalgette and George Humphries, *The Law Relating to County Councils* (1888).
9. J. W. Bazalgette, Graduate I.C.E., "On Reclaiming Land from the Sea, with plans illustrative of works in Loughs Swilly and Foyle", *Min. Proc. I.C.E.* **Vol. I** (1837-40, Session 1840), p. 41.
10. J. W. Bazalgette's Graduate Membership certificate, Muniment Room, I.C.E. He was proposed by John MacNeill and supported by Joshua Field and James Simpson.
11. J. W. Bazalgette's Transfer to Membership certificate, Muniment Room, I.C.E. His form includes the signatures of Josiah Parkes, Charles Hutton Gregory, Gerge Parker Bidder and John U. Rastrick.
12. *Cassell's Saturday Journal*, 30 August 1890, p. 1160.
13. *London Exhibited* (John Weale, London, 1851), p. 821.
14. Parliamentary Papers (London, 1847-48), XXXII, p. 3.

SIR JOSEPH WILLIAM BAZALGETTE (1819-1891)

15. *11 & 12 Vict*, c. 112, "An Act to Consolidate, and continue in force for Two Years . . . the Metropolitan Commission of Sewers" (4 September 1848).
16. Ibid.
17. Op. cit. (13).
18. *The Engineer*, 20 April 1894, p. 335.
19. Obituary of Mr. Frank Forster, *Min. Proc. I.C.E.*, **Vol. XII** (1852-53), p. 157.
20. *Civil Engineer and Architect's Journal*, Vol. **15** (1852), p. 160.
21. Haywood's overtime schedule, Engineer's Main Drainage Papers, 2.2, City Corporation Record Office.
22. Engineer's Main Drainage Papers, 1.16, City Corporation Record Office.
23. Op. cit. (21).
24. Ibid.
25. *18 & 19 Vict.*, "An Act for the Better Management of the Metropolis" (14 August 1855).
26. *The Engineer*, 28 March 1856, p. 169.
27. Ibid.
28. Minute of 1 January, M.B.W. Minutes, 1856, p. 7.
29. *The Observer*, 6 January 1856.
30. Ibid.
31. Minute of 14 January, M.B.W. Minutes, 1856, p. 15.
32. *The Observer*, 20 January 1856.
33. Minute of 25 January, M.B.W. Minutes, 1856, pp. 19-20. The candidates were:

J. W. Bazalgette	1, Greek Street, Soho.
R. Rawlinson	34, Parliament Street.
Thomas Cox	Croydon.
D. Henry	Borough Engineer's Office, Wolverhampton.
W. Y. Freebody	9, Duke Street, Westminster.
Richard Dixon	Rotherham, Yorkshire.
John Alfred Ikin	14, Chatham Place, Blackfriars.
Wm. Trego	1, Chelsea Villas, West Brompton.
G. R. Booth	9, Portland Place, Wandsworth Road.

34. Ibid., p. 20. The voting was as follows:

Candidate	Show of Hands		
	1	2	3
Mr. Bazalgette	31	32	32
Mr. Henry	15	12	0
Mr. Rawlinson	12	10	0
Mr. Booth	1	0	0

35. Parliamentary Papers (London, 1889), XXIX, *Final Report of the Royal Commission appointed to Inquire into Certain Matters connected with the working of the Metropolitan Board of Works* (8 April 1889), p. 327, Bazalgette's evidence of 13 July 1888.
36. *Cassell's Saturday Journal*, 30 August 1890, p. 1161.
37. Op. cit. (35), p. 327.
38. To help with this extensive parliamentary business, Bazalgette was assisted by the consulting engineering firm of Law & Chatterton. See Obituary of Mr. Henry Law, *Min. Proc. I.C.E.*, Vol. **CXLII** (1899-1900 Part IV), p. 362.
39. Op. cit. (35), p. 341, evidence of 13 July 1888.
40. Ibid., p. 474, evidence of 14 August 1888.
41. Ibid., p. 321, evidence of 10 July 1888.
42. Ibid., p. 363, evidence of 17 July 1888.

SIR JOSEPH WILLIAM BAZALGETTE (1819–1891)

43. Ibid., p. 322, evidence of 10 July 1888.
44. Ibid., p. 335, evidence of 13 July 1888.
45. *Cassell's Saturday Journal*, 30 August 1890, p. 1161.
46. M.B.W. Minutes, 1856, p. 35.
47. Report to M.B.W. (6 April 1858), "The Main Drainage of the Metropolis, presented by Messrs. Bidder, Hawksley, and Bazalgette, in accordance with a resolution of the Board, dated 23 November 1857".
48. Minute of 15 October, M.B.W. Minutes, 1858, p. 666.
49. *21 & 22 Vict.*, c. 104, "An Act . . . to extend the powers of the Metropolitan Board of Works for the Purification of the Thames and the Main Drainage of the Metropolis" (2 August 1850).
50. This sewer, 12 miles long, runs from Kensal Green, under Bayswater Road and Oxford Street and then through Clerkenwell, to the River Lea. The contractor was Thomas Brassey, who used steam cranes to lift the excavated material.
51. Minute of 12 May, M.B.W. Minutes, 1859, p. 310.
52. Bazalgette's salary and travelling allowances progressed as follows:

Date	Salary £	Expenses £	Total £
14 January 1856	1000	—	1000
15 June 1859	1000	200	1200
19 July 1861	1200	200	1400
21 November 1862	1500	200	1700
13 October 1865	2000	200	2200*

*Remained so until Bazalgette retired in 1889.

The Assistant Engineers' salaries progressed as follows:

Date	Salary (£)	Expenses (£)	Total (£)
February 1856	350	50	400
July 1859	350	150	500
November 1861	500	200	700
October 1865	800	200	1000

53. Minute of 25 February, M.B.W. Minutes, 1859, p. 115. Twenty-five designs were sent in and considered by the Committee.
54. Minute of 30 July, M.B.W. Minutes, 1857, p. 16. The Committee members requested, and were supplied (in June) with photographs of the competition drawings.
55. Engineer's Eighth Annual Report (to the M.B.W.), 4 July 1864.
56. *The Engineer*, 29 July 1864, p. 71.
57. Minute of 7 April, M.B.W. Minutes, 1865, p. 434.
58. Ibid.
59. Ibid. 269 guests sat down to lunch in the covered reservoir. The guests included the Dean of St. Paul's, Edwin Chadwick, Professor Airey, William Haywood and J. R. M. McClean (President of the Institution of Civil Engineers).
60. Minute of 6 July, M.B.W. Minutes, 1866, p. 891.
61. Op. cit. (35), p. 235.

62. Ibid., p. 363.
63. Ibid., p. 330. The first sludge-vessel was built at Barrow, at a cost of £16,353, arrived in the Thames in June 1887 and was named the *Bazalgette*.
64. The idea of embanking the Thames goes back certainly to the fourteenth century. Sir Christopher Wren conceived a plan to embark the Thames from Temple Gardens to the Tower after the Great Fire of 1666. In the 1860s, Parliamentary Committees and Commissions discussed plans submitted by the following engineers: Fowler, McClean, Hemans, Page, Shields, Gisborne and Bazalgette. Parliament had allocated the Coal Duty revenue to pay for the works.
65. *Cassell's Saturday Journal*, 30 August 1890, p. 1160.
66. The north bank between Westminster and Blackfriars was indented with coal wharves and river access to the City Gas Works was also a problem. The Duke of Northumberland owned extensive property in the area, which he eventually conveyed to the M.B.W.
67. Report of the Select Committee on the Thames Embankment Bill, 23 June 1862, Bazalgette in evidence on 28 May 1862.
68. Derek Hudson, *Munby: Man of Two Worlds*, (London, Abacus, 1974), p. 188, "Friday 15 April 1864".
69. The contractors were:
    Contract No. 1, Westminster to Waterloo Bridge: Mr. Furness.
    Contract No. 2, Waterloo Bridge to Temple: Mr. Ritson.
    Contract No. 3, Temple to Blackfriars, plus the whole roadway: Mr. Webster.
70. Op. cit. (35), p. 329.
71. *26 & 27 Vict.*, c. 75, "An Act for the Embankment of part of the River Thames, on the South Side thereof . . ." (29 July 1863).
72. *31 & 32 Vict.*, c. 135, "An Act to enable the Metropolitan Board of Works to Embank the River Thames between the Royal Hospital at Chelsea and Battersea Bridge", (13 July 1868).
73. Edward Bazalgette, "The Victoria, Albert, and Chelsea Embankments of the River Thames", *Min. Proc. I.C.E.*, Vol. **LIV** (1877-78 Part IV), pp. 1-24 and ibid., p. 30, J. W. Bazalgette in discussion on 9 April 1878.
74. Op. cit. (35), p. 337.
75. Ibid.
76. Ibid., p. 321.
77. *The Engineer*, 21 May 1880, p. 373.
78. Ibid.
79. Parliamentary Papers (London, 1854), XIV, *Report from the Select Committee on Metropolitan Bridges*. The Committee considered that, although it may become necessary to build additional bridges, it would be better, first, to ascertain the effect of freeing the bridges from tolls.
80. *The Artizan*, Vol. **XII**, No. CXLIII, 1 December 1854, p. 34.
81. *40 & 41 Vict.*, c. 99, "An Act to provide for throwing open for the free use of the public certain Toll Bridges within the Metropolis" (12 July 1877).
82. *Cassell's Saturday Journal*, 30 August 1890, p. 1161.
83. Op. cit. (35), p. 337.
84. Ibid.
85. Report to M.B.W. (26 May 1880): "Metropolitan Bridges, Report by Sir J. W. Bazalgette, CB, as to repairs urgently Necessary".
86. Bazalgette's bridge designs were done in conjunction with his son Edward, who was by then "Assistant Engineer (Metropolitan Bridges)".
87. Minute of 17 February, M.B.W. Minutes, 1882, p. 266.
88. *Collins' Illustrated Guide to London* (1877), p. 104.
89. These were all high-level, fixed bridges and the designs included a parallel lattice girder, a braced cantilever and a lattice arch design.
90. Bazalgette's report to the M.B.W. (15 March 1877): "Proposed Bridge over the Thames below London Bridge", p. 14.
91. Parliamentary Papers (London, 1879), *Select Committee on the "Tower High Level Bridge (Metropolis) Bill"*, p. 127. Bazalgette in evidence on 20 May 1879.

SIR JOSEPH WILLIAM BAZALGETTE (1819-1891)

92. *50 & 51 Vict.*, c. CLXXII, "An Act for Enabling the Metropolitan Board of Works to make a new means of Communication across the River Thames by means of a Tunnel at Blackwall" (8 August 1887).
93. Op. cit. (35), p. 331.
94. Ibid.
95. Ibid.
96. Ibid., p. 336.
97. Ibid., p. 339.
98. Ibid., p. 336.
99. Ibid., p. 334.
100. Ibid., p. 323.
101. R. H. Parsons, *Early Days of the Power Station Industry*, (Cambridge, C.U.P., 1939), pp. 4-5.
102. Op. cit. (35), p. 336.
103. Minute of 12 August, M.B.W. Minutes, 1859, p. 593.
104. The premises in Greek Street soon became inadequate and were sold, by auction, for £6400 in August 1861 (M.B.W. Minutes, 1861, p. 630). The M.B.W. acquired a new site in Spring Gardens on which it built a new office building.
105. H. P. Boulnois (1846-1927) became Engineer to Exeter (1874), Poetsmouth (1883) and Liverpool (1890) Corporations and Inspector with the Local Government Board in 1897.
106. H. Percy Boulnois, *Reminiscences of a Municipal Engineer*, (St. Bride's Press, London, 1920), p. 29.
107. Ibid., p. 30. (See also A. W. Skempton, "Portland Cements, 1843-1887", *Transactions of the Newcomen Society*, Vol. **35** (1962-63), p. 117).
108. Ibid.
109. Ibid., p. 31.
110. Ibid., p. 34.
111. Minute of 4 July, M.B.W. Minutes, 1862, p. 521.
112. Minute of 23 September, M.B.W. Minutes, 1864, p. 941.
113. Minute of 12 July, M.B.W. Minutes, 1867, p. 882.
114. Minute of 6 November, M.B.W. Minutes, 1868, p. 1224.
115. Edward Bazalgette's Membership certificate, 8 April 1879, Muniment Room, I.C.E.
116. Minute of 21 January, M.B.W. Minutes, 1869, p. 46.
117. Minute of 24 September, M.B.W. Minutes, 1875, p. 281.
118. Printed report "To the Corporation of Maidstone" dated "Spring Gardens, London, S.W., July 1877", Maidstone Corporation Minute Book, Maidstone Council Offices.
119. *40 & 41 Vict.*, c. 138, "An Act for Authorizing the construction of a Bridge over the River Medway at Maidstone; and for other purposes" (2 August 1877).
120. Letter, J. W. Bazalgette to Maidstone Corporation, 7 August 1879, suggesting his form of advertisement for tenders (which was adopted), Maidstone Corporation Minute Book, p. 179, Maidstone Council Offices.
121. Para. 3 of printed letter "The Mayor to Mr Ellis", 10 September 1877, Maidstone Corporation Minute Book, p. 195, Maidstone Council Offices.
122. Bridge Committee Report, 30 October 1877, Maidstone Corporation Minute Book, p. 179, Maidstone Council Offices. The four tenders were: William Webster, £22,300; Henry Lee and Sons, £24,500; Ball and Gammon, £28,800 an Mr. Hill, £29,875.
123. See *South Eastern Gazette*, 9 August 1879, p. 4, for full description, speeches, etc.
124. Minute of 8 June 1881, Maidstone Corporation Minute Book, p. 407, Maidstone Council Offices.
125. Enclosure of 27 November, Report dated "Spring Gardens, 25 November 1863", M.B.W. Minutes, 1863, p. 1069. Copies of London and to each of the Vestries and District Boards of the Metropolis. At the next meeting of the Board on 4 December, The Vestry of Shoreditch requested a copy for each member of the committee and the Chelsea Vestry requested 60 copies.
126. Sir Joseph's Presidential Address was given on 8 January 1884 and dealt with "engineering works which promote the health and comfort of the inhabitants of large cities". He took a global view and produced statistics of population and death-rate per 1000 for 45 large cities. (*Min. Proc. I.C.E.*, Vol. **LXXVI** (1883-84), Part II), pp. 1-69).
127. Minute of 8 February, M.B.W. Minutes, 1889, p. 326.

SIR JOSEPH WILLIAM BAZALGETTE (1819-1891)

128. Minute of 15 February, M.B.W. Minutes, 1889, p. 423.
129. *The Engineer*, 15 February 1889, p. 141.
130. Minute of 9 July, London County Council Minutes, 1889, p. 540. The L.C.C. could not match Bazalgette's continuity of service. Joseph Gordon died on 9 November 1889 and was replaced by C. Dunscombe who, in turn, was replaced by A. R. Binnie (later Sir Alexander) in 1890—three engineers in little over a year.
131. *Cassell's Saturday Journal*, 30 August 1890, pp. 1160-61: "Representative Men at Home—Sir Joseph Bazalgette, C.B., at Wimbledon".
132. Ibid.
133. Ibid.
134. Ibid.
135. Sir John Coode on the death of Sir J. W. Bazalgette, *Min. Proc. I.C.E.*, Vol. **CV** (1890-91 Part III), p. 106.
136. Ibid.
137. *The Engineer*, 20 March 1891, p. 231.
138. *The Times*, 16 March 1891.
139. During 1855, the last year of the Metropolitan Commission on Sewers, Bazalgette was drawn into conflict with John Grant, one of his District Engineers. The issue concerned the relative merits of glazed pipe sewers and brick sewers. Bazalgette preferred brick sewers and Grant preferred pipes. Grant published a report on 30 March 1855, Bazalgette replied with a report on 30 April and then, on 8 May, there appeared the "Rejoinder of Mr. J. Grant to Mr. Bazalgette's remarks". Bazalgette was annoyed and, on 10 May, wrote to the Commissioners thus:

    "Though quite prepared on any fitting occasion when required to meet Mr. Grant's observations . . . the tone of that document is such that I feel I should best consult what is due to me personally and officially, and to the position I occupy, by at present refraining from any counter observations whatever.

    Nevertheless, Bazalgette selected Grant as one of his Assistant Engineers at the M.B.W. and, on Grant's death in 1888, he described Grant as:

    "A man on whose word I could depend thoroughly an a most careful painstaking officer with reference to his work. He has introduced into the Metropolis some of the best work that has ever been constructed in public works, and he has been the means of improving the manufacture of cement for which the country is greatly indebted to him". (Op. cit. (35), p. 338).
140. Charges of corrupt practice were made against the Board in 1886 with regard to the allocation of building sites in Shaftesbury Avenue. A Royal Commission was appointed, in 1888, to "Inquire into Certain Matters connected with the working of the Metropolitan Board of Works". They questioned many Officers and Members of the Board, including Bazalgette, who was asked 661 questions between 10 July and 14 August 1888. They made their report without exposing any misconduct of those interviewed. They had, however, received a letter from the Architect's Assistant for Valuation stating that he "had failed in his duty to the Board" but was too ill to be interviewed. The Metropolitan Board of Works was wound up under the powers of the Local Government Act (*51 & 52 Vict*, c. 41 (13 August 1888)). The London County Council came into office on 21 March 1889.
141. *Cassell's Saturday Journal*, 30 August 1890.
142. The memorial, with bronze bust, is by George Simmonds and was unveiled in 1901.

DISCUSSION

Mr. Peter Skilton asked whether Bazalgette had investigated the possible damage to the Thames embankment service ducts which might arise from the roots of the trees lining the embankments? **Dr. Smith** replied that Bazalgette seemed to have foreseen the problem; the tree roots were contained in special pockets within the Portland cement concrete, which were provided as part of the design.

Mr. Geoffrey Binnie said that within a year of Bazalgette's retirement his place had been taken by Alexander Binnie (the speaker's grandfather), who was responsible for completing Bazalgette's unfinished schemes as well as for building the Blackwall Tunnel. In relation to the paper, Mr. Binnie wondered whether George Roe, who had been one of Bazalgette's Assistant Engineers, was any relation of John Roe, who had developed the egg-shaped sewer? Dr. Smith said he thought the two Roes were related, but could not confirm that without further checking.

A visitor asked whether Bazalgette was known to have introduced hot-air central heating in any public building? Such heating had been used in 1890 in a Wimbledon house built by Bazalgette for his daughter which, unusually for its day—had no fireplaces. Dr. Smith said such matters would have been the responsibility of the Architect's Department of the Board and not of Bazalgette.

SIR JOSEPH WILLIAM BAZALGETTE (1819-1891)

Mr. John Butler asked which firm had been responsible for making the 8½ feet diameter metal pipes used to carry the Northern Outfall Sewer across railway lines? Dr. Smith said that Rothwell's of Bolton were responsible for the metal pipes used on the Northern Outfall contract.

Mr. E. F. Clark said that, as mentioned in the paper, Bazalgette had not originated the concept of intercepting sewers but that the idea went back to before Frank Forster's time having, apparently, been suggested by one John Martin, an artist, in 1834. Mr. Clark went on to say that the political moves which preceded the final enabling Act of August 1858 were even more complex than Dr. Smith had indicated. No fewer than 5 pairs of sites for the outfalls were considered, those furthest down the river being below Tilbury and Gravesend respectively. When the Chairman of the Metropolitan Board of Works, in the words of a later official report '. . . enquired whether Her Majesty's Government would be prepared to contribute to the additional cost . . .', the First Commissioner stated that '. . . in his opinion, Parliament would refuse to make any contribution to the works in question'. The resulting deadlock was only resolved by a change of Government to one in which Disraeli became Chancellor of the Exchequer. He enunciated the principle that 'He who pays the fiddler has a right to call the tune'. As a result, the Metropolitan Board of Works, emboldened by the report submitted on 6 April 1858 by Bazalgette, Hawksley and Bidder, went ahead with the scheme which forms the basis of the present arrangements. Dr. Smith replied that it was probable that the concept of intercepting sewers could be traced back even earlier than 1834 but, by 1851, it was firmly established as a realistic engineering project with proper design drawings and so on.

Rear-Admiral Derek Bazalgette said that his great-grandfather would have been delighted with Dr. Smith's paper about him. The family had few details about Sir Joseph but it was known that he was small, very asthmatic and probably rather irascible. He was obviously a 'workaholic' who had time only to work on London—and produce children. He was not infallible; papers in the family's possession recorded controversies he had encountered over a scheme for draining Windsor Castle. Although Queen Victoria had made him a Companion of the Bath, he had incurred the hostility of the Commissioner of Lands and Revenues, and of the citizens of Windsor. This seemed to suggest that Sir Joseph's scheme was, on that occasion, less than perfect.

Mr. Robert Cargill asked how well Bazalgette's works were surviving. Would the intercepting sewers need replacing soon? Dr. Smith said that the basic system was a palimpsest which had been worked on by successive generations of engineers. It was difficult to say which portions of it were the original work of the 1860s and which were more recent. The Thames Water Authority had a rolling programme of maintenance, upgrading and renewals going on all the time, but the original works had certainly stood up remarkably well to the passage of time.

Mr. David Perret asked whether Bazalgette's road-building activities had been, in a sense, part of a more general scheme of urban improvement? All the large London schemes seemed to be aimed at large slum concentrations but, under the legislation of the 1870s, the M.W.B. had no obligation to rehouse those who were displaced. Dr. Smith replied that artisans' housing was, indeed, one of Bazalgette's responsibilities, discharged jointly with the Architect's Department, but space had not permitted the development of that theme in the current paper.

A visitor asked why the outfall positions had been chosen at Crossness and Beckton? Recent observations suggested that, on the Beckton side at least, the Thames current flowed upstream whatever the state of the tide; perhaps the flow testing had not been done so thoroughly on the northern side? Dr. Smith replied that he supposed the most important consideration had been that the sewage should not back up as far as the Houses of Parliament!

Mr. J. S. Allen (President) proposed a vote of thanks to Dr. Smith, which was carried with acclaim.

15

Eugène Belgrand (1810–1878): civil engineer, geologist and pioneer hydrologist

George Atkinson

INTRODUCTION

Geology stands to civil engineering in the same relation as faith to works and the 'works' of the engineer should be based on the 'faith' of the geologist.

Boyd Dawkins, FRS, quoted by Robert F. Leggett in *Geology and Engineering* (McGraw-Hill, New York, 1939).

Belgrand

Fig. 1. Eugène Belgrand (1810–1878), civil engineer, geologist and pioneer hydrologist

François-Eugène Belgrand (1810–1878), creator, for Paris, of a unique system of water supply and of sewers which are still a tourist attraction, is one of nineteenth century's most interesting yet least known figures—a civil engineer, geologist and pioneer hydrologist. Born and brought up in a small Burgundian town on the edge of the forest of Othe, Belgrand was a countryman, though spending the last twenty five years of his life in Paris. A true provincial, who studied and understood rural France, particularly the waters of the Upper Seine basin, he might today be called a 'green'.

I was drawn to my study of a man hardly known in France, a country which usually goes out of its way to honour outstanding citizens, by an interest, dating back many years, in the very different way French civil engineers were trained, compared with their fellows in the English-speaking world in the nineteenth century, a difference which still exists. It is an interest first stimulated by the writings of Stanley B. Hamilton, a distinguished President of the Society and one-time colleague at Building Research Station.

Ecole polytechnique.

Fig. 2. Entrance to the Ecole polytechnique, rue Descartes, Paris

Ecole des ponts et chaussées.

Fig. 3. Entrance to the Ecole des ponts et chaussées, rue des Saints-Pères, Paris

My interest was revived by Denis Smith's description of the life and works of Joseph Bazalgette (1819–1891),[1] a near contemporary of Belgrand whose work for London mirrored in a number of ways that of Belgrand for Paris, but whose professional training was very different. Bazalgette began his engineering career like other British civil engineers last century as an Articled Pupil. Belgrand after studying mathematics and latin at a leading Paris lycée, won, in a highly competitive examination, a place in the École polytechnique, there to receive a scientific education unparalleled in Europe at the time. Graduating with honours, he went on to the École des ponts et chaussées where, as an 'élève' in the state Corps des ponts et chaussées,[2] he acquired the theoretical side of civil engineering, spending two summer periods in the field gaining practical experience.

Passing out, Belgrand, like other student engineers in his 'promotion',[3] was posted to a provincial public works services as *ingénieur aspirant*, to become, after two years in the service, an *ingénieur ordinaire 2e classe* in charge of an *arrondissement*.[4] From then on his career, like most other members of the Corps, was to be in the service of the state until retirement or, in the case of Belgrand, death.

My particular interest in Belgrand's life and works was aroused when I found, in the archives of the Institution of Civil Engineers, an early volume of his major publication. *Les Travaux Souterrains de Paris*,[5] presented by him to the Institution in 1875. Fortunately the Institution also has a set of the *Annales des Ponts et Chaussées*,[6] in which most of Belgrand's

EUGÈNE BELGRAND (1810–1878): CIVIL ENGINEER

DÉCRET IMPÉRIAL

Portant organisation du Corps des Ingénieurs des Ponts et Chaussées.

Au quartier-général impérial du Pont-de-Brique, près Boulogne, le 7 Fructidor an XII.

NAPOLÉON, par la grâce de Dieu et par les constitutions de la République, EMPEREUR DES FRANÇAIS; sur le rapport du ministre de l'intérieur; le Conseil d'état entendu, DÉCRÈTE ce qui suit :

TITRE I.er

Formation du Corps des Ingénieurs des Ponts et Chaussées.

ART. I.er

Le corps des ingénieurs des ponts et chaussées sera composé, à l'avenir, de cinq cent trente-sept individus divisés en grades de la manière qui suit :

5 inspecteurs généraux,
15 inspecteurs divisionnaires,
2 adjoints,
134 ingénieurs en chef,
306 ingénieurs ordinaires,
15 aspirans.

477 ingénieurs,
60 élèves.

537.

Fig. 4. Napoleonic decree of 1804 reorganising the Corps des ponts et chaussées

engineering papers appear, while the London Library and the Royal Society both have a set of the *Comptes Rendus de l'Académie des Sciences*[7] where many of his scientific papers are to be found.

Belgrand's scientific and technical writings show the width of his interests, some of his engineering works—like the Arcueil acqueduct—being Roman in scale. His feeling for natural phenomena, combined with the respect in which he was held by his contemporaries, became apparent as I continued my study. They show him to be not only engineer of key services in one of the two principal cities of nineteenth century Europe but a pioneer in the science of hydrology.

Yet Belgrand is not well known in France, and is all but unknown outside his native land. It is not possible to discuss here why this may be so, nor to deal in any depth with the physical, political, and social environment within which he worked. To examine the contributions made by colleagues and assistants like Adolphe-Auguste Mille (1812–1894), Alfred-Augustin Durand-Claye (1841–1888), and Georges Lemoine (1841–1922), or his relations with Georges Haussmann (1803–1891), his patron and friend, and Jean-Charles Adolphe Alphand (1817–1891), Belgrand's younger colleague in the technical services of the City of Paris, who took over responsibilities for the City's water and sewers on Belgrand's death, are other matters still to be explored.

This paper is, therefore, but an introduction to the achievements of one of the nineteenth century's most able, least well-known civil engineers. Within limits of space, the paper looks in the main at three phases in Belgrand's career: as district engineer at Avallon (Yonne) in the 1840s; as director of Paris water and sewers from 1854 until his death in 1878; and his contributions as a pioneer hydrologist to the French Academy of Sciences after his election in August 1871.

EUGÈNE BELGRAND (1810–1878): CIVIL ENGINEER

FAMILY AND EARLY LIFE

François-Eugène Belgrand was born on 23 April 1810, at Evry-le-Châtel [Aube].[8] He was the eldest of a family of six boys, two of whom were also students at the École polytechnique. Their father, Antoine François Belgrand, a *propriétaire* at Arc-en-Barrois [Haute-Marne], died when the family were young, and Eugène and his brothers were brought up by their mother, Mme. Marie Madeleine Eugénie Truchy, a courageous, wise person, whose father M. Truchy, a local landowner, was a man of some substance. Belgrand, as the eldest, assumed a responsibility beyond his years—an event which is said to have made a significant contribution to his character development as a youth.

Living in a small town surrounded by woods and looking over meadow land through which flowed a number of small rivers, forming the head waters of the Yonne, for the first twelve years of his childhood Eugène Belgrand benefited from a rural life in one of the most beautiful regions of France. From long walks through the forests covering the Langres plateau, he gained at an early age knowledge of the ground of the upper Seine-Yonne basin which was to become a life-long interest and a major influence on his career.

In 1822, at the age of twelve, Belgrand entered the Collège de Chaumont,[9] a former Jesuit college with a good educational reputation. He came with a strong background in the classics, French literature and mathematics, to be reinforced by rigorous study in a college where Jesuit traditions still prevailed.

The education in mathematics at Chaumont and at Pension Mayer, Paris, facilitated Belgrand's entry as a boarder into the oldest and most prestigious of the Paris *lycées*, the Lycée Louis-le-Grand. Despite its reputation of being 'a cross between a Jesuit seminary and a military barracks', Louis-le-Grand gave young Frenchmen a secondary education, particularly in science and mathematics, of an academic standard unparalleled within the western world. In Belgrand's case, he was able to join the special class in mathematical analysis, a fellow student being the gifted mathematician, Evariste Galois (1811–32)[10]

THE STUDENT ENGINEER

Two years or so at a Lycée like Louis-le-Grand was an essential route for entry by examination into the prestigious École polytechnique into which Belgrand made his entry in 1829 at the age of 19, the same year as a life-long colleague, friend and biographer, Léon Lalanne. It was a period of unrest in France, as it was throughout Europe. For many young Frenchmen, and especially those at the École polytechnique, the restoration of a Bourbon king had been a bitter blow. The Bourbon years were a time of increasing disillusionment, leading in 1830 to open revolt against the reactionary government of Charles X, last of the Bourbons.[11]

During the 'Three Glorious Days' of July 1830, Belgrand and other students of his *promotion* moved by a speech by the physicist Arago during a *séance* of the Académie des Sciences in which anti-Bourbon sentiments were inserted on an eulogy on Fresnel, took to the streets in a mass demonstration against the killing, during an attack on the barracks of the Swiss Guard, of a fellow student, Vanneau.[12] Forty eight hours later the School was closed, and did not reopen for three and a half months.[13] Adolphe-Auguste Mille [1812–1894], who was in the 1832 *promotion* at the École polytechnique, mentioned later that Belgrand, moved by the occasion, never wished to recall that day.[14]

Having graduated with honours at the École polytechnique, Belgrand, now aged 22 years, entered the École des ponts et chaussées[15] as an *élève ingénieur* in 1832. A feature of training at the School was that student engineers spent the summer months in the field, attached usually to a *départemental* service.[16] The service to which the young Belgrand was attached in his first year at the School was that of the Côte-d'Or, whose chief town is Dijon. Among the Dijon 'départemental' archives is the original instruction, sent by the 'Directeur général des ponts et chaussées et des mines, Division du personnel, Missions d'élèves' to the Côte d'Or *préfet*, dated 26 April 1832. It stated that Belgrand, '*élève ingénieur* de 3eme classe', would arrive at Dijon on 11 May to be employed on works relating to a bridge at Pontarlier, that he had Ffr75.50 for travelling expenses and Ffr50 as a first instalment of his Ffr100 field allowance, that the *département* was to pay the student a monthly salary of Ffr150 from local funds, and ensure he returned to Paris by 20 November.

It happened that Belgrand was not sent to Pontarlier, but put in charge of work on a small bridge across the Brenne river at Vitteaux-en-Auxois.[17] During the work he witnessed an incident which had a major impact on his career. Following an exceptionally severe rain storm lasting less than an hour, rainwater streaming down every slope of the valley caused the level of the Brenne to rise rapidly. In an instant, river water flooded over the bridge parapets. It was evident that the design rules used had not taken into account local conditions, About the incident, he wrote later:

> I began these studies, so to speak, since the beginning of my career. It was in 1832 when I was a young engineer attached to the Côte d'Or 'département' in charge of the construction of a small bridge. The waterway openings had been calculated according to the empirical rule we had been taught the preceding winter at the École des ponts et chaussées which only took account of the slopes of watercourses and the heights of the land bordering a valley and nothing about the nature of the terrain, the calculated dimensions being well below that needed. I concluded that there was a gap in our studies, and, since that time, I am convinced that a civil engineer cannot just be a surveyor, he must also be a geologist.[18]

THE YOUNG DISTRICT ENGINEER

> Cette rigide observation des devoirs est un des titres des Corps des ponts et chaussées à l'estime et, on peut le dire, à la reconnaissance publique. Il est rare qu'on en parle, tant la chose est naturelle et dictée par le sentiment de la dignité professionelle. (Lalanne, pp. 343–344).

On 1 May 1834 Belgrand, aged 24, was posted to Puy de Dôme as the engineer in charge of the Clermont-Ferrand *arrondissement*. In the autumn of the same year his appointment in the

CLERMONT-FERRAND.

'Fig. 5. Clermont-Ferrand, Belgrand's first post as a district engineer

'Corps des ponts et chaussées' as an *ingénieur aspirant* was confirmed. Named after France's oldest and highest extinct volcano, the Puy de Dôme *département* in the Auvergne region was, and still is, one of France's most interesting regions for the geologist. As a young district engineer, his responsibilities, like those of his fellows throughout provincial France, were varied but hardly mind-stretching. He was probably the most highly educated young man in the district, yet much of the work was mundane. Furthermore, he had to work with, and gain the confidence of, both the local bureaucrats in the Clermont Ferrand prefecture and his own works staff who thought they knew—and generally did know—more about the practical aspects of civil engineering than a young man fresh from Paris. In particular he had to gain the respect and confidence of a key grade—that of the *conducteur* the non-commissioned officer in the local public works service.

To his great regret, this left Belgrand with little spare time to continue his studies of the effects of geology on water regime in a region of France of great interest from the standpoint of hydrology. Writing in January 1835 to his life-long friend Lalanne, he expressed this regret: 'Judge my misery. During the eight months I have been in the centre of this magnificent country, I still haven't had time for painting!'[19] He did, however, find some compensation in assisting an 'ingénieur des mines' in running a public course on chemistry.

Having served two years as a probationer, Belgrand was promoted to *ingénieur ordinaire de 2e classe* on 15 June 1836, having been posted on 8 June to the Côte d'Or[20] as district engineer in charge of two *arrondissements*—Semur-en-Auxois and Châtillon-sur-Seine—taking up residence at Montbard, a small town in the Burgundian uplands on the Brenne river, located between the two *arrondissements*. He had already spent time in the Côte d'Or *département* during his first student *mission* in 1832, and was now in a part of France with which he was familiar since childhood. An experienced walker, a skilled draughtsman with excellent eye-sight and assured memory, he had more time to observe the effects of local terrain and topography on rainfall and river flow. Here he was fortunate in that ground conditions in the two *arrondissements* were very different—at Semur impermeable lias clay over the granite of the Morvan and at Châtillon permeable oolitic limestone. While at Montbard Belgrand spent six months in Rome during the winter of 1837–38 looking after a seriously ill brother, a time which allowed him to carry out a study of the system of Roman aqueducts about which he was later to write.[21]

DISTRICT ENGINEER AT AVALLON

> Si je fus étonné de rencontrer dans cet homme grand, chauve, dans l'extérieur campagnard n'annoncait pas l'intelligence supérieure, un géologue et un hydrologue des plus forts, il me parut l'être tout autant d'avoir, dans son Préfet, un adopte, ou tout au moins un curieux des deux sciences auxquelles il consacrait ses loisirs. Une sympathie mutuelle. (G. Haussmann, *Memoires, 'Les Services d'Ingénieurs'*, p. 1123.).

On 1 April 1839, Eugène Belgrand, aged 29 and recently married,[22] was transferred to the Yonne *département* as district engineer at Avallon.[23] The transfer marked an important phase in his career. He was to design and see completed his first water scheme. He met, and impressed, Georges Haussmann, soon to be *préfet de la Seine* and creator of modern Paris. And he saw the first of many papers on hydrology published.[24] The river Yonne, after which the *département* is named, is one of the major tributaries of the Seine, and during his stay in the Yonne *département* at Avallon, and for a short time at Auxerre, the principal town of the *département*, Belgrand found time to continue his study of the hydrology of the Seine basin which was to be a subject of interest for the rest of his life.

At Avallon, Belgrand was able to put into practice the understanding acquired as a young engineer of the link between local geology and water regime. Sited on a rocky promontory 100 metres above a deep valley through which ran a small river, the Cousin, Avallon suffered annually when its wells dried out. There had been schemes to pump water from the Cousin river, but none had so far proved practical or economic, and, on the 14 August 1843, the municipal council asked their new district engineer to join a committee set up to investigate a reliable supply of water for the town.[25]

Having discarded the idea of pumping water from the Cousin river on practical grounds as well as cost, Belgrand sought what now would be called a 'green' solution. Investigating the wooded slopes across the Cousin valley, he found a suitable source of water, at a sufficient height to supply Avallon by gravity, in the Étang Minard about 4km from Avallon. The scheme, prepared for consideration by the municipality, involved constructing a 3.6 km long masonry conduit leading to a cylindrical reservoir sited at Alleux in open country across the Cousin valley from Avallon. The reservoir was a masonry structure, 16 m in diameter, built in Roman cement, within which 500 cubic metres of water could be stored under a thin 'shell' roof. The stored water was taken across the deep Cousin valley by means of a siphon, formed from cast-iron pipes, each 2.50 m in length, and 162 mm internal diameter, to water-points within the walled town of Avallon.

Features of the scheme were the 88 m deep siphon, the world's deepest at the time, constructed with cast-iron pipes tested to a pressure equivalent to a 270 m head of water, the special jointing of the pipes, an extensive use of roman cement obtained from a local cement works at Vassy, the 'shell' roof covering the reservoir, and the 30 m single span bridge, carrying the pipeline over the Cousin river.[26]

To obtain technical approval for Belgrand's novel design from the ministry of public works, and loan sanction from the interior ministry, Belgrand's plans, specification and estimate of cost were sent to Paris, first for technical checking by M. Mary, '*ingénieur en chef* in the service of the city of Paris', and, then, for final approval by the Conseil général des ponts et chaussées. Only minor modifications were required, like an increase in reservoir capacity from 500 to 600 cubic metres, and larger compensation for owners of land through which the pipeline passed. By the end of 1844, all the administrative and technical issues having been resolved, construction under Belgrand's supervision was commenced. Tenders for the specialist iron work were sought from Paris firms, accepted during 1847 and, soon after, the work had been completed.

In the summer of 1850, a historic but little known event occurred, namely, a meeting between Georges Haussmann, at the time 'préfet' of the Yonne 'département', and Eugène Belgrand, the engineer who was to give Paris a modern water supply and system of sewers.[27] For Haussmann, appointment as 'préfet' of the 'département' of Yonne, based on Auxerre, was but a stepping stone to the ultimate prize, 'préfet de la Seine' and virtual dictator of Paris. He had been given the post by Louis Napoleon, then Prince-President, to keep an eye on socialist agitators obstructing Louis Napoleon's rise to power as Emperor, achieved a year later in a *coup d'état*. However, Haussmann had also a more conventional task as *préfet*—to visit small towns in the *département*, like Avallon, show the flag, settle administrative disputes, and, a key task, to sound local opinion on behalf of Paris.

In the course of such a visit it was usual for the *préfet* to inspect new municipal schemes like the Avallon water supply. During that visit, he spent two days with Belgrand, during which he was impressed both with the scheme and with its designer, finding, beneath the provincialism of the man 'un géologue et un hydrologue des plus forts'[28] During their tour of inspection, the two men discovered a mutual interest in the relationship between geology and water supply, and established *une sympathie mutuelle* which was last during the years when Haussmann and Belgrand worked together in Paris. Having developed an interest in the influence of drinking water on the health of a local population from an earlier post in the Pyrénées, Haussmann inquired about the source of the constant supply of fresh, clear water for the town's public fountains. Belgrand surprised at such technical questions from an administrator gave Haussmann clear and detailed explanations.

EUGÈNE BELGRAND (1810–1878): CIVIL ENGINEER

THE SEINE NAVIGATION SERVICE

On 1 June 1852, Belgrand was posted to Paris in the service of Seine Navigation.[29] The posting happened at a critical stage in 19th century French history. It was a turning point in Belgrand's career as a civil engineer bringing him in direct contact with senior colleagues in the Corps des ponts et chaussées. It also gave him the opportunity to initiate a few years later one of the major civil engineering works of the nineteenth century as well to consolidate his pioneering hydrological studies. With the posting to take charge of Section III of Seine Navigation between Paris and Rouen came promotion to *ingénieur en chef des ponts et chaussées*, and the separate responsibility of organising a new hydrométrique service for the whole Seine river basin. How far the posting and promotion was helped by a recommendation made by Haussmann when Yonne *préfet* we do not know.[30] Although Haussmann did not become *préfet de la Seine* until a year later, he was already in touch with Napoleon III, and may have planted the idea of a comprehensive hydrometric service in the minds of Napoleon III and his ministers.

Fig. 6. Hôtel de Ville, Paris, Belgrand's office before its destruction by the Commune in 1871

Belgrand's stay in Seine Navigation was short. We do know that one of his engineers, the younger Emmery,[31] was building locks and a barrage at Meulan, and that the posting gave Belgrand the opportunity to extend his knowledge of the Seine below Paris which was to prove useful later when he decided to discharge the main *collecteur* for his new main drainage scheme downstream at Asnières.

DIRECTEUR DES EAUX ET DES ÉGOUTS: THE PARIS WATER SUPPLY

De l'eau, de l'eau. "Je voudrais que toutes les maisons de Paris eussent de l'eau comme celles de Londres" Voltaire 1767

When Haussmann became 'préfet de la Seine' on 23 June 1853, the Paris water supply differed little from that at the beginning of the century.[32] As yet there was no piped domestic supply. Water was drawn from 'borne-fontaines'—public water points located at cross-roads and markets—and sold to individual households by 'porteurs d'eau'. Most of the water came from the Seine, increasingly polluted by streams within the urban area, the largest the Bièvre an open sewer. Napoleon I had tried to improve the water supply by building a canal to bring water from the Ourcq, a tributary of the Marne. Begun in 1808, the canal was not completed until the 1830s. By that time the Ourcq water had itself become polluted by the waste emptied into the canal by 1500 barge families.

Meryon.
La pompe Notre-Dame.
Eau-forte.
Cl. Giraudon.

Fig. 7. Before Belgrand's two schemes, most Paris water was pumped directly from the Seine

For Haussmann provision of pure water was a religion. During his short period as *préfet de la Yonne* he had become aware of Belgrand's qualities as a water engineer. Together, at the Auxerre prefecture, the two men had studied the geology and hydrology of the Seine basin, planning a new supply of water for Paris.[33] Belgrand was now in Paris working in the Seine Navigation Service, but it took time to arrange with the public works ministry his transfer to the city of Paris. Meanwhile, Haussmann had to persuade Louis Napoleon to abandon the idea of giving the Paris water supply to a private concessionaire, and to entrust public water and sewerage services to the municipality.[34]

In April 1854, Haussmann was able to get Belgrand to investigate sources of potable water for Paris.[35] The time allowed was short. Fortunately Belgrand, having studied the hydrology of the Seine basin for fifteen years, had a thorough knowledge of its water resources, and was able to recommend the more promising without much further detailed investigation. He reported that the most abundant springs were in Burgundy (Tonnerrois), Champagne Pouilleuse (south of the Aisne and at the head of the Aube and Yonne) and, less abundant, in the Eure valley. While waters from the sources in Burgundy were of better quality, their distance—up to 250 km from Paris—was judged to be uneconomic.[36] Subject to further investigation, Belgrand opted for the waters of the Somme-Soude, closer to Paris, as the source meeting specified requirements for early exploitation.

THE SOMME-SOUDE[37]

The Somme-Soude, a small river in Champagne, proved a failure. Its abandonment on 26 May 1859, partly the result of political pressure from electors of the Seine-et-Marne *département* where the source was located, and partly its inadequacy during a long period of drought was the cause of one of the few serious criticisms made by Haussmann of Belgrand's work, unjustified in view of Belgrand's warning on the need for further gauging of the stream's flow before a final decision.

THE DHUIS[38]

First of two successful schemes, the Dhuis still provides Paris with a supply of fresh, potable water from springs emerging from 'terrains tertaires lacustres' which lay above *marnes vertes de Montmartre*. The Dhuis is a tributary of the Surmelin, which joins the Marne river 10 km east of

Fig. 8. Map of the Seine basin showing lines of the Dhuis and Vanne aqueducts

Fig. 9. One of the cast iron siphon-bridges on the Dhuis aqueduct

EUGÈNE BELGRAND (1810–1878): CIVIL ENGINEER

Château-Thierry. It was chosen as a source because, being at an altitude of 128 metres, its waters could be delivered by gravity to the heights of Ménilmontant in the east of Paris at an altitude of 108 metres. The aqueduct pipes are ovoid in section, then an innovation, with a 1m.40 channel. It is 114 km in length, 10 km being in tunnel, and crosses streams and low ground by means of 17 km of 1 metre diameter cast iron pipes forming siphons. Precautions were taken at source to limit calcium carbonate deposits in the pipeline.

Dhuis water supplies a large reservoir of novel design at Ménilmontant in the 20th *arrondissement*. Built in 1865 as a two-storey structure, the upper storey able to hold up to 100,000 cubic metres of potable Dhuis water, supplying those sections of Paris, particularly in the north-east, at a higher altitude and which were, at the time of its construction, without any supply. The lower storey holds 31,000 cubic metres of non-potable water for washing streets etc pumped from works on the Marne at Saint-Maur.

THE VANNE[39]

The Vanne scheme supplies water to the Montsouris reservoir in the south of Paris, serving those parts of the city lying at a lower level. Work was started in 1867, a legal order—'declaration d'utilité publique'—to exploit springs supplying water to the river, compensate watermills etc and construct a 173 km aqueduct to Paris having been granted on 19 December 1868. Work was interrupted during the Franco-Prussian War and the scheme was not completed until 1874.

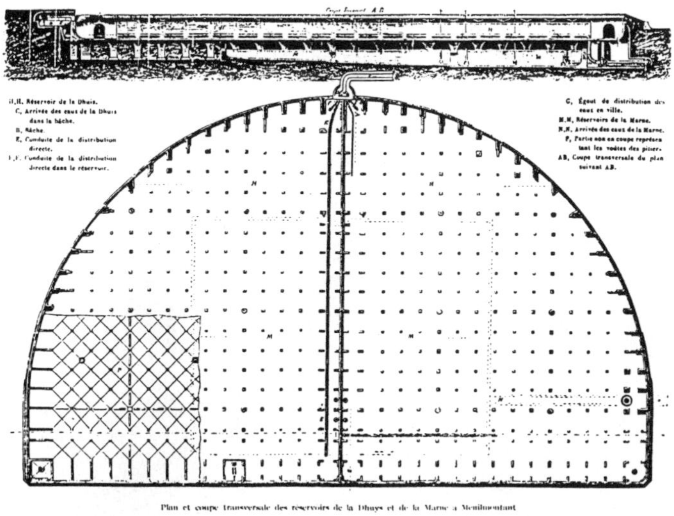

Fig. 10. The Ménilmontant two-level reservoir, Paris: upper level Dhuis potable water, lower level non-potable Marne water pumped from St Maur installation

The Vanne, a tributary of the Yonne, rises in open country 14 km west of Troyes (Aube). Before entering the Yonne *département*, it receives water from an important stream, the Noisel. After passing through a series of small *communes* where further springs are located, the river reaches the Yonne to the immediate south of Sens. Within its catchment area of 965 sq.km—665 sq.km permeable chalk and 300 sq.km red *limon de plateaux* mixed with boulders surface—streams are rare. Due to the high permeability of the chalk rainwater feeds directly into an exten-

EUGÈNE BELGRAND (1810–1878): CIVIL ENGINEER

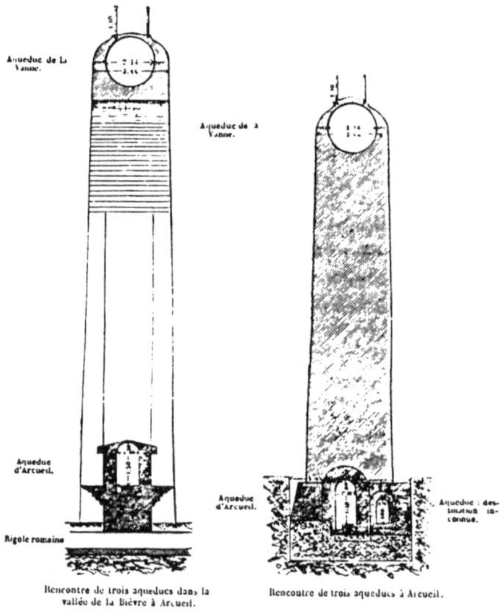

Fig. 11. Vanne aqueduct, Arcueil, sited over ruins of Roman aqueduct and Marie de Medici aqueduct

Fig. 12. Montsouris reservoir, Paris, serving the Vanne system

sive network of aquifers so that, although the area has one of the lowest rainfalls in the Seine basin—601 mm/year, compared with 2158 mm/year on the higher parts of the Morvan uplands to the south, among the smaller Seine tributaries the Vanne is least affected by periods of drought.

EUGÈNE BELGRAND (1810–1878): CIVIL ENGINEER

The valley itself is in the form of a trough whose complex geology is described in Volume IV *of Les Travaux Souterrains de Paris.*[40]

Drawing on springs in the valley, the aqueduct runs along the left bank of the river to a point 6 km east of Sens[41] where it crosses the Vanne close to the TGV line to Lyon. From high ground east of Sens, the aqueduct reaches to the Yonne just south of the small town of Pont-sur-Yonne where it crosses the river, the Paris-Sens railway and the N6 highway on a high, arcaded structure, the first of a number built in 'béton aggloméré du système Coignet'.[42] After entering high ground on the left bank of the Yonne, it crosses smaller valleys on similar arcarded structures before reaching the Loing to the immediate west of Moret-sur-Loing (Seine-et-Marne) where it crosses the river and canal by another high arcaded structure to enter higher land leading to the Forest of Fontainebleau, the crossing of which because of extremely complex ground conditions challenged the ingenuity of Belgrand and his engineers.[43] The last major work after passing close to Orly airport is the 990 m long arcaded structure built across the Bièvre valley at Arcueil on the site of earlier gallo-romano and renaissance structures. Finally the Vanne aqueduct reaches the Montsouris reservoir just within the former line of fortifications in the south of Paris.

Larger in area—3 hectares—and capacity—275,000 cubic metres—than the Ménilmontant, the Montsouris reservoir is in the form of a rectangle 265 m by 136 m, its 3 m 55 high upper storey holding 110,000 cubic metres at an altitude of 80 m, and the 5 m 50 high lower storey 165,000 cubic metres at an altitude of 74 m 50. Both Montsouris and Ménilmontant reservoirs are vaulted and protected by a 400 mm layer of earth from frost and excessive evaporation, a technique adapted from the much smaller reservoir Belgrand built at Avallon in the 1840s.[44] Before the water leaves the Vanne valley, it is raised to the 117 m 17 level by three sets of water-powered turbines. It then falls by gravity over 173 km to the 80 m level at Montsouris, 41 km 90 being underground and 16 km 23 in siphons formed from paired 1 m 10 diameter cast iron pipes, the fall being between 100 and 130 mm/km, except in the siphons where a fall of 550 mm/km was adopted.

NON-POTABLE WATER, THE ST-MAUR INSTALLATION

Just before the Marne joins the Seine upstream from Paris the river takes a large loop. To provide a short cut for navigation, a start was made during the Empire (1809) to build a canal in tunnel across the narrow neck of the loop at St-Maur. When the canal was completed a number of watermills were built along its banks, using the flow of the water as a source of power. Although the municipal authorities would have liked to use this power to pump water from the Marne, the land on which the mills stood had been alienated in 1822, and it was not until 1864 that the mills were taken over at a cost of Ffr5.5 million. Belgrand was then able to build the St-Maur pumping station supplying Marne water to the lower non-potable level of the Ménilmontant reservoir. Later additional pumping facilities were introduced to supply a new reservoir at one of the highest locations in Paris at Belleville. This source of non-potable water is still used for street cleansing, public fountains and gardens.

THE PARIS SEWERS[45]

> Great and manifold as are the evils entailed by the scarcity of water in the bedrooms and kitchens of Paris, there is another deficiency greater and infinitely worse in effects. The want of drains and sewers is the great defect of all the cities in France, and a tremendous defect it is.
>
> Frances Trollope, 'Paris and the Parisians in 1835'

When Eugène Belgrand took charge of Paris sewers in 1854, they were in poor condition. On the Right Bank the situation was as follows. A line of high ground ran from Belleville in the east to Chaillot in the west, and, in earlier years, water from the ridge collected in *le ruisseau de Ménilmontant*, a stream which ran at the foot to discharge into the Seine below Chaillot. As Paris

EUGÈNE BELGRAND (1810–1878): CIVIL ENGINEER

spread towards the stream, the stream was used as a tip for rubbish. Silted up, in heavy rain there was flooding, particularly in the area around the Marais, a one-time marsh where, in the seventeenth century, nobility built their residences. Many attempts were made to drain the area with limited success. Poorly constructed and filled with domestic and industrial waste, the drains were frequently blocked.

After the 1830–1831 epidemic of Asiatic cholera, a progressive *préfet*, Comte Claude-Philibert Rambuteau, and his engineer, Henri-Charles Emmery, initiated a series of reforms. One, diverting blocked sewers directly into the Seine, only put off large scale reconstruction of Right Bank drainage for more than a decade; another, closing the disgraceful Montfauchon depôt, where night-soil, animal carcases, and household rubbish were dumped, had long term benefits.[46] In 1851, a start was made to cover the Ménilmontant stream and turn it into the *collecteur du Ceinture*. Modern sewers were constructed under rue de Rivoli and boulevard Sébastopol. All, however, discharged directly into the Seine within Paris, and it was not until Georges Haussmann became *préfet de la Seine* in June 1853 that there was a coordinated programme of sewer construction.

Having set up a separate Service des eaux et égouts, Haussmann sent Adolphe Mille,[47] a contemporary and engineer colleague of Belgrand, on a mission to Britain to report on urban sanitation reforms in London and Scotland.[48] His report was the basis of Haussmann's June 1855 proposals to the Paris municipal council in which the principles of a reform programme were set out, Belgrand and his engineers were entrusted with the task of up-grading older sewers, continuing the Rivoli and Sébastopol projects, and constructing further sewers under new streets to improved designs, all connected, through *collecteurs élementaires*, to *grand galeries* where the water channels would be wide enough to carry boats used for cleansing and removal of silt. Mille undertook improvements to handling night soil and the introduction of water-borne sanitation.

Conditions for sewer construction were different either side of the Seine. On the Right Bank, where the land was low-lying and it was difficult to ensure a sufficient fall particularly for the *grand galeries* parallel with the Seine, Belgrand adopted a radical solution. Instead of discharging the sewers directly into the Seine within the city, he took advantage of the fact that, below Paris, the Seine made a series of S-bends. By constructing a new *collecteur général* across

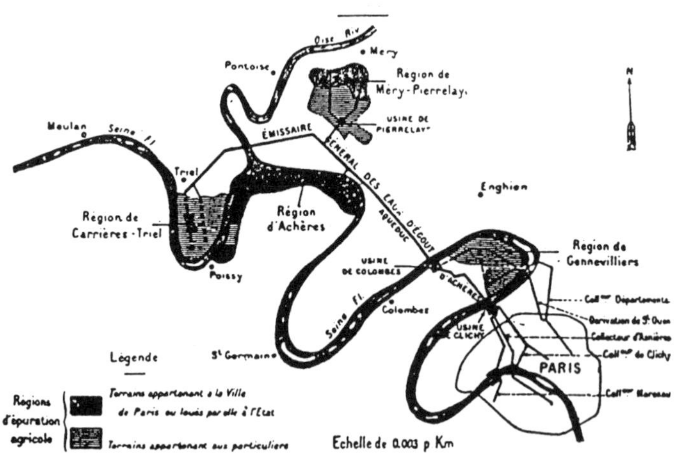

Fig. 13. Belgrand took advantage of the Seine bends below Paris to take his main collector first to Asnières (using de Clichy), later extended to serve the Gennevilliers sewage farm and later to Achères and Trueil

EUGÈNE BELGRAND (1810–1878): CIVIL ENGINEER

the peninsular formed by the first of these bends, a fall in level of 2 m 50 was achieved. His plan had two advantages: by better handling of flood water, it reduced the risk of flooded cellars in older districts on the Right Bank; and it facilitated control of silt and its removal.

The new *collecteur* started at Concorde in the centre of Paris. Picking up existing sewers and *collecteurs*, it passed in a tunnel under the Monceau ridge and through open country beyond the line of fortifications to an outfall at Asnières on the Seine 7 km downstream from Paris. The

Grand égout collecteur.

Fig. 14. Main collector showing pipes carrying potable and non-potable water and cart for collection of silt

total length of the Asnières *collecteur* was 5182 m, of which 2828 m was in tunnel. Gigantic in size, it had a 3 m 50 wide and 1 m 35 deep drainage channel, contained within a 5 m 60 wide and 4 m 40 high vaulted chamber, with 0.90 m inspection passages on both sides of the channel. Started in 1855–56, construction went on during 1857–58, the Asnières *collecteur* was brought into operation on 10 March 1859.

The situation on the Left Bank differed on two accounts: the Latin Quarter around the Sorbonne, although not low-lying, was densely built-up; and the Bièvre, a stream of some importance, was heavily polluted. Belgrand discarded an earlier scheme which would have only covered-in the Bièvre, allowing it still to discharge into the Seine. Instead he constructed a new Left Bank

EXCURSION IN THE PARIS SEWERS.

Fig. 15. During the nineteenth century the Belgrand's sewers attracted tourists

collecteur, picking up the Bièvre sewer. The original plan was to connect it with the Asnières *collecteur* at Concorde by means of a siphon under the Seine. However, this part of the plan had to be modified when it proved impractical to connect sewers serving quartiers west of Concorde with the Asnières collecteur. An alternative scheme which took the Left Bank *collecteur* further downstream to pont d'Alma was adopted. Taken under the river by a siphon at this spot, sewer water was carried in a second general *collecteur* located to the west of the Asnières *collecteur*. Picking up sewers in Grenelle, Passy and Auteuil, the second *collecteur* joined the main *collecteur* in the suburb of Levallois-Perret to discharge into the Seine at Asnières. Work on the first section of the Bièvre *collecteur* started in 1860, but, because of design changes and difficulties in the Sorbonne area, it was not completed until November 1869. Belgrand took personal interest in the Alma siphon, inventing a technique for clearing silt collecting in the two pipes by means of a large wooden ball forced through by the weight of water behind.[49]

In contrast to Belgrand's achievements in the construction of sewers, the story of improvements to public sanitation was less happy. At the start of the nineteenth century, domestic waste was still thrown into the street. Domestic sanitation depended on cesspits which were, from time to time, emptied at night by contractors in *tonneaux* the product being taken, with domestic waste, carcases of animals, etc, to the Montfauchon depôt. After its closure in 1846 night-soil and domestic waste, still collected at night, was taken to be a depôt on the Ourcq canal at La Villette, and carried by barge to a 'model' installation in the Forest of Bondy, 7 km beyond the line of fortifications, where *poudrette* was made from human manure as fertiliser . . .

During the first half of the century, there had been small improvements in public sanitation; but the practice of cesspit use and night-soil collection was to prevail for many years. Any change depended on solving two distinct but related problems: namely, ending night-soil collection, and replacing it with a water-borne system; and utilisation of the resulting sewage. Obstacles to change were on the one hand commercial and financial and on the other technical. Belgrand's two contributions—a new and much improved system of sewers, and a vast new system for water supplies—helped; but it was for other engineers, particularly Adophe Mille and his younger colleague, Alfred Durand-Claye, to ensure the replacement of night-soil collection by a system of

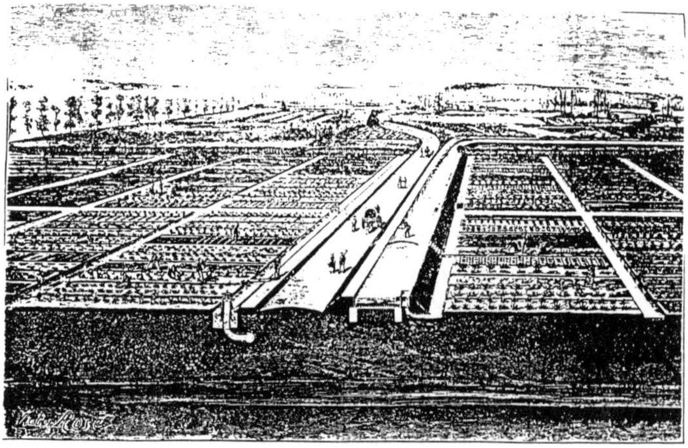

10. Sewage farming at Gennevilliers in the 1870s.

Fig. 16. Adolphe Mille's experimental sewage farm at Gennevilliers

water-borne sanitation, thus finally remedying a sanitary condition complained of by Mrs. Trollope half a century earlier.

TWO HORRIBLE YEARS: 1870–1871

After the achievements of the 1860s, the year 1870 and the first half of 1871 were joyless. At the beginning of the year, Georges Haussmann was dismissed, and before a new administration in the Hôtel de Ville had had a chance to settle in, Emperor Napoleon III declared war on Prussia. After a series of defeats, the Emperor surrendered at Sedan on 1 September. France was invaded and on 19 September Paris was surrounded by the Prussians. Belgrand, with colleagues in the Service of the Ville de Paris and the *département* de la Seine, remained in the besieged city, sharing its deprivations, helping in its defence and, in the case of Belgrand, ensuring a supply of water, and since the Bondy installation was in Prussian hands, disposal of night-soil and domestic waste.[50]

The Armistice negotiated early in 1871 was not to be the final stage in the tragedy. For Belgrand, the worst was yet to come. The Commune, which took charge of Paris at the end of March 1871, having replaced Belgrand in charge of water and sewers, in the bloody days of retreat before the troops of the new French Republic, set fire to the Hôtel de Ville, destroying his library and records. For Belgrand, there were two compensations: on 1 January he had been awarded the Grand Cross of a Commander of the Legion d'Honneur for services to the people of the besieged city, and on 28 August he was elected to the Académie des Sciences as one of seven *membres libres*.

THE ACADEMY OF SCIENCES[51]

Election to membership of the Academy of Sciences on 28 August 1870 as one of seven *membre libres* was possibly the most gratifying event in Belgrand's career. His first contact with the Academy had been in 1830 as a student at the École polytechnique. His first paper on hydrology had been sponsored by Arago, the Academy's permanent secretary. And, after becoming an academician, he was to contribute some thirty papers as well as taking an active part in the Academy's activities. The Academy's rooms at *l'Institut* on Quai Malaquais were only a few minutes walk from Belgrand's residence in rue Bonaparte, while his office, established after the Hôtel de Ville fire[52] in the Luxembourg Palace, and the École des ponts and chaussées in rue des Saints Pères were both close at hand.

Living in such a convenient location, Belgrand was fortunate in being able to direct his engineers' work, slowed down by financial stringency following the recent war, to give attention to the Seine hydrometric service, now under the daily control of Georges Lemoine, and yet have time to develop his hydrological studies. That Eugène Belgrand was to die at the age of 68, only eight years after becoming an active and valued member of the Academy, is regrettable. But few men are as fortunate to have enjoyed such a rich and varied life, and to have seen much of the engineering work for which he was responsible completed, and his scientific work published and appreciated by his peers.

THE SEINE HYDROMETRIC SERVICE[53]

In 1852, even before posting to Paris, Belgrand's pioneer work on the Seine basin hydrology was becoming recognised.[54] And, when a hydrométrique service for the Seine river basin was planned, clearly he was the right man to organise it. The service dates officially from 5 February 1854, when the Minister of Agriculture, Commerce and Public Works authorised its establishment, entrusting its management to Belgrand, then in charge of Seine navigation from Paris to Rouen. The service's main task was to record rainfall and river water levels throughout the Seine basin. Initially two stations were set up, at Montbard on the Brienne and at Saint Dizier on the Marne, with telegraphic communication with Paris, to give a 24-hour flood warning. From 1 May 1854, variations in the levels of the Seine and its tributaries were recorded without break, the results being published by the ministry. Belgrand remained responsible for the service for 24 years until his death in 1878.

The importance Belgrand placed on the service suggests it was very much his brainchild and, while the observations made served practical functions like advanced warnings of floods, he regarded the data collected of equal importance for his long term study of the hydrology of the Seine basin. Writing three years after its establishment, Belgrand reported that regular observations of variations in the levels of the Seine and its tributaries were being made at 49 stations since 1 May 1854. There had, however, been a problem with the design of the rain gauges and comparative trials were in progress at sites at Dijon and on the Normandy coast. In the meantime he was having to make do with rainfall data from ten sites, all in the Yonne *département*.

The data already collected on the state and flow of waters made it possible to compare local geology and water regimes at different locations in the Seine and its major tributaries the Yonne, the Marne and the Oise. It showed that a very heavy fall of rain in the Morvan, the hard rock area at the head of the basin, would be noticed by a high level at Paris three or four days later and that a mass of water—*les eaux tormentils*—from such a impermeable terrain would precede that from permeable terrain—*les eaux tranquilles*—further downstream some days later. Examining 38 reports of exceptionally high levels of the Seine at Paris, Belgrand was able to draw lessons from the terrible floods which struck France in the early days of June 1875.[55] Comparing water levels predicted with those actually observed, he suggested that his predictions would have been even closer if there had been communication between those in the field and Belgrand at Paris, pointing out that although official mail reached his office later than ordinary letters, experience from two stations in telegraphic communication would have enabled predictions to be made twenty-four hours earlier.

In the early years of the service Belgrand had part-time assistance from two engineers in the Seine navigation service, Rozat de Mandres and Collignon. In 1866, he was joined full-time by a young *ingénieur des ponts et chaussées*, Georges Lemoine.[56] Although the principal task of the service was to collect data for agriculture and navigation and, especially, for flood warning, from an extensive knowledge of the different ground conditions in the upper Seine basin, Belgrand and Lemoine were able to 'tune' their predictions of flood conditions, checking their methodology against earlier recorded floods. Jointly, they published two papers in the *Comptes Rendus* on predictions of variations in water flows in the Seine during 1874 and 1875, as well as annual summaries of observations made by the Service Hydrométrique between 1854 and 1876.

So well was the service organised under Belgrand that his successors had nothing essential to add. From the beginning of 1855, Belgrand was able to forecast 3 days ahead the level of flood water at Paris, a result of rises in water levels upstream; and, in 1872, Lemoine was able to predict high water levels in the Seine at Monterau (2 days ahead); Aisne at Pont Aven (3 days ahead); Marne at Damery, near Épernay (3 days ahead); and, Oise at Compiègne (4 days ahead). On Belgrand's death in 1878, Lalanne, Director of the École des ponts et chaussées, was given overall responsibility for the service. Lemoine was in charge of its daily management, taking over full responsibility when Lalanne retired from public service three years later, at a time when the service came under a Flood Warning Commission, having the wider task of establishing hydrological and flood warning services at a national level.

A PIONEER HYDROLOGIST

The river basin, bounded by its drainage divide and subject to surface and sub-surface drainage under gravity to the ocean or to interior lakes, forms the logical area unit for hydrological

EUGÈNE BELGRAND (1810–1878): CIVIL ENGINEER

studies. Rosemary J. More, *Water, Earth and Man: a synthesis of hydrology* . . . (London, Methuen, 1969).

That Eugène Belgrand was a pioneer in the science there is no doubt. Yet a hundred and twenty two years after his first paper on the hydrology of the upper Seine basin was published in the proceedings of the French Académie des Sciences, neither it, nor his key paper[57] were mentioned in an important collection of papers on hydrology and associated topics, *Water, Earth and Man*. There have been great advances in the science of hydrology during the last hundred years, especially in areas like data collection and modelling. Rosemary J. More, from whose paper, 'The Basic Hydrological Cycle', the heading to this section was taken, presented in a complete form the components of the hydrological cycle, and their inter-relationship as a preliminary to a discussion on computer and analogue modellings. Belgrand's studies were confined to the Seine basin, and were directed to the relationship between terrain, rainfall and river flow, and vegetation, yet, except in techniques, there are no fundamental differences between Belgrand's work and that of more recent hydrologists.

Belgrand's work was based on personal knowledge of the extensive network of streams and rivers which forms the Seine basin over a period of sixty years—longer if account is taken of his youth at Evry-le-Châtel and Chaumont. It was based on an experience of the physical terrain through which they flowed, and, after he took charge of the Seine hydrometric service, on its observations and on historical data, especially on major Seine floods. Together with his assistant, Georges Lemoine, he was able to create a real life 'analogue model' from which they were able to predict exceptionally high river levels at Paris and elsewhere in the Seine basin, validating predictions against hydrometric observations.

The extent of Belgrand's scientific work is remarkable, taking into account that, for much of the time, he was responsible for major civil engineering projects as *directeur des eaux et égouts de la Ville de Paris*. Starting with his 1847 paper on the hydrology of the upper Seine, Belgrand wrote 55 papers on hydrology and related matters. Some were short contributions to Monday meetings of the Académie des Sciences. Others were reports of observations from the Service Hydrométrique, possibly jointly with Lemoine and others were devoted to hydrological principles and their applications. Papers range from a contribution on the nature of eddies in flowing waters, presented to a meeting of the Académie des Sciences a few days before his sudden death, to more general reviews of the water systems of France, Greece and Egypt. Twenty-four papers on related meteorological topics were published in the *Annuaire de la Société météorologique de France*. A number, published in the *Annales des ponts et chaussées*, took on a somewhat different form for his engineering colleagues. Belgrand's monumental publication: *Les Travaux Souterrains de Paris* was not only an account of his engineering works. It also brought together much of the work.

BELGRAND'S DEATH

Eugène Belgrand died in Paris early in the morning of 8 April 1878 at his home, 21 rue Bonaparte on the Left Bank. He was in his 68th year, still active in the service of the City of Paris, and in full charge of his faculties. Seven days before he had contributed a paper on the nature of eddies in moving streams to a meeting of the Academy of Sciences.[58] In announcing his death at the usual Monday meeting of the Academy a few hours later, the president, M. Fizeau, said that Belgrand's death was unexpected and that he died a few hours after being taken suddenly ill.[59] He was with his family at rue Bonaparte a few steps away from the Academy, and, until his actual death, was in full possession of his senses. After the announcement, the session was adjourned.

The funeral took place at his parish church, Saint Germain-des-Prés, in the presence of his wife and members of his family. His coffin was placed in the crypt of Saint Germain-des-Prés, later to be taken to Châtillon-sur-Seine. At the funeral, General Favé read a tribute on behalf of the Academy, to be followed by M. Daubrée on behalf of the Academy's *section du minéralogie et géologie* of the Academy.[60] There were obituaries in a number of Paris journals, including one by his colleague, Mille, in the *Revue scientifique*.[61] Later, Léon Lanne, inspecteur général des ponts et chaussées, a fellow student and lifelong friend then in retirement at Étretat, contributed a lengthy obituary to the *Annales des ponts et chaussées*, to which was attached a list of Belgrand's publications, prepared by M. Vialey, civil engineer and a relative of the Belgrand family.[62]

On Belgrand's death, Jean-Charles Adolphe Alphand, who was already technical director of works, City of Paris, took over the *Direction des eaux et égouts*, adding water and sewers to his many duties including as organiser of the 1878 *Exposition Universelle*, Paris. Later in the year he reported to the Paris municipal council on the arrangements for continuing Belgrand's work.

LAST WORDS

> Il y a en effet, dans M. Belgrand, deux hommes éminents, un homme de science et un ingénieur, tous deux tels qu'on ne saurait dire quel est celui des deux qu'il l'emporte sur l'autre. (Discours de M. Le Général Favé au nom de l'Académie des Sciences. 8 avril 1878).[63]

The life and works of Eugène Belgrand raise a number of questions. First, there are the comparative contributions of Haussmann and Belgrand to the concept of supplying Paris with pure water from distant sources for domestic uses while using non-potable water, taken from the Seine and the Marne rivers, for industrial uses, street cleansing etc. Secondly, there is the question of responsibility for the unsuccessful Somme-Soude scheme. Was it unsuccessful because the scheme was rushed by Haussmann? Or was Belgrand's survey inadequate, leading to an error of technical judgement? Thirdly, how far were the criticisms (mainly for over-spending and poor management made by Alphand on taking over Belgrand's office on his death in 1878) justified or were they caused by Alphand's jealousy at his colleague's academic and scientific achievements?

In debating these questions, the personalities of the two men (Haussmann and Belgrand), their education, social background, careers and standing among colleagues and in public opinion are important. Haussmann was an exceptionally able administrator, but he was also very much a Louis Napoleon man.[65] In nineteenth century France, *préfets* had to be politicians as well as administrators; and Haussmann's career before he was posted to Paris showed him to be a man able to influence local politics in the interest of the government in Paris, while keeping an eye on local politicians. His posting to the Seine prefecture, for example, owed much to the support he managed for the Prince President, Louis Napoleon, when he was *préfet* of the Gironde *département*. Being a political *préfet*, however brillant as an administrator, had risks as Haussmann found when, in 1870, he was dismissed by Napoleon III's last chief minister, Emile Ollivier, a politician jailed as a young man for a few days when Haussmann was *préfet* of the Var 'département'.

Haussmann, after studying law at the Sorbonne, got to the top in the prefectorial service by maintaining links with persons of influence in Paris. Belgrand and fellow engineers, once over the hurdle of entry to the École polytechnique, had a career in the state service which was more or less planned with the expectation, if they showed special merit, of becoming an *inspecteur général* or, if reasonably competent, an *ingénieur en chef* in charge of a 'départemental service'. Benefiting from the esprit de corps which existed between polytechniciens and members of the Corps des ponts et chaussées, they were able to pursue special interests side by side with normal engineering responsibilities. Publication in technical and scientific journals won respect among colleagues, but hid achievements from the general public.[64] As a result many books have been written about Haussmann, none about Belgrand or his colleague Mille.

NOTES AND REFERENCES

1. Denis Smith, 'Sir William Joseph Bazalgette (1819–1891) Engineer to the Metropolitan Board of Works', *Trans. Newcomen Soc.* Vol. 59 (1986–87), pp. 89–110.

EUGÈNE BELGRAND (1810–1878): CIVIL ENGINEER

2. The Corps des ponts et chaussées founded in 1716 and reorganised during the Revolution was given its nineteenth century structure by the *Décret Imperial portant organisation des Corps des Ponts et Chaussées*, signed on 24 August 1804 by Napoleon at Boulogne ready to invade England.
3. The name given to a group of students passing through the School together.
4. A district, the basic administrative unit of government in France.
5. Described in Appendix 3.
6. Listed in Appendix A1.
7. Listed in Appendix B2.
8. Evry-le-Châtel lies in Burgundy south east of the Forêt d'Othe in wooded country at a height of 145 m above meadow land formed by the confluence of two small rivers, the Armance and Bernon, tributaries of the Yonne.
9. Chaumont in Haute-Marne occupies a rocky platform separating the Marne and its tributary the Suize. Lalanne has described Belgrand's life as a youth: 'Ce n'était pas uniquement au collège de Chaumont qu'il avait acquis l'instruction littéraire, le goût des arts, le sens profond de la nature dont il a donné tant de preuves. Jouissant des charmes d'une vie campagnarde jusqu'au douze ans, après des longues promenades dans les forêts qui couvrent les plateaux du bassin de l'Ource (a tributary of the Seine) . . . Il avait donc déjà l'esprit ouvert aux principaux objets de l'enseignement lorsqu'il entra au collège; et sa constitution vigoureuse développée par la vie rustique lui permit de suivre ses études sans relâche, aussi bien que de travailler jusqu'à son dernier jour sans paraître en ressentir aucune fatigue'. Lalanne, pp. 338–339.
10. 'C'est dans la classe de mathématiques spéciales du collège Louis-le-Grand qu'il recut d'un excellent professeur, feu M. Richard, une instruction qui ne se bornait pas aux matières du programme restreint dont on se contentait alors pour l'admission à l'École'. Lalanne, p. 339.
11. Eugène Belgrand entered the École polytechnique at a time of general unrest. In July 1830 the people of Paris led by students of the École polytechnique and supported by the commercial middle class rose in revolt against Charles X, the last of the Bourbons, who abdicated and was replaced by Louis Philippe, a constitutional monarch.
12. 'Ce fut un beau jour que le vendredi 9 juillet, alors que le canon des Invalides, vers quatre heures de soir, vint annoncer de sa voix la plus éclante à Paris qui osait à peine y croire, la chute d'Alger l'imprenable. . . . C'es sous l'influence de la double émotion de ce triomphe militaire et de la lutte entrainée entre le régime parlamentaire et le pouvoir royal que s'achevait l'année scolaire des deux promotions de 1828 et 1829, lorsque les ordonnances de 26 juillet 1830 vinrent, comme un coup de foudre, arrêter subitement le cours de nos études'. Lalanne described the atmosphere during a eulogy on Fresnel read by the physicist Arago attended by many École polytechnique students: 'Des salles réitérées d'aplaudissements, dont les polytechniciens donnaient le signal, acceuillirent certains passages qui, mettant en relief les opinions libérales de Fresnel et la noble abnégation avec laquelle il les professait, paraissaient une allusion brûlante aux préoccupations du moment'. Lalanne, pp. 339–341.
13. 'On rentrait à l'École avec le coup d'émotions causées à la fois par la séance académique et par l'effervescence qui commencait à se manifester dans Paris. Quarante-huit heures après, l'École était vide et ne devait se rouvrir que trois mois et demi plus tard. Des élèves marchaient à la tête de combattants dont étaient les chefs improvisés.' Lalanne, pp. 341.
14. Although Belgrand participated in the July event, 'il ne parlait jamais de ces circonstances si émouvantes'. Mille, p. 1191.
15. Founded in 1747, after a somewhat chequered time during the Revolution the School was reorganised under Napoleon, its organisation, course content, etc., being set out under the Napoleonic Corps des ponts et chaussées decree. It was later installed in a building in rue Saint-Pères.
16. The 1804 decree required students to spend their summer months gaining experience in the field: 'Les élèves pourront être envoyés en campagne dans le cours de floréal (april—may) de chaque année . . . Il recevront . . . le traitement des aspirans . . . Les élèves ainsi envoyés en dehors seront tenus d'être rentrés à l'école le 1(er) frimaire (novembre): . . ., para. 65.

EUGÈNE BELGRAND (1810–1878): CIVIL ENGINEER

17. 'En 1832, simple élève ingénieur, dit-il lui-même dans la préface de son livre *La Seine*, j'étais chargé de surveiller la construction d'un petit pont de trois arches sur la Brenne, à Vitteaux-en-Auxois. Pendant mon séjour dans cette ville, je fus témoin d'u de ces phénomènes qui ne s'effacent pas de la mémoire. À la suite d'une forte pluie d'orage, qui dura moins d'une heure, je vis les eaux pluviales ruisseler de toutes parts à la surface des coteaux qui bordent la vallée. En un instant, la Brenne éprouva une crue énorme, qui s'éleva audessus des niveaux des parapets du pont que nous construisons. Ce pont en remplacait un autre beaucoup plus petit, qui avait été emporté par une crue. On avait calculé le nouveau débouché d'après les règles les plus larges admises alors, et cependant il était évident pour moi qu'on était resté bien au dessous des dimensions nécessaires'. Mille p. 1991 quoting from 'la préface de son livre *la Seine*'.
18. 'Je fus bientôt connuit à cette conclusion, que ce double phénomène était dû à la nature du sol. . . . J'en conclus qu'il y a avait une lacune dans nos études, et, dès cette époque, je fus convaincu qu'un ingénieur des ponts et chaussées doit être non-seulement "géomètre", mais aussi "géologue"'. Mille, p. 1191.
19. 'Juge de ma misère; m'écrivait-il le 5 janvier 1835, depuis huit mois au milieu de plus magnifique pays je n'ai pas encore eu le temps de m'occuper de peinture!' Lalanne, p. 343.
20. There are in the Côte d'Or archives two notices from the Directeur-Général des Ponts et Chaussées to the Côte d'Or préfet respectively dated 25 May 1836 announcing Belgrand's promotion from 15 June, and 26 May announcing his transfer from Puy de Dôme to Côte d'Or on 6 June 1836.
21. '. . . les premières données du livre (*Les Aqueducs romaine*) dataient d'un voyage fait à Rome; de 1837 à 1838, M. Belgrand avait accompagné en Italie un de ses frères gravement malade; il y resta six mois'. Mille, p. 1194.
22. In 1839 Belgrand married Mlle de Brosse 'appartenant à l'une des meilleures familles de pays'. Lalanne, p. 343.
23. Avallon (Yonne), a small town on the edge of the Morvan uplands stands on a promontory above a precipitous winding gorge 100 m deep at the bottom of which flows the small Cousin river.
24. 'Études hydroliques: Dans les granits et les terres juraissique formant la zone supérieur du bassin de la Seine' (Avallon 1 février 1846), *Annales*, 1846. II, No. 153, pp. 129–183.
25. There are a number of papers in the Côte d'Or archives between the Avallon mairie and other authorities on the proposed water supply.
26. 'Sur l'établissement d'une conduite destinée à améner les eaux du ru d'Aillon à Avallon, suivi de quelques considérations sur les ouvrages en ciment romain' (Avallon 1 decembre 1849), *Annales*, 1850. II, No. 238, pp. 412–451.
27. 'Vers 1850, M Haussmann devint préfet de l'Yonne et visita Avallon. Comme administrateur, il avait dans le Midi cherché aussi à livrer des eaux vives aux populations. Il fut frappé de la justesse des idées de jeune ingénieur et de la hardiesse des travaux qu'il avait sous les yeux. Il s'en souvint quand il fut préfet de la Seine, en en 1854 il demandait à M, Belgrand de déterminer les sources qu'on pouvait amener à Paris'. Mille, p. 1191.
28. Georges Haussmann described his meeting with Belgrand and their visit to the Fabrique de Ciment de Wassy in his *Mémoires*, (1979), reprint, Vol. 1, Ch. IV., 'Les Services d'Ingénieurs', p. 112.
29. There is a copy of the notice transferring Belgrand to the Seine Navigation in the Yonne archives. The Service was organised in three sections: the first section upstream from Paris, the second the stretch through Paris including the bridges, and the third, for which Belgrand was responsible, the stretch below Paris. Promoted *ingénieur en chef des ponts et chaussées 2e classe*, he was also made responsible for maintenance of hydrological data for the whole Seine basin. G Haussmann, *Mémoires*, (1979), reprint, Vol. 1, Ch. IV., 'Les Services d'Ingénieurs', p. 103.
30. '. . . que mes notes très favorables avaient contribué à faire nommer ingénieur en chef après mon départ d'Auxerre pour Bordeaux.' G Haussmann, *Mémoires*, (1979), reprint, Vol. 1, Ch. IV., 'Les Services d'Ingénieurs', p. 114.
31. *Annales*, 1855 II, pp. 351–352. Henri-Charles-Léopold Emmery de Sept Fontaines [1815–1879], whose father had been founder and first editor of the *Annales des ponts et chaussées*.

EUGÈNE BELGRAND (1810–1878): CIVIL ENGINEER

32. For Paris water before 1854 see Belgrand *Les Travaux Souterrains de Paris* Vol. III 'Les Eaux: Première Section Les Anciennes Eaux' (Paris, 1882) and G Haussmann, *Mémoires*, (1979), reprint, Vol. 2, Ch. IX, 'Service des Eaux', pp. 21–28.

33. 'Il eut avec moi des longs entretiens, au cours desquels nous recherchâmes, dans le Département de l'Yonne, au moyen de la carte géologique . . . C'est encore à Auxerre que prit forme cette théorie de M. Belgrand, d'apparence paradoxale dont la justesse est aujourd'hui confirmée par l'expérience: les sources abondantes et pérennes se rencontrent dans les contrées arides, où les couches perméables du sol ont une grande épaisseur, où l'eau des pluies et des neiges, absorbées profondement, reparait seulement dans quelques ravins offrant au sous-sol imperméable de rares points d'affleurement, comme en Champagne'. G Haussmann, *Mémoires*, (1979), reprint, Vol. 1, Ch. IV., 'Les Services d'Ingénieurs', p. 113.

34. G Haussmann, *Mémoires*, 1979 reprint, Vol. 2, Ch. IX, 'Service des Eaux', p. 42.

35. '. . . en 1853 . . . Je le décidai facilement à se charger d'un Service d'Études ayant pour objet l'exploration complète des sources du Bassin de la Seine, le jaugeage de ces sources et l'analyse de leurs eaux'. G Haussmann, *Mémoires*, (1979), reprint, Vol. 1, Ch. IV., 'Les Services d'Ingénieurs', p. 114.
'Il tait si bien prêt que, recevant en avril la mission de rechercher les sources qui pourraient alimenter Paris, il répondait dès le mois du juillet par la belle carte du bassin de la Seine, carte qui est un modèle de géologie appliquée et d'hydrologie exprimée par la couleur. Les considérations éclairaient l'examen, les deductions qui en sortaient parurent, dès 1854, dans le Mémoire du préfet au Conseil municipal. ©e Mémoire était la programme des améliorations d'assainissement qu'il s'agissait d'entreprendre. . . .'. Mille, p. 1191–1192.

36. Belgrand summarised the investigation later in 'Hydrologie—Sur les conditions qu'on a dû chercher à réaliser dans le choix de sources destinée à l'alimentation de la Ville De Paris', 21 avril 1873, *Comtes Rendus*, Tome 76.1 pp. 989–997.

37. For one version of this less happy project, G Haussmann, *Mémoires*, (1979), reprint, Vol. 2, Ch. XI, pp. 128–133.

38. Fully described by Buffet from Belgrand's papers in *Les Travaux Souterrains de Paris* Vol, IV, 'Les Eaux': Deuxième Section: Les Eaux Nouveaux' (Paris, 1882), there is a shorter description of the scheme in Lalanne, pp. 353–354.

39. Again fully described by Buffet from Belgrand's papers in *Les Travaux Souterrains de Paris* Vol. IV, there is a shorter description of the scheme in Lalanne, pp. 355–357. Haussmann gives the administrative background to the Dhuis and Vanne schemes in his *Mémoires*, 1979 reprint, Vol. 2, Ch. XI, pp. 133–151.

40. pp. 149–153.

41. At Sens, a Roman town, the remains of a gallo-roman aqueduct drawing water from springs in the Vanne valley have been discovered.

42. Structures along the aqueduct include the 3757 metre long 'béton aggloméré' arcade crossing the valley of the Yonne, the masonry arcade at Guy, and the arcade across the Loing at Moret, also in 'béton aggloméré'. The 900 metres long arcade across the Bièvre valley at Arcueil, south of Paris was the last of three aqueducts built at this spot. The first, of which only the ruins of masonry piers may be seen today, is what is left of a Roman aqueduct, erected in the time of Emperor Julian the Apostate (361–363 AD) to carry water from springs at Rungis to the Baths of Gallo-Roman Lutetia, the ruins of which form part of the Cluny museum on boulevard St Michel. The second, 400 yards long, consisting of twenty dressed stone arches, was built between 1613 and 1624 by the architect Salomon de Brosse to convey Rungis waters, about twelve miles distant, to the Luxembourg palace which de Brosse designed for Maria dei Medici at the beginning of the 17th century. The third built by Belgrand between 1868 and 1872 for the Vanne aqueduct consists of seventy seven 40 metres high arches using the de Brosse arcade as foundations. Buffet described the use of 'béton aggloméré' in Vol. IV 'Les Eaux': Deuxième Section: Les Eaux Nouveaux' p. 284 sq.

43. There is a description of the difficulties met with crossing the Forest of Fontainebleau in *Les Travaux Souterrains de Paris*, Vol. IV, Lalanne pp. 353–354. Belgrand gave an account of some aspects of the

work in *Comtes Rendus* 'Hydrologie—Sur la perméabilité du sables de Fontainebleau', 21 juillet 1873, Tome 77.2 pp. 178–182.

44. Belgrand, 'Sur l'établissement d'une conduite destinée à améner les eaux du ru d'Aillon à Avallon, suivi de quelques considérations sur les ouvrages en ciment romain' (Avallon 1 decembre 1849) *Annales*, 1850. II, No. 238, pp. 448–449.

45. Haussmann gives the history and administrative background to Paris sewers in his *Mémoires*, (1979), reprint, Vol. 2, Ch. X, pp. 98–110. The last volume of *Les Travaux Souterrains de Paris*, Vol. V, 'Deuxième Partie: Les Égouts' (Paris, 1887), is devoted to the work of Belgrand and his engineers. See also Mille, p. 1191 and Lalanne, pp. 357–361.

46. *Galignani's New Paris Guide* 1838 pp. 240–241 described the place 'where most of the night-carts of Paris are emptied. A raised causeway of stone advances between two black and deep pools; along the edges of this the carts, which are enormous tuns placed on wheels are arranged, and empty their contents into a shelving trough placed a little below the causeway, from whence, after much racking and examination, they fall into the upper pond. Men remain here whole days searching for money, jewels, and other valuable articles, which may chance to be found in the sewers, and are sometimes very successful in their search. The contents of the upper pool drain into the second, and from thence into three others successively; the water escapes, and the solid sediment is cut out and used for manure. The sides of the upper pool are slaughter houses for horses, where all the worn-out animals of the capital are brought . . .'.

47. Adolphe-Auguste Mille (1812–1894) civil engineer and environmentalist, 'Sur le service des vidanges publiques de la Ville de Paris' *Annales*, 1854/1 Mémoire No. 71 pp. 129–157.

48. 'Préfecture du Département de la Seine. Rapport sur le mode d'assainissement des villes en Angleterre et en Ecosse presenté à M. le Préfet de la Seine par M. Mille, ingénieur des ponts et chaussées' (Paris, 1853) pp. 33; Mille, 'Sur le mode d'assainissement des villes en Angleterre', *Annales*, 1855/1, Mémoire No. 108, pp. 129–170.

49. Belgrand, 'Note sur l'égout collecteur dit de la Bièvre et la siphon de l'Alma' *Annales*, 1869. II, Note 241, pp. 425–457.

50. Mille, p. 1193 and Lalanne, p. 365–366; also Belgrand, 'Note: Conservation de l'eau des réservoirs de la Dhuis, pendant le siège de Paris', 30 decembre 1872, *Annales* Tome 75.2 pp. 1840–1841.

51. Founded in 1666 four years after the Royal Society, the Academy was suppressed for a time during the Revolution, reorganised in 1803 by Napoleon who strengthened the divisions for mathematical and physical sciences, each having 30 members in five sections but suppressed the division for moral and political science. In 1816 a new category of membership—'académicien libre'—was created for men like Belgrand whose talents could not easily be accommodated in any particular section. The class of correspondent for distinguished foreigners was extended. The *Comtes rendus* were introduced by Arago in 1835 as records of the Academy's weekly Monday meetings. At the time of Belgrand's election as an 'académicien libre', corresponding members included Henry Fairbairn and Henry Moseley. The Academy is one of the component bodies of the 'Institut de France', meeting in the seventeenth century building designed by Le Vau on the Quai Malaquais a short distance from Belgrand last Paris residence in rue Bonaparte.

52. 'Sous la Commune . . . le plus grand mal, c'est que son cabinet fut brûlé lors de l'incendie de l'Hôtel de Ville, et que là périrent les matériaux déjà accumulées pour l'histoire des eaux de Paris'. Mille, p. 1193.

53. Created in 1854 by the Ministry of Commerce and Public Works its task was to record rainfall, snow and levels of water in the Seine and its tributaries. Belgrand was responsible for analysing and publishing summaries of the observations, as well as using the data collected for flood prediction and hydrological studies, assisted by a young engineer Georges Lemoine who was to take over responsibility for the service on Belgrand's death and later to have a distinguished career as a scientist-civil engineer-educationalist. Belgrand, 'Service hydrométrique du bassin de la Seine: Mémoire sur les observations hydrométriques, du 1 mai 1854 au 30 avril 1855; sur la qualité des eaux de sources du basin' *Annales*, 1857. I pp. 257–307.

EUGÈNE BELGRAND (1810–1878): CIVIL ENGINEER

54. Belgrand's hydrological studies were becoming more widely known by the early 1850s. For example, in *Annales*, 1853. II p. 183 there is a references to his 1846 investigations into the impermeability of certain terrains: 'L'hydrologie est une science toute nouvelle, et qui doit attirer l'attention des ingénieurs'. His reputation had been enhanced by the publication of his 'Carte hydrolique et géologique du bassin de la Seine'.

55. Belgrand, 'Hydrologie—Perturbations atmospheriques de la saison chaude de l'année 1875. Note sur le groupe de pluies du 21 au 24 juin 1875; crue de la Garonne; désastre de Toulouse; inondations du midi de la France', 29 novembre 1875, Tome 81.2 pp. 1017–1025; 6 decembre 1875, pp. 1082–1086; 13 decembre 1875, pp. 1168–1175.

56. Georges Lemoine (1841–1922) was born at Tonnerre (Yonne) on 26 January 1841 and like Belgrand, whom he joined in 1866, was familiar with the upper Seine from his early youth. In 1872 he developed a technique for predicting flood levels for the Seine at Montereau 2 days ahead; of the Aisne at Pontavent 3 days ahead; of the Marne at Damay near Épernay 3 days ahead; and on the Oise at Compegène 4 days ahead. On Belgrand's death he managed the hydrometric service under Lalanne, taking over full responsibility after Lalanne's retirement in 1881 when he was appointed 'ingénieur en chef'. Later he left the service for a distinguished educational career as professor of chemistry at the École polytechnique, being elected in turn a member of the chemistry section of the Academy of Sciences in 1899, and becoming in turn the Academy's vice-president and for a short time its president before his death in 1922.

57. Belgrand, 'Études hydroliques: Dans le bassin de la Seine entre les limites jurassiques et Paris' (Avallon 2 avril 1851), *Annales*, 1852. I, No. 24, pp. 1–228.

58. Belgrand, 'Hydrologie—Étude sur la tourbillon des cours d'eaux'. *Comptes Rendus de la Séance de lundi 1 avril 1878*, Tome 96, 1878. 1, pp. 798–801.

59. 'Il prenait encore la parole le lundi de sa dernière semaine; à propos de la discussion des tourbillons, causes des tempêtes dans l'atmosphère, il explique par des tourbillons les mouvements de propagations et de transport des eaux courantes. Le samedi, il dinait chez M. Chasles, qui réunissait comme de coutume chaque annèe, une groupe de ses collèques de l'Institut. C'est au retour, dans la nuit du samedi au dimanche 8 avril, qu'il était pris d'une obstruction d'intestine, et qu'il s'éteignait, trente-six heures après, la pensée encore tournée vers son service: c'était le 8 avril à dix heures du matin. Il avait soixante-huit ans. Il succombait en pleine jouissance de ses facultés, de son autorité, de sa grande considération. Mille, p. 1194.

60. Général Favé, 'Discours (sur M Belgrand) au nom de l'Académie des Sciences', followed by M. Daubrée, 'Discours (sur M Belgrand) au nom de la Section de Minéralogie et de Géologie' in *Comtes Rendus des Séances de l'Académie des Sciences*, Vol LXXXVI No. 14, pp. 842–850.

61. A. Mille, 'M. Belgrand', *Revue scientifique*, 15 juin 1878, pp. 1190–1194.

62. Léon Lanne, 'Notice sur la vie et les travaux de M. E. Belgrand' *Annales des Ponts et Chaussées*, 1881. 2 pp. 337–387.

63. Two recent French publications give a balanced view of Haussmann's work at 'préfet de la Seine': Henri Malet, *Baron Haussmann et La Rénovation de Paris* (Les Editions Municipales, Paris, 1973), and Jean des Cars and Pierre Pinon, *Paris-Haussmann* (Paris, Editions du Pavilion de l'Arsenal, 1991).

64. The answer is to be found in the following comment in Belgrand's letter to the Institution of Civil Engineers in July 1875: 'Les journaux ne sont pas occupés de ces publications, ils sont absorbés par la politique et ne songent guères à la science de l'Ingénieur'.

SOURCES

There is no book in English or French devoted to the life and works of Eugène Belgrand and little else in English publications except for short notes on his Paris works by Harold Clunn, *The Face of Paris* (Simpkins Marshall, London, 1933) and in early editions of the guides such as Karl Baedeker, *Paris and its Environs* (Leipzig). Donald Reid, *Paris sewers and sewermen, realities and representations* (Harvard University Press, London, 1993) is about the life of sewermen rather than the engineering of Paris sewers. There is much in English on Georges Haussmann and Napoleon III but it is about the rebuilding of Paris, its architecture and urban planning, or on the politics of the period.

Sources in French include the key obituary by Léon Lanne, 'Notice sur la vie et les travaux de M. E. Belgrand' *Annales des Ponts et Chaussées*, 1881, 2 pp. 337–387; and shorter obituaries by A. Mille, 'M. Belgrand', *Revue scientifique*, 15 juin 1878, pp. 1190–1194, Général Favé, 'Discours (sur M Belgrand) au nom de l'Académie des Sciences', followed by M. Daubrée, 'Discours (sur M Belgrand) au nom de la Section de Minéralogie et de Géologie' in *Comtes Rendus des Séances de l'Académie des Sciences*, Vol. LXXXVI, No. 14, pp. 842–850, Comte de Franqueville. 'M. Belgrand' in *La première siécle de l'Institut de France* (Paris, 1895–97), Vol. II, p. 35, and E Franceschini, 'Francois-Eugène Belgrand' in *Dictionnaire de Biographie Francaise* (Paris: 1936) Vol. 5, p. 95.

Other sources are the contributions made by Belgrand and colleagues to *Annales des ponts et chaussées* (Appendix A), and *Comtes Rendus des Séances de l'Académie des Sciences* (Appendix B), Georges Haussmann, *Mémoires du Baron Haussmann. Grands Travaux de Paris*, (Guy Durier, Paris, 1979) and Eugène Belgrand, *Les Travaux Souterrains de Paris*, (Paris: 1872–1887) (Appendix C). Relevant documents, including official reports, have been found in the 'départemental' archives of Côte d'Or and Yonne, and in the Ville de Paris archives and bibliothèques administrative et historique.

Sources relating to Paris water and sewers in the nineteenth century include: Horace Say, 'Travaux dirigés pas les ingénieurs', pp. 371–428 in *Études sur l'administration de la Ville de Paris et du Département de la Seine* (Paris, 1846); M. Blerzy, 'L'eau de la Seine' in *Études sur les travaux publics*, (Paris, 1873); Gaston Cadoux, 'Les eaux de Paris', pp. 39–64 in *La Vie des Grandes Capitales. Études comparatives de Londres–Paris–Berlin–Vienne–Rome* (Paris 1913); Robert Auzelle, 'Les Eaux de Paris', in *La Vie Urbaine*, Jul—Dec 1953, No. 3–4, pp. 291–293; Louis Réau, 'Les eaux de Paris' in *L'Oeuvre du Baron Haussmann. Préfet de la Seine (1853–1870)*, (Presse Universitaires de France, Paris, 1954) pp. 131–137; Henri Malet, 'Les travaux utilitaires—Les eaux—les égouts de Paris', pp. 258–268 in *Le Baron Haussmann et La Rénovation de Paris* (Les Editions Municipales, Paris, 1973); A Brunot and R Coquand, 'L'équipement de Paris: l'eau et l'assainissement' in *Histoire du Corps des Ponts et Chaussées* (CNRS, Paris, 1982) pp. 286–7, Pierre Pinon, 'Les réseau techniques de l'eau salubre, limpide et fraîche, . . . et des égouts' in *Paris-Haussmann* (Paris, Editions du Pavilion de l'Arsenal, 1991), pp. 150–160; and Bertrand Lemoine, 'Le Paris des ingénieurs. Belgrand, l'homme e l'eau', pp. 23–26, and Sabine Baries and André Guillerme, 'Paris sens dessus dessous', pp. 114–129 in *Paris d'Ingénieurs* (Paris, Editions du Pavilion de l'Arsenal, 1995).

APPENDIX 1

The Technical Papers of Eugène Belgrand

Most of the technical writings of Eugène Belgrand, other than books such as *Les Travaux Souterrains de Paris*, were published in the *Annales des Ponts et Chaussées**, a few being technical versions of scientific paper listed in Appendix 2.

'Études hydroliques: Dans les granits et les terres juraissique formant la zone supérieur du bassin de la Seine' (Avallon 1 février 1846) 1846, II, No. 153, pp. 129–183.

'Sur l'établissement d'une conduite destinée à améner les eaux du ru d'Aillon à Avallon, suivi de quelques considérations sur les ouvrages en ciment romain' (Avallon 1 décembre 1849), 1850 II, No. 238, pp. 412–451.

'Études hydroliques: Dans le bassin de la Seine entre les limites jurassiques et Paris' (Avallon 2 avril 1851), 1852, I, No. 24, pp. 1–228.

EUGÈNE BELGRAND (1810–1878): CIVIL ENGINEER

'Hydrolique: De l'influence des forêts sur l'écoulement des eaux pluviales', 1854, I, No. 61, pp. 1–27.

'Recherches statistiques sur les matériaux de construction employés dans le département de la Seine. Rapport sur le mémoire de M. l'Ingénieur Michaelot en date du 20 Octobre 1853', 1855, II, No. 124, pp. 189–212.

'Écluse et barrages de Meulun', 1855, II, Chronique, pp. 351–353.

'Service hydrométrique du bassin de la Seine: Mémoire sur les observations hydrométriques, du 1 mai 1854 au 30 avril 1855; sur la qualité des eaux de sources du basin', 1857, I, No. 168, pp. 257–307.

'Débit de la Seine pendant l'étiage de 1858: Communication de MM. Lagallisserie & Belgrand', 1858, II, Chronique, pp. 221–222.

'Notice sur la régime de la pluie dans le bassin de la Seine', 1865, II, No. 107, p. 30–58.

'Étude sur le régime de cours du bassin de la Seine pendant les crues du mois de septembre 1866', 1868, II, No. 195, pp. 235–312. (with G Lemoine).

'Note sur l'égout collecteur dit de la Bièvre et la siphon de l'Alma', 1869, II, Note 241, pp. 425-457.

'Le bassin parisien aux âges antéhistorique par M. Belgrand', 1870, I, Bibliographie, pp. 193–202.

'Note sur l'état probable des eaux courants du bassin de la Seine dans l'été et l'automne de 1870', 1870, I, No. 255, pp. 532–546. (with G Lemoine).

'Appareils de jaugeage des eaux', 1870, II, Chronique, pp. 95–99.

'De mélange des eaux courantes au confluence de deux cours d'eau; moyens de l'éviter', 1872, I, Chronique, pp. 133–135.

'Hygiène Publique—De l'action de l'eau sur les conduites de plomb', 1873/2 pp. 280–297.

'Hydrologie—Sur les conditions qu'on a dû chercher à réaliser dans le choix de sources destinée à l'alimentation de la Ville De Paris', 1873, I, Chronique, pp. 315–326. (extrait de *Comtes Rendus* 21.4. 1873).

'Étude de la grande crue de la Seine en mars 1876; en commun avec M. Lemoine', 1877:I Note 30. pp. 435–466.

*The full title is *Annales des Ponts et Chaussées**. Mémoires et Documents relatif à l'Art des Constructions et au Service de l'Ingénieur: Lois, Décrets, Arrêtés et Autres Actes concernant l'Administration des Ponts et Chaussées. It was published annually in two volumes at the Librarie des Ponts et Chaussées et Mines, Quai des Augustins 42, près la rue des Grands Augustins, Paris. There is an index, 'Tables Généraux', for 1831–1910.

APPENDIX 2
THE SCIENTIFIC PAPERS OF EUGÈNE BELGRAND

The scientific writings of Eugène Belgrand were mostly published in the *Comptes Rendus de l'Académie des Sciences.* His first paper, sponsored by Arago, the mathematical physicist, at the time the Academy's *secrétaire perpétuel*, was published in 1847. For the next twenty years, his scientific papers were published in the *Annuaire de la Société météorologie de France* or the *Bulletin de la Société géologique de France.* When elected in 1871 Belgrand became a regular contributor to the *Comptes Rendus* which recorded the proceedings of the regular monday meetings of the 'Académie des Sciences' contributing twenty nine papers—some short contributions to a discussion, most five or more pages. Many relate to hydrology, including reports on river flows and flood predictions in the Seine basin; others range over a wide area from Roman aqueducts to public works in the USA, and the effects on health of using lead pipes for carrying potable water.

EUGÈNE BELGRAND (1810–1878): CIVIL ENGINEER

His last paper, on the nature of eddies in rivers, was presented to the Academy a week before his sudden death on 8 April 1878.

*Comptes Rendus de l'Académie des Sciences**

'Études hydroliques dans les granits et les terres juraissique formant la zone supérieur du bassin de la Seine', Avallon 1.12.1846. Mémoire présenté à l'Académie des Sciences par Arago. 5 mai 1847, Tome 24.1, pp. 787–788.

'Geologie: Le bassin parisien aux âges ante-historiques', 16 mai 1870, Tome 70.1, pp. 1080–1087.

'La Seine. Études sur la régime de la pluie, des sources, des eaux courants; applications diverses à l'art de l'Ingénieur et à l'Agriculture', 16 mai 1870, Tome 71.2, pp. 886–895.

'Essai sur les aqueducs romains', 16 mai 1870, Tome 73.1, pp. 170–175.

'Note: Conservation de l'eau des réservoirs de la Dhuis, pendant le siège de Paris'. 30 decembre 1872, Tome 75.2, pp. 1840–1841.

'Note sur les crues de la Seine et de ses affluents', 9 decembre 1872, Tome 75.2, pp. 1584–1594.

'Deuxième Note sur les crues de la Seine et de ses affluents', 16 decembre 1872, Tome 75.2, pp. 1676-1677.

'Hydrologie: La Seine. Études hydroliques', 12 mai 1873, Tome 76.1, pp. 1172–1174.

'Hygiène Publique—De l'action de l'eau sur les conduites de plomb', 10 novembre 1873, Tome 77.2, p. 1055–1062.

'Hydrologie—Des sources du bassin de la Seine', 10 mars 1872, Tome 76.1, pp. 610-617.

'Geologie Agricole: La Seine; études hydroliques, seconde partie: application à agriculture', 19 mai 1873, Tome 76.1, pp. 1240–1247.

'Travaux publics—Note sur les travaux publics des États Unis d'Amérique', 11 août 1873, Tome 77.2, pp. 417-423.

'Hydrologie—Sur les conditions qu'on a dû chercher à réaliser dans le choix de sources destinée à l'alimentation de la Ville De Paris', 21 avril 1873, Tome 76.1, pp. 989–997.

'Hydrologie—Sur la crue de la Seine le 23 janvier 1873', 27 janvier 1873, Tome 76.1, pp. 189–190.

'Hydrologie—Sur la perméabilité du sables de Fontainebleau', 21 juillet 1873, Tome 77.2, pp. 178–182.

'Hydrologie—Abaissement probable du débit des eaux courants du bassin de la Seine dans l'été et l'automne de 1874', 14 juin 1873, Tome 78.1, pp. 1526–1532 (with G Lemoine).

'Remarques relatives à la Communication de M. Bobiérre sur les conditions dans les quelles le plomb attaqué par l'eau', 2 février 1874, Tome 78.1, pp. 318–321.

'Remarques relatives à une communication de M. Prost sur les trépidations du sol de Nice' 2 février 1874, Tome 78.1, pp. 336–337.

'Hydrologie—Sur le service hydrométrique du bassin de la Seine', 30 mars 1874, Tome 78.1, pp. 870–878 (with G Lemoine).

'Hydrologie—Abaissement probable du débit des eaux courants du bassin de la Seine dans l'été et l'automne de 1875', 14 juin 1873, Tome 80.11, pp. 1438–1440 (with G Lemoine).

'Hydrologie—Coup d'oeil d'ensemble sur le regime des principales rivières du Nord, du Centre et du Midi de la France', 18 janvier 1875, Tome 80.1, pp. 147–154.

'Hydrologie—Perturbations atmospheriques de la saison chaude de l'année 1875. Note sur le groupe de pluies du 21 au 24 juin 1875; crue de la Garonne; désastre de Toulouse; inondations du

EUGÈNE BELGRAND (1810–1878): CIVIL ENGINEER

midi de la France', 29 novembre 1875, Tome 81.2, pp. 1017–1025; 6 decembre 1875, pp. 1082–1086; 13 decembre 1875, pp. 1168–1175.

'Sur la crue de la Seine de février et mars 1876', 13 mars 1876, Tome 82.1, pp. 596–598.

'Hydrologie—Sur le débit et de le crue de la Seine de 17 mai 1876; réponse à une communication de M. Boussiugault', 20 mars 1876, Tome 82.1, pp. 659–660.

'Hydrologie—Sur la crue de la Seine et sur les moyens de préserver Paris des débordements du fleuve', 8 mai 1876, Tome 82 1876.1, pp. 1086–1092.

'Hydrologie—Étude sur la tourbillon des cours d'eaux', 1 avril 1878. Tome 96 1878.1, pp. 798–801.

*The full title is: *Comptes Rendus Hebdomadairement des Séances de l'Académie des Sciences publiés conformement à une décision de l'Académie en dater du 13 Juillet 1835 par MM les Secrétaires perpétuels* (Paris, Gauthier-Villars, Imprimeur-Libraire, Quai des Augustins 55.).

Bulletin de la Société géologique de France

'Carte agronomique et géologique de l'arrondissement d'Avallon, notice sur cette carte' Mémoire présenté, le 7 septembre 1849, au Conseil général de l'Yonne et insérée, par son ordre, dans "L'Annuaire de l'Yonne" 1850, pp. 212-256; 1851, pp. 168–262, 1851.

'Sur les depôts d'alluvions de la vallée de la Seine', 1863.

'Note sur les terrains quaternaires du Bassin de la Seine', 1864.

'Compte rendu de la course de la Société géologique de France sur les sabliéres à Paris', 1867.

'Sur le tuf de Contehian, Hérault', 1868.

'Sur les callioutis de Riege et Saint Quinieu', 1868.

'Sur les plantes fossilisées quaternaires dans le bassin de la Seine', 1868.

'Note sur l'ancienne historie de la Seine', 1868.

'L'âge des tourbes dans la vallée de la Seine', 1869.

'Note complémentaire sur le limon du plateau', 1870.

'Note sur la présence des stries à la surface d'une table de grés de Fontainebleau, dans la localité dites la Padole', 1870.

'Sur divers objets trouvés dans les tranches de l'aqueduc de la Vanne', 1870.

'Note sur l'ouvrage intitulé "La Seine–Études hydroliques"', 1874.

Annuaire de la Societé météorologique de France

'Étude des lois pour régissent les cours d'eaux', 1853.

'Hydrologie. De l'influence des forêts sur l'écoulement des eaux pluviales Paris 1 octobre 1853, avec un complément imprimé à la suite', 1853.

'De la simultanéité des pluies qui producent les crues des grands cours d'eau du plateau central de la France/de la Seine, de la Loire, du Saône et de la Meuse', 1854.

'Des relations qui existent entre la mode d'écoulement des eaux pluviales à la surface du sol et la culture des praries', 1855.

'Service hydrométrique du bassin de la Seine. Mémoire sur les observations hydrométriques, du 1 mai 1854 au 30 avril 1855 et sur la qualité des eaux de sources', 1856.

'Note sur les crues d'eau du bassin de la Seine en 1856', 1856.

'Note sur l'averse tombe à Paris le 21 mai 1857', 1857.

'Observations du service hydrométrique du bassin de la Seine jusqu'en 1855; études sur les sources du bassin de la Seine', 1857.

EUGÈNE BELGRAND (1810–1878): CIVIL ENGINEER

'Essai des diverses eaux dans le bassin de la Seine', 1858.

'Note sur l'hydrométre', 1858.

'Sécheresses du 1857 et 1858', 1858.

'Note sur les puits de Passy', 1864.

'Températures des eaux distribuer dans Paris (Suite de mémoire de 1859)', 1859.

'Températures des eaux distribuer dans Paris (Suite de mémoire de 1859)', 1860.

'Deux notes sur les puits de Passy, études des nappes souterraines', 1861.

'Les grands débordements de la Seine à Paris', 1864.

'Notice sur la régime de la pluie dans le bassin de la Seine', 1865.

'Résumé des observations centralisées en 1867 par la Service Hydrologique de la bassin de la Seine, au common avec M Lemoine', 1867.

'Note sur les observations ozonométrique organisées à Paris d'après les ordres de M. le Préfet de la Seine' (with G Lemoine), 1868.

'Résumé des observations centralisées en 1867 par la Service Hydrologique de la bassin de la Seine, au common avec M Lemoine', 1868.

'Étude sur la crue de septembre 1866, au common avec M Lemoine', 1868.

'Note sur la régime des pluies et des cours d'eau dans le bassin de la Seine à l'époque quaternaire', 1869.

'Résumé des observations centralisées en 1868 par la Service Hydrologique de la bassin de la Seine' (with G Lemoine), 1869.

'Résumé des observations centralisées, pendant les années 1869 et 1870, par le service hydrométrique du bassin de la Seine', 1871.

'Hydrologie du département de l'Yonne: études sur les régimes des cours d'eau et les cultures', 1851.

Other publications

'Monographie des eaux de sources de la banlieue de Paris' *La Science*, 1855.

'Résumé des Tomes I et IV; *Bulletin Météorologique spécial de l'Association scientifique de France.*

'Observations de 1872–1873', *Atlas pour les observations pour 1872, 1873 & 1874,* (with G Lemoine).

'Résumé des tomes III et IV', *Bulletin météorologique spécial de l'Association scientifique de France.*

'Observations de 1874', *Atlas de l'Observatoire pour 1875* (with G Lemoine).

'Notice sur les Crues de Principales Rivières en France en mars 1876', *Atlas de L'Observatoire Météorologique* Anné 1876.

'Publications des observations du service hydrométrique du bassin de la Seine':

Cours d'eau . . . . 1854–1876 (with G Lemoine)
Pluie . . . . 1861–1876 (with G Lemoine)

APPENDIX 3

Les Travaux Souterrains de Paris: 1872–1887

To record his work and serve as a guide to other engineers, Belgrand started work on a monumental publication *Les Travaux Souterrains de Paris* in 1873. In five volumes of text in large 8vo and five volumes of plates in atlas format, publication was not completed until 1887, nine years after Belgrand's death. Vol. I. *La Seine: Régime de la pluie, les sources des eaux courantes* (Paris, 1872–1875) pp. 622, is devoted to the hydrology of the Seine basin. Vol. II. *Les Eaux. Introduction. Aqueducs Romaines* (Paris, 1875) pp. 239, included material collected when Belgrand was in Rome (1837-38). Vol. III. *Les Eaux: Premiere Section: Les Anciens Eaux* (Paris, 1877, pp. 779), is devoted to the history of Paris water. The remaining two volumes were published posthumously from notes left by Belgrand. Vol. IV. *Les Eaux: Deuxième Section: Les Eaux Nouveaux* (Paris, 1882) pp. 596, which was edited by M. Buffet, one of Belgrand's engineers, assisted by M. Mourot, Belgrand's secretary, from papers in Belgrand's office, covers work on Paris water supplies in the early part of the nineteenth century as well as work carried out under Belgrand's direction. Vol. V. *Deuxième Partie: Les Égouts. Troisième Partie: Les Vidanges* (Paris, 1887) pp. 117, was to have been edited by Edouard Couche who had worked with Belgrand during the last years of his life, but completed by M. Humblot, another of Belgrand's engineers, on Couche's death. It covers works on sewers and sanitary services.

Eugène Belgrand presented the first three volumes of *Les Travaux Souterrains de Paris* and some other of his papers to the Institution of Civil Engineers on 8 July 1875. The publications, together with a letter in which he apologises for the letter being in French—'Ne connaissant pas la langue anglaise.'—are in the ICE archives.

Name and Place Index

Please note: references to illustrations and tables are indicated [in this index] by italic page numbers, and notes by suffix 'n' (e.g. 26n.35 means page 26, note 35).

Abbey Mills pumping station (Britain), 328, *336*
Academy of Sciences (France), 368n.51
  Belgrand's papers, 363, 371–3
  members, 361, 368n.51, 369n.56
Adams, Julius W., 295
Alatri (Italy), Roman siphon, 10, 26n.35
Albert, Prince (Consort to Queen Victoria), 229
Alcalà (Spain), 116, *126*, 130
Alcantará (Spain), Roman bridge, 21, 22, 23, 27n.59
Alcantarilla (Spain), Roman dam, 22
Alexander, Robert, 197
Allen, Horatio, 186
Allhusen, Christian, 163
Almuñecar (Spain), Roman siphon, 10
Alphand, Jean-Charles Adolphe, 347, 364
American Society of Civil Engineers, 192
Amwell (Britain), 50, 51, *54*, 65, 66, 70
Angell, Lewis, 253
Angitia (Italy), Roman siphon, 10
Arago, François, 348, 361, 365n.12, 368n.51
Arcueil aqueducts (France), 347, *356*, 357, 367n.42
Arles (France), Roman siphon, 10, 26n.30
Armstrong, William George, 155, 157, 161, *176*, 180
Armstrong, W. G., & Company, 163
Arnott, Dr Neil, 278
Ashby, T., 9, 44n.10
Ashton & Frost engines, 31
Aspendos (Turkey), Roman siphon, 10, 26n.38
Atkinson, Robert, 157
Atkinson, R. W., 226
Aubrey, John, 51, 68
Austin, Henry, 260n.3, 264, 276 & n.53, 278n.58, 279, 284, 285n.81, 287
Avallon (France), 350–1

Backhouse, Rowland, 75
Backhouse, Samuel, 49, 66, 75
Baker, Moses N., 311, *313*
Balfe, M.W., 191, 192n.26
Ball, Edward, 60
Baltimore (USA), 204, 309
Barking (Northern Outfall) sewerage works (Britain), 328
Barr(e)y, Richard, 90, 107, 108, 109, 112, 131
Bartlemew, John, 71, 72
Bateman, John Frederic, 155, 161, 162, 163, 166, 168, *176*, 266
Bateman, Robert, 75
Bazalgette, Charles Norman (son of Sir Joseph), 322
Bazalgette, Edward (son of Sir Joseph), 322, 332, 340n.86
Bazalgette, Sir Joseph, 244, 245, 321–43
  in conflict with John Grant, 279, *280*, 342n.139
  as consulting engineer, 331–2
  death, 334
  on design of sewers, 244, 245, 260, 281n.63, 284–5, 286 & n.82, 294–5, 326–8
  drainage reports, 336–7
  early engineering career, 322
  family background, 321–2
  health problems, 322, 343
  inspection of London pipe sewers, 279, *280*
  memorial to, 334, 342n.142
  at Metropolitan Board of Works, 279, 324–31, 333–4
  at Metropolitan Commission of Sewers, 279, 281n.63, 323–4
  as President of ICE, 333, 341n.126
  retirement, 333–4
  river crossings work, 330–1
  salary, 323, 325, 339n.52
  street improvement work, 331
  Thames embankment work, 328–9
Bazalgette, Joseph William, Jnr (son of Sir Joseph), 322
Beckton Outfall sewerage works, 328, 343
Beighton, Henry, 81, *92*, 104, 105
Belginum (Germany), Roman force pump, 38
Belgrand, Eugène, 345–75
  admitted to Academy of Sciences, 361
  background, 345–6
  death, 363
  as district engineer: in Côte d'Or *département*, 350; in Puy de Dôme

Belgrand, Eugène continued
département, 349–50; in Yonne *département*, 350–1
early education, 346, 348
family background, 348
funeral, 363–4
as hydrologist, 362–3
Paris sewerage system, 357–61
Paris water-supply, 353–7
on Roman aqueducts, 12, 17, 350, 371
scientific publications, 371–4
Seine hydrometric service organized by, 361–2
in Seine Navigation Service, 352, 366n.29
as student engineer, 346, 348–9
technical publications, 346–7, 370–1, 374–5
Bellamy, Christine, 287n.91
Bellingham (manager of Farlington Waterworks), 145
Bentham, Jeremy, 262
Berwick, Thomas John, 164
Bidder, George Parker, 287n.90, 323, 326, 327, 337n.11
Binnie, Sir Alexander, 342 & n.130
Binnie, G.M., 244
Binnie, Geoffrey (Sir Alexander's grandson), 342
Blackman, D.R., 44n.10
Blackwall Tunnel (Britain), 330–1, 342
Blomfield, R., 6
Boone, Stephen, 71
Borough of Portsmouth Waterworks Company, 148, 152, 248, 257
Boston (USA), 317n.6
Bottrell, Henry & Richard, 147, 148
Boulnois, H. P., 331–2, 341n.105
Boultham (Britain), water works, 31, *32*
Boulton & Watt engines, 135, 138
Bracebridge Heath (Britain), reservoir, 31, *32*
Brading Harbour (Isle of Wight), 76
Bramble, Benjamin, 152
Bramwell, F. J., 329
Bramwell, Sir Frederick, 253
Brassey, Thomas, *335*, 339n.50
Brayshaw, M., 257
Brenne river (France), 349, 350
Bridgnorth (Britain), 89–90, *91*
Bristol (Britain), 90, 93
Britain, sanitation in, 225, 229
Bronx River (New York, USA), 183, *184*
Brooklyn (USA), 293, 295
Broxbourne (Britain), 65, 66, 70
Brunel, Isambard Kingdom, 244, 323
Buck, George, 135, 136, 138
Buffalo (USA), 317n.6
Bulmer, Bevis, 48
Burdett, Edward, 93, 94, 95, 96
Burr, Aaron, 183
Bush Hill (Britain), aqueduct, 66, 67

Cadiz (Spain), Roman siphon, 10
Cail, Richard, 161
Cambridge (Britain), 225
Campagna (Italy), aqueduct bridges, 21, 33
Canterbury, Archbishop of, 95
Casher, Edward, 146, 147, 148, 152, 153n.18
Catcleugh Reservoir (Britain), 155, *156*, 172–5, 178
rainfall statistics, 175, 180n.72
Catskill Aqueduct (USA), 193
Cay, Dr Jabez, 106
Chadwell (Britain), springs, 50, 51, 70
Chadwick, Edwin, 259–88
arterial-venous concept of sewer design, 262, 274
attitude to engineers, 259, 266, 268, 277, 279
on blockage of sewers, 283
and Croydon sewerage system, 278
disagreement(s) with Hawksley, 266–7, 271, 285
expulsion from General Board of Health, 287
at General Board of Health, 244, 262, 267, 273
on manual cleaning out of sewers, 265, 284n.76
at Metropolitan Commission of Sewers, 272, 274
at Metropolitan Sanitary Commission, 267–8, 269, 271–2
pipe sewers advocated by, 260, 276, 287n.90, 294
and Poor Law Commission, 241, 262, 267
Sanitary Report, 241, 259, 262
Chandler, Charles E., 303, 318n.32
Charles I, King of Britain & Ireland, 77
Charles II, King of Britain & Ireland, 105
Chase Manhattan Bank (USA), 183
Chaumont (France), 348, 365n.9
Cherchel (Algeria), Roman bridge, 21
Chesbrough, E. S., 295, 298
Cheshunt (Britain), 57, 64, 65, 66, 69, 70
Chicago (USA), 291, 293, 295
Choisy, Auguste, 13
Clerkenwell (Britain), 65, 66
Cleveland (USA), 291
Clinton, De Witt, 185

Cobbett, William, 145
Coleby (Britain), *41*, 43
Cologne (Germany), Roman aqueducts, 33
Colt Crag Reservoir (Britain), 155, *156*, 164, *165*, 166, *167*, 177
  repairs, 168–9
Colthurst, Christopher, 55
Colthurst, Edmund, 48, 49, 50, 51, 55, 62, 64
Compton, Sir William, 66
Coode, Sir John, 334
Cooper, Edmund, 323, 325, 327
Copley, Sir Godfrey, 81–2, 101–2, 111
Cousin (France), 350
Cowie, Morgan, 260n.3, 270–1, 272, 274–5
Cox, Thomas, 278, 338n.33
Craven, Alfred W., 192–3
Crescent City, *see* New Orleans
Cresy, Edward, 260n.3, 264
Crossness (Southern Outfall) sewerage works (Britain), 328, 329, *335*, *336*, 343
Croton Lake Reservoir (USA), 186, 191
Croton Maid, The (boat), 191
Croton River (USA), 183, 186
Croton Water System (New York, USA), 183–93
  completion, 191–3
  construction details of aqueduct, 187
  dam, 186, *186*
  Distributing Reservoir, 190, *191*
  gatehouse, 186, *187*
  High Bridge, 189–90, 192, *193*
  Jumel Tunnel, 190
  Mill River Crossing, 188, *188*
  physical setting, 183–5
  plans for aqueduct, 185–9
  Receiving Reservoir, 190
  Saw Mill River Crossing, 189
  Sing Sing Kill Bridge, 187–8
Croydon (Britain), 278, 284
Ctesibius of Alexandria, 38
Cubitt, Sir William, 244, 323, 324

Darge, John, 48
Dawkins, Boyd, quoted, 345
Deacon, G. F., 252
Deal (Britain), 93–6
Defoe, Daniel, 81, 84, 109, 110
De Mun, Louis, 201
Dennett, Daniel, 97
Deptford pumping station (Britain), 326, *336*
Derby (Britain)
  All Saints' (church), 80, 81, 82
  Company of Mercers, 87–8
  Gunpowder Mill, 85
  St Michael's Church, 87, 89
  silk mills, 82, 86–7
  water conduits, 86, 88
  water-supply, 85–9
Detroit (USA), 291
Dhuis river (France), 354–5
Diaper, Henry St John, 152, 248
Dilke, O. A., 2, 5
Dodd, Ralph and Barrodall, 112, 133, 135
Donaldson, George, 260n.3, 278, 281, 323
Dorchester (Britain), Roman aqueduct, 35
Douglass, David B., 185
Drachmann, A. G., 2
Drainage Construction Company (USA), 301
Drayton Basin reservoir (Britain), *139*, 141, 255
Dunscombe, C., 342n.130
Durand-Claye, Alfred, 347, 360

Eastney Pumping Station (Britain), 253, *254*
East River (New York, USA), 183, *184*
Eddy, Harrison P., 293, 314, 320n.60
Edmonton (Britain), 65, 66
Edo (*later* Tokyo, Japan)
  drainage system, 223
  governance, 232
  mortality statistics, 233–4
  population, 211, 212, 213–14
  waste disposal system, 221–3
  water-supply systems, 216–18; Kanda system, 216, 217, 218, 227; Tama River system, 216–17
Eitai Island (Japan), 222–3
Elizabeth I, Queen of England & Ireland, 48
Elkesley (Britain), water works, 31
Elswick (Britain), 180
Emmery, Henri-Charles-Léopold, 352, 358, 366n.31
Enfield (Britain), 57, 65, 66, 67, 70
Engineer, The, 324, 334
Engineering News, 301
Engineering Record, 320n.57
Engledue, Dr W. C., 151, 152
Erie Canal (USA), 185
Etwall Manor (Britain), 113
Exeter (Britain), 96–8, 283n.72

Farlington Waterworks (Britain), 135–6, 138, *139*, 141, 145–6, *154*, 240–1, 247–8, 255
  amalgamation with Portsea Island Waterworks, 147, 247
  sea-wall breach affecting, 147–8, 247
Ferdinand, King of Spain, 118
Field, Joshua, 327, 337n.10

Fiennes, Celia, 84, 86, 93, 96, 97, 98, 108
Finer, S. E., 261, 285n.79
Finley, M. I., 2, 4
Flinn, M. W., 260n.2
Forbes, R. J., 8
Ford, R. W., 152
Forster, Alfred L., 168
Forster, Frank, 323, 343
Forster, John Reay, 168, 169, 171, *176*
Fox, William, 164
France, sanitation in, 228–9
Frank's Island (USA), lighthouse, 202 & n.22
Frontinus, 5, 6, 14, 18, 20, 25n.19
Fuller, John, 59–60
Fulton, Robert, 200, 201
Furness, George, 333, 340n.69

Gallatin, Albert, 201
Galois, Evariste, 348
Galton, Douglas, 329
Garrard, Sir John, 71
Gilbert, Henry, 99, 100
Gordon, Joseph, 334, 342n.130
Gotto, Edward, 260n.3
Graff, Frederick (assistant to Latrobe), 196n.3
Graff, Frederick C. (investor), 199 & n.11
Grainger, Richard, 155, 157
Grant, George, 146
Grant, John, 260n.3, 276, 279, *280*, 323, 325, 327, 329, 342n.139
Great London Drainage Company (Britain), 324
Greatorex, J. E., 255
Great Yarmouth (Britain), 112
Gregory, Charles Hutton, 337n.11
Griscom, John H., 225
Guadaira river (Spain), 116
Guthrie, William Dyce, 260n.3, 264

Hadley, John, 83, 101, 104
Haga, Godfrey, 199
Hale, J. L., 260n.3, 270n.36, 272
Hallington reservoirs (Britain), 155, *156*, 161–71, 177–8
  rainfall statistics, 162, 163, 179n.33
Hamilton, Stanley B., 346
Hampshire Telegraph, 249, 251
Hardwick/Hargate Manor (Britain), 113
Harlem River (New York, USA), 183, 189
Harrison, John Thornehill, 281
Hartley, Sir Charles, 329
Hartley, Jesse, 244
Hartsholme (Britain), 31, *32*
Hassall, Revd Thomas, 65
Hatton, Edward, 103
Haussmann, Georges, 347, 350, 351, 352, 353, 354, 358, 361, 364
Havant (Britain), springs, 248, 255
Hawkshaw, John, 253
Hawksley, Charles, 168, 174, 175, *176*, 181
Hawksley, Thomas, *176*
  disagreement(s) with Chadwick, 266–7, 271, 285
  and sewer design, 244, 246, 260, 271, 323, 326, 327
  and Tyneside reservoirs, 155, 172, 174
Haywood, William, 323, 324
Hazen, Allen, 314
Hennock, E. P., 240
Henzell, Charles George, 170, 171, 174
Hering, Rudolf, 299–300, 301, 302, 304–5, 307, 308, 309, 310, 312, 319n.52
Hermany, George, 298
Hero of Alexandria, 44n.12
Hertford (Britain), 50
Highbury (Britain), aqueduct, 75
Hill, George Henry, 163, 164, 169, *176*
Hillen, Henry J., 98, 99
Hobbs, John, 113
Hoddesdon (Britain), 66, 70, 77
Holborn & Finsbury Sewers Commission (Britain), 264
Holloway (Britain), 65, 66
Hone, Philip, 188, 191
Honnecourt, Villard de, 5
Horne, Thomas, 64
Hornsey (Britain), 65, 66
Hosking, William, 264
Houghton, John, 86
Howell, Henry H., 174
Hudson River (USA), 183, *184*
Hughes, Edward, 56, 60, 61
Hutton, William, 87
Hyde, Lawrence, 49, 75

Ingelbert, William, 50, 51
Ingen (plough used on New River), 68
Institution of Civil Engineers
  members, 244, 260 & n.3, 322, 337n.11
  papers presented to, 170, 244, 252, 281, 322
  Presidents, 245, 253, 268n.28, 333, 334, 339n.59, 341n.126
  publications in archives, 345–7, 369n.64, 375
  reaction to Chadwick's attacks, 277
Irving, Washington, 188
Isle of Dogs Pumping Station (Britain), *336*

Islington (Britain), 51, 65, 66, 98

James I, King of England & Scotland, 49, 57, 59, 60, 61, 69–70, 74, 76, 77
James II, King of England, Ireland & Scotland, 93
Japan, urban population, 211, 213–15
Jefferson, Thomas, 197
Jenkins, Rhys, 82
Jervis, John Bloomfield, 185, 190, 193
Joahnes/Jones, Howell, 64
Johnson, Sir Thomas, 81
Jucef Abu Jacub, 115, 118
Jurassic ridge (Britain), 29, 40, *41*

Kanazawa (Japan), 215
Kay-Shuttleworth, James, 262n.10
Keith-Lucas, B., 239–40
Kielder Reservoir (Britain), 181
King's Lynn (Britain), 98–9
Kingston (Hants, Britain), 149, 150
Klemm, F., 8
Korschelt, O., 226n.22
Kyoto (Japan), population, 211, 214–15

Lacey, Aver(e)y, 68, 72
Lalanne, Léon, 348, 362, 364, 369n.56
Landport (Britain), 149, 150
Landport & Southsea Improvement Commissioners, 152, 248–51, 257
  dissatisfaction with, 251
  drainage scheme, 249, *250*
  powers transferred to Town Council, 251
  road improvements, 248–9
Langstone Harbour (Britain), 253, *254*
Latrobe, Benjamin, 196–210
  on African slaves' behaviour, 206
  as architect, 195, 200, 202–3, 204, 207–8
  bankruptcy, 204
  burial place, 209
  death, 209
  on mosquitoes, 206–7
  on New Orleans architecture, 204, 205
  on New Orleans people, 204–5
  steamboat schemes, 200, 201, 208
  on treatment of black slaves, 205
  waterworks schemes, 196–9, 203–4, 208–9
Latrobe, C. S., 309
Latrobe, Henry (son of Benjamin), 198, 199, 201–2, 203
Law, Henry, 333, 338n.38
Lawton, Benjamin, 161
Lee, William, 260n.3, 285n.81
Leeds (Britain), 99–101
Leicester (Britain), 44n.8, 225
Lemoine, Georges, 347, 361, 362, 363, 368n.53, 369n.56
Lenox (USA), 297, 309
Leslie, James, 272n.40, 281
Leslie, John, 275
Lewyn, William, 60–1, 64
Lincoln (Britain)
  contours, *30*
  location and geology, 29, *41*
  medieval water-supply, 29, *30*, 31
  nineteenth century water-supply, 31, *32*
  population, 31, 36
  possible Roman inverted siphons, 39, 40, 43, 44, 45n.21
  Roaring Meg aqueduct, *33*, *34*, 35–6, *37*, 43
  Roaring Meg spring, 26n.37, 35, 40
  as Roman *colonia*, 31
  Roman water-supply, 10, 26n.37, 29, 31, 33–44
  spring line, 29, *30*
  twentieth century water-supply, 31, *32*
  water towers, 31, *32*, 34, 44
  wells, 29, 43
Lincoln Water Company, 31
Lindley, William, 266
Liverpool (Britain), 82, 244
Livingston, Robert, 200, 201
Lockwood, Charles, 192 & n.27
London (Britain)
  Albert Embankment, 329
  Aldersgate conduit, 47
  Battersea Bridge, 330
  Chelsea Embankment, 329
  City Corporation, 48, 57, 59, 76–7
  Commission of Sewers, 264, 323
  Dowgate conduit, 47, 74
  Great Fire (1666), 78, 103, 340n.64
  Hammersmith Bridge, 330
  London Bridge Water Works, 47, 72, 81, 83, *92*, 102–5
  Marchant's Water Works, 82, 83, 101–2
  plague, 77
  pollution of water supply, 225
  population, 47, 211
  Putney Bridge, 330
  St Thomas's Hospital, 329
  sewerage system, 294–5, 323–43
  Tower Bridge, 330
  underground railway, 324
  Victoria Embankment, 328–9, 331, 333, 334
  Waterloo Bridge, 188, 330
  water-supply system, 47–77, 98, 101–5, 227

London (Britain) continued
  see also New River scheme
London County Council (Britain), 328, 331, 334, 342n.130
Lovick, Thomas, 260n.3, 272, 279, 323, 325, 327, 334
Lowbridge, Richard, 97
Lyon (France)
  Roman aqueducts, *16*, 17, 33
  Roman siphons, 10, 14–15, *16*, 17–20, 21–2, 40; Beaunant siphon, *16*, *18*, 19, 22; Cotte-Chally siphon, *16*, 27n.54; d'Ecully siphon, *16*, 21; Grange-Blanche siphon, *16*, 21; header tanks for, *18*, 20; St Genis siphon, *16*, 19, 20, 21; St Irénée siphon, *16*, 19, 21; Soucieu siphon, *16*, 17, 18, *18*, 19, 21, 40, *40*; Tourillons siphon, *16*, 22; Tupinier siphon, *16*, 27n.54; weight of lead used, 18, 40; width of venter bridges, 20

Macclesfield (Britain), 105
McGuire, J. B., 162, 163
McNeill, Sir John, 322, 337n.10
McQueen, Robert, 200n.15
Macritchie, Charles, 310
Maidstone Bridge (Britain), 332–3
Manchester (Britain), 286n.86
Manhattan Company (USA), 183
Manhattan Island (New York, USA), 183, *184*, 225
Marchant, Hugh, 102
Mark, Jacob, 199
Market Rasen (Britain), 40, *41*
Marly (France), 81, 105
Marne river (France), 353, 354
Massachusetts (USA), 310
Massachusetts Board of Health (USA), 311
Medworth, Joseph, 260n.3, 272, 273n.46
Memphis (USA), 297–9, 304–5, 311
Ménilmontant reservoir (Paris, France), 355, *355*, 357
Mérida (Spain), Roman bridge, 21, 23
Metcalf, Leonard, 293
Metropolitan Board of Works (London, Britain), 279, 324–31, 333–4
  bridges and other river crossings, 330–1
  corruption charges, 342n.140
  Engineer, 279, 325
  flood prevention duties, 329–30
  main drainage works, 326–8
  payment procedures, 326
  principal contracts, *335–6*
  street improvements, 331
  tender procedures, 325–6
  Thames embankments, 328–9
Metropolitan Commission of Sewers (London, Britain), 267, 272, 279, 281, 323–4
Metropolitan Sanitary Commission (London, Britain), 267–8, 269, 271
Metz (France), Roman bridge, 23
Middleton, Hugh, *see* Myddelton, Sir Hugh
Mille, Adolphe, 347, 358, 360, 364
Minchin, Thomas, 145
Mississipi River (USA), 195, 201–2, 293, 297, 310
Montagu, Sir Henry, 49, 55, 62, 75
Montauzan, Germain de, 18, 21
Montbard (France), 350
Montsouris reservoir (Paris, France), 355, *356*, 357
More, Rosemary J., 363
Morice, Peter, 103
  see also Morris
Moritz, L. A., 2
Morris, George P. (American poet), 191, 192n.26
Morris, Peter (London), 47, 83, 103
Morris, Thomas (grandson of Peter), 103
Morse, Edward S., 230–1, 232–3
Mowlem (contractor), *335*
Myddelton, Sir Henry, 55
Myddelton, Sir Hugh, 47–78, 98
  agreement with King Charles I, 77
  allowance(s) paid to, 63–4
  amount spent on New River, 73–4
  appointment by City, 51–2, 55
  baronetcy, 76
  burial place, 78
  charitable disposition, 61, 65
  death, 77
  disbursement books for New River scheme, 52, *53*, 55–7, 60–1, 62–4, 67–70
  family background, 52
  formation of New River Company, 74–5
  loan for New River scheme, 73
  as Member of Parliament, 52
  memorial to, *54*
  paid by King James I, 73, 74
  in partnership with: Edmund Coulthurst, 52, 55; King James I, 59–60, 61; others, 61–2
  portrait, *46*, 77
  at reading of New River Bill, 49, 52
  recognition by City of London, 76–7
  signature, *78*
Myddelton, Robert (Hugh's brother), 49, 52

Myddelton, Sir Thomas (Hugh's brother), 64, 75

Nance, Andrew, 146, 147, 148, 152, 153n.18
Napoleon I, 353
Napoleon III, 351, 353, 361, 364
Narni (Italy), Roman bridge, 21, 23
Nevill, Sir Henry, 62, 71, 75
Newcastle & Gateshead Water Company
  formation, 155, 180
  reservoirs, 155–78; Catcleugh, 155, *156*, 172–5, 178; Colt Crag, 155, *156*, 164, *165*, 166, *167*, 168–9, 177; East Hallington, 155, *156*, 161–3, 177; Little Swinburn, *156*, 164, *165*, *166*, 168, 177–8; West Hallington, 155, *156*, 169–71, 178; Whittle Dean, 155, *156*, 157–61, 177
Newcastle-upon-Tyne (Britain), water-supply, 106–7, 155–82
Newington (Britain), 65, 66
New Orleans (USA)
  Charity Hospital, 202
  Customhouse, 197
  French Opera House, 202
  Louisiana State Bank building, 208
  St Louis Cathedral, 207–8
  waterworks system, 196–201, 203, 207, 208–9
Newport (RI, USA), 302
New River Company, 74–5, 78
New River Head (Britain), 51
  construction of Round Pond and water house, 71
  river first let into Round Pond, 64
New River (London, Britain)
  book of disbursements, 52, *53*
  compensation for land-owners, 66
  delays, 56–7, 59
  expenditure on, 52, *53*, 55–7, 60–1, 62–4, 67–70, 73–4
  fall/gradient/slope, 66
  first proposed, 48
  flooding due to, 69, 70
  inadequacy, 75, 227
  income from, 74
  King James I as partner, 59–60, 61
  land-owners affected by, 65–6
  legislation covering, 49–51, 57
  length, 66, 98
  Myddelton's appointment, 52, 55
  new (7–year) agreement, 59
  opposition to, 56–9
  partners, 49, 52, 59–60, 61–2
  progress of work, 65
  rent paid to king instead of partnership share, 77
  Sorocold's water-engine used, 98
  surveying, 62–3, 66–7, 76
  water-distribution system, 71–3
  water rates charged, 74
  workmen: nationalities, 68; numbers, 56, 68; rates of pay/piecework, 68–9
New York State Board of Health (USA), 302
New York (USA)
  Central Park, 190, 296
  New Croton Aqueduct, 183, 193
  Old Croton Water System, 183–93
  Public Library, *191*, 192n.26
  sewerage system, 312
  water-supply system, 183–93
Nichol, John, 310
Nicholson, Robert, 159, 161, *176*
Nicholson, William, 133, 135
Nineveh (Iraq), water-supply system, 8
Noisel stream (France), 355
Nonius Datus, 25n.14
Norfolk (VA, USA), 302
Northumberland, Duke of, 157, 340n.66
Norwich (Britain), 107–8
Nottingham (Britain), 108–9, 267

Observer, The (London), 325
Odessa (Russia), 333
Okubo Togoro Tadayuki, 216
Ollivier, Emile, 364
Olmsted, Frederick Law, 190
Orr, Benjamin G., 204, 209
Osaka (Japan)
  population, 211, 214
  waste disposal system, 218–21
Otby Top (Britain), 40, *41*
Owen, T. E., 152

Page, Thomas, 278, 279, 281, 285
Paris (France)
  Bièvre stream, 353, 357, 359–60
  École polytechnique, *346*, 348, 361, 365n.11,12, 369n.56
  Hôtel de Ville, *352*, 361
  Ourcq canal, 353, 360
  reservoirs, 355, *355*, *356*, 357
  sewers, 225, 357–61
  water supply, 225, 352–7
Parkes, Josiah, 337n.11
Parkes, Richard, 64, 72
Parnell, William, 56, 62, 64, 67

Partridge, Bellamy, 186
Peacock, S. E., 251–2
Peet, Henry, 82
Penn, John, 327
Pergamon (Asia Minor), siphon, 39
Philadelphia (USA)
  sewerage system, 299, 312, 320n.56
  water-supply system, 196, 291
Philbrick, Edward S., 304
Philip, HRH The Prince, 9
Phillips, John, 260n.3, 268–9, 275, 285n.81
Pierson, G. S., 306, 319n.39
Pinkney, David H., 225
Pitt, Sir William, 76
Plommer, H., 2, 6
Plummer, George, 145
Pocantico River (USA), 188
Pointon, V. F. J., 257
Pond, Edward, 62–3, 64
Pont du Gard (France), Roman aqueduct bridge, 10, 21, 23, 27n.59, 33, 40
Portsdown Hill (Britain), 136, 147, 241, 248
Portsea Island (Britain)
  cholera, 150, 151
  geology, *132*
  map(s), *134*, *137*, *142*, *238*, *250*, *254*
  pre-1800 water-supply, 131–2
Portsea Island Waterworks Company, 135, 136, 138, 143, 146, 240
  amalgamation with Farlington Waterworks, 147, 247
  White Swan Waterworks, *132*, 133–5, *140*, 146, 149, *238*, 255
Portsmouth (Britain)
  administrative developments, 251, 256
  Borough of Portsmouth Waterworks Company, 148, 152, 248, 257
  compared with national trends, 256–7
  daily demand for water, 138, 149
  death rate, 150
  Dockyard, 133, 143, 145, 148
  drainage schemes, 151, 246–7, 253–5
  house-building activity, 149
  Improvement Commissioners, 136, 143, 147, 149, 239–40, 242–4, 256
  local Board of Health, 150, 151
  map(s), *137*, *142*, *144*
  population, 133, 138, 148, *239*
  St Helen's Road, *134*
  Town Council, 152, 251–5
  United Portsmouth, Portsea & Farlington Waterworks Company, 147–8, 149, 152, 247
  water-supply: nineteenth century, 132–52, 240–1, 247–8, 255–6; pre-1800, 109, 131–2
Portsmouth & Arundel Canal Company (Britain), 146
Pullman (IL, USA), 306
Pyrke, Jonathan, 97

Quick, Joseph, 255
Quincy (IL, USA), 310

Rafter, George, 311
Rambuteau, Comte Claude-Philibert, 358
Rammell, Thomas, 260n.3, 284
Ranger, William, 246, 260n.3, 278 & n.57
Rastrick, John U., 337n.11
Rawlinson, Robert, 150–2, 244–5, 245–6, 247, 260n.3, 272n.40, 276, 281, 283, 284 & n.76, 303, 325, 338n.33,34
Reid, William, 159
Rendel, J. M., 244
Rennie, John, 266, 330, 333
Rey Pastor, Alphonso, 22–3
Rider, William, 93, 94, 95, 96
Ridley, John, 157
Rigby, William, 164, 168
Riseholme (Britain), *41*, 43
Ritson (contractor), *335*
River Aire (Britain), 100, 101
River Avon (Britain), 90
River Colne (Britain), 49
River Derwent (Britain), 82, 86, 88, *114*
River Exe (Britain), 97
River Frome (Britain), 35
River Lea/Lee (Britain), 48, 49, 58, 75, 98, 339n.50
River Leen (Britain), 108, 109
River North Tyne (Britain), 181
River Pont (Britain), *156*, 161
River Rede (Britain), *156*, 172
River South Tyne (Britain), *156*, 181
River Thames (Britain)
  embankments, 328–9, 340n.64, 342
  sewage discharge into, 225, 326, 328
  water intake from, 47–8, 101–5, 225
River Trent (Britain), 109
River Tyne (Britain), *156*, 161, 180
River Witham (Britain), 29, *30*, *32*
Rodez (France), Roman siphon, 10, 26n.29
Rodrigues, J., 229
Roe, George, 323, 342
Roe, John, 260n.3, 262n.8, 264, 269, 272, 273 & n.46, 276, 283, 342
Rolt, Tom, 29

Rome (Italy)
  aqueducts, 6, 8, 33, 36, 44n.10, 131, 350
  Pantheon, 5, 8, 9
  siphon, 10, 26n.33
Roosevelt, Nicholas, 200
Roquefavour (France), bridge, 22, 27n.55
Roriczer, Mathias, 5
Rotherhithe (London, Britain), 82
Rothwell & Co., *336*, 343
Rungis (France), springs and aqueduct, 367n.42

St Malo (France), force pump, 38, *39*
St-Maur (France), 357
Saldae (Algeria), Roman aqueduct tunnel, 25n.14
Salisbury, Robert, Earl of, 59, 60
Sanitary Engineer, 303, 304
Saracole, George, *see* Sorocold, George
Savory, Thomas, 81
Saw Mill River (USA), 189
Schramke, T., 186, 187, 190, 191
Schuylkill River (USA), 196
Sedgwick, William T., 311
Segovia (Spain), Roman aqueduct bridge, 21, 23, 33
Seine hydrometric service (France), 361–2, 368n.53
Seine Navigation Service (France), 352, 366n.29
Seine river, 105, 225, 352–62
Seine river basin (France), *354*, 361, 363
Seracoal/Seracoll, George, *see* Sorocold, George
Seville (Spain)
  new waterworks, 130
  water-collection system, 115–17, *127–8*; length of tunnel, 117; volume of water, 117
  water conduit, 117–18
  water-distribution system, 118–30, *127–9*; Carmona Gate reservoir, 122, *128*; dividing-boxes, 124–5, *129*; gauge box for, 119, *120*, 122; unit of measure used, 119, *120*
Shaftesbury, 7th Earl, 242
Sheffield (Britain), 109–11
Sheppard, John, 243
Silchester (Britain), Roman force pump, 36, 38
Simpson, James, 155, 157, 159, *176*, 179n.19, 244, 286, 337n.10
Smallman, James, 200n.15
Smith, Arthur, 105
Snygge, Sir George, 60
Soame, Richard, 103, 104, 107, 108
Somme-Soude river (France), 353, 354
Sopwith, Thomas, 155, 157, *176*
Sorocold, George (?grandfather), 80, 113
Sorocold, George, 79–112
  accident, 81
  as engineer (generally), 81–3
  family background, 79–80
  first (recorded) job, 82
  marriage, 80
  origin of surname, 79
  in partnership with Richard Barr(e)y, 90, 107, 108, 109, 112, 131
  patent for sawing machine, 83
  as water-supply engineer, 83–112, 131
  wife, 80, 81
Southampton (Britain), 241, 246
Southgate (Britain), 65
Southsea (Britain), 150, 152, 243, 251
  see also Landport & Southsea Improvement Commissioners
Spital Brook (Britain), 65, 66
Sprotborough (Britain), 111–12
Staley, Cody, 306, 319n.39
Stamford (CT, USA), 302
Stein, Albert, 209
Stephenson, Robert, 244, 323, 324, 327
Stone, May N., 226
Strong, George Templeton, 191
Surmelin river (France), 354
Surracolde, George, *see* Sorocold, George
Swinburn reservoirs (Britain), *156*, 163–9, 177–8
Swinnerton, Sir John, 64
Switzer, Stephen, 102
Symons, George James, 174, 175

Tokugawa (Japan), mortality statistics, 233
Tama River (Japan), 216–17
Taylor, Peter (MP for Portsmouth), 132, 147
Taylor, Richard (pipe-layer in London), 72
Termini Imerese (Italy), Roman siphon, 10, 26n.36
Terrey, William, 110, 111
Thames Conservancers, 329
Theobalds (Britain), 65
Theobalds Park (Britain), 57, 69
Thom, Robert, 264
Thompson, E. A., 2
Thompson, F. H., 35–6
Thoresby, Ralph, 80, 84, 89, 90, 93, 98, 100–1, 105, 107, 109, 111, 112
Times, The (London), 274, 275, 334

Tokugawa Shogunate (Japan), urban population growth in, 213–15
Tokyo (Japan), 211
  water quality, 226
  see also Edo
Toledo (Spain), Roman aqueduct bridge, 22–3, 26n.27, 27n.59
Tone, John Furness, 161, 179n.19
Tonkin, William, 147, 148
Torreblanca (Spain), 117
Tottenham (Britain), 283n.72
Toulmin Smith, Joshua, 288
Tower, F. B., 191
Towne, Henry R., 302
Towns Improvement Company (Britain), 267
Trollope, Frances, quoted, 357
Tyne Salmon Conservators, 172
Tyneside towns, reservoirs supplying, 155–78

United Portsmouth, Portsea & Farlington Waterworks Company, 147–8, 149, 152
  amount of water supplied by, 149, 152, 248
  debts, 148
  formation, 147, 247
  take-over by Borough of Portsmouth Water Company, 148, 152, 248

Vanne river (France), 355–7
Vetch, James, 260n.3
Victoria, Queen of Great Britain, 333, 343
Vitruvius, 3, 5, 6, 9, 10, 11–15, 24, 44n.16, 189

Wacher, J. S., 39, 40
Waddell (contractor), *335*
Waddington (Britain), *41*, 43
Walker, James, 268n.28, 328
Ware (Britain), 65, 67, 70, 75
Waring, George E.
  criticisms of his designs, 304–5, 306
  patents, 301
  personality, 300–1
  professional ethics, 301
  as sanitary reformer, 296–7, 311
  as self-publicist, 293, 300–1
  separate sewers advocated by, 297, 304, 309, 315
  sewerage systems designed by, 297–9, 304–5, 309
Warner, William, 93, 94, 95, 96
Washington (USA)
  Capitol building, 195, 199, 202
  sewerage system, 312
  White House, 195, 202
Watanabe (Japan), 220
Webb, Sidney & Beatrice, 239, 240
Webster, William, 329, 333, *335*, *336*, 340n.69, 341n.122
Westminster Sewers Commission (Britain), 264, 266, 268, 269, 275
Whalley, Peter, 110
Whitacres, Miles, 60
White, Canvass, 185
White, K. D., 2
Whitmore, Sir William, 89
Whittle Dean reservoirs (Britain), 155, *156*, 157–61, 177
  rainfall statistics, 162, 163, 179n.33
Whittle Dean Water Company, 155
Wicksteed, Thomas, 264, 270n.35, 281, 282, 286, 324
Williams, Benezette, 306
Williams, J. Butler, 260n.3, 264
Williams, Son & Willington, *335*
Windsor Castle (Britain), 343
Wirksworth (Britain), 112
Wolf River (USA), 298
Wolley, William, 86–7
Wood Green (Britain), 65, 67
Woolwich ferry (Britain), 331
Wormley (Britain), 57, 65, 66, 67
Wormwood Scrubs Prison (London, Britain), 181
Wren, Sir Christopher, 78, 340n.64
Wright, Edward, 48, 55–6, 63
Wynne, Sir John, 52n.21, 60, 63

Yarnold, William, 106
Yonne river (France), 350, 355

Subject Index

Please note: references to illustrations and tables are indicated [in this index] by italic page numbers, and notes by suffix 'n' (e.g. 27n.45 means page 27, note 45).

Acts of Parliament
  Bazalgette's involvement, 325
  Bristol water-supply, 90
  Clause Acts, 241, 242
  Deal water-supply, 93–6
  Exeter water-supply, 96–8
  Highways Act(s), 243
  Improvement Commissioners, 239
  Lincoln water-supply, 31
  Local Government Act(s), 251, 342n.140
  Metropolis Toll Bridges Act (1877), 330
  on mortgage foreclosure, 146
  Newcastle upon Tyne water-supply, 106, 107
  New River (London), 50, 51
  Norwich water-supply, 107–8
  Nottingham water-supply, 108–9
  Nuisance Removal and Diseases Prevention Act, 150, 277
  Portsmouth water-supply, 132, 135, 152, 240, 248
  Public Health Act(s), 151–2, 241–2, 246, 256
  sewers commissions, 323
  Thames embankments, 328–9
  Thames flood prevention, 329
  Tyneside water-supply, 155, 157, 162, 164, 169, 172, 174, 175
agricultural uses of sewage, 219, 222, 228, 230, 236, 263, 295, 328, 360
animal-powered pumps, 38, 39, 99
aqueducts
  Croton Aqueduct, 186–92
  fall/gradient/slope required, 20, 27n.45, 66, 187, 357
  New River scheme (Britain), 67, 75
  Paris water-supply, 354, 355, *356*, 357, 367n.42
  Roman, 6, 8, 21, 22, 33, 36, 44n.10, 117–18, 131, 350
Assyrian engineering, 7–8

back drainage system, 244–5
bacteriology, 311
bamboo pipes, 217
bathing facilities, 192, 231
Bill of Mortality, 102
boiling of water, 230
book-leaf marl, 170
brewers, water used by, 74, 143, 227
brick built sewers, 43, 245, 253, 264, 269, 281, 282, 285, 286, 287, 359
  size, 43, *245*, 253, 287, 359
brickmaking, onsite, 181
bridges, 21, 22–3, *23*, 187–8, 189–90, 330, 332–3
  see also aqueducts

calcium carbonate deposits in pipes, 40, 355
canals, 185, 353, 357
carters, 131, 147, 353
cast-iron pipes, 112, 124, 133, 135, 157, 159, 162, 170, 171, 172, 190, 192, 198, 351, 355, 357
  fracture of, 181
  size, 135, 157, 159, 162, 170, 171, 172, 190, 192, 198, 355, 357
castle towns (in Japan), 215, 218, 224
catch basins (in sewers), 308
cement, *see* hydraulic...; Portland...; Roman cement
cesspits/cesspools, 151, 224, 225, 245, 262, 290–1, 360
chain-of-buckets/pots water-lift devices, 26n.37, 36
chalk, 40, 355
cholera, 150, 151, 155, 161, 180, 181, 234, 358
chopsticks, 230
coffer-dams, 190
concrete
  aqueduct laid on, 187
  reservoirs repaired using, 168, 169
  sewer laid on, 253
  strength of, 9
  use by Romans, 4, 5, 8
consulting engineers, 152, 155, 157, 161, 172, 253, 322, 323, 326, 331–3
consumption per day (of water), 8, 36, 117, 138, 155, 157, 163, 172, 255
  per person, 36, 138, 291
Copley Medal, 102
cost considerations
  New River (London), 73–4

cost considerations continued
  pavement/road construction, *252*
  reservoir construction, 168, 171, 179n.40
  sewerage systems, 282, 298, 305–7, 314
crossing of rivers/streams, 67
Ctesibian pump, 26n.37, 38
cypress, *see* wooden pipes

daily demand, *see* consumption of water
daimyos (in Japan), 214n.3
  night-soil contracts, 222
  residences in Edo (Tokyo), 213–14
dams, 22, 43, 172, 186
  see also embankments
death rates, 150, 233, 237
dilution capacity (of streams/rivers, for sewage), 312, 316
discoloration of water, 141, 180, 216
diseases
  cholera, 150, 151, 155, 161, 180, 181, 234, 358
  enteric infections, 228, 234–5
  factors affecting, 181, 206, 225, 227–8, 229, 234–5, 262, 278, 292, 302–3, 311–12
  plague, 77, 234
  typhoid fever, 31, 229, 311, 312, 316
  yellow fever, 196, 203, 206, 208, 209, 293
distribution systems, 71–3, 118–25, 130, 136, 138, 255
donkey mills, 39
drainage systems, 43, 223, 244–5, 253–5, 306, 308–9, 358–60
  see also sewers; storm water sewers

earth closets, 297
earthenware pipes
  fifteenth century, 124
  in Roman systems, 8, 12, 14, *33*, 35
  size, *33*, 35
eating utensils, 230
elm, *see* wooden pipes
embankments
  collapse/slippage during construction, 70, 170, *171*, 186
  reservoir, 159, *160*, 162, *163*, *165–7*, 168, *169*, 170–1, *173*, 175, 186
  river, 69, 70, 328–9
  see also dams
engineers, difference in training (Britain/France), 346, 348–9

fall/gradient/slope
  aqueducts, 20, 27n.45, 66, 187, 357
  sewers, 266, *271*, 272, *274*, 283 & n.72, 326, 359
fecal pollution (of water sources), 149, 151, 225, 245, 291, 316n.2, 326
'filth theory' (of disease), 292, 296, 302, 311
filtration
  filter bed, 31, 180, 217
  mechanical, 186
fire-fighting, water for, 77, 143, 183, 291
fishery interests, compensation to, 172–3, 180
fish passes, 172
flash (for river crossing), 67
flooding, 307, 358
  during construction work, 70, 186
flood prevention (London, Britain), 329–30
flood warning service (France), 361–2
flushing dams (in sewers), 264
flush tanks (in sewers), 297, 298, 302, 304, 305
force pumps, Roman, 26n.37, 36, 38–9
fountains and waterfalls, 36, *42*, 191, 192n.26, 198, 227
Fourneyron turbines, 15

garbage, *see* rubbish disposal
General Board of Health (Britain), 150, 242, 244, 246, 262, 273, 275, 277, 287
geological faults, effect on reservoirs, 166, 168
germ theory of disease, 303, 311
gravitational water-supply systems, Roman construction, 8–9
Greek aqueducts, 8
Greek technology, 4, 6, 8, 10
grit-removing traps, 284

health, *see* General Board of Health; local boards...; Medical Officer of Health; National Board of Health; state boards...
health problems
  effect of water quality, 150, 181, 225
  see also diseases
Health of Towns inquiries/reports, 241, 264, 266
Highways Board(s) (Britain), 243–4
hot-air central heating, 342
human excrement
  disposal in storm water sewers, 290, 294
  use as fertilizer, 227–8, 234–5, 294, 360
  see also fecal pollution; night soil
hydraulic cement/mortar/concrete, 8, 159, 181, 187
hydrology, 157, 174, 351, 362–3

Improvement Commissioners (Britain), 136, 143, 147, 149, 239–40, 242–4
intercepting sewers, 305, 323, 343
inverted siphons
  features, *11*
  meaning of term, 7
  Roman, 10–24, 39–40
  see also siphons
investors
  New Orleans waterworks project, 199, 204
  New River (London) scheme, 49, 61–2, 74–5
  Portsmouth waterworks, 133, 135, 143, 146, 148
irrigation schemes, 217, 218

landfills, 223
landowners affected by
  New River scheme (London), 57, 65–6
  Tyneside reservoirs, 157, 164, 180
lead pipes, 47, 48, 72, 73, 89, 101, 107, 109, 124, 125
  bursting strength, 17
  in Roman systems, 8, 12, 18, 40
  size, *16*, 17, 39, 40, 125
life expectancy, 233, 234, 236
lighthouses, 201–2
lime mortar, 159
limestone, 116, 166, 168, 181
local Board(s) of Health (Britain), 150, 151, 242, 246
  Town Council acting as, 251–5
Local Government Act(s) (Britain), 251, 342n.140
Local Government Board (Britain), 244, 287n.91

macadam pavements, 249, 252
  costs, *252*
  effect on drainage systems, 261
maintenance
  aqueducts, 67
  canals/rivers/watercourses, 57, 70
  public roads, 150, 224, 248–9, 251–3
  reservoirs, 168–9, 177
  rivers, 57, 70, 223, 329
  Roman venter bridges, 20
  seawalls, 147–8
  sewers, 265, 283–4, 286, 287, 304, 358, 359, 360
  water-distribution systems, 73
manhole access (in sewers), 244, 284, 298, 304–5
measurement systems, urban water-supply, 119, *120*, 122, 125, 130
mechanical power, Romans' attitude to, 3
mechanization, Romans' attitude to, 4, 39
Medical Officer of Health, 256
Metropolis Toll Bridges Act (1877), 330
military personnel, use in engineering projects, 5, 25n.14
modern analytical techniques, application of, 15, 24
mole-catchers, 57, 70
mortality statistics, 150, 233–4, 237
mosquitoes, 206–7

National Board of Health (USA), 298, 299, 301, 305
network, meaning of term, 261n.5
Newcomen engine, 99
night soil
  commercial trade in Japan, 219–20
  disposal of (in Europe), 219, 240
  ownership rights, 219, 240
  treatment before spreading on fields, 228
  use as fertilizer, 219, 222, 228, 230, 236, 360
  see also human excrement
Nuisance Removal and Diseases Prevention Act (1846), 150, 277

oak, *see* wooden pipes
open channels, 22, 117, 216
  compared with siphons, 21
  fall/gradient/slope required, 20, 27n.45
  flow calculations by Romans, 20, 25n.19
overflow drains (for storm water), 305, 308–9

paja (Spanish measure of water-supply), 119
  dimensions, 119, *120*
  discharge equivalents, 122
patents, 83, 301
paving improvements, 239, 248–9
  construction details, 249, 252
  effect on drainage systems, 261, 307
  in London, 331
  and water-pipe laying, 136
payment procedures, 73, 326
pine, *see* wooden pipes
pipe-boring machinery, 71, 83, 87, 111, 207
pipe resistance formula, 19
pipes
  bamboo, 217
  cast-iron, 112, 124, 133, 135, 157, 159, 162, 170, 171, 172, 190, 192, 198, 351, 355, 357

pipes continued
earthenware, 8, 12, 14, *33*, 35, 40, 124, 217
joints in, 12, *33*, 35, 111
lead, 8, 12, 17, 18, 39, 40, 47, 48, 72, 73, 89, 101, 107, 109, 111
masonry, 39
size of: cast-iron pipes, 135, 157, 159, 162, 170, 171, 172, 190, 192, 198, 355, 357; earthenware pipes, *33*, 35; lead pipes, *16*, 17, 39, 40, 125; wooden pipes, 72
stone, 8, 39, 47, 217
wooden, 8, 47, 71–2, 83, 93, 97, 107, 111, 196, 198, 216, 217
pipe sewers, 245, 249, 253, 269, 270, 272–3, 279–82, 298
design table for, *274*
effect of surface run-off, 275–6, 285, 308
fall/gradient/slope required, 266, *271*, 272, *274*, 283 & n.72
size, 249, 253, 270, *274*, 276 & n.53, 278, *280*, 283, 286 & n.86, 298
structural integrity, 286
trial data, 272–3
plague, 77, 234
Poor Law Commission, 241, 262, 267
population statistics, 31, 36, 47, 133, 138, 148, 211, 213–15, *239*
Portland cement, 181, 332
port towns
in Britain, 131–54, 237–58
in Japan, 215
pressurized pipes
bends in, 12
joints in, 12, *33*, 35, 111
leakage from, 124, 255
Roman, 8–9, 39
strength of materials required, 9
testing of, 17, 190, 351
privies, 151, 229, 290–1
public baths, *42*, 231
Public Health Act (Britain), 151–2, 241–2, 246, 256
Government buildings exempt from, 247
Portsmouth's rejection of, 151–2, 246–7, 256–7
public works, concept used by Romans, 4
puddled clay, 159, 166, 168, 175, 181, 182, 286
puddling (procedure), 175, 181–2
pumped drainage systems, 151, 253, 326
pumping
by animal-powered pumps, 38, 39, 99
compared with road transport, 38, 44n.11
by force pump, 26n.37, 36, 38–9
by Newcomen engine, 99
by steam engine, 31, 135, 138, 190, 196, 198, 241, 253, 255, *336*
by water turbines, 357
by waterwheel, 38, 47, 82, 83, 86, 87, 89, 90, *91–2*, 97, 98, 99, 102, 104, 110, 131
by wind-power, 98, 99

railways
Bazalgette's work in Britain, 322, 324, 332
construction materials transported by, 163, 171, 175
Jervis's work in USA, 185, 193
water transported by, 31
rainfall statistics, 162, 163, 175, 179n.33, 180n.72
rain gauges, design, 362
rational model for sewerage design choice, 293–4, 300, 315
development retarded by Waring's system, 293, 301, 307
endogenous factors affecting, 304–10, 315
exogenous factors affecting, 300–3, 315
increasing use, 312, 315
and problems from separate systems without storm drains, 314–15
and water pollution, 310–14, 315–16
regulation of use of water-supply, 125, 130, 192, 218
reservoirs
in American cities, 186, 190, *191*, 196, 198
for British towns and cities: Bristol, 90; Derby, 87, 89; Great Yarmouth, 112; Lincoln, 31, *32*; Portsmouth, 136, *139*, 141, 255; Sheffield, 110; Tyneside, 155–78
for French towns and cities, 351, 355
geological faults affecting, 166, 168
Roman and Moorish, 8, 122
Tyneside impounding catchement reservoirs, 155–78
rise-and-fall waterwheel, 83, 86, 101, 104, 131
river embankments, 69, 70, 328–9
river maintenance, 57, 70, 223, 329
river transport, 208
roads
building of new ones in London, 331
construction costs, *252*
maintenance of, 150, 192, 224, 248–9, 251–2

road transport (of water), 38, 44n.11, 131
Roman bridges, 21, 33
  height limits, 22, 23
Roman cement, 8, 159, 181, 351
Roman engineering, 1–24
  developments emphasized, 4, 33, 43–4
  difficulties in studying, 2, 5
  failures, 43
  historians on, 2
  and mechanization, 4
  political aspects, 9
  sources of techniques used, 4
  structural engineering, 5
  technical notebooks, 5
  and water-power, 3
Roman siphons, 10–24
Roman tunnels, 43, 116
row houses, 192
Royal Commission on Health of Towns, 241, 262
rubbish disposal, 222–3
running water, self-purification of, 310, 312, 316

sakkieh/saquieh (chain of buckets) water-lift device, 36, 131
samurai class (in Japan), 214n.3
  in government positions, 232
  numbers in Edo (Tokyo), 213
sanitary reform, 151, 237–375
Sanitary Report (by Edwin Chadwick), 241, 259, 262
sanitary sewers, 263, 275, 285, 289, 289–90, 295, 297, 304–7, *313*, 314
  storm water affecting, 275–6, 285, 307–9, 314
sanitation, Japanese compared with European cities, 224–35
self-purification theory (for running water), 310, 312, 316
semi-circular arch, use by Romans, 4, 27n.62
Separate System of Sewerage (by Staley & Pierson), 306, 319n.39
sewage end-use, sewer design affected by, 263, 275, 294, 295
sewage treatment facilities, 227–8, 264, 309, 310, 313, 314, 360
sewage velocity, factors affecting, 264, 269, 270–1
sewerage systems, advantages, 292–3
'sewer gas' theory (of disease), 296, 302, 303, 311, 315
sewer outlets, positioning of, 225, 309–10, 343
sewers
  access manholes for, 244, 284, 298, 304–5
  American, 289–320
  arterial-venous concept, 262, 274
  blockages in, 278, 283, 284, 323
  brick sewers, 43, 245, 253, 264, 269, 281, 282, 285, 286, 287, 359
  British, 253–5, 259–88
  Chadwick's approach, 263, 264, 269, 276, 287
  combined systems, 270, 275–6, 285, 289, 294–6, 305, 306, 312–13, *313*, 315, 358–61
  Commission(er)s of, 50, 242, 264, 265–6, 268, 269, 272, 274, 275, 323–4
  design choices: brick/pipe, 260–1, 269, 282–3, 287–8; combined/separate systems, 275–6, 285, 293–316
  egg-shaped, 253, 264, 269
  fall/gradient/slope required, 266, *271*, 272, *274*, 283 & n.72, 326, 359
  French, 357–61
  intercepting, 305, 323, 343
  maintenance of, 265, 283–4, 286, 287, 358, 359, 360
  manual cleaning out of, 265, 284n.76, 287
  pipe sewers, 245, 249, 253, 269, 270, 272–3, 279–82, 287–8, 298
  purpose of, 275, 285, 286
  Rawlinson's designs, 244–5, 281, 286
  Roman, 43
  self-flushing, 269
  separate systems, 263, 275, 285, 289–90, 295, 297, 304–7, 312, *313*, 315
  size of: brick sewers, 43, *245*, 253, 287, 359; design considerations, 270, *274*, 275–6; pipe sewers, 249, 253, 270, *274*, 276 & n.53, 278, *280*, 283, 286 & n.86, 298; problems arising from downsizing, 278
  storm water, 43, 253, 275–6, 285, 290, 306, 307–9
  trial data, 272–3
  ventilation of, 276, 287, 298
shaduf water-lift device, 36, 44n.14, 131
Shinto religion, 230–1
shipping, water for use on board, 143
siphons
  air-pressure devices in, 13, 14–15
  bends in, 12
  bridges for, *11*, 23, 354
  compared with open channels, 21
  fall/slope required, 20
  nineteenth century, 190, 351, 357

siphons continued
  Roman, 10–24, 39–40
  seventeenth century, 217
  venters in, 11, 13
  Vitruvius on, 9, 11–15, 24
  see also inverted siphons
slaves, 205–6
sludge vessels, 328, 340n.63
slums, 233, 343
springs, 29, 47, 50, 51, 70, 116, 117, 133, *139*, 141, 216, 248, 255
  pollution of, 147
stand-pipes, 161, *189*
state boards of health (USA), 302, 311, 316
statistics
  mortality, 150, 233–4, 237
  population, 31, 36, 47, 133, 138, 148, 211, 213–15, *239*
  rainfall, 162, 163, 175, 179n.33, 180n.72, 356
steam engines, 31, 135, 138, 190, 196, 198, 200, 207, 241, 253, 255, *336*
steam-powered riverboats, 200, 201, 208
stone pipes, 8, 39, 47
stopcocks, 73
storm water sewers, 43, 253, 275–6, 285, 290, 306, 307–9
stream disposal (of sewage), 310, 312, 316
street cleaning, 150, 192, 224, 291, 308, 355, 357
street improvements, *see* paving improvements
street lighting, 88, 249, 331
structural engineering, Roman, 5
surface run-off, sewers affected by, 275–6, 285, 307–9
surveying
  Croton Aqueduct (New York, USA), 185
  in Japan, 217
  New River (Britain), 62–3, 66–7
  Roman, 5, 33
  sewers, 275
swallow-holes, 168
systems approach to sanitary systems, 261, 263–4
  meaning of term, 261n.5
  opposition to, 281–2, 288

telegraphic communications, 362
tender procedures, 325–6
tenements, 233
triple system of sewerage, 285n.81
tunnels
  London, 330–1
  New York water system, 186, 187, 190
  Paris, 355, 357
  Roman, 43, 116
  Tyneside reservoirs, 159, 161, 174, 181
typhoid fever, 31, 229, 311, 312, 316

urban growth
  in Britain, 133, 148, 237
  effect on drainage systems, 261, 291, 292, 312
  in Japan, 211, 213–15, 236
  in USA, 193, 208, 291
urban water-supply
  fifteenth/sixteenth century, 115–30
  nineteenth century, 31, 131–210, 240–1, 255–6, 351–7
  regulation of use, 125, 130, 192, 218
  Roman, 7–24, 31–44, 116–17
  sixteenth/seventeenth century, 47–77, 85–112, 216–18
  water rates charged, 74, 93, 141, 145, 191, 197
urine
  collection of, 220, 222
  use as fertilizer, 219 & n.12

ventilation
  of aqueducts, 116–17, 187
  of sewers, 276, 287, 298

war
  American projects affected by, 199, 200–1, 203
  British economy affected by, 133, 143, 145
  French projects affected by, 355, 361
waste disposal systems
  in Japan, 218–24
  water carriage system, 292, 360–1
  see also sewers
water closets, 151, 225, 244, 246, 297
  effect on drainage systems, 261, 292
water-lifting devices
  Archimedean screw, 36
  bucket-wheel, 36
  chain of buckets/pots, 26n.37, 36
  see also pumping
Water Police (in New York, USA), 192
water-power, Romans' attitude to, 3
water rates, 74, 93, 141, 145, 191, 197
water towers, Roman, 33, 35, 44
water turbine, pumping by, 357
waterwheel
  pumping by, 47, 82, 83, 86, 87, 89, 90, *91–2*, 97, 98, 99, 102, 104, 110, 131, 357

waterwheel continued
  rise-and-fall waterwheel, 83, 86, 101, 104, 131
  Romans' use of, 3, 38
waterworks
  nineteenth century, 31, 130, 135–6, *139–40*, 196, 198, 241, 248, 255, 291
  sixteenth/seventeenth century, 47, 48, 71, 101–5
wells
  bored (artesian), 31, 180, 241
  pollution of, 149, 151, 245, 291
  sunk, 29, 43, 94, 96, 98, 131, 168, 171, 240
wolds (in Britain), 40, *41*
wooden aqueducts, 67
wooden pipes, 8, 47, 97, 107, 198, 216, 217
  boring of, 71, 83, 87, 111, 207
  cypress pipes, 201, 207 & n.40
  elm pipes, 71–2, 83, 87, 93
  oak pipes, 196
  red pine pipes, 217
  size, 72
workmanship, poor, effects of, 159, 278
workmen
  drinking habits, 159, 188
  wage rates, 69, 157, 188

yellow fever, 196, 203, 206, 208, 209, 293